Henry William Jeans

Nautical Astronomy and Navigation

Henry William Jeans

Nautical Astronomy and Navigation

ISBN/EAN: 9783337396213

Printed in Europe, USA, Canada, Australia, Japan

Cover: Foto ©berggeist007 / pixelio.de

More available books at **www.hansebooks.com**

NAUTICAL ASTRONOMY

AND

NAVIGATION.

PART I.

CONTAINING RULES FOR FINDING THE LATITUDE AND LONGITUDE,
AND THE VARIATION OF THE COMPASS.

With numerous Examples.

Designed for Beginners.

———•———

By H. W. JEANS, F.R.A.S.

LATE MATHEMATICAL MASTER AND EXAMINER AT THE ROYAL NAVAL COLLEGE, PORTSMOUTH

AUTHOR OF A WORK ON
" PLANE AND SPHERICAL TRIGONOMETRY;" " HANDBOOK OF THE STARS;" " PROBLEMS IN
ASTRONOMY, NAVIGATION, ETC. WITH SOLUTIONS :"
FORMERLY MATHEMATICAL MASTER IN THE ROYAL MILITARY ACADEMY, WOOLWICH ; AND AN EXAMINER OF OFFICERS
IN THE MERCHANT-SERVICE IN NAUTICAL ASTRONOMY, ETC.

THIRD EDITION.

LONDON:
LONGMANS, GREEN, READER, AND DYER.
1876.

PREFACE.

———•———

In the present edition the Author has adapted the rules not only to the Tables of Dr. Inman (the most comprehensive and useful yet published), but also to those in more general use, such as Riddle's, Norie's, &c. The student will therefore find now no difficulty in working out the examples contained in the book by any of the above tables. In the last edition of the Author's work on *Plane and Spherical Trigonometry*, rules are also given depending on the common tables of sines, &c., as well as on the table of haversines contained in Inman's Tables.

Langstone House, Havant,
May 23 1876.

CONTENTS.

NAVIGATION.

NAUTICAL ASTRONOMY.

NAVIGATION OR PLANE SAILING.

CHAPTER I.

(1.) Two distinct methods are used for navigating a ship from one place to another: the first is an application of the common rules of Plane Trigonometry, the necessary angles and measurements being supplied by means of the compass and log-line; the second and more exact method requires a knowledge of the rules of Spherical Trigonometry and of the principal definitions and facts in Astronomy, the necessary data being obtained by astronomical observations taken usually with the sextant. The latter method is for this reason called *Nautical Astronomy*; the characteristic name of the former being *Navigation* or *Plane Sailing*.

Before we proceed to give the rules in navigation for finding the *place* of the ship, that is, its latitude and longitude, we will reprint, for the sake of reference, the definitions and terms in pp. 143-44 of *Navigation*, Part II., and also some trigonometrical and nautical problems taken for the most part out of the author's *Trigonometry*, Part I. These problems are intended to serve as a useful introduction to navigation; at the same time they will show the student that a knowledge of the rules in plane trigonometry is nearly all that will be required to enable him to understand and work out the problems and examples in navigation or plane sailing.

Definitions in Navigation.

(2.) The following are the principal terms in Navigation : the definitions of these terms, like those in Nautical Astronomy, must be thoroughly understood and committed to memory.

Course.
Distance.
Departure.
True difference of latitude.
Meridional difference of latitude.
Difference of longitude.
Middle latitude.

B

Definitions of the preceding terms.

The course is the angle which the ship's track makes with all the meridians between the place left and the place arrived at.

The distance is the spiral line made by the ship's track in describing the course between the place left and the place arrived at.

The departure is the sum of all the arcs of parallels of latitude drawn between the place left and the place arrived at, through points indefinitely near to one another taken on the distance, and intercepted between the meridians passing through those points.

The true difference of latitude is the arc of a meridian intercepted between the parallels of latitude drawn through the place left and the place arrived at.

The meridional difference of latitude is the value in minutes of a great circle of the line on a Mercator's chart, into which the true difference of latitude has been expanded.

The difference of longitude is the arc of the terrestrial equator intercepted between the meridians passing through the place left and the place arrived at.

The middle latitude is the mean of the latitudes (supposed of the same name) of the place left and the place arrived at.

These definitions will be clearly understood by means of the following diagrams.

Let P represent the pole of the earth, TZ an arc of the equator, PT the meridian passing through a known place G, as Greenwich, A and F two other places on the earth (considered as a sphere), PU, PZ, their meridians.

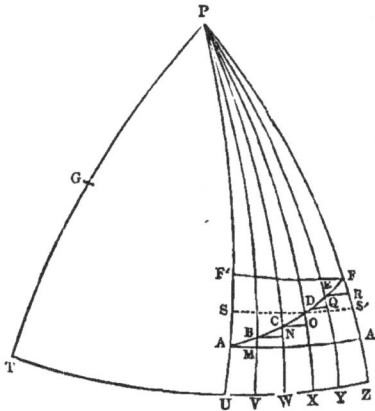

Through the points A and F suppose a curve line AF to be drawn, cutting all the intermediate meridians PV, PW, PX, &c., *at the same angle;* that is, making the angle PAB=PBC= PCD=&c. Then this common angle PAF is called the *course.* The arc AF* is the *distance.* Draw the parallels of latitude AA′, FF′; then, since the latitude of A is the arc AU, and the longitude of A the arc TU, and the latitude of F is the arc FZ, and the longitude of F is the arc TZ; therefore the difference, or, as it is usually called, the *true difference of latitude,* between A and F is the arc AF′ or A′F, and *the difference of longitude* between A and F is the arc of the equator UZ. Again, suppose the intermediate meridians PV, PW, &c., to be

---

* A F is sometimes called the *rhumb line,* sometimes the *loxodromic curve,* sometimes the *equiangular spiral.*

drawn through points B, C, D, &c., taken on the line AF indefinitely near to each other; and through the points A, B, C, D, &c., the arcs of parallels of latitude AM, BN, CO, &c. On this supposition (namely that the points A, B, C, &c., are indefinitely near to each other) the elementary triangles ADM, BCN, CDO, &c., may be considered without any error to be right-angled *plane triangles*. *The departure* between A and F=A M+B N+C O+ . . . E R, the points A, B, C, &c., being supposed to be indefinitely near to each other.

If a parallel of latitude s s' be drawn through the middle of A F', then the arc of the meridian S U is called the mean or *middle latitude* between A and F. It is manifest that the arc s s' will be nearly equal to A M+B N+ . . . D Q+E R, the departure, A and F being supposed to be on the same side of the equator. For short distances s s' is substituted without any practical error for the departure, and one of the principal rules in Navigation is deduced from it.

(3.) There are two kinds of charts; the Plane chart, and Mercator's chart.

The Plane chart.

The plane chart is a representation of the earth's surface, considering it as a plane. When a small portion of the surface is concerned, this mode of representing it will lead to no practical error; hence coasting charts are usually constructed in this manner, in which the different headlands, light-houses, &c., are laid down according to their bearings.

Mercator's chart.

The chart used at sea for marking down the ship's track, and for other purposes, is called Mercator's chart. It exhibits also the surface of the earth *on a plane;* but the meridians are drawn perpendicular to the equator, and therefore the arcs AM, B'B, &c., of parallels of latitude intersected between any two meridians are increased to am, $b'b$, &c., and become equal to one another and to line uv, and therefore to the intercepted arc U V of the equator. If we wish to make the figures (supposed to be very small) $ambb'$, $b'bcc'$, &c., on the chart similar to AMBB', B'BCC', &c., of the globe, it is evident we must increase the sides bm, bc, &c., in the same proportion as am, $b'b$, &c. (that represent AM, BB', &c.), have been increased. Let us therefore suppose the straight lines am, $b'b$, $c'c$, &c., have been drawn at such a distance from each other that the above similarity of figure is preserved (and

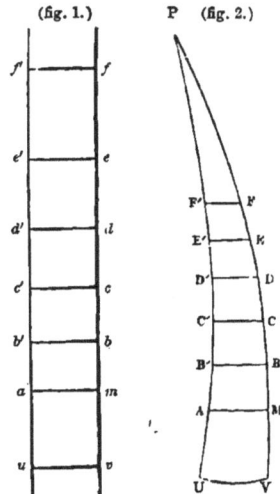

(fig. 1.) P (fig. 2.)

this can only be done by supposing the surfaces $ambb'$, $b'bcc'$, &c., *indefinitely* small, so that the surfaces AMBB', B'BCC', &c., may be considered as plane surfaces). Then a representation of the earth's surface, or any part of it, so constructed, is called a *Mercator's chart*.

The straight line mf, into which MF, the true difference of latitude between M and F, has been expanded, is called the *meridional difference of latitude* between M and F, and the values of bv, cv, &c., in minutes, are called the *meridional parts* of B, C, &c.: hence the *meridional difference of latitude* between two places is the difference of the meridional parts for the two places.

The method of constructing a Mercator's chart and laying down a ship's track thereon will be given hereafter (see art. 16).

From these definitions and principles are deduced the following

ELEMENTARY RULES IN NAVIGATION.

Rule (a). *To find the true difference of latitude*, having given the latitude from and latitude in.*

(1.) When latitude from and latitude in have *like names*, that is, are both north or both south.

Under the latitude from, put down the latitude in, take the difference and reduce the same to minutes; place N. or S. against the result according as the latitude in is north or south of the latitude from ; the remainder is the *true difference of latitude*.

(2.) When latitude from and latitude in have *unlike names*, that is, one north and the other south.

Take the sum of the two latitudes, reduce it to minutes, and attach N. or S. thereto, according as the latitude in is north or south of the latitude from ; the result is the *true difference of latitude*.

EXAMPLES.

1. Find the true difference of latitude, having given latitude from 42° 10′ N., and latitude in 50° 48′ N.

lat. from 42° 10′ N.

lat. in   50 48 N.

$\overline{8\quad38}$

60

∴ T. D. lat. $\overline{518}$ N.

2. Find the true difference of latitude, having given latitude from 3° 42′ N., and latitude in 2° 40′ S.

lat. from 3° 42′N.

lat. in   2 50 S.

$\overline{6\quad32}$

60

∴ T. D. lat. $\overline{392}$ S.

* The latitude of the place left is called the latitude *from*, the latitude of the place arrived at is called the latitude *in*.

Find the true difference of latitude in each of the following examples:

	Lat. from	Lat. in	Answers.
3.	33° 42′ N.	40° 40′ N.	∴ T. D. lat. = 418 N.
4.	40 40 N.	33 42 N.	.. = 418 S.
5.	3 42 S.	1 40 N.	.. = 322 N.
6.	3 8 S.	14 42 S.	.. = 694 S.
7.	68 48 N.	38 30 N.	.. = 1818 S.
8.	14 14 N.	0 0	.. = 854 S.

Rule (*b*). *To find the meridional difference of latitude,* having given the latitude from and latitude in.

Take the meridional parts for the two latitudes from the table of meridional parts: subtract if the names be alike, and add if the names be unlike, the result is the *meridional difference of latitude;* N. or S. being attached thereto according as the latitude in is north or south of latitude from.

EXAMPLES.

9. Find the meridional difference of latitude, having given latitude from 42° 10′ N., and latitude in 50° 48′ N.

    lat. from   42° 10′ N.
    lat. in   50 48 N.
    mer. parts...2795·2 N.
    mer. parts...3549·8 N.
    mer. diff. lat. ...754·6 N.

10. Find the meridional difference of latitude, having given latitude from 3° 42′ N., and latitude in 7° 32′ S.

    lat. from   3° 42′ N.
    lat. in   7 32 S.
    mer. parts ...222·2 N.
    mer. parts ...453·3 S.
    mer. diff. lat. ...675·5 S.

Find the meridional difference of latitude in each of the following examples:

	Lat. from	Lat. in	Answers.
11.	34° 42′ N.	33° 15′ N.	M. D. lat. = 104·9 S.
12.	14 14 N.	30 14 N.	.. = 1041·7 N.
13.	84 10 N.	80 30 N.	.. = 1681·5 S.
14.	2 8 S.	3 10 N.	.. = 318·1 N.
15.	4 5 N.	4 5 S.	.. = 490·4 S.
16.	0 0	2 45 N.	.. = 165·1 N.

Rule (*c*). *To find the middle latitude,* having given the latitude from and latitude in.

The names being supposed to be *alike*, that is, both north or both south.

Add together the two latitudes, and take half the sum; the result is the middle latitude.

When the names are unlike, the mid. lat. (which is seldom required but for obtaining the departure) should be found by means of a table; but in this case it may perhaps be as well to avoid the use of the middle latitude in any of the common problems in navigation.

EXAMPLES.

17. Find the middle latitude, having given latitude from 3° 42′ N., and latitude in 13° 52′ N.

$$
\begin{array}{ll}
\text{lat. from} & 3°\ 42′\,\text{N.} \\
\text{lat. in} & 13\ \ 52\ \text{N.} \\
\hline
& 2)\overline{17\ \ 34} \\
\text{mid. lat.} & \overline{8\ \ 47}\ \text{N.}
\end{array}
$$

Find the middle latitude in each of the following examples :

	Lat. from	Lat. in	Answers.
18.	38° 42′ N.	30° 30′ N.	mid. lat. 34° 36′ N.
19.	62 17 S.	62 30 S.	.. 62 23½ S.

Rule (d). *To find the difference of longtitude,* having given the longitude from and longitude in.

(1.) When the longitude from and longitude in have *like names;* that is, are both east or both west.

Under longitude from put longitude in, take the difference, and reduce the same to minutes; place E. or W. against the remainder according as the longitude in is east or west of longitude from; the remainder will be the difference of longitude.

(2.) When the longitude from and longitude in have *unlike names;* that is, one east and the other west.

Take the sum of the two longitudes, reduce it to minutes, and attach E. or W. thereto according as the longitude in is east or west of the longitude from; the result is the true difference of longitude.

NOTE. If the difference of longitude found by this rule exceed 180° it must be subtracted from 360°, and the remainder brought into minutes must be considered the difference of longitude, with the contrary letter attached to it.

20. Find the difference of longitude, having given the longitude from=110° 42′ W., and longitude in=100° 42′ W.

$$
\begin{array}{ll}
\text{long. from} & 110°\ 42′\,\text{W.} \\
\text{long. in} & 100\ \ 42\ \text{W.} \\
\hline
& 10\ \ \ 0 \\
& 60 \\
\hline
\end{array}
$$

∴ diff. long. $\overline{600}$ E.

21. Find the difference of longitude, having given long. from 12° 10′ E., and long. in 2° 45′ W.

$$
\begin{array}{ll}
\text{long. from} & 12°\ 10′\ \text{E.} \\
\text{long. in} & 2\ \ 45\ \text{W.} \\
\hline
& 14\ \ 55 \\
& 60 \\
\hline
\end{array}
$$

∴ diff. long. $\overline{895}$ W.

Find the difference of longitude in each of the following examples :

	Long. from	Long. in	Answers.
22.	33° 40′ E.	40° 10′ E.	Diff. long. 390 E.
23.	104 0 W.	110 30 W.	.. 390 W.
24.	2 45 W.	3 30 E.	.. 375 E.
25.	0 0	4 10 W.	.. 250 W.
26.	3 10 W.	3 10 E.	.. 380 E.
27.	179 0 E.	179 0 W.	.. 120 E.

Rule (e). *To find the latitude in,* having given the latitude from and true difference of latitude.

(1.) When the latitude from and true difference of latitude have *like* names.

To the latitude from, *add* the true difference of latitude (turned into degrees and minutes, if necessary) ; the sum will be the latitude in, of the same name as the latitude from.

(2.) When the latitude from and true difference of latitude have *unlike* names.

Under the latitude from put the true difference of latitude (in degrees and minutes, if necessary) ; take the less from the greater ; the remainder, marked with the name of the greater, is the latitude in.

EXAMPLES.

28. Find the latitude in, having given the latitude from 42° 30′ N., and true diff. lat. 342′ N.

    60)342′ N.
    T. D. lat. 5° 42′ N.
    lat. from 42 30 N.
    lat. in   48 12 N.

29. Find the latitude in, having given the latitude from 42° 30′ S., and true diff. lat. 342′ N.

    60)342′ N.
    T. D. lat. 5° 42′ N.
    lat. from 42 30 S.
    lat. in   36 48 S.

Find the latitude in, in each of the following examples :

	Lat. from	T. D. lat.	Answers.
30.	30° 10′N.	182′N.	Lat. in 33° 12′N.
31.	3 2 S.	190 N.	.. 0 8 N.
32.	2 48 S.	368 N.	.. 3 20 N.
33.	2 48 S.	288 N.	.. 2 0 N.
34.	4 48 N.	288 S.	.. 0 0
35.	0 10 N.	228 N.	.. 3 58 N.

Rule (*f*). *To find the longitude in,* having given the longitude from and the difference of longitude.

(1.) When the longitude from and diff. long. have *like names.*

To the long. from, *add* diff. long. (turned into degrees, if necessary) ; the sum will be long. in, of the same name as long. from.

(2.) When the long. from and diff. long. have *unlike names.*

Under long. from put diff. long. (in degrees and minutes, if necessary) ; take the less from the greater ; the remainder, marked with the name of the greater, is the long. in.

NOTE. If the long. in, found as above, exceed 180°, subtract it from 360°, and attach to the remainder the contrary name to the one directed in the Rule.

EXAMPLES.

36. Find the long. in, having given long. from 38° 42′ W., and diff. long. 384·5′ W.

$$60)384·5$$
$$6° \; 24·5′ \; W.$$

long. from 38° 42′ W.
diff. long.   6   24·5 W.
long. in   45   6·5 W.

Find the longitude in, in each of the following examples :

	Long. from	Diff. long.	Answers.
37.	62° 32′ E.	1000·5′ W.	long. in 45° 51·5′ E.
38.	2 30 E.	126·6 E.	. . 4 36·6 E.
39.	3 40 W.	220·0 E.	. . 0 0
40.	0 0	100·4 W.	. . 1 40·4 W.
41.	179 59 W.	2·0 W.	. . 179 59·0 E.

NOTE. The Nautical problems following (42 to 68) may be omitted by the student, unless he is acquainted with the practical rules in Plane Trigonometry ; proceeding at once to the construction of the Mariner's Compass, p. 20, and then to the corrections in Plane Sailing, p. 30.

(4.) It is shown in any work on Trigonometry that if in the right-angled triangle
P C N the angle P C N (opposite the perpendicular P N) be denoted by A, then,

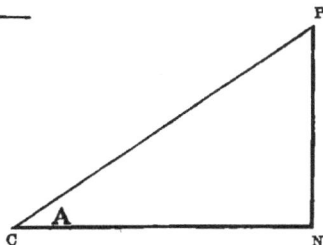

$$\frac{\text{Perp.}}{\text{Hyp.}} \text{ or } \frac{PN}{CP} = \text{sine of the angle A or } PN = CP \sin. A.$$

$$\frac{\text{Hyp.}}{\text{Perp.}} \text{ or } \frac{CP}{PN} = \text{cosecant of the angle A or } CP = PN \text{ cosec. A.}$$

$$\frac{\text{Perp.}}{\text{Base}} \text{ or } \frac{PN}{CN} = \text{tangent of the angle A or } PN = CN \text{ tan. A.}$$

$$\frac{\text{Base}}{\text{Perp.}} \text{ or } \frac{CN}{PN} = \text{cotangent of the angle A or } CN = PN \text{ cot. A.}$$

$$\frac{\text{Hyp.}}{\text{Base}} \text{ or } \frac{CP}{CN} = \text{secant of the angle A or } CP = CN \text{ sec. A.}$$

$$\frac{\text{Base}}{\text{Hyp.}} \text{ or } \frac{CN}{CP} = \text{cosine of the angle A or } CN = CP \text{ cos. 'A.}$$

These six equations or formulæ enable us to work out all the ordinary
questions in Navigation involving right angles. Before, however, the student proceeds with the following problems he should make himself also acquainted with the common rules in Plane Trigonometry (see the author's
Trigonometry, Part I. Rules 1, 2, and 3).*

NAUTICAL PROBLEMS.

42. Wishing to find the height of a tower, I observed with a sextant
the angle of elevation of its top above the level of my eye to be 32° 14'. I

* In this edition the rules are adapted to the common table of sines, &c., as well as
to the table of haversines ; but preference will be given to the use of the latter, as it
diminishes considerably the labour of calculation.

then measured the distance from the place of observation to the base of the tower, and found it to be 142 feet. Required the height of the tower.

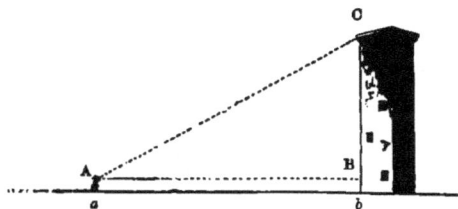

Let CB* represent the tower, A the place of observer. Draw the horizontal line A B at a height above the ground ab equal to the height of the eye, and join A C.

Then in right-angled triangle A B C are given A B $=142$, angle C A B $=32°$ 14′ and D $=90°$: to find C D, the height of the tower, above the horizontal line A B.

(Mark the figure in the usual way. See *Trigonometry*, Part I., rule for right-angled plane triangles.)

Calculation.

Since $\dfrac{CB}{AB}=$ tan. A.

     ∴ CB $=$ A B tan. A.

∴ log. C B $=$ log. A B $+$ log. tan. A $- 10$.

     (*Trig.* Part I. art. 31.)

log.   A B......2·152288

,, tan. A......$\overline{9·799717}$

,,   C B......$\overline{1·952005}$

∴ C D $=89·5$ feet.

To the value of C B must be added the height of the eye A a : the result will be the height of the tower required.

43. To find the height of a tower, I observed the angle of elevation of its top above the level of my eye (supposed to be 5 feet above the ground) to be 47° 56′. I then measured the distance from the place of observation to the base of the tower, and found it to be 190·4 feet. Required the height of the tower.   *Ans.* 216 feet.

44. On the opposite bank of a river to that on which I stood is a tower known to be 216 feet high : with a pocket sextant I ascertained the angle between a horizontal line drawn from my eye (supposed to be 5 feet above the ground) and its top to be 47° 56′. Required the distance across the river from the place where I stood to the bottom of the tower.

Let c b (fig. to Prob. 1) represent the height of tower $=216$ feet ; A a the

---

* The figures or diagrams of the following problems are not drawn accurately to scale ; the student should endeavour to draw them as neatly as he can by the eye, so as to indicate the form without regard to the exact value of the sides and angles in the problems to which they refer. Problems will be given hereafter (see art. 9), which are solved not only by logarithms but by *construction ;* that is, by using mathematical instruments. The practice of using instruments thus obtained will form a proper introduction to the construction of charts, and the tracing the ship's track thereon.

height of spectator's eye=5 ft.; and AB or ab width of river. Suppose AB parallel to horizontal line ab: join CA, then CAB the angle of elevation= 47° 56', and BC=height of tower−5=211 feet are given : to find AB or ab the width of river.

Since $\dfrac{AB}{BC}$=cot. CAB

∴ AB=BC cot. CAB

∴ log. AB=log. BC+log. cot. CAB−10

    =log. 211+log. cot. 47° 56'−10

Calculation.

log. 211............2·324282

 ,, cot. 47° 56' 9·955453

 ,, AB ....2·279735

∴ AB=190·4

width of river.

45. On the opposite bank of a river to that on which I stood is a tower known to be 94·5 feet high : with a pocket sextant I ascertained the angle between a horizontal line drawn from my eye (supposed to be 5 feet above the ground) and its top to be 32° 14'. Required the distance across the river from the place where I stood to the bottom of the tower.

Ans. 142 feet.

46. Being ordered to place a target at 500 yards from the ship, and knowing that the height of the truck above the water-line was 213 feet : it is required to find what angle the height will subtend on my sextant when I am at the required distance (before allowing for index correction).

Let BC represent the ship's mast, A the required place of target : then the angle BAC is the angle which must be read off on the sextant (supposing it to have no index correction).

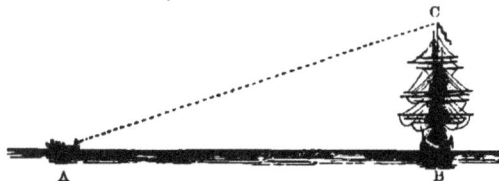

In right-angled triangle are given the side BC=213 feet, AB=500 yards =1500 feet, and B=90°, to find the angle A.

Since tan. A=$\dfrac{CB}{AB}$

∴ log. tan. A−10=log. CB−log. AB

∴ log. tan. A=10+log. CB−log. AB

Calculation.

log. CB+10...12·328380

 ,, AB .... 3·176091

 ,, tan. A .... 9·152289

∴ angle on sextant=8° 5'.

47. Sailing in company with another ship, and being ordered to keep at the distance of 500 yards from her, and knowing that the height of her mast

above the hammock-nettings was 198 feet, it is required to find what angle on my sextant will indicate the proper distance (see fig.).

Ans. 7° 31′ 15″.

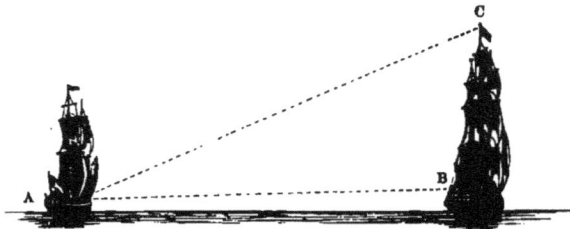

48. Being on Southsea Common, and wishing to find my distance from a ship at anchor at Spithead, I observed with a sextant the angle between the ship and the steeple of St. Thomas's Church to be 72° 42′. I then walked 500 yards in a direct line towards the church, and again took the angle between the ship and the steeple, and found it to be 82° 45′. Required my distance from the ship at each observation.

Let c be the ship at anchor, A my first station, B the second, and P the church. Then the angle CAP=72° 42′, the angle CBP=82° 45′ and the line AB=500 yards, are given : to find BC and AC my distances at each observation. Since CBP=82° 45′ ∴ adjacent angle ABC=97° 15′, and the angle ACB=10° 3′ (since the three angles of a plane triangle are together equal to 180°).

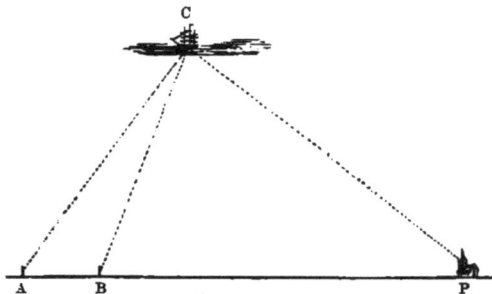

(1.) To find BC (by *Trigonometry*, Part I. Rule 11).

In triangle AB=500   log. AB......... 2·698970
            CAB=72° 42′   „ sin. A...... 9·979895
            ACB=10   3   12·678865
BC: AB : : sin. CAB : sin. ACB   „ „ ACB... 9·241814
                                  „ „ BC.... 3·437051
                            ∴ BC=2735 yards.

(2.) To find AC.

In triangle, AB = 500   log. AB .... 2·698970
   ABC = 97° 15' „ sin. ABC... 9·996514
   ACB = 10 3   12·695484
AC : AB : : sin. ABC : sin. ACB „ „ ACB... 9·241814
          „ AC .... 3·453670
         .·. AC = 2841 yards.

49. Two ships sailing in company, in order to determine nearly their distance from an object c on the shore, are separated from each other two nautical miles AB (fig. somewhat similar to the one to last problem), the angle is then observed from each ship between the object and the other ship: at A the angle is 85° 10', at B the angle ABC is 82° 45'. Required the distance of each ship from the object c on shore.

Ans. 9·478 and 9·52 miles.

50. To determine the height of a lighthouse c on the summit of a cliff on the seashore, I observed the angle of elevation CAD of its top above the level sand to be 26° 40'; then, measuring AB = 200 yards on the sand in a direct line towards it, I again observed the angle of elevation CBD of its top, and found it to be 33° 30'. Required the height CD of the lighthouse above the shore.

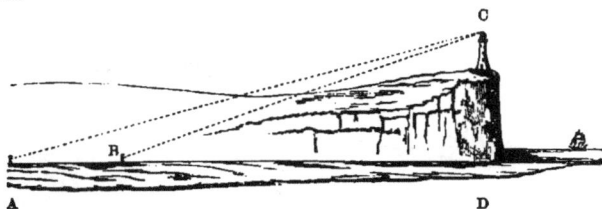

(1.) Find CB.

Given AB = 200   In oblique-angled triangle ABC,
 A = 26° 40'   CB : AB : : sin. A : sin. ACB
 ABC = 146 30 log. AB........... 2·301030
.·. ACB = 6 50   „ sin. A....... 9·652052
            11·953082
       „ „ ACB ... 9·075480
       „ CB .... 2·877602
       .·. CB = 754·4 yards.

(2.) Find CD.

In right-angled triangle BCD are given CB (its log. 2·877602); angle CBD = 33° 30' and D = 90°, to find CD.

Since $\dfrac{CD}{CB}$=sin. CBD

$\quad$.˙. CD=CB sin. CBD

.˙. log. CD=log. CB+log. sin. CBD—10

log. CB	2·877602
,, sin. CBD...	9·741889
,, CD..........	2·619491

$\quad$.˙. CD=416·3 yards
$\qquad$ the height of lighthouse.

51. Standing in for the land, I observed the summit of a lofty mountain near the shore. I took the angle of elevation of the peak, and found it to be 12° 25′; after having run 3½ miles directly towards it, I again took the angle of elevation, which was 30° 13′. Required the height of the mountain, and its distance from the second station (see fig.).

$\qquad$ *Ans.* Height 1·24 miles, BD=2·13 miles.

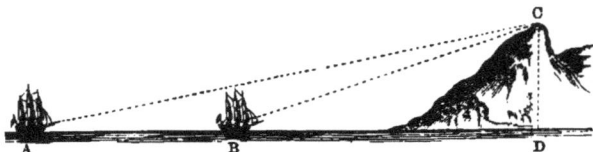

52. To find the distance between two ships at anchor at C and D, I measured on the shore a base-line AB=735 yards, and with a sextant observed the following angles :

At A,
$\quad$ CAD=63° 30′
$\quad$ DAB=35   10

At B,
$\quad$ DBC=80° 16′
$\quad$ CBA=28   20

Required the distance CD.

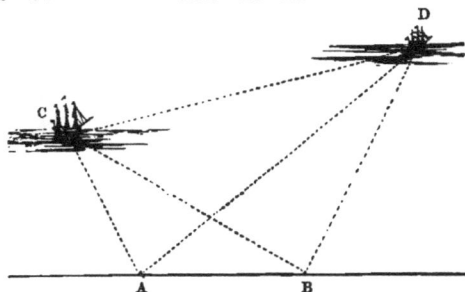

$\quad$ (1.) Find AC.$\quad$ (2.) Find AD.$\quad$ (3.) Find CD.
$\qquad$ (1.) To find AC.

In triangle ACB,
$\quad$ AB=735
$\quad$ CBA=28° 20′
$\quad$ CAB=63   30+35° 10′
$\qquad$ =98   40
and .˙. ACB=53

AC : AB : : sin. CBA : sin. ACB.

log. AB.........	2·866287
,, sin. CBA...	9·676328
	12·542615
,, sin. ACB...	9·902349
,, $\qquad$ AC...	2·640266

$\qquad$.˙. AC=436·8.

(2.) To find AD.

In triangle ABD,

AB=735

DAB=35° 10'

ABD=28 20+80° 16'

=108° 36'

and ∴ ADB=36° 14'

AD : AB : : sin. ABD : sin. ADB.

log. AB......... 2·866287

,, sin. ABD... 9·976702

12·842989

,, sin. ADB... 9·771043

,,   AD... 3·071346

∴ AD=1178·5.

(3.) To find CD (by Rule 4, second method, *Trig.* Part I.).

In triangle ACD,

AC= 436·8

AD=1178·5

∴ AC+AD=1615·3

AC−AD= 741·7

CAD=63° 30'

∴ ½CAD=31 45

log. (AD+AC)...3·208253...log. (AD+AC)...3·208253

,, (AD−AC)...2·870226   ,, sin. ½CAD...9·721162

0·338027   2·929415

,, tan.½CAD...9·791563   ,, sin. arc.......9·904757

,, tan. arc. ...10·129590   ,, CD............3·024658

∴ the distance CD=1058·3 yards.

53. To determine the distance between two redoubts C and D by which the entrance into a harbour is defended, a boat is placed at A with its head towards a tree seen at E (produce the line AB to some point E) in the direction AB, and the angles CAD=22° 17' and DAE=48° 1' were observed. The boat is then moved to B, a distance of 1000 yards, directly towards the tree, and the angles CBD=53° 15' and DBE=75° 43' are observed. Required the distance between the redoubts C and D.   *Ans.* 1290 yards.

As the two following problems are of great use in Marine Surveying, we will solve them by logarithms, and also by a geometrical construction. In Problem 98 of the author's volume of Astronomical Problems, analytical solutions of the same problems are also given.

54. Wishing to determine the position of a sunken rock at the entrance of a bay, and the water being smooth, I anchored a boat upon it, and measured with a sextant the angles which three objects, A, B, and C, on the shore subtended at the boat. They were as follows: the angle between A and the object B to the right was 26° 27', and the angle between B and the object C to the right was 34° 12'. On my chart of the bay I carefully measured with compasses the distances between the three objects, and found AB=5 miles, BC=6 miles, and AC= 7 miles. Required the distance of the rock from A, B, and C.

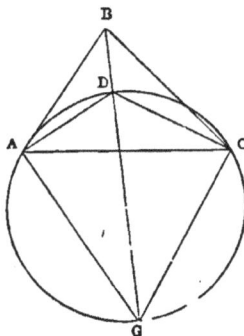

(1.) *By Construction.*

By means of a scale of equal parts make the triangle ABC, having the side AB=5, BC=6, and AC=7. At the point C, on the side of AC farthest from the boat, make the angle ACD=26° 27′, the angle observed between the other two objects A and B; and at the point A, on the farthest side also from the boat, make the angle CAD=34° 12′, the angle between the other two objects B and C: produce the sides AD and CD till they meet in D. Then describe a circle to pass through the three points A, D, and C; and the position of the rock will be somewhere in the circumference of that circle. To find that position, join BD, and produce it to the circumference in G; then G will be the station sought, or the position of the rock.

For, the angles in the same segment of a circle being equal (*Euclid*, b. iii.), therefore AGB=ACD=26° 27′, and CGB=CAD=34° 12′; and these were the angles observed at the boat. Hence G must be the position of the boat; and GA, GB, and GC measure respectively the distance of the rock from each of the objects A, B, and C.

(2.) *By Trigonometry.*

Assume any point G to be the position of the boat, and let A, B, and C be the objects. Describe roughly a circle passing through the three points A, G, C. Join GA, GB, and GC. Then GA, GB, and GC are the distances required. Draw AD, CD to the point of intersection D. Then, by Geometry, since the angles in the same segment of a circle are equal, ∴ CAD=CGB=34° 12′, and ACD=AGB=26° 27′.

[1.] Find AD, having given in the triangle ADC the side AC=7, the angle ACD=26° 27′, and ADC=180°−(34° 12′+26° 27′)=119° 21′.

[2.] Find angle BAC, having given the three sides of the triangle ABC.

[3.] Find angle ABD, having given AB, AD, and angle BAD (=BAC−CAD).

[4.] Find GA, GB, and angle BAG, having given in the triangle ABG the side AB and the angles AGB and ABG.

[5.] Find GC, having given in the triangle AGC the side AC, the angle AGC, and the angle CAG=(BAG−BAC).

Calculation.

[1.] To find AD.

AC : AD :: sin. ADC : sin. DCA

AC=7	0·845098	AB=5
ADC=119° 21′	9·648766	BC=6
DCA= 26 27	10·493864	AC=7
	9·940338	
	0·553526	

∴ AD=3·577

[2.] To find angle BAC.

7	9·154902
5	9·301030
2	0·602060
6	0·301030
8	9·359022
4	BAC=57° 7′ 15″
4	CAD=34 12 0
2 ∴ BAD=	22 55 15

[3.] To find angle ABD.

$$\text{AB}+\text{AD} : \text{AB}-\text{AD} :: \tan. \tfrac{1}{2}(\text{ADB}+\text{ABD}) : \tan. \tfrac{1}{2}(\text{ADB}-\text{ABD})$$

AB=5			0·153205
AD=3·577			10·692995
8·577			10·846200
1·423			0·933335
BAD= 22° 55′ 15″			9·912865
180			39° 17′ 15″
157 4 45			78 32 22
78 32 22		ADB=117 49 37	
		ABD= 39 15 7	

[4.] To find GA, GB, and the angle BAG.

AB=5	AB : GA :: sin. AGB : sin. ABG	
AGB= 26° 27′	AB : GB :: sin. AGB : sin. BAG	
ADG= 39 15	0·698970	0·698970
65 42	9·801201	9·959711
180	10·500171	10·658681
BAG=114 18	9·648766	9·648766
	0·851405	1·009915
	GA=7·103	GB=10·23

[5.] To find GC.

	AC : GC :: sin. AGC : sin. CAG
AC=7	0·845098
AGC= 60° 39′	9·924491
BAG=114 18	10·769589
BAC= 57 7	9·940338
∴ CAG= 57 11	0·829251
	∴ GC=6·75

55. Sailing in a deep and unknown bay, I suddenly found the soundings decrease; and suspecting I was on a coral reef, I hauled off, having anchored a boat on it, from which the following angles were taken between three remarkable objects, A, B, and C, that appeared on the distant shore, namely: between A and B, a high-pointed rock to the right of A, the angle was 116° 40′; between B and C, a bluff summit to the right of B, the angle was 112° 30′. I obtained afterwards the distances between the three points A, B, and C, by measuring on the shore convenient base-lines, and taking angles, as pointed out in Example, p. 14. The distance from A to B was 5·75 miles, from B to C 7·5 miles, and from A to C 8·25 miles. It is required to find the position of the reef.

C

(1.) *By Construction.*

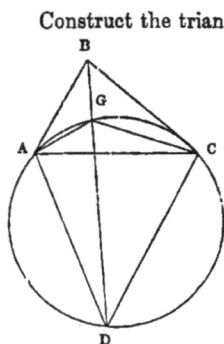

Construct the triangle ABC from the scale of equal parts, by taking AB=
5·75, BC=7·5, and AC=8·25. At the point C, in
the straight line AC, make ACD=63° 20′, the
supplement of the angle subtended by the other
two points A and B. Again, at the point A, in the
straight line AC, make CAD=67° 30′, the supplement of the angle subtended by the other two points
B and C : produce the lines AD and CD to meet in
the point D. About the triangle ADC describe a
circle ; then the place of the reef will be somewhere
in the circumference of this circle. To find it, join
BD ; and the point of intersection G is the position
of the reef required.

For since the angles in the same segment of a circle are equal (*Euclid*,
b. iii.), therefore AGD=ACD=63° 20′; therefore the angle AGB=116° 40′.
Again, CGD=CAD=67° 30′ ; therefore the angle BGC=112° 30′. And these
were the angles observed at the boat; therefore G must be the place of the
boat, or position of reef.

(2.) *By Trigonometry.*

Assume any point G as the position of the reef, and let A, B, and C be the
objects on shore. Describe a circle passing through the three points A, G,
and C. Join BG, and produce it to meet the circle in D. Join GA and GC.
Then GA, GB, and GC are the distances required. Join AD and CD. Then,
by Geometry, since the angles in the same segment are equal, ∴ angle DAC
=DGC=180°−DGC=180°−112° 30′=67° 30′, and angle ACD=AGD=180°
−AGB=180°−116° 40′=63° 20′.

[1.] Find AD, having given in the triangle ADC, AC=8·25, angle ACD=
63° 20′, and angle ADC=180°−(67° 30′+63° 20′)=49° 10′.

[2.] Find the angle BAC, having given the three sides of the triangle
ABC.

[3.] Find the angle ABD, having given AB, AD, and the angle BAD=
(BAC+CAD).

[4.] Find GA, GB, and the angle BAG, having given in the triangle ABG
the side AB and the angles AGB and ABG.

[5.] Find GC, having given in the triangle AGC the side AC, the angle
AGC, and the angle CAG=BAC−BAG.

Calculation.

[1.] To find AD.

$$AC : AD :: \sin. ADC : \sin. DCA$$

AC=8·25		0·916454
ADC=49° 10′		9·951159
DCA=63 20		10·867613
		9·878875
		0·988738
		∴ AD=9·744

[2.] To find the angle BAC.

AB=5·75	8·25		9·083546
AC=8·25	5·75		9·240332
BC=7·5	2·50		0·698970
	7·5		0·397940
	10·0		9·420788
	5·0	BAC= 61° 46′ 15″
	5·0	CAD= 67 30 0
	2·5	BAD=129 16 15

[3.] To find the angle ABD.

$$AD+AB : AD-AB :: \tan. \tfrac{1}{2}(ABD+ADB) : \tan. \tfrac{1}{2}(ABD-ADB)$$

AD= 9·744	0·601408
AB= 5·750	9·675890
15·494	10·277298
3·994	1·190164
180° 0′ 0″	9·087134
BAD=129 16 15	6° 58′
50 43 45	25 22
25 21 52	∴ ABD=32 20

[4.] To find GA, GB, and the angle BAG.

AB=5·75	AB : GA :: sin. AGB : sin. ABG	
AGB=116° 40′	AB : GB :: sin. AGB : sin. BAG	
ABG= 32 20	0·759668	0·759668
149 0	9·728227	9·711839
180	10·487895	10·471507
BAG= 31 0	9·951159	9·951159
	0·536736	0·520348
	∴ GA=3·44	∴ GB=3·31

[5.] To find GC.

AC=8·25	AC : GC :: sin. AGC : sin. CAG	
AGB=116° 40′	BAC=61° 46′	0·916454
BGC=112 30	BAG=31 0	9·708882
229 10	∴ CAG=30 46	10·625336
360		9·878875
∴AGC=130 50		0·746461
		∴ GC=5·58

The problems about to follow require a knowledge of the several points of the Mariner's Compass. We will therefore first show how the compass card is constructed and an expeditious method of learning the bearings of its points and quarter-points from the meridian. These points must be thoroughly known and committed to memory.

THE COMPASS.

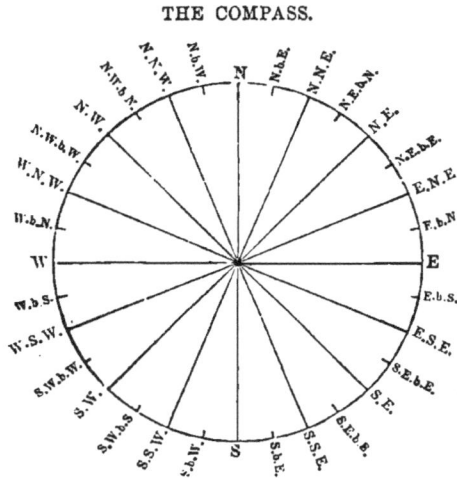

The compass card is represented above: each quadrant is divided into eight equal parts, called *points;* each point therefore contains the eighth part of 90°, or 11° 15′. The four cardinal points are the North, South, East, and West points; the intermediate points are formed and named as follows:

The middle point between N. and E. is..............................N.E.
(formed simply by putting these letters together).
    Similarly :
The middle point between N. and N.E. is..........................N.N.E.
    „   „   E. „ N.E. is..........................E.N.E.
    Again, one point from N. towards E. is N. by E. or ....N.b.E.
    „   „   E.   „   N. is E. by N. or ....E.b.N.
    „   „   N.E. „   N. is N.E. by N. or .........N.E.b.N.
    „   „   N.E. „   E. is N.E. by E. or .........N.E.b.E.

The other three quadrants of the compass are divided and named in a similar manner.

Before the student proceeds further, he should form as neatly as he can, in the manner pointed out above, and without the aid of instruments, the above compass, writing it down *several times,* until he is thoroughly acquainted with its construction and the 32 parts or points into which it is thus divided.

(6.) *Bearings or Angular Distances by Compass.*

The points of the compass are frequently referred to with respect to their position or bearing to the *right* or *left* of the cardinal point towards which the spectator is looking: thus, suppose the compass card to represent the horizon, and the spectator to be placed in the centre of the card and looking towards the north, then any point as N.E. is said to be 4 points to the right of N. (written thus—4 r. N.): E.b.N. is 7 points right of N. or 7 r. N. If the spectator is looking towards the east, then N.E. is 4 left of E. or 4 l. E., E.b.N. is 1 left of E. or 1 l. E., and so on.

EXAMPLES.

56. Required the bearings of the following points—first, from the north ; second, from the east :

N.N.E   N.E.b.N.   N.b.E.   N.N.W.   N.W.   W.b.N.

Answer.

Bearings from North......2 r. N.	N.N.E. or 3 r. N.	N.E.b.N. or 1 r. N.	N.b.E. or 2 l. N.	N.N.W. or 4 l. N.	N.W. or 7 l. N.	W.b.N. or
East ....6 l. E.	5 l. E.	7 l. E.	10 l. E.	12 l. E.	15 l. E.	

57. Required the bearings of the following points from the north, east, south, and west respectively.

S.b.E.   S.E.b.S.   S.E.b.E.   S.S.W.   W.b.S.   E.S.E.

Answer.

Bearings from	S.b.E. or	S.E.b.S. or	S.E.b.E. or	S.S.W. or	W.b.S. or	E.S.E. or
North......15 r. N.	13 r. N.	11 r. N.	14 l. N.	9 l. N.	10 r. N.	
East ...... 7 r. E.	5 r. E.	3 r. E.	10 r. E.	15 r. E.	2 r. E.	
South ...... 1 l. S.	3 l. S.	5 l. S.	2 r. S.	7 r. S.	6 l. S.	
West ...... 9 l. W.	11 l. W.	13 l. W.	6 l. W.	1 l. W.	14 l. W.	

58. Required the compass bearings of the following points :

2 r. N.   5 l. N.   3 r. S.   12 r. S.   5 r. E.   4 l. W.

Answer.

2 r. N. or N.N.E.	3 r. S. or S.W.b.S.	5 r. E. or S.E.b.S.
5 l. N. or N.W.b.W.	12 r. S. or N.W.	4 l. W. or S.W.

Each point of the compass, moreover, is subdivided into *quarter-points*, and named from the adjacent points ; thus $2\frac{1}{2}$ points to the right of north is N.N.E.$\frac{1}{2}$E.; $7\frac{3}{4}$ points to the left of north is W.b.N.$\frac{3}{4}$W., or rather W.$\frac{1}{4}$N.

59. Required the bearings of the following points—first, from the north; second, from the east:

N.N.E.$\frac{1}{4}$E.   N.$\frac{3}{4}$E.   E.b.N.$\frac{1}{2}$N.   N.E.$\frac{1}{2}$N.   N.W.$\frac{1}{2}$W.   N.$\frac{3}{4}$W.

Answer.

Bearings from	N.N.E.$\frac{1}{4}$E. or	N.$\frac{3}{4}$E. or	E.b.N.$\frac{1}{2}$N. or	N.E.$\frac{1}{2}$N. or	N.W.$\frac{1}{2}$W. or	N.$\frac{3}{4}$W. or
North...	2$\frac{1}{4}$ r. N.	$\frac{3}{4}$ r. N.	6$\frac{1}{2}$ r. N.	3$\frac{1}{2}$ r. N.	4$\frac{1}{2}$ l. N.	$\frac{3}{4}$ l. N.
East.....	5$\frac{3}{4}$ l. E.	7$\frac{1}{4}$ l. E.	1$\frac{1}{2}$ l. E.	4$\frac{1}{2}$ l. E.	12$\frac{1}{2}$ l. E.	8$\frac{3}{4}$ l. E.

60. Required the bearings of the following points—first, from the south; second, from the west:

S.b.E.$\frac{1}{2}$E.   S.E.b.S.$\frac{1}{4}$S.   S.S.E.$\frac{3}{4}$E.   S.$\frac{3}{4}$W.   W.S.W.$\frac{1}{4}$S.   W.S.W.$\frac{1}{4}$W.

Answer.

Bearings from	S.b.E.$\frac{1}{4}$E. or	S.E.b.S.$\frac{1}{4}$S. or	S.S.E.$\frac{3}{4}$E. or	S.$\frac{3}{4}$W. or	W.S.W.$\frac{1}{4}$S. or	W.S.W.$\frac{1}{4}$W. or
South.....	1$\frac{1}{2}$ l. S.	2$\frac{3}{4}$ l. S.	2$\frac{3}{4}$ l. S.	$\frac{3}{4}$ r. S.	5$\frac{3}{4}$ r. S.	6$\frac{1}{4}$ r. S.
West......	9$\frac{1}{2}$ l. W.	10$\frac{3}{4}$ l. W.	10$\frac{3}{4}$ l. W.	7$\frac{1}{4}$ l. W.	2$\frac{1}{4}$ l. W.	1$\frac{3}{4}$ l. W.

61. Required the compass bearings of the following points:

2$\frac{1}{4}$ r. N.	1$\frac{3}{4}$ l. N.	10$\frac{1}{4}$ r. S.	7$\frac{3}{4}$ l. N.	3$\frac{1}{2}$ r. S.	3$\frac{1}{2}$ l. S.
6$\frac{1}{2}$ r. S.	10$\frac{1}{2}$ l. S.	14$\frac{3}{4}$ r. N.	8 r. N.	8 l. S.	15$\frac{1}{2}$ r. N.

Answers.

2$\frac{1}{4}$ r. N. or N.N.E.$\frac{1}{4}$E.   10$\frac{1}{4}$ r. S. or W.N.W.$\frac{1}{4}$N.   3$\frac{1}{2}$ r. S. or S.W.$\frac{1}{4}$S.

6$\frac{1}{2}$ r. S. or W.S.W.$\frac{1}{2}$W.   14$\frac{3}{4}$ r. N. or S.b.E.$\frac{1}{4}$E.   8 l. S. or East.

1$\frac{3}{4}$ l. N. or N.N.W.$\frac{1}{4}$N.   7$\frac{3}{4}$ l. N. or W.$\frac{1}{4}$N.   3$\frac{1}{2}$ l. S. or S.E.$\frac{1}{2}$S.

10$\frac{1}{2}$ l. S. or E.N.E.$\frac{1}{2}$N.   8 r. N. or East.   15$\frac{1}{2}$ r. N. or S.$\frac{1}{2}$E.

(7.) Attached to the compass card, and coinciding with the line N.S., is a magnetic bar of steel, by means of which the card, when balanced on a fine point near its center, will indicate the compass bearing or direction of any object beyond it. Thus, the compass being placed near the helm, the bearing of the ship's head is seen at once, and the direction in which the ship is steered is readily noted.

The Log-line.

(8.) The log is a flat piece of thin wood of a quadrantal form, loaded in the circular side with lead sufficient to make it swim upright in the water; to this is fastened a line about 150 fathoms long, called the *log-line*, which is divided into certain spaces called *knots;* the length of each knot is supposed to be the same part of a nautical mile (about 6080 feet) that half a minute is of an hour; hence 1 knot$=\dfrac{6080}{120}=51$ feet nearly. If, therefore, 1 knot runs out in half a minute (shown by a half-minute glass), the rate of the ship is supposed to be 1 mile an hour; if 2 knots, the rate is 2 miles an hour, and so on. The length of the knot is very rarely so much as 51 feet, and the hour-glass used is not always a half-minute glass: various modifications of the two instruments are made, to render this method of measuring the ship's way tolerably correct; these will be more clearly understood in the use of the instruments themselves.

NAUTICAL PROBLEMS SOLVED BY TRIGONOMETRY AND ALSO BY CONSTRUCTION.

(9.) It is proved in *Navigation*, Part II. p. 151, that the *distance, true difference of latitude, departure,* and *course* between any two places on the earth may be correctly represented by the three sides and one of the angles of a right-angled plane triangle; and that the *meridional difference of latitude* and *difference of longitude* by two sides of a triangle which is similar to the same right-angled plane triangle. Thus, let A and B be the two places, AB a straight line joining them, and AC that part of the meridian passing through A that is intercepted between A and a straight line BC drawn through B perpendicular to AC: then

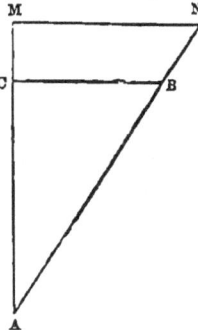

AC will represent the true difference of latitude.

AB	„	„	distance.
BC	„	„	departure.
angle CAB	„	„	course from A to B.

Again, if AC be produced to M, so that AM may be equal to the meridional difference of latitude between A and B, and MN be drawn parallel to CB to meet AB produced to N : then

AM will represent the meridional difference of latitude, and MN the difference of longitude between the two places A and B.

The line AN is not used in navigation.

We thus see that questions in navigation or plane sailing may be much simplified by considering the above six terms as forming parts of two similar right-angled plane triangles connected together as shown in the above figure; for then we can solve nearly all the questions in plane sailing by the simple application of the rules in Trigonometry for right-angled plane triangles.

We will proceed to exemplify this by means of a few useful problems in sailing, and will at the same time show how these problems may be solved by *construction;* that is, by measuring with mathematical instruments the several lines and angles given by the problem, limiting ourselves at present to questions that require a knowledge only of the several parts of the smaller triangle A B C.

62. A ship from latitude 47° 30′ N. has sailed S.W.b.S. 98 miles : find by construction, and by calculation, the latitude in and departure.

(1.) *By Construction.*

Let A represent the point the ship departed from, A D the meridian, and

A p, drawn at right angles to it, the parallel of latitude of the ship. At the point A, with a chord of 60°, describe the quadrant m p, and cut off mc=S.W.b.S. or 33° 45′=the course ; and through c draw a line A B. From a scale of equal parts take A B=98 miles, the distance ; and through B draw B D parallel to A p, meeting A D in D. Then B is the place the ship has arrived at, A D is the difference of latitude, and B D is the departure. If A D

and B D are measured by the same scale of equal parts, it will be found that the difference of latitude A D is about 81 miles, and the departure B D about 54 miles. The figure may be more easily laid off by means of a protractor (see any work on Practical Geometry).

(2.) *By Trigonometry.*

In the right-angled triangle A B D are given the course D A B=33° 45′, and distance A B=98 miles ; to find the difference of latitude A D, and departure B D.

By Rule, p. 9, $\frac{A D}{A B}$=cos. D A B ∴ A D=A B cos. D A B

By same Rule, $\frac{B D}{A B}$=sin. D A B ∴ B D=A B sin. D A B.

Reducing these formulæ to logarithms, we have :

log. A D=log. A B+log. cos. D A B−10
log. B D=log. A B+log. sin. D A B−10

A B=98   D A B=33° 45′

log. A B.........1·991226	log. A B.........1·991226
,, cos. D A B...9·919846	,, sin. D A B...9·744739
,, A D..........$\overline{1·911072}$	,, B D..........$\overline{1·735965}$

∴ A D=81·5=1° 21′ 30″S.   ∴ B D=54·4

Lat. from...47   30   0 N.

Lat. in......$\overline{46\quad 8\quad 30}$ N. and dep. 54·4 W.

The Traverse Table.

Questions involving two or more parts of a right-angled plane triangle are often very easily solved by means of a table called the Traverse Table, which contains the difference of latitude and departure calculated for any course and distance, so that when any two of these quantities are given the other two may be found *by inspection:* thus in the last example are given the course and distance to find diff. lat. and departure. Entering the table, therefore, with the course S.W.b.S=3 points and distance 98 miles, we find the corresponding diff. lat. and dep. to be 81·5 and 54·4 respectively.

63. A ship from latitude 20° 30′ N. has sailed W.S.W. 120 miles: find by construction, by logarithms, and by the traverse table the latitude she is in and the departure she has made.

Ans. Lat. in. 19° 44′ 6″ N., dep. 110′·9.

When a ship has described more than one course during the day, and it is required to show by a diagram the latitude she has arrived at, we may proceed as in the following example.

64. A ship in latitude 47° 30′ N. sailed N.N.W. 90 miles, and E.b.S. 60 miles : find latitude in and departure.

(1.) *By Construction.*

Let A represent the place the ship left, and with any convenient radius describe the circle NWSE to represent the horizon of the ship. Draw two diameters NS and WE at right angles to each other. Let NS represent the meridian, and WE the parallel of latitude the ship departed from. To mark off the several courses and distances during the day, we may proceed as follows :

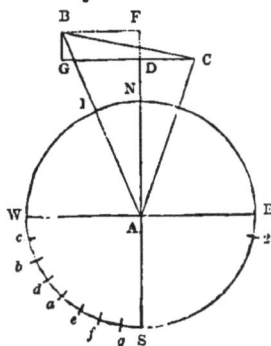

[1.] Divide one of the quadrants, WS, into eight equal parts, to form a scale of points. This may be done by bisecting WS in a, and then Wa in b, and Wb in c: the other points, *defg*, may then be readily filled in.

[2.] Mark off on the circumference the several courses, thus: take N_1= N.N.W., or two points from the scale in WS ; and E_2=E.b.S., or one point from the east.

[3.] Through A_1 draw the straight line AB, and make AB=90 miles by a scale of equal parts; and through B, parallel to a line passing through A_2, draw BC=60 miles. The point C represents the place the ship has arrived at. Join AC, and through C draw CD parallel to WE, meeting AN produced in D. Then AD is the difference of latitude, and DC the departure made good during the day. Also the angle CAD and line AC represent the direct course and distance from A to C.

If we measure AD by the scale of equal parts, we shall find the difference of latitude AD about 71 miles to the north, and the departure DC about 24 miles to the east of the place the ship left. The latitude arrived at is found thus :

$$\begin{array}{ll}
\text{Lat. A} & 47^\circ \ 30' \ \text{N.} \\
\text{Diff. lat.} & 1 \ \ \ 11 \ \ \text{N.} \\
\text{Lat. in} & \overline{48 \ \ \ 41} \ \text{N. and dep. 24 E.}
\end{array}$$

(2.) *By Trigonometry and Traverse Table.*

To calculate the difference of latitude and departure between A and C, we must proceed as follows :

Through B draw BF parallel to WE, meeting the meridian produced in F. Then in the right-angled triangle ABF are given the course BAF=2 points, and distance AB=90 miles, to calculate AF the diff. lat. and BF the departure. Again, through B draw BG parallel to the meridian NS; and through C draw CG parallel to WE, meeting BG in G. Then in the triangle BGC are given the course GBC=7 points, and BC=60 miles, to calculate BG =FD the diff. lat. and GC the departure. By performing the calculation, we find that AF=83·2 miles to the north, and BG=11·7 to the south; so that the diff. lat.=83·2 N.—11·7 S.=71·5 miles. Similarly may be found GC=58·8 to the east, and FD or DG=34·4 to the west; so that the departure =58·8 E.—34·4 W.=24·4 E.

This method of computing the diff. lat. and departure separately for every course is in practice avoided by making use of the Traverse Table. The diff. lat. and departure, when taken out of the table, are arranged under proper heads in the following form :

Points.	Courses.	Dist.	Diff. lat.		Dep.	
			N.	S.	E.	W.
2	N.N.W.	90	83·2	—	—	34·4
7	E.b.S.	60	—	11·7	58·8	—
			83·2	11·7	58·8	34·4
			11·7		34·4	
			71·5 N.		24·4 E.	

$$\begin{array}{ll}
\text{Lat. from} & 47^\circ \ 30' \ \ \text{N.} \\
\text{Diff. lat.} & 1 \ \ \ 11·5 \ \text{N.}=71·5' \ \text{N.} \\
\text{Lat. in} & \overline{48 \ \ \ 41·5} \ \text{N. and dep. 24·4. E.}
\end{array}$$

65. A ship in latitude 50° 30′ N. has sailed during the day N.N.E. 100 miles and W.b.S. 70 miles: what latitude is she in, and what departure has she made? 　　　　　*Ans.* Lat. in 51° 48′·7 N., dep. 30′·4 W.

66. A ship from latitude 50° 48′ N. has sailed during the day on the following courses. Required the latitude in, and departure, and the direct course and distance.

1. S.E..........40 miles.	4. N.W.b.W....30 miles.
2. N.E.........28　„	5. S.S.E.........36　„
3. S.W.b.W...52　„	6. S.E.b.E......58　„

(1.) *By Construction.*

Let A be the place sailed from, and N W S E the horizon of the ship. Draw the meridian N S, and parallel of latitude W E.

Divide one of the quadrants into eight equal parts for a scale of points, as in the last example, and by means of this scale mark off the circumference the several courses, viz. S_1=S.E., N_2=N.E., S_3=S.W.b.W., N_4=N.W.b.W., S_5=S.S.E., and S_6=S.E.b.E.

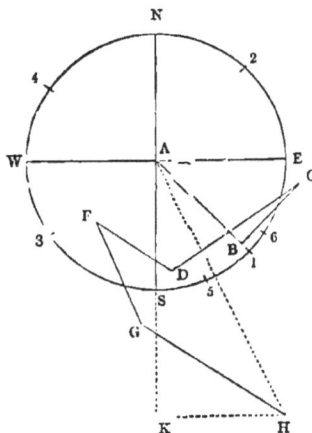

Through A draw A B=40 by a scale of equal parts; through B, and parallel to A_2, draw B C=28 miles; through C, and parallel to A_3, draw C D=52 miles; through D, and parallel to A, draw D F=30 miles; through F, and parallel to A, draw F G=36 miles; and lastly, through G, and parallel to A_6, draw G H=58 miles. The point H is the place the ship has arrived at. Join A H, and through H draw H K parallel to W E and meeting the meridian N S produced in K. Then A K is the difference of latitude, and K H the departure made good during the day. Also the angle K A H represents the direct course, and the line A H the direct distance from A to H.

If we measure A K by the scale of equal parts, we shall find the difference of latitude A K about 86 miles to the south, and the departure K H about 42 miles to the east of the place the ship left. The latitude arrived at is found thus:

Lat. A.............50° 48′ N.
Diff. lat.......... 1　26 S.
Lat. in...........49　22 N. and dep. 42′ E.

(2.) *By Trigonometry and Traverse Table.*

The diff. lat. and departure for each course and distance may be computed as in Example, p. 25. But to avoid this tedious operation, the several quantities may be taken out of the Traverse Table by inspection, as follows:

Points.	Courses.	Dist.	Diff. lat.		Dep.	
			N.	S.	E.	W.
4	S.E.	40	—	28·3	28·3	—
4	N.E.	28	19·8	—	19·8	—
5	S.W.b.W.	52	—	28·9	—	43·2
5	N.W.b.W.	30	16·7	—	—	24·9
2	S.S.E.	36	—	33·3	13·8	—
5	S.E.b.E.	58	—	32·2	48·2	—
			36·5	122·7	110·1	68·1
				36·5	68·1	
			Diff. lat.=86·2		42·0=dep.	

Lat. from.........50° 48′ 0″ N.
Diff. lat.......... 1 26 12 S.=86·2 S.
Lat. in............49 21 48 N. and dep. 42′ E.

To calculate the direct course from A to H, or the angle KAH, and the direct distance AH, we have in the plane right-angled triangle AKH, AK= 86·2, and KH=42; to find the course KAH, and distance AH.

[1.] To find the direct course KAH, tan. $KAH = \dfrac{KH}{AK}$

∴ log. tan. KAH—10=log. KH—log. AK
or log. tan. KAH=10+log. KH—log. AK

KH=42

AK=86·2

    log. KH+10....11·623249
     „ AK.......... 1·935507
     „ tan. KAH... 9·687742
∴ the direct course, or KAH=S.25° 58′ 30″ E.

[2.] To find the distance AH.

$\dfrac{AH}{AK}$=sec. KAH, or AH=AK sec. KAH

∴ log. AH=log. AK+log. sec. KAH—10
   log. AK...............1·935507
    „ sec. KAH—10..0·046247
    „ AH...............1·981754
∴ distance, or AH=95·8 miles.

The two following examples, taken out of that valuable old work, Robertson's *Elements of Navigation,* are given for practice in construction, and for the singularity of the form of the diagrams resulting from the several courses and distances.

67. A ship sails from a place in latitude 40° N. on the following courses. Required the latitude arrived at (by Construction).

Course.			Course.		
1	S.E.b.S.	29'	13	West	62'
2	N.N.E.	10	14	North	10
3	E.S.E.	50	15	West	8
4	E.N.E.	50	16	South	10
5	S.S.E.	10	17	West	62
6	N.E.b.N.	29	18	South	7
7	West	25	19	E.¾S.	62
8	S.S.E.	10	20	South	110
9	W.S.W.½W.	42	21	W.N.W.½W.	42
10	North	110	22	N.N.E.	10
11	E.¾N.	62	23	West	25
12	North	7			

Ans. The ship returns to the place sailed from.

68. Two ships, A and B, part company in lat. 31° 31′ N., and meet again at the end of two days, having run as follows:

The ship A.			The ship B.		
Course.			Course.		
1	N.N.E.	96'	1	N.N.W.	96'
2	W.S.W.	96	2	E.S.E.	96
3	E.S.E.	96	3	W.S.W.	96
4	N.N.W.	96	4	N.N.E.	96

Required the latitude arrived at, and the direct course and distance of each ship (by Construction).

Ans. Direct course due north, and distance 104 miles. Lat. in 33° 15′ N.

CORRECTIONS IN PLANE SAILING.

(10.) Three corrections are sometimes necessary to be applied to the course steered by compass, to reduce it to the true course; and the converse. These are called :

      (1.) The correction for variation of the compass.

      (2.) The correction for deviation of the compass.

      (3.) The correction for leeway.

(1.) *The Correction for Variation of the Compass.*

(11.) The magnetic needle seldom points to the true north. Its deflection to the east or west of the true north is called the *variation of the compass;* it is different in different places, and it is also subject to a slow change in the same place. The variation of the compass is ascertained at sea by observing the magnetic bearing of the sun when in the horizon, or at a given altitude above it. From this observation the true bearing is found by rules given in nautical astronomy. The difference between the true bearing and the observed bearing is the variation of the compass.

The method of correcting the course for variation will be more readily understood by means of a few examples.

Suppose the variation of the compass is found to be two points to the east, that is, the needle is directed two points to the right of the north point of the heavens; then the N.N.W. point of the compass card will evidently point to the true north, and every other point on the card will be shifted round two points. If, therefore, a ship is sailing *by compass* N.N.W., or, as it is expressed, the compass course is N.N.W., her true course will be north; that is, *two points to the right of the compass course.* In a similar manner it may be shown that, when the variation is two points westerly, the true course will be *two points to the left of the compass course.* Hence this rule :

To find the true course, the compass course being given.

        Easterly variation allow to the right.

        Westerly   ,,   ,,   left.

From the preceding considerations it will be easy to deduce the converse rule, namely :

To find the compass course, the true course being given.

        Easterly variation allow to the left.

        Westerly   ,,   ,,   right.

69. Find the true course, having given the compass course N.W.$\frac{1}{2}$W. and variation 3$\frac{1}{4}$W.

	pts.	qrs.	
Compass course......	4	2	left of N.
variation...............	3	1	left.*
true course...........	7	3	left of N.=W.$\frac{1}{4}$N.

70. Find the compass course, having given the true course W.$\frac{1}{4}$N. and variation 3$\frac{1}{4}$W.

	pts.	qrs.	
True course	7	3	left of N.
variation.............	3	1	right.
compass course......	4	2	left of N.=N.W.$\frac{1}{2}$W.

Find the true course in each of the following examples:

	Compass course.	Var.	Answers.
71.	N.N.E.	2$\frac{1}{4}$W.	N.$\frac{1}{4}$W.
72.	N.W.	1$\frac{3}{4}$E.	N.N.W.$\frac{1}{4}$W.
73.	S.W.$\frac{3}{4}$W.	1$\frac{1}{2}$E.	W.S.W.$\frac{1}{4}$W.
74.	S.	2W.	S.S.E.
75.	W.	2$\frac{1}{2}$E.	N.W.b.W.$\frac{1}{2}$W.

Find the compass course in each of the following examples:

	True course.	Var.	Answers.
76.	N.N.E.$\frac{1}{2}$E.	$\frac{1}{4}$W.	N.N.E.$\frac{3}{4}$E.
77.	N.	1$\frac{1}{2}$E.	N.b.W.$\frac{1}{2}$W.
78.	S.S.W.	2W.	S.W.
79.	S.W.	0	S.W.
80.	N.b.W.$\frac{1}{4}$W.	1$\frac{1}{4}$W.	N.

(2.) *The Correction for Deviation of the Compass.*

(12.) This correction of the compass arises from the effect of the iron on board ship on the magnetic needle, in deflecting it to the right or left of the plane of a great circle called the *magnetic meridian.* The increased quantity of iron used in ships has caused this correction to be attended to now more than formerly, as its effects and magnitude have become more perceptible. The amount of the deviation arising from this local cause varies as the mass of iron changes its position with respect to the compass. When a fore and aft line coincides with the direction of the magnetic meridian, the iron in the ship may be supposed to be nearly equally distributed on both sides of the needle, and its effect in deflecting the needle may be inappreciable. In other positions of the ship with respect to the magnetic meridian, the iron

---

* When names are alike (that is, both left or both right), *add :* when unlike, *subtract,* marking remainder with the name of the greater.

may produce a sensible deflection of the needle ; and this deflection or deviation will in general be the greatest when the ship's head points to the east or west.

Various methods are used to determine this correction. The one usually adopted is to place a compass on shore, where it may be beyond the influence of the iron of the ship, or any other local disturbing force, and to take the bearing of the ship's compass, or some object in the same direction therewith ; at the same time, the observer on board takes the bearing of the shore compass ; then if 180° be added to the bearing at the shore compass, so as to bring it round to the opposite point, the difference between the result and the bearing at ship's compass will be the amount of the deviation of the compass for that position of the ship.

The ship is then swung round one or two points, and a similar observation made ; and thus the local deviation found for a second position of the ship. This being repeated for every point or two points of the compass, the deviation is thus known for all positions of the ship. A table, similar to the one below, is then formed, and the courses corrected for this deviation by the following rules ; which resemble those already given for correcting for variation.

Deviation of Compass of H.M.S. ——, for given positions of the ship's head.

Direction of ship's head.		Deviation of compass.	Direction of ship's head.		Deviation of compass.
		nearly			nearly
N.	E.	2° 45′ or $\frac{1}{4}$ pt.	S.	W.	3° 0′ or $\frac{1}{4}$ pt.
N.b.E.	E.	4 57 or $\frac{1}{2}$,,	S.b.W.	W.	4 20 or $\frac{1}{2}$,,
N.N.E.	E.	7 30 or $\frac{3}{4}$,,	S.S.W.	W.	5 0 or $\frac{1}{2}$,,
N.E.b.N.	E.	9 0 or $\frac{3}{4}$,,	S.W.b.S.	W.	6 7 or $\frac{1}{2}$,,
N.E.	E.	10 0 or $\frac{3}{4}$,,	S.W.	W.	7 0 or $\frac{1}{2}$,,
N.E.b.E.	E.	10 55 or 1 ,,	S.W.b.W.	W.	7 27 or $\frac{3}{4}$,,
E.N.E.	E.	10 40 or 1 ,,	W.S.W.	W.	7 50 or $\frac{3}{4}$,,
E.b.N.	E.	9 55 or $\frac{3}{4}$,,	W.b.S.	W.	8 20 or $\frac{3}{4}$,,
E.	E.	8 50 or $\frac{3}{4}$,,	W.	W.	8 50 or $\frac{3}{4}$,,
E.b.S.	E.	7 15 or $\frac{3}{4}$,,	W.b.N.	W.	8 10 or $\frac{3}{4}$,,
E.S.E.	E.	5 35 or $\frac{1}{2}$,,	W.N.W.	W.	6 50 or $\frac{1}{2}$,,
S.E.b.E.	E.	3 40 or $\frac{1}{4}$,,	N.W.b.W.	W.	5 40 or $\frac{1}{2}$,,
S.E.	E.	1 50 or $\frac{1}{4}$,,	N.W.	W.	4 50 or $\frac{1}{2}$,,
S.E.b.S.	E.	0 20 or 0 ,,	N.W.b.N.	W.	3 20 or $\frac{1}{4}$,,
S.S.E.	W.	0 56 or 0 ,,	N.N.W.	W.	1 40 or 0 ,,
S.b.E.	W.	2 20 or $\frac{1}{4}$,,	N.b.W.	E.	1 10 or 0 ,,

(13.) *To find the true course,* having given the compass course and the deviation.

Easterly deviation allow to the right.
Westerly   ,,   ,,   left.

81. Correct the compass course W.b.S. for deviation ¾W. (known from table, p. 32).

pts. qrs.
Compass course.......7   0 right of S.
deviation.........,......0   3 left.
true course 6   1 right of S., or W.S.W.¼W.

82. Correct the compass course N.W.½W. for deviation ½W. (from deviation table, p. 32), and also for variation of compass 3¼W.

pts. qrs.
Compass course 4   2 l. N.
deviation...........0   2 l.
variation 3   1 l.
                      3   3 l.
true course   8   1 l. N.
                      16
or true course 7   3 r. S.=W.¼S.

Find the true course in each of the following examples, by correcting for deviation from table, p. 32, and for variation :

	Compass Course.	Var.	Answers.
83.	N.N.E.	2¼W.	N.½E.
84.	N.W.	1¾E.	N.N.W.¾W.
85.	S.W.¾W.	1½E.	S.W.b.W.¾W.
86.	S.	2W.	S.S.E.¼E.
87.	W.	2½E.	W.N.W.¼W.
88.	W.¾N.	1½W.	W.S.W.½W.

(14.) *To find the compass course,* having given the true course and deviation.

Easterly deviation allow to the left.
Westerly   „   „   right.

NOTE.—The true course should first be corrected for variation (if any) by Rule, p. 30 (which is similar to the above), so as to get a compass course nearly, and then this course for deviation, from table, p. 32.

89. Required the compass course, the true course being W.S.W.¼W., variation 0, and deviation ¾W. (see table).

pts. qrs.
True course 6   1 r. S.=compass course nearly.
deviation...........0   3 r.
compass course ...7   0 r. S., or W.b.S.

D

90. Required the compass course, the true course being S.W., variation of compass $2\frac{1}{4}$E., and deviation as in table, p. 32.

	pts.	qrs.	
True course	4	0 r.	S.
variation	2	1 l.	
compass course nearly ...	1	3 r.	S., or S.b.W.$\frac{3}{4}$W.
deviation...................	0	2 r.	
compass course............	2	1 r.	S. = S.S.W.$\frac{1}{4}$W.

Required the compass course in each of the following examples (for deviation, see table, p. 32) :

	True course.	Var.	Answers.
91.	N.$\frac{1}{2}$E.	$2\frac{1}{4}$W.	N.N.E.
92.	N.N.W.$\frac{3}{4}$W.	$1\frac{3}{4}$E.	N.W.
93.	S.W.b.W.$\frac{3}{4}$W.	$1\frac{1}{2}$E.	S.W.$\frac{3}{4}$W.
94.	S.S.E.$\frac{1}{4}$E.	2W.	S.
95.	W.N.W.$\frac{1}{4}$W.	$2\frac{1}{2}$E.	W.
96.	W.S.W.$\frac{1}{2}$W.	$1\frac{1}{2}$W.	W.$\frac{3}{4}$N.

(3.) *The Correction for Leeway.*

(15.) This correction is the angle which the ship's track makes with the direction of a fore and aft line : it arises from the action of the wind on the sails, &c. not only impelling the ship forwards, but pressing against it sideways, so as to cause the actual course made to be to *leeward* of the apparent course,

as shown by the fore and aft line. The amount of leeway differs in different ships, depending on their construction, on the sails set, the velocity forwards, and other circumstances. Experience and observation, therefore, usually determine the amount of leeway to be allowed.

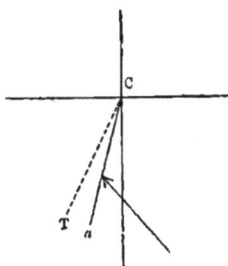

The method of correcting for leeway will be best seen by the following example :

Suppose the apparent course is S.S.W.$\frac{1}{2}$W., and leeway two points, the wind being S.E., required the correct course.

Draw two lines at right angles to each other towards the cardinal points of compass, and a line, as ca, to represent (roughly) the course of the ship, and another to represent the direction of the wind (as the arrow in fig.) ; then it will be seen that the corrected course, as cT, will be to the *right* of the apparent course ; *the observer being always supposed to be at the center c, and looking towards the cardinal point from whence the course is measured;* hence

pts. qrs.

Apparent course......2   2 r. S.

leeway 2   0 r.

corrected course......4   2 r. S.=S.W.½W.

EXAMPLES.

Correct the following courses for leeway, so as to find the true courses:

	Apparent course.	Wind.	Leeway.	Answers.
97.	N.N.E.	W.N.W.	1½	N.E.½N.
98.	N.W.	N.N.E.	2	W.N.W.
99.	E.S.E.	S.	2½	E.½N.
100.	E.	N.b.E.	¾	E.¾S.

Correct the following compass courses for deviation, variation, and leeway, so as to find the true courses. The deviation is found in table, p. 32, and the variation of compass is supposed to be in each example 2½ W.

	Course.	Wind.	Leeway.	Answers.
101.	N.W.¼W.	W.S.W.	2½	N.W.¾W.
102.	S.E.¼E.	E.N.E.	2¼	S.E.½E.
103.	W.¼S.	S.S.W.	2	W.S.W.½W.
104.	N.¾W.	W.b.N.	1½	N.b.W.¾W.

These examples may be worked out in the following manner:

pts. qrs.

Ex. 101.   Compass course.........4   1 l. N.

deviation........0   2 l.

variation 2   2 l.

3   0 l.

7   1 l. N.

leeway 2   2 r.

true course.......... ...4   3 l. N.=N.W.¾W.

In the preceding examples the courses, both true and compass, are corrected for variation and deviation by a formal rule. The student, however, should also know how to make these corrections by means of a construction, as in the following examples:

105. Given the true course=N. 42° 28′ E., and the variation of the compass=1½ points easterly; construct a figure to show the compass course.

Construction.

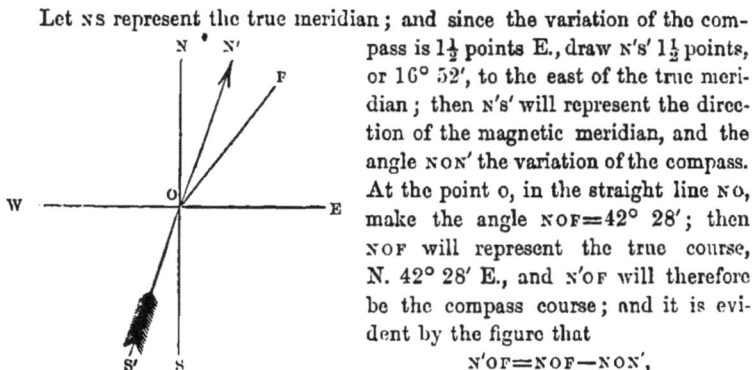

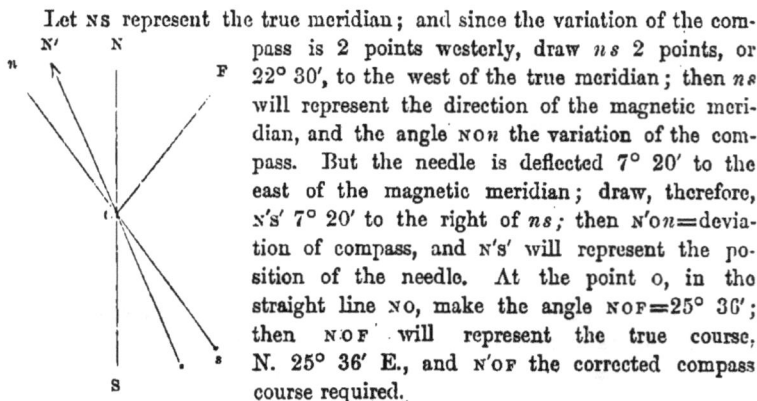

Let NS represent the true meridian; and since the variation of the compass is 1½ points E., draw N′S′ 1½ points, or 16° 52′, to the east of the true meridian; then N′S′ will represent the direction of the magnetic meridian, and the angle NON′ the variation of the compass. At the point o, in the straight line NO, make the angle NOF=42° 28′; then NOF will represent the true course, N. 42° 28′ E., and N′OF will therefore be the compass course; and it is evident by the figure that

$$N′OF = NOF - NON′,$$

or compass course=true course − variation
$$= 42° 28′ - 16° 52′ = 25° 36′;$$

and since this angle is to the right of the magnetic meridian,

∴ the compass course=N. 25° 36′ E.

106. Given the true course=N. 25° 36′ E., the variation=2 points W., and deviation on account of local attraction=7° 20′ E.; to find the corrected compass course (by construction).

Construction.

Let NS represent the true meridian; and since the variation of the compass is 2 points westerly, draw ns 2 points, or 22° 30′, to the west of the true meridian; then ns will represent the direction of the magnetic meridian, and the angle NOn the variation of the compass. But the needle is deflected 7° 20′ to the east of the magnetic meridian; draw, therefore, N′S′ 7° 20′ to the right of ns; then N′On=deviation of compass, and N′S′ will represent the position of the needle. At the point o, in the straight line NO, make the angle NOF=25° 36′; then NOF will represent the true course, N. 25° 36′ E., and N′OF the corrected compass course required.

By the figure, N'OF = NOF + (NON − NON')

$$= 25°\ 36' + (22°\ 30' − 7°\ 20')$$
$$= 25\ \ 36 + 15\ \ 10$$
$$= 40\ \ 46$$

∴ the corrected compass course = N. 40° 46′ E.

By practical rule (p. 34),

True course.................25° 36′ r. N.
variation ....22 30 r.
compass course nearly.....48 6 r. N.
deviation .... 7 20 l.
∴ compass course ....40 46 r. N. = N. 40° 46′ E.

107. Given the true course, S. 15° 58′ E.; variation of compass, $2\frac{1}{4}$ points W.; deviation, 4° 20′ W. Construct figure, and find compass course.
Ans. Compass course, S. 13° 40′ W.

CONSTRUCTION OF A MERCATOR'S CHART.

(16.) A Mercator's chart represents the surface of the earth as a plane (p. 3), and is constructed as follows:

Draw at the bottom of a sheet of paper a straight line to represent the most southern parallel of latitude required for the chart; divide it into equal parts, as degrees, &c., regulating the length of each degree according to the number required in the chart and the size of the paper: or if the chart is to be drawn to any given scale, as one inch or ·7 of an inch, &c., make the length of each degree of longitude on the scale one inch or ·7 inch accordingly. The line so drawn at the bottom of the paper we may call the *longitude line;* at each extremity of this line erect a perpendicular: these perpendiculars are the *graduated meridians,* on which must be marked the length of each degree of latitude.

To obtain the linear measure of the degrees of latitude.

(17.) Write down on a slip of paper, in a vertical column, the degrees of latitude which the chart is to contain, beginning with the highest degree: take out from the Table of Meridional Parts the meridional parts for each degree, and write them down opposite their corresponding latitudes; take the successive differences between the first and second, second and third, &c., of these meridional parts, and thus make a second vertical column. Then, to find the points on the graduated meridians through which each parallel of latitude is to be drawn, transfer these meridional differences of latitude to the graduated meridians, by measuring along the longitude line at the bottom of the chart the number of minutes, &c. contained in each meridional difference of latitude taken in order,* making a dot on the graduated meridians at the extremity of each measure ; connect these dots by straight lines : these will be the parallels of latitude required. The intermediate meridian lines are then to be filled in, by drawing lines through the divisions of the base-line, or through every fifth degree, or through as many as may be considered sufficient; a compass should then be drawn on the chart (or more than one, if the chart is large) ; this will be useful to determine the bearings of different points, or for more conveniently finding the latitude and longitude of the ship when her course and distance run are given. To construct the compass, take some convenient intersection of a meridian and parallel as a center, and describe a circle with any suitable radius ; mark the points of the circumference cut by the meridian with the letters N. and S., and complete the compass by inserting the other points.

To lay down upon the chart a point whose latitude and longitude are given.

(18.) Lay the edge of a ruler (or doubled edge of paper) along the given parallel of latitude ; measure off the degrees, &c. between the given longitude and the longitude of the nearest meridian line drawn on the chart; apply this difference to the edge of the ruler in the proper direction, and the point on the chart whose latitude and longitude are given will be found.

EXAMPLE.

108. Let it be required to lay down on the chart a point whose latitude is 50° 48′ N., and longitude 22° 10′ W.

Place the edge of the ruler over latitude 50° 48′ N. in the chart, and with a pair of compasses, or otherwise, take 2° 10′ (the difference between 22° 10′ and 20°, assuming that the meridian line of 20° is the nearest on the chart), and lay it along the ruler towards the left from the meridian of 20° : the position of the required point will then be determined.

In this manner a ship's daily track is usually pricked off; for the latitude and longitude being known at noon, her place is given at that time, and the entire track during the voyage can be seen by connecting, by straight lines, the places or points on the chart thus found.

---

* This may be easily done by placing along the graduated longitude line the doubled edge of a piece of paper, and transferring to it the required lengths ; or by taking the proper distance by a pair of compasses.

To copy a chart on a different scale.

(19.) Having drawn the meridians and parallels as pointed out in p. 37, find the latitude and longitude of the prominent points in the chart, and transfer these points to the new chart, p. 38; then sketch in neatly with the hand the outline of the coast between the assumed points, and insert all the other necessary parts of the chart, as rocks, shoals, islands, &c., as accurately as possible.

(20.) *To find the course and distance between two given places on the chart.*

1. *To find the course.* Place the edge of a parallel ruler over the two places on the chart, and keeping one part of the parallel ruler firm, move the other till the edge passes through the center of the compass described on the chart : the edge thus lying on the compass will point out the course between the two places. It also may be found by means of the small semicircular protractor contained in most cases of mathematical instruments in the following manner. Place the straight edge of the protractor against the edge of the ruler as it lies upon the two places, and slide it along till the center of the protractor is on one of the meridian lines ; then the course will be seen on that point in the circumference of the protractor through which the meridian line passes. A rectangular protractor will determine this with equal facility.

2. *To find the distance.* The distance is found (nearly) by transferring the space or interval between the two places as it appears on the chart to the side line, or graduated meridian, *as nearly opposite the two places as possible :* the degrees, &c. (turned into minutes) which this space measures on the graduated meridian will be the distance required. If the places have the same latitude, the distance is found more accurately as follows : Take half the space or interval between them ; apply it to the graduated meridian above and below the parallel on which the places are situated : the difference between the degrees of the extreme points (turned into minutes) will be the distance required (nearly). If the places have the same longitude, it is evident the difference of their latitudes (or the sum, if they are on different sides of the equator) will be the distance between the two places.

To find the latitude and longitude in by the chart, having given the course and distance from a given place.

(21.) Lay down the course on the chart in the manner pointed out in pages 25-27, or by any other method suited to the instruments at hand. To the line (or edge of the ruler) thus lying in the direction of the

course, apply the distance run (turned into degrees and minutes if necessary), measured from that part of the graduated meridian which is adjacent
to the given place, and to that to which the ship is sailing : this distance so
taken along the line or edge of the ruler from the place sailed from will
determine the position of the ship on the chart, that is, its latitude and
longitude in.

(22.) The following is an example of constructing a Mercator's chart
and of tracing the ship's track thereon.

109. Construct a Mercator's chart on a scale of one inch extending from
54° N. to 58° N., and from long. 178° E. to 178° W., and lay down thereon
the ship's track, namely the several true courses and distances from the following sailings, thus forming a track chart :

Compass courses and dist. on each course.	S.	E.b.S.	N.b.E.	N.$\frac{1}{4}$W.	N.W.	S.W.b.W.
	90′	100′	65′	60′	80′	75′

Variation of the compass, 1 point E.

Correcting compass courses to get the true courses, we have

True courses and distance.	S.b.W.	E.S.E.	N.N.E.	N.$\frac{1}{2}$E.	N.W.b.N.	W.S.W.
	90′	100′	65′	60′	80′	75′

First, to construct the chart within the given limits.

This is done by following the directions given in Arts. 16, 18, pp. 37,
38, as follows :

1. At the bottom of the paper draw the longitude line, and divide it into
four equal parts each one inch long, to contain the four degrees of longitude,
namely, from 178° E. to 178° W., and erect the graduated meridians at each
extremity of the longitude line.

2. Subdivide each degree of the longitude line into 6 or any other convenient number of equal parts : if into 6, as in the diagram, then each sub-
division will be 10′. This longitude line may now be used as a scale of
equal parts from which to set off any longitude distances on the chart ; but
it will be better to make a scale of equal parts on a separate piece of paper
by drawing a straight line and dividing it into several equal parts each one
inch, and one of these equal parts again to subdivide into 6 or more equal
parts.

3. Write down the meridional parts for each degree of latitude so as to
get the M. D. lat. for each degree, beginning with the highest degree : thus

lat. 58...merid. parts...4294
  ,, 57... ,, ...4183...M. D. lat.=111′ between 57° and 58°
  ,, 56... ,, ...4074... ,, =109 ,, 56 ,, 57
  ,, 55... ,, ...3968... ,, =106 ,, 55 ,, 56
  ,, 54... ,, ...3865... ,, =103 ,, 54 ,, 55

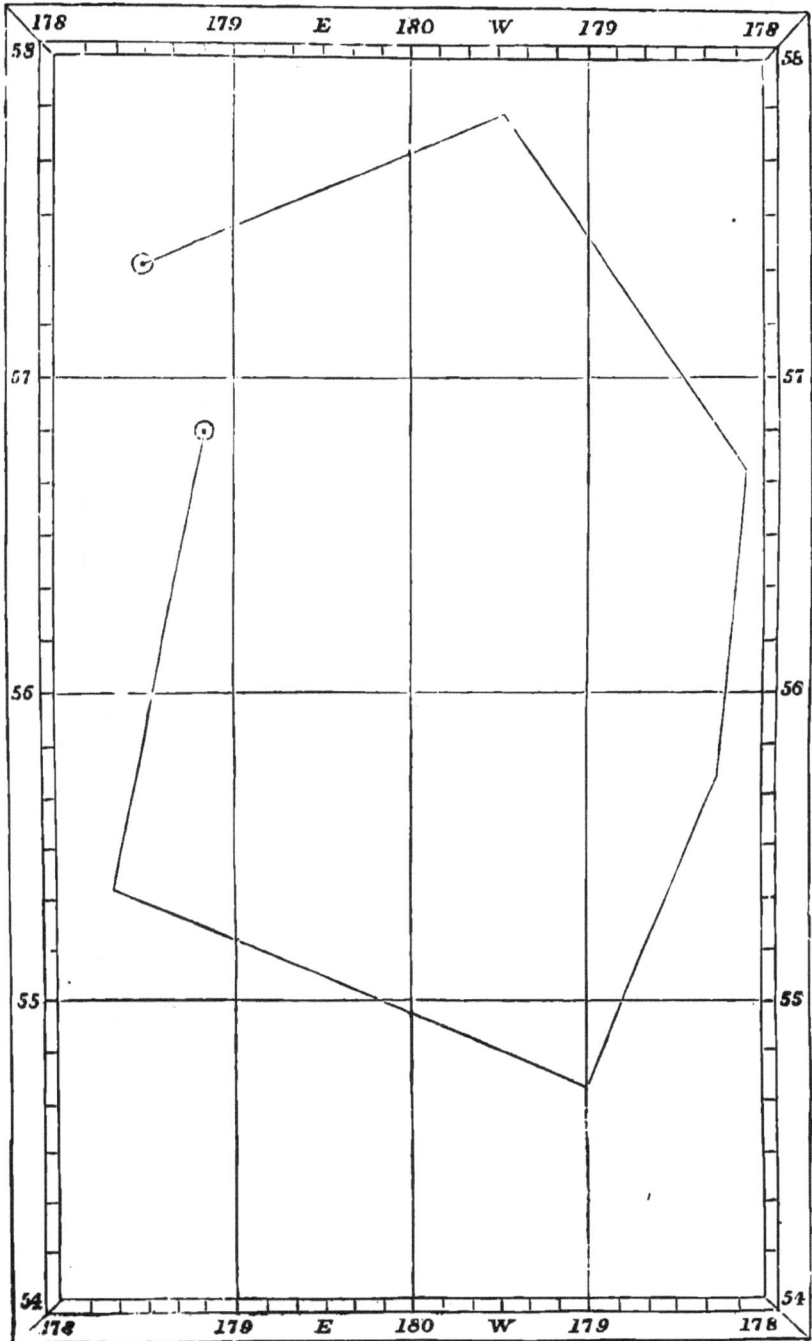

Ans. Lat. in 57° 21' N., long. in 178° 31' E.

4. Transfer the quantities 103, 106, 109, 111, taken off the scale or from the longitude line to the graduated meridians; draw the parallels of latitude and intermediate meridians and mark them with their proper degrees (see preceding page), and the chart will be ready to lay down on it the track of the ship, that is, the several true courses and distances made during the day. This is done as follows:

Second, to lay down the ship's track on the chart.

1. Find the point on the chart corresponding to the place the ship sailed from: thus

Lay the edge of a ruler (or a doubled edge of a piece of paper) over lat. 56° 54' N.; and since the longitude is 178° 50' E.—that is, 10' to the left of 179° E.—with a pair of compasses or otherwise take a distance of 10' and lay it along the ruler to the left from meridian 179° E.; make a small dot at the point, and the position of the ship, namely lat. 56° 54' N. and long. 178° 50' E., is determined.

2. Draw the several lines to represent the ship's track: thus (Art. 21)

From the point thus found draw a line S.b.W. (or 1 point west of meridian 179° E.) and equal to 90', remembering to take all the *distances from that part of the graduated meridians adjacent to the respective courses*. From the southern extremity of this line draw a line E.S.E. 100', and proceed in the same manner to lay down the several other courses given in the question; when it will be found that the ship has arrived at a place in lat. 57° 21' N. and long. 178° 31' E.

3. Finish the chart off neatly by rubbing out pencilled and superfluous lines, and surrounding it with parallel lines, &c., as a boundary (see track chart on preceding page).

Fundamental Formulæ for Plane Sailing.

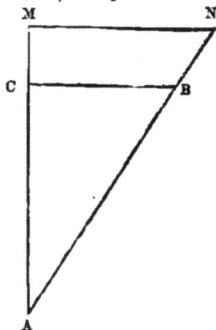

(23.) In p. 23 it is shown that the six terms—distance, tr. diff. latitude, departure, meridional difference of latitude, diff. long., and course—may be correctly represented by two similar right-angled plane triangles A B C, A M N.

If any two or more of these terms contained in either of the triangles are known, the others may be found by means of the trigonometrical ratios (Art. 4, p. 9).

Thus, in triangle A B C, let A B the distance and B A C the course be given: to find B C the departure, and A C the true diff. lat.

By p. 9, $\dfrac{BC}{AB}$=sin. A or $\dfrac{dep.}{dist.}$=sin. course $\therefore$ dep.=dist. sin. course (1).

„ $\dfrac{AC}{AB}$=cos. A or $\dfrac{T.\ D.\ lat.}{dist.}$=cos. course

$\therefore$ T. D. lat.=dist. cos. course (2).

In triangle AMN, given AM the M. D. lat. and MAN the course, to find MN the diff. long.

By p. 9, $\dfrac{MN}{AM}$=tan. A or $\dfrac{diff.\ long.}{M.\ D.\ lat.}$=tan. course

$\therefore$ diff. long.=M. D. lat. tan. course (3).

To obtain the formulæ for proving the rules in parallel sailing, and middle lat. sailing, we must proceed as pointed out in p. 163, *Nav.* Part II. These formulæ are

In parallel sailing......dist.=diff. long. × cos. lat..........(4).
In middle lat. sailing...dep.=diff. long. × cos. mid. lat....(5).

Collecting these Navigation formulæ* for the sake of reference,

Departure = distance × sin. course..................(1).

True diff. lat. = distance × cos. course ....(2).

Diff. long. = meridional diff. lat. × tan. course...(3).

In parallel sailing, distance = diff. long. × cos. lat............(4).

In mid. lat. sailing,

departure (nearly) = diff. long. × cos. mid. lat....(5).

(This latter formula is only approximately correct: see *Nav.* Part II. p. 145.)

* The mathematical student will no doubt use these formulæ and figure in p. 42 to solve questions in Navigation, rather than follow the formal rules contained in the following pages. These rules, he will see, are, in fact, simply the above formulæ expressed in words.

RULES IN NAVIGATION.

Rule 1. *To find the course and distance from one place to another,* having given the latitudes and longitudes of the two places (by using meridional parts, called Mercator's method).

(1.) Find true difference of latitude, meridional difference of latitude, and difference of longitude; reduce the true difference of latitude and difference of longitude to minutes, attaching thereto the proper letters. Rules (*a*), (*b*), (*d*).

(2.) *To find the course.* From the log. difference of longitude (increased by 10) subtract the log. mer. diff. latitude; the remainder is the log. tan. course, which find in the tables, and place before it the letter of the true difference latitude, and after it the letter of the difference longitude, to indicate the direction of course. At the same opening of the tables, take out the log. secant course.

(3.) *To find the distance.* Add together log. secant course and log. true difference latitude; the sum (rejecting 10 in the index) will be the log. distance, which find in the tables.

110. Required the course and distance from A to B.

lat. A 45° 15′ N.		long. A 35° 26′ W.	
„ B 47 10 N.		„ B 32 15 W.	

M.P.

lat. A...45° 15′ N.	3051·2 N.		long. A...35° 26′ W.	
„ B...47 10 N.	3217·4 N.		„ B...32 15 W.	
1 55	M. D. lat. 166·2 N.		3 11	
60			60	

T. D. lat. 115 N.   diff. long. 191 E.

log. diff. long. +10..12·281033   log. sec. course 0·182767

„ M. D. lat....... 2·220631   „ T. D. lat...2·060698

„ tan. course......10·060402   „ dist.........2·243465

∴ course N. 48° 58′ E.   ∴ distance 175′.

Required also the *compass* course in the above example: var. of compass being 2 points W., and deviation on account of local attraction as in table (p. 32). See Rule, p. 33.

	pts.	qrs.	
True course......48° 58′ r N. or 4	1	r. N.*	
variation............................2	0	r.	
compass course nearly.............6	1	r. N.=E.N.E.¼E.	
deviation............................1	0	l.	
∴ compass course.................5	1	r. N.=N.E.b.E.¼E.	

---

* Degrees are converted into points, or the converse, by means of the table for that purpose in the nautical tables.

Examples in Navigation are usually worked without attaching to each logarithm taken out its name or designation, as in the following example :

111. Required the course and distance from A to B.

lat. A...51° 31′ N.   long. A...0° 6′ W.

„ B...54 33 N.   „ B...3 5 E.

		M.P.	
51° 31′ N.		3618	0° 6′ W.
54 33 N.		3921	3 5 E.
3 2	M. D. lat.	303	3 11
60			60

diff. lat. 182 N.   diff. long. 191 E.

    12·281033   0·072650

    2·481443   2·260071

    9·799590   2·332721

∴ N. 32° 13′ 30″ E.=course. 215·1=dist.

Required the course and distance from A to B in each of the following examples, by Rule 1 or Mercator's method.

	Lat. from and lat. in	Long. from and long. in.	Answers. Course and distance.
112.	lat. A 49° 52′ S.	long. A 17° 22′ W.	course N. 26° 36′ E.
	lat. B 42 13 S.	long. B 11 50 W.	dist. 513·3 miles.
113.	lat. A 49 10 N.	long. A 29 17 W.	course N. 37° 48′ W.
	lat. B 56 45 N.	long. B 39 5 W.	dist. 576 miles.
114.	lat. A 50 48 N.	long. A 1 10 E.	course N. 41° 55′ W.
	lat. B 52 35 N.	long. B 1 25 W.	dist. 144 miles.
115.	lat. A 58 24 N.	long. A 4 12 W.	course N. 32° 34′ E.
	lat. B 63 17 N.	long. B 2 13 E.	dist. 347·6 miles.
116.	lat. A 2 37 N.	long. A 110 42 W.	course S. 75° 12′ W.
	lat. B 0 0	long. B 120 36 W.	dist. 614·4 miles.
117.	lat. A 3 30 N.	long. A 33 40 E.	course S. 42° 32′ E.
	lat. B 4 10 S.	long. B 40 42 E.	dist. 624 miles.

Required also the compass courses in Examples 115, 116, and 117, the variation of compass being 2 points E., and deviation as in table (p. 32). See Rule, p. 33.

ANSWERS.

115. compass course N.½E. nearly.

116. „ „ S.W.b.W.¼W. nearly.

117. „ „ E.S.E.¼E. „

Rule 2. *To find the latitude and longitude in, having given the course and distance* (by Mercator's method).

(1.) *To find latitude in.* Add together log. cos. course* and log. distance, the sum (rejecting 10 in the index) will be log. true difference latitude, which find in the tables ; reduce to degrees and minutes, and place the letter N. or S. against it, according as course is northward or southward.

(2.) Apply true difference latitude to latitude from, so as to get the latitude in. Rule (*e*).

(3.) *To find longitude in.* Take out the meridional parts for the two latitudes, and get M. D. lat. Rule (*b*).

(4.) Add together log. tangent course and log. meridional difference latitude ; the sum (rejecting 10 in the index) will be the log. difference longitude, which find in the tables ; reduce to degrees and minutes, and place the letter E. or W. against it, according as the course is eastward or westward.

(5.) Apply difference longitude to longitude from, so as to get longitude in. Rule (*f*).

EXAMPLES.

118. Sailed from A, N. 37° 10′ E., 472·6 miles ; required the latitude and longitude in.

lat. A 27° 20′ N.   long. A 25° 12′ E.

log. cos. course 9·901394		log. tan. course 9·879740	
„ dist ....2·674494		„ M. D. lat. 2·641474	
„ T. D. lat. 2·575888		„ diff. long. 2·521214	
∴ T. D. lat. 376·6′		diff. long. 332·1′	
or 6° 17′ N.	M.P.	or 5° 32′ E.	
lat. from 27 20 N.......1706 N.		long. from 25 12 E.	
„ in 33 37 N.......2144 N.		„ in 30 44 E.	
M. D. lat. 438			

119. A ship in latitude 27° 0′ S. and longitude 123° W. sailed S.S.E.½E. (or S. 28° 7′ 30″ E.) 150 miles : required the latitude and longitude in.

9·945430		9·727957	
2·176091		2·176091	
2·121521		1·904048	
6,0)13,2·3		80′·1	
diff. lat....... 2° 12′ 18″ S.	M.P.	1° 20′ 6″ E.	
lat. from.....27 0 0 S.	1683 S.	123 0 0 W.	
„ in........29 12 18 S.	1833 S.	121 39 54 W.	
	150	Long. in.	

* Take out, at same opening of tables, log. tan. course, and place it a little to the right.

Required the latitude and longitude in, by Rule 2 or Mercator's method, in each of the following examples, having sailed from A as follows :

	Course and dist. from A.		Lat. A.	Long. A.	Lat. in.	Long. in.
120.	N. 26° 36′ E.	513·5′	49° 52′ S.	17° 22′ W.	42° 13′ S.	11° 50′ W.
121.	S. 48 58 W.	175·2	47 10 N.	32 15 W.	45 15 N.	35 26 W.
122.	N. 29 10 E.	373·4	52 10 N.	17 32 W.	57 36 N.	12 15 W.
123.	S. 37 7 E.	370·0	70 14 S.	25 30 E.	75 9 S.	38 5 E.
124.	N. 47 47 E.	272·4	50 15 S.	15 10 E.	47 12 S.	20 16 E.

Rule 3. *To find the course and distance* (middle latitude method), having given the latitudes and longitudes of the two places.

(1.). Find the true difference latitude, middle latitude, and difference longitude by Rules (*a*), (*c*), (*d*).

(2.) *To find the course.* Add together log. cos. mid. lat. and log. diff. long., and from the sum subtract log. true difference latitude ; the remainder is the log. tan. course, which find in the tables, and mark it with the same letters as the true difference latitude and difference longitude. From the same opening take out the log. secant of course.

(3.) *To find distance.* To the log. secant course just found add the log. true difference latitude ; the sum (rejecting 10 in index) will be the log. distance.

EXAMPLES.

125. Required the course and distance from A to B, by middle latitude method.

lat. A 50° 25′ N.   long. A 27° 15′ W.

„ B 47 12 N.   „ D 30 20 W.

lat. A 50° 25′ N.........50° 25′ N.	long. A 27° 15′ W.
„ B 47 12 N.........47 12 N.	„ B 30 20 W.
3 13 2)97 37	3 5
60 mid. lat. 48 48	60
T. D. lat. 193 S.	diff. long. 185 W.
log. cos. mid. lat. 9·818681	log. sec. course 0·072849
„ diff. long. ... 2·267172	„ T. D. lat...2·285557
12·085853	„ dist.........2·358406
log. T. D. lat. ... 2·285557	∴ dist. 228·2′
„ tan. course... 9·800296	∴ course S. 32° 16′ W.

Required the course and distance from A to B in each of the following examples, by middle latitude method :

	Lat. from and lat. in.	Long. from and long. in.	Answers. Course and dist.
126.	lat. A 49° 52′S.	long. A 17° 22′W.	N. 26° 40′ E.
	lat. B 42 13 S.	long. B 11 50 W.	513·6
127.	lat. A 21 15 S,	long. A 0 30 W.	S. 14° 37′ E.
	lat. B 30 27 S.	long. B 2 10 E.	570·5
128.	lat. A 60 15 S.	long. A 14 55 E.	S. 32° 50′ E.
	lat. B 65 36 S.	long. B 22 30 E.	382

Rule 4. *To find the latitude and longitude in* (by middle latitude method), having given the course from a given place, and distance.

(1.) *To find latitude in.* Add together log. cos. course* and log. distance ; the sum (rejecting 10 in the index) is the log. true difference latitude, which find from tables, and mark N. or S. according as the course is northward or southward.

Apply true difference latitude (turned into degrees and minutes, if necessary) to the latitude from, and thus get latitude in. Rule (*e*). Find the middle latitude. Rule (*c*).

(2.) *To find longitude in.* Add together log. sin. course, log. distance, and log. secant middle latitude; the sum (rejecting 20 in the index) is the log. difference longitude, which find in tables, and mark E. or W. according as the course is eastward or westward. Apply the difference longitude (in degrees and minutes) to the longitude from, and thus get longitude in. Rule (*f*).

129. Sailed from A, S. 37° 10′ W., 472·6 miles ; required lat. in and long. in (by middle lat. method).

lat. A 27° 20′ S.　　　　　long. A 25° 12′ W.

log. cos. course...9·901394	log. sin course...9·781134
„ dist. ....2·674494	„ dist...........2·674494
„ T. D. lat......2·575888	„ sec. mid. lat.0·064531
∴ T. D. lat. 376·6′	„ diff. long....2·520159
or 6°17′S.	∴ diff. long. 331·3′
lat. from 27 20 S.	or 5°31′W.
„ in 33 37 S.	long. from 25 12 W.
2)60 57	„ in 30 43 W.
mid. lat. 30 28	

---

* Take out at the same opening log. sin. course, and put it down a little to the right.

Required the latitude and longitude in, by middle latitude method, in each of the following examples, having sailed from A as follows:

	Course and dist. from A.	Lat. A.	Long. A.	Answers. Lat. in.	Long. in.
130.	N. 25° 42′ W. 427·3′	64° 10′ N.	40° 15′ W.	70° 35′ N.	48° 17′ W.
131.	S. 48 58 W. 175·2	47 10 N.	32 15 W.	45 15 N.	35 26 W.
132.	N. 34 48 W. 383·7	50 25 N.	3 40 E.	55 40 N.	2 24 W.

PARALLEL SAILING.

Rule 5. *To find the course and distance*, having given the latitude of the two places, and their longitudes.

(1.) Find the difference longitude.

(2.) The *course* is evidently due east or due west, according as the longitude in is to the east or west of longitude from.

(3.) *To find the distance.* Add together log. cos. latitude and log. difference longitude ; the sum (rejecting 10 in index) is the log. distance, which find in the table.

133. Required the course and distance from A to B.

lat. A...80° N.   long. A...3° 50′ E.
 ,, B...80 N.   ,, B...6 10 W.

long. from........3° 50′ E.   log. cos. lat.......9·239670
 ,, in...........6 10 W.   ,, diff. long.....2·778151
        10 0   ,, dist...........2·017821
        60   .·. dist. 104·2′.
     600 W.

    .·. the course is west.

134. Required the *compass* course and distance from A to B.

lat. A ......50° 48′ N.   long. A ......100° 0′ E.
 ,, B ......50 48 N.   ,, B ......101 0 E.

Variation of the compass two points E., and deviation as in table, p. 32.

			pts. qrs.
long. A...100° E.	9·800737	True course...............	8 0 r. of N.
,, B...101 E.	1·778151	variation..................	2 0 l.
1	1·578888	compass course nearly	6 0 r. of N. = E.N.E.
60	37·9=dist.	deviation	1 0 l.
60 E.		compass course.........	5 0 r. of N.
		or N.E. by E.	

E

Required the true course and distance from A to B in each of the following examples:

	Lat. A and B.	Long. A.	Long. B.	Answers. Course and dist.
135.	70° 10′ S.	15° 10′ E.	22° 15′ E.	East 144·2′
136.	50 48 N.	5 0 W.	5 0 E.	East 379·2
137.	50 10 N.	40 25 W.	50 10 W.	West 374·7
138.	48 10 N.	100 0 W.	110 0 W.	West 400·2
139.	75 13 N.	15 20 E.	0 0 E.	West 234·7
140.	80 15 N.	179 0 E.	176 0 W.*	East 50·8

Rule 6. *To find the longitude in*, having given the course and distance, and latitude and longitude from.

Add together log. sec. lat. and log. distance; the sum (rejecting 10 in the index) will be the log. difference longitude. Find the natural number thereof, and turn it into degrees, and mark it E. or W. according as the course is E. or W. Apply difference longitude to longitude from, and thus find longitude in.

The latitude in is the same as the latitude from.

EXAMPLE.

141. Sailed from A due east 1000 miles, required the latitude and longitude in. Lat. A...32° 10′ S.; long. A...28° 42′ W.

lat. in=lat. from=32° 10′ S.
log. sec. lat. ....0·072372
„ dist. ....3·000000
„ diff. long. ....3·072372
∴ diff. long. 1181′, or 19° 41′ E.
long. from ....28 42 W.
∴ long. in .... 9 1 W.

Required the latitude and longitude in, in each of the following examples:

	Course and dist.	Lat. from.	Long. from.	Answers. Lat. in.	Long. in.
142.	East 492·5′	52° 10′ N.	0° 29′ W.	52° 10′ N.	12° 54′ E.
143.	East 1752	60 0 N.	5 10 W.	60 0 N.	53 14 E.
144.	East 560	57 32 N.	13 5 W.	57 32 N.	4 18 E.
145.	West 740	60 0 N.	50 0 W.	60 0 N.	74 40 W.

* In this example it is evident we must modify the general rule; for the diff. long. is never considered to be greater than 180°. When, therefore, the above rule gives the diff. long. greater than 180°, subtract it from 360°, and apply thereto a contrary letter to the one directed by the rule; the result will be the diff. long. to be used.

Application and use of formulæ in page 43.

(24.) The preceding rules are the principal ones used in Navigation. It would be easy for the mathematical student to make for himself others, by means of the relations between the several terms course, dist., dep., &c., as shown by the formulæ and diagram in p. 43 : he would find then no difficulty in solving a great variety of problems similar to the following :

146. Sailed from A, in long. in 3° 10′ W., 300 miles due east, and altered my longitude 10 degrees; required the latitude and longitude in.

Thus, by form (4)...dist.=diff. long. × cos. lat.

∴ cos. lat. $=\dfrac{\text{dist.}}{\text{diff. long.}}=\dfrac{300}{600}=\frac{1}{2}$ ∴ lat. in $=60°$, and long. in $=6°$ 50′ E.

147. Wishing to make a small island, I took the ship to windward of it in the same latitude with the island, namely 50° 48′ N. The longitude of the ship by chronometer was 20° 35′ W., and the long. of the island was 23° 50′ W. What was my distance from the island?

In this example of parallel sailing we have given lat. 50° 48′, and diff. long. 3° 15′, or 195′, to find distance.

By form (4)...dist.=diff. long. × cos. lat.

log. diff. long..........2·290035
  ,, cos. lat............9·800737
  ,, dist.............. ..2·090772 ∴ dist. 123·2 miles.

148. What course must be steered so that the departure may be one-third the distance?

In fig. p. 42, we have given the relation between the departure CD and distance AB; that is

$$\frac{\text{dep.}}{\text{dist.}}=\text{sin. course}$$

and by the question, $\dfrac{\text{dep.}}{\text{dist.}}=\frac{1}{3}$

∴ sin. course $=\frac{1}{3}$ and course $=19°$ 28′.

149. Sailed between the N. and E. 100 miles, and altered my latitude 1° 10′ : required the course.

In fig. p. 42, AC=T. D. lat.=1° 10′=70′ ; AB=distance=100′

and cos. course $=\dfrac{\text{AC}}{\text{AB}}=\dfrac{70}{100}=\dfrac{7}{10}$ ∴ course=N. 45° 35′ E.

To find the course and distance from one place to another, as from A to B, having given T. D. lat., mid. lat., and diff. long. By fig. p. 42.

150. Find course and distance from A to B.

lat. A......58° 24′ N. long. A....4° 12′ W.
,, B......63 17 N. ,, B....2 13 E. (Ex. 115).

M. P.

lat. A...58° 24′ N.	4339·8	long. A...4° 12′ W.
,, B...63 17 N.	4942·6	,, B...2 13 E.
4 53	M. D. lat....602·8=AM	6 25
60		60
T. D. lat......293=AC		diff. long......385=MN

To find the course.

(By fig.)...tan. course=$\dfrac{MN}{AM}$

| log. MN+10...12·585461 |
| ,, AM......... 2·780173 |
| ,, tan. course 9·805288 |
| ∴ course=N. 32° 34′ E. |

To find the distance.

distance AB=AC sec. course.

| log. AC............ 2·466868 |
| ,, sec. course...10·074293 |
| ,, dist........... 2·541161 |
| ∴ distance=347′·6. |

To find the latitude and longitude in, having given the latitude and longitude from, the course, and distance. By fig. p. 42.

151. Required the latitude and longitude in, having sailed from A, in lat. 52° 10′ N., long. 17° 32′ W. (see fig.), N. 29° 10′ E., 373·4 miles. (See Ex. 122.)

In triangle CAD, CA=AB cos. A, or T. D. lat.= dist. × cos. course ; from which T. D. lat. may be found, and therefore M. D. lat. and lat. in.

In triangle AMN, MN=AM tan. A, or diff. long.=M. D. lat. × tan. course ; from which diff. long. is found, and therefore long. in.

To find T. D. lat.

By fig., T. D. lat.=dist. × cos. course.
log. dist............2·572174
,, cos. course...9·941117
,, T. D. lat......2·513291
∴ T. D. lat.=326
or 5° 26′ N.
lat. from......52 10 N....3681·5
,, in.........57 36 N....4249·3
M. D. lat.= 567·8

To find diff. long.

By fig., diff. long.=M.D. lat. × tan. course.
log. M. D. lat..........2·754195
,, tan. course ....9·746726
,, diff. long. ....2·500921
,, diff. long. ......316·9
5° 17′ E.
long. from...17 32 W.
,, in......12 15 W.

Some useful examples may be worked out in middle latitude sailing and parallel sailing by means of the two figures in p. 154, *Nav.* Part II.: thus

152. Required the course and distance from A to B by middle latitude sailing : lat. A=58° 24′ N., lat. B.=63° 17′ N., long. A=4° 12′ W., long. B 2° 13′ E. (Ex. 115).

In the figures let A and D be the two places ; then U z=diff. long., A c=T. D. lat., s u the middle lat. ∴. s s₁=departure (nearly)=c D. (See *Nav.* Part II. p. 150.)

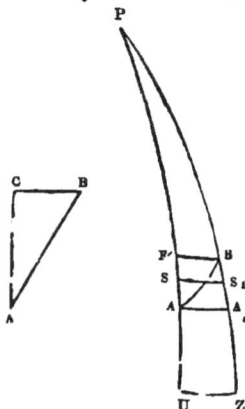

1. $\dfrac{SS_1}{UZ}$=cos. SU, or dep.=diff. long. × cos. mid. lat.

2. $\dfrac{CB}{AC}$=tan. A, or dep.=T. D. lat. × tan. course.

Equating, we have
T. D. lat. tan. course=diff. long. × cos. mid. lat.

∴. tan. course=$\dfrac{\text{diff. long.} \times \text{cos. mid. lat.}}{\text{T. D. lat.}}$, which determines the course ;

and $\dfrac{AB}{BC}$=sec. A, or distance=T. D. lat. × sec. course ;

whence distance is known.

lat. A......58° 24′ N.	58° 24′ N.	long. A... ..4° 12′ W.
,, B......63 17 N.	63 17 N.	,, B......2 13 E.
4 53	2)121 41	6 25
60	mid. lat....60 50 30″	60
T. D. lat. 293 N.		diff. long. 385 E.

To find course.   To find distance.

By above formula,

tan. course=$\dfrac{\text{diff. long.} \times \text{cos. mid. lat.}}{\text{T. D. lat.}}$ dist.=T. D. lat. × sec. course.

log. cos. mid. lat.... 9·687842   log. sec. course...0·074616
,, diff. long......... 2·585461   ,, T. D. lat......2·466868
12·273303   ,, distance ......2·541484
,, T. D. lat......... 2·466868   ∴. dist.=348′.
,, tan. course...... 9·806435 ∴. course=N. 32° 38′ E.

NOTE. ss₁, used above as the departure, is the departure *nearly :* this method, therefore, is only approximately correct, but it will be found near enough for all practical purposes.

153. Required the latitude and longitude in, having sailed from A in lat. 52° 10′ N. and long. 17° 32′ W., N. 29° 10′, E. 373·4 miles (Ex. 122), by middle latitude method (see last figures).

1. $\dfrac{AC}{AB}$ = cos. A .·. T. D. lat. = dist. × cos. course,

which determines diff. lat. and .·. the latitude in.

2. $\dfrac{CB}{AB}$ = sin. A .·. departure = dist. × sin. course.

3. $\dfrac{UZ}{SS_1}$ = sec. su .·. diff. long. = dep. × sec. mid. lat.

$\qquad\qquad\qquad$ = dist. × sin. course × sec. mid. lat.

To find T. D. lat.	To find diff. long.

T. D. lat. = dist. × cos. course.

$\qquad$ diff. long. = dist. × sin. course × sec. mid.lat.

log. dist. ....2·572174	log. distance ......2·572174
,, cos. course.....9·941117	,, sin. course...9·687842
$\overline{\qquad 2\cdot513291}$	,, sec. mid. lat. 0·240149
log. T. D. lat. = 326	$\overline{\qquad 2\cdot500165}$
$\overline{\qquad 5°\ 26'\ \text{N}.}$	.·. diff. long. = 316′
lat. from ...52 10 N.	or $\qquad\overline{5°\ 16'\ \text{E}.}$
,, in.......57 36 N.	long. from...17 32 W.
$\overline{\quad 2)109\ \ 46}$	,, in......12 16 W.
mid. lat. ...54 53	

P

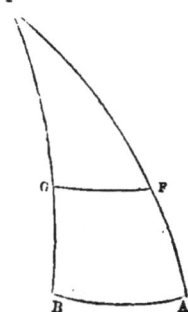

154. Find the course and distance from G to F : lat. G = lat. F = 50° 48′ N., long. G = 5° W., long. F = 5° E. (Ex. 136). This is an example in parallel sailing.

$\qquad$ long. G.... 5° W.

$\qquad$,,   F... 5 E.

$\qquad\qquad\overline{10} = 600' = $ diff. long.

GB = 50° 48′, and GF is the arc of the parallel described by ship. Then $\dfrac{GF}{AB}$ = cos. GB,

or dist. = diff. long. × cos. lat.

log. diff. long....2·778151

$\quad$,, cos. lat......9·800737

$\quad$,, dist. ....$\overline{2\cdot578888}$

The course is evidently *East.*

.·. distance = 379·2.

155. Required the latitude and longitude in, having sailed due east 560 miles from a place G (see fig.) in lat. 57° 32′ N. and long. 13° 5′ W. (Ex. 144).

lat. G B=57° 32′, dist. G F=560
and A B=diff. long. required

By fig., $\dfrac{AB}{GF}$=sec. G B

∴ diff. long.=dist. sec. course

log. dist.......2·748188
 ,, sec. lat...0·270180
 ,, diff. long. $\overline{3\cdot018368}$
∴ diff. long.=1043
 or $\overline{17° 23′}$ E.
long. from......13 5 W.
∴ ,, in......... $\overline{4\ 18}$ E.

The examples from 110 to 145 may be worked in a similar manner (by making a figure to suit each case) as Examples 146 to 155.

EXAMPLES.

By Construction and Traverse Sailing.

A. A ship sails from A, in latitude 24° 32′ N., on the following courses and distances : required latitude in and direct course and distance. (1.) S.W.b.W. 45′; (2.) E.S.E. 50′; (3.) S.W. 30′; (4.) S.E.b.E. 60′; (5.) S.W.b.S$\frac{1}{4}$W. 63′.   *Ans.* Lat. 22° 3′, south, 149·2′.

B. A ship sails from A, in lat. 28° 32′ N., on the following courses and distances: required lat. in and direct course and distance. (1.) N.W.b.N. 20′; (2.) S.W. 40′; (3.) N.E.b.E. 60′; (4.) S.E. 55′; (5.) W.b.S. 41′; (6.) E.N.E. 66′.   *Ans.* Lat. 28° 32′ N., east, 70·2′.

C. Since yesterday at noon we have run the following courses : required diff. lat. and departure, and direct course and distance. (1.) S.W.b.S. 20′; (2.) W. 16′; (3.) N.W.b.W. 28′; (4.) S.S.E. 32′; (5.) E.N.E. 14′; (6.) S.W. 36′.   *Ans.* Diff. lat. 50·7′, dep. 50·7′, S.W., 71·7′.

EXAMPLES.

By Construction and Trigonometry.

A′. Two ships, A and B, sail from two islands bearing the one from the other N.E. and S.W., their distance being 76′. A sails S.b.E., and B E.b.S.: at last they meet. How far has each sailed ?
        Ans. A sails S.b.E. 68·4′, B sails E.b.S. 68·4′.

B′. Coasting along shore, a headland bore N.E.b.N.; then, having run E.b.N. 15′, the headland bore W.N.W.: required the distance from headland at each observation.   *Ans.* ,8·5′ and 10·8′.

C′. Yesterday noon we were in lat. 33° 15′ N., and bound to a port in latitude 28° 35′ N., lying 196′ to the west; and this day at noon we were in lat. 30° 20′ N., having made departure 168′ west: required the direct course and distance to the port.   ·   *Ans.* S. 14° 55′ W., dist. 108·8′.

(25.) *To find the place of the ship at noon*, that is, its latitude and longitude, having given the latitude and longitude at the preceding noon, the compass courses, and distances run in the interval, the deviation of the compass for each course on account of local attraction, the variation of the compass, the leeway, the velocity and direction of current (if any), &c., constitutes what is called the Day's Work.

The Day's Work.

Rule 7. (1.) Correct each course for variation, deviation, and leeway; thus get the true courses, and arrange the same in a tabular form, as in the example, p. 28. Add together the hourly distances sailed on each course, and insert the same in table opposite the true course.

(2.) Take out of the traverse table the true difference latitude and departure for each course and distance, putting them down in the columns headed with the same letters as in course. Previously to opening the traverse table, fill up the columns of true difference latitude and departure not wanted by drawing horizontal lines; this will frequently prevent mistakes.

(3.) If the ship does not sail from a place whose latitude and longitude are known, her bearing and distance from some near object, as a church-spire, &c., must be ascertained, and also its latitude and longitude. Then the ship is supposed to sail from this known object to her anchorage, her course being the opposite to the bearing of the object from the ship. This course must be corrected like the rest for variation and deviation, and inserted in the table as an actual course, with the distance of the object as a distance.

(4.) If a current sets the ship in any ascertained direction, and with a known velocity, these also may be conceived to be an independent course and distance, and must be corrected for variation, and should be for deviation also, if the latter correction is appreciable, which is rarely the case.

(5.) *To find the latitude in.* The quantities in the four columns of true difference latitude and departure being added up separately, the difference between the north difference of latitude and south difference of latitude, with the name of the greater, will give the true difference of latitude made at the end of the day. The departure is found in a similar manner. Apply true difference latitude to latitude from, so as to obtain the latitude in.

(6.) *To find the longitude in.* Add together log. sec. mid. lat. and log. departure, the result (rejecting 10 in the index) is the log. difference longitude. Find this in table, and thus the longitude in is found.*

The following example, worked out in detail, will perhaps be sufficient to explain the operations directed in the above general rule.

EXAMPLE.

156. April 27th, 1852, at noon. A point of land in latitude 36° 30′ S. and longitude 110° 20′ W. bore by compass E.b.N.½N. (ship's head being S.E. by S. by compass), distant 14 miles; afterwards sailed as by the following log account; required the latitude and longitude in, on April 28th, at noon.

Hours.	Knots.	Courses.	Winds.	Lee-way.	Devi-ation.	Remarks.
1	2·5	S.W.½W.	S.b.E.	2¼	½ l.	P.M.
2	3·4					
3	2·3					
4	3·2	W.b.S.½S.	S.b.W.	2½	¾ l.	
5	4·4					
6	2·3					Variation of com-
7	2·3	W.b.N.¾N.	S.W.	2	¾ l.	pass 1¾ E.
8	3·3					
9	4·0					
10	5·4					
11	4·2					
12	4·4					
1	3·3	N.W.½W.	W.b.S.¾S.	2½	½ l.	A.M.
2	3·3					
3	3·5					
4	4·2					A current set the
5	6·3	W.b.S.	S.½W.	1½	¾ l.	ship the last 8
6	3·7					hours, by com-
7	2·5					pass, E.½S., 2
8	5·0					miles an hour.
9	5·2	S.W.	S.b.E.	2½	½ l.	
10	3·4					
11	6·3					
12	5·4					

(1.) The column in the above table headed deviation should be formed from the general table of deviations (p. 32) previously to correcting courses.

---

* Or thus: To find diff. long., add together log. M. D. lat. and log. dep., and from the sum subtract log. T. D. lat.; the remainder is the log. diff. long., which find in the tables.

Thus, in the first course in the preceding table, the ship's head is S.W.$\frac{1}{2}$W.; looking in the deviation table, we see that the corresponding correction is $\frac{1}{2}$ W. or $\frac{1}{2}$ l, and so for the others.

(2.) Form a table such as below, by writing down the headings, points, courses, &c., over the seven columns which are to be filled in with the corrected courses, &c.

Points.	Courses.	Distance.	Diff. lat.		Departure.	
			N.	S.	E.	W.

(3.) *To correct the courses.*

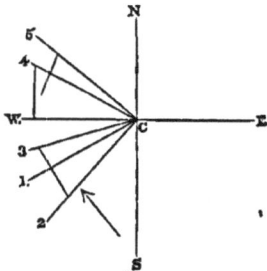

The courses are more readily corrected by drawing two lines at right angles, to represent the N., S., E., and W. points of the compass, and then a line to represent (roughly) the compass course of the ship. The direction in which the correction for leeway must be applied will then be easily seen.

After some experience in correcting courses, they can be made mentally, and the diagram dispensed with.

To correct the departure course which is W.b.S.$\frac{1}{2}$S. (the opposite bearing to E.b.N.$\frac{1}{2}$N.).

Draw a line roughly in the fig. W.b.S.$\frac{1}{2}$S. as c 1 ; it is then seen that

pts. qrs.

Compass course............6 2 r. S.

variation............1 3 r.

ship's head S.E.b.S. ∴ dev. 0 0   (See table, p. 32.)

—— 1 3 r.

true course ....8 1 r. S.

or 7 pts. 3 qrs. left of N., or W.$\frac{1}{4}$N. dist. 14′.

Insert this course and distance in table below.

Points.	Courses.	Dist.	Diff. lat.		Departure.	
			N.	S.	E.	W.
7¾	W.¼N.	14·0	0·7	...	...	14·0
8	W.	8·2	...	...	...	8·2
6	W.N.W.	9·9	3·8	...	...	9·2
8¼	N.W.¾N.	23·6	19·0	...	...	14·1
¾	N.¾W.	14·3	14·1	...	...	2·1
6½	W.b.N.½N.	17·5	5·1	...	...	16·7
7¾	W.¼S.	20·3	...	1·0	...	20·3
5¾	S.E.b.E.¾E.	16·0	...	6·8	14·5	...
			42·7	7·8	14·5	84·6
			7·8			14·5
		T. D. lat. 34·9 N.			Dep. 70·1 W.	

First Course.—S.W.½W.

. Draw a line in fig. S.W.½W. as c 2; then

		pts. qrs.
Compass course............		4 2 r. S.
variation ....1 3 r.		
deviation..........0 2 l.		
	——	1 1 r.
		5 3 r. S.
leeway (wind S.b.E)......		2 1 r.
true course.................		8 0 r. S., or due W. 8·2′.

The distance 8·2′ is found by adding up the hourly distances until the course is altered, at 4 o'clock. Insert this course and distance in the table.

Second Course.—W.b.S.½S.

Draw a line in fig. W.b.S.½S. as c 1.

		pts. qrs.
Compass course...........		6 2 r. S.
variation ....1 3 r.		
deviation ....0 3 l.		
	——	1 0 r.
		7 2 r. S.
leeway		2 2 r.
true course		10 0 r. S.
		or 6 0 l. N.
		=W.N.W.9·9′.

Insert this course and distance in the table.

Third Course.—W.b.N.¾N.	*Fourth Course.*—N.W.½W.

Draw a line W.b.N.¾N. as c 4.　　Draw a line N.W.½W. as c 5.

	pts.	qrs.			pts.	qrs.	
Compass course............6	1 l.	N.		Compass course............4	2 l.	N.

Compass course............6　1 l. N.
variation ....1　3 r.
deviation ....0　3 l.
　　　　　　　　1　0 r.
　　　　　　　5　1 l. N.
leeway......................2　0 r.
true course.................3　1 l. N.
　　　　or N.W.¾N. 23·6'.

Compass course............4　2 l. N.
variation..........1　3 r.
deviation ....0　2 l.
　　　　　　　　1　1 r.
　　　　　　　3　1 l. N.
leeway......................2　2 r.
true course.................0　3 l. N.
　　　　or N.¾W. 14·3'.

Insert course and distance in table.　　Insert course and distance in table.

Proceed with the 5th and 6th courses in the same manner, thus :

Fifth Course.　　　　　　　　　*Sixth Course.*

　　pts. qrs.
W.b.S. 7　0 r. S. as c 3.
1　3 r.
0　3 l.
——　　1　0 r.
　　　　8　0 r. S.
　　　　1　2 r.
　　　　9　2 r. S.
　　or 6　2 l. N.
=W.b.N.½N. 17·5'.

　　pts. qrs.
S.W.　4　0 r. S.
1　3 r.
0　2 l.
——　　1　1 r.
　　　　5　1 r. S.
　　　　2　2 r.
　　　　7　3 r. S.
　　or W.¼S. 20·3'.

　　　　　　　　pts. qrs.
Current course E.½S....7　2 l. S.
variation ....1　3 r.
true course...............5　3 l. S.
　　　or S.E.b.E.¾E. 16·0'.

Previously to opening the traverse table to take out the difference latitude and departure corresponding to each course and distance in the above
table, fill the columns not wanted : thus in the first course W.¼N. the N.
and W. columns will be wanted ; fill up the S. and E. columns by drawing
a line under S. and E.   In the second course W., the three columns N., S.,
and E., will not be wanted ; fill them up with lines.   In the same manner
proceed with the other course .

(4.) *To find difference latitude and departure* for each course and distance, by traverse table.

Enter traverse table, and take out the difference latitude and departure corresponding to $7\frac{3}{4}$ points, and distance 14·0. (Look out rather $7\frac{3}{4}$ points and 140 distance, the diff. lat. and dep. for which are 6·9 and 139·8; and move the decimal points one place to the left,) and put down the results to the nearest tenth, which are ·7 and 14·0. Insert them in the spaces left unmarked under N. and W.

The second course being due W. 8·2', the departure will be 8·2 (the same as the distance).

With third course 6 points and distance 9·9 (looking for 99, and making the proper change in decimal points), the diff. lat. is 3·8' and dep. 9·2'.

In a similar manner find difference latitude and departure for the other courses.

When the four columns are added up, it appears that the ship has sailed N. 42·7' and S. 7·8'; therefore upon the whole the true difference latitude is 34·9' N.; and her departure has been 14·5' E. and 84·6' W.; hence the departure made good in the 24 hours is 70·1 W.

(5.) *To find the latitude in,* apply the true difference latitude to the latitude from, in the usual manner, to obtain the latitude in.

(6.) *To find the longitude in.** With the latitude from and latitude in, find middle latitude. Add together log. secant mid. lat. and log. departure; the result (rejecting 10 in index) is the log. difference longitude, which, found in the tables, and applied to the longitude from, gives the longitude in. Thus:

To find latitude in.	To find longitude in.
T. D. lat.... 0° 34′ 54″ N.	log. sec. mid. lat...0·093148
lat. from....36 30 0 S.	„ departure......1·845718
„ in. ......35 55 6 S.	„ diff. long.......1·938866
2)72 25 6	∴ diff. long. 87′
mid. lat......36 12 33	or 1° 27 W.
	long. from....110 20 W.
	„ in.......111 47 W.

---

* Or thus : To find long. in (by inspection).

Since $\dfrac{\text{dep.}}{\text{dist.}} = \text{sin. course}$

and $\dfrac{\text{dep.}}{\text{diff. long.}} = \text{cos. mid. lat.} = \text{sin. complement mid. lat.}$

If, therefore, the traverse table is entered with complement of mid. lat. as a course and with the given departure, the distance corresponding thereto will be the difference of longitude nearly.

NAUTICAL ASTRONOMY.

———◆———

CHAPTER II.

(26.) NAUTICAL Astronomy teaches the method of finding the *place* of a ship, that is, its latitude and longitude, by means of astronomical observations.

(27.) The following are the principal terms in Nautical Astronomy; they are more fully explained in *Navigation*, Part II., to which the student is referred: they are inserted in this place for the sake of reference. The definitions of these terms should be thoroughly understood and carefully committed to memory.

> True place of a heavenly body.
> Apparent place of a heavenly body.
> Axis of the earth.
> Terrestrial equator.
> Poles of the earth.
> Axis of the heavens.
> Celestial equator.
> Poles of the heavens.
> The ecliptic.
> Obliquity of the ecliptic.
> True latitude of spectator.
> Reduced or central latitude of spectator.
> Meridians of the earth.

True zenith.
Reduced zenith.
Visible horizon.
Rational horizon.
Poles of the horizon.
Vertical circles, or circles of altitude.
Celestial meridian.
North and south points.
Prime vertical.
East and west points.
Circles of declination.
Circles of latitude.
Right ascension of a heavenly body.
Declination of a heavenly body.
Longitude of a heavenly body.
Latitude of a heavenly body.
Altitude of a heavenly body.
Azimuth or true bearing of a heavenly body.
Amplitude of a heavenly body.
Hour-angle of a heavenly body.
Solar year.
Sidereal year.
Mean solar year.
Sidereal day.
Apparent solar day.
Mean sun.
Mean solar day.
Sidereal time.
Apparent solar time.
Mean solar time.
Equation of time.
Sidereal clock.
Mean solar clock or chronometer.

Definitions of the preceding Terms in Nautical Astronomy.

(28.) To a spectator on the earth, the sun, moon, and stars seem to be placed on the interior surface of a hollow sphere of great but indefinite magnitude. The interior surface of this sphere is called the *celestial concave*, the center of which may be supposed to be the same as that of the earth.

(29.) The heavenly bodies are not in reality thus situated with respect to the spectator; for they are interspersed in infinite space at very different distances from him; the whole is an optical deception, by which an observer, wherever he is placed, is induced to imagine himself to be the center

of the universe. For let us suppose the elliptical figure p q p_1 q_1 to represent the earth, P Q P_1 Q_1 the celestial concave, and m a heavenly body. Then a spectator at A, not being able to estimate the distance of m, would imagine it to be in the celestial concave at M.

This figure will enable us to explain the terms *true* and *apparent* *place* of a heavenly body. The body m viewed from the surface of the earth would appear to a spectator A to be at M in the celestial concave: but if it could be seen from the center of the earth c, the point occupied by m would be M_1, the extremity of a line drawn from the center c of the earth through the heavenly body to the celestial concave. M is called the *apparent* place, and M_1 the *true* place of the heavenly body m.

(30.) The *axis of the earth* is that diameter about which it revolves: the *poles* of the earth are the extremities of the axis.

(31.) The *terrestrial equator* is that great circle on the earth that is equidistant from each pole.

(32.) A spectator on the earth, not being sensible of the motion by which in fact he describes daily a circle from west to east with the spot on which he stands, views in appearance the heavens moving past him in the opposite direction, or from east to west. The sphere of the fixed stars, or, as it is more usually called, the *celestial concave*, thus appears to revolve from east to west round an imaginary line which is the axis of the earth produced to the celestial concave: this line is therefore called the *axis of the heavens*.

(33.) The *poles of the heavens* are the extremities of the axis of the heavens.

(34.) The *celestial equator* is that great circle in the celestial concave which is perpendicular to the axis of the heavens; or it may be defined to be the terrestrial equator expanded or extended to the celestial concave. The poles of the celestial equator and the poles of the heavens are therefore identical.

(35.) While the earth thus performs its daily revolution, it is carried with great velocity from west to east round the sun, and describes an elliptic orbit once every year. This *annual* motion of the earth round the sun causes the latter body, to a spectator on the earth, insensible of his own change of place, to appear to describe a great circle in the celestial concave from west to east. This may be explained by a figure. Let AbA_1 be the earth's orbit, s the sun, and $cdel$ the celestial concave; then, to a spectator at a the sun is seen at a point c in the celestial concave : but when the earth has arrived at b the spectator (not being sensible of his motion from a to b) imagines the sun to be at d, and thus it would seem to have described the arc c d in the time the earth actually moved from a to b. It appears from this, that when the earth has arrived again at a, the sun will again be at c, having described one complete circle in the celestial concave among the fixed stars. The great circle thus described by the sun is called the *ecliptic*.

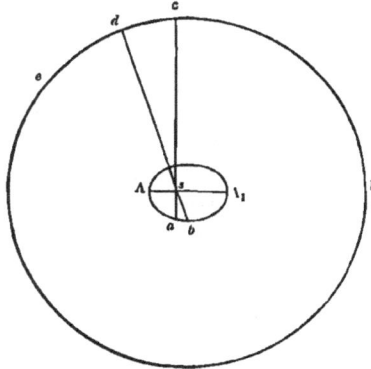

(36.) The axis of the earth as it is thus carried round the sun continues always parallel to itself, and it may be assumed without any sensible error, on account of the smallness of the earth's orbit (small, when compared with the distance of the heavenly bodies), to be always directed to the same points in the celestial concave, namely, the *poles* of the heavens.

(37.) From observation, the celestial equator is found to be inclined to the ecliptic at an angle of about 23° 28′. This inclination of the equator to the ecliptic is called the *obliquity of the ecliptic*. The axis of the earth, therefore, which is perpendicular to the equator, is inclined to the ecliptic, or, as it is in the same plane, to the earth's orbit, at an angle of 66° 32′.

(38.) In consequence of the whirling motion of the earth about its axis, the parts near the equator, which have the greatest velocity, acquire thereby a greater distance from the center than the parts near the poles. By actual measurement of a degree of latitude in different parts of the earth, it has been computed that the equatorial diameter is longer than the axis or polar diameter by 26 miles : the former being about 7924 miles; the latter about 7898 miles, and that the form of the earth is that of an *oblate spheroid*. (Note, p. 123.) It is usual, however, in drawing the figure of the earth to exaggerate very much its ellipticity; this is done for the sake of drawing the lines about the figure with greater clearness; for if it were constructed according to its true dimensions, the line pp_1 (fig. art. 29) (being only about the $\frac{1}{300}$th part of itself less than qq_1) would appear to the eye of the same length as qq_1, and thus the figure that more nearly resembles the earth would be a sphere.

(39.) If $A G$, a perpendicular to the earth's surface, be drawn passing through A, the angle $A G q$ formed by the line $A G$ with the plane of the equator is the *latitude*, or true latitude of the point A.

(40.) If $A C$ be a line drawn from A to C, the center of the earth, then the angle $A C q$ is called the reduced or central latitude of the point A. The difference between the true and reduced latitude is not great: it is, however, of importance in some of the problems in Nautical Astronomy. This correction has accordingly been calculated, and forms one of the Nautical Tables.

(41.) Sections of the earth passing through the poles, as $p A p_1$, are called *meridians* of the earth. If the earth is considered as a sphere (which it is very nearly), the meridians will be circles: on this supposition, moreover, the perpendicular $A G$ would coincide with $A C$, and the latitude of a place on the surface of the earth may, on this supposition, be defined to be the arc of the meridian passing through the place, intercepted between the place and the equator. If $G A$ be produced to meet the celestial concave at z, the point z is the zenith of the spectator at A. If $C A$ be produced to the celestial concave at z', then z' is called the *reduced* zenith of the spectator at A. The point opposite to z in the celestial concave is called the *Nadir*. In the figure the terrestrial equator qq_1 is extended to the celestial concave, and therefore $Q c Q_1$ is the plane of the *celestial* equator.

By means of this figure we may define the zenith, reduced zenith, latitude, and reduced latitude, as follows:

(42.) The *zenith* is that point in the celestial concave which is the extremity of the line drawn perpendicular to the place of the spectator, as z.

(43.) The *reduced* zenith is that point in the celestial concave which is the extremity of a straight line drawn from the center of the earth, through the place of the spectator, as z'.

(44.) The *latitude* of a place A on the surface of the earth is the inclination of the perpendicular $A G$ to the plane of the equator: thus the angle $A G Q$ is the latitude of A. The arc $z Q$ in the celestial concave measures the angle $A G Q$; hence $z Q$, *or the distance of the zenith from the celestial equator, is equal to the latitude of the spectator*.

(45.) The *reduced latitude* of the place A is the inclination of $z'C$ or $A C$ to the plane of the equator: or it is the angle $A C Q$ or arc $z' Q$, which measures the angle. Since the curvature of the earth diminishes from the equator to the poles, the reduced latitude $z'Q$ must be always less than the true latitude zQ, and therefore the difference zz′ must be subtracted from the true latitude to get the reduced latitude.

The formula for computing the difference between the true and reduced latitude of any place is investigated in *Navigation*, Part II.

(46.) The *visible horizon* is that circle in the celestial concave which touches the earth where the spectator stands, as *h*Ar; and a circle parallel to the *visible* horizon, and passing through the center of the earth, is called the rational horizon : thus HOR is the rational horizon. These two circles, however, form one and the same great circle in the celestial concave : thus n and r in the figure must be supposed to coincide. This may be readily conceived, when we consider that the distance of any two points on the surface of the earth will make no sensible angle at the celestial concave ; therefore either of these two circles is to be understood by the word horizon. The *poles* of the horizon of any place are manifestly the zenith and nadir.

(47.) Great circles passing through the zenith are called *circles of altitude* or *vertical* circles. That circle of altitude which passes through the poles of the heavens is called the *celestial meridian*. The points of the horizon through which the celestial meridian passes are called the *north* and *south* points. A circle of altitude at right angles to the meridian is called the *prime vertical*. This last circle cuts the horizon in two points called the *east* and *west* points. The east and west points are manifestly the poles of the celestial meridian.

(48.) Since the horizon and celestial equator are both perpendicular to the celestial meridian, the points where the horizon and celestial equator intersect each other must be 90° distant from every part of the meridian (Jeans' *Trig*. P. II. art. 65) ; that is, the celestial equator must cut the horizon in the east and west points.

(49.) The ecliptic (art. 35) is divided into twelve parts, called signs, which receive their names from constellations lying near them. These divisions or signs are supposed to begin at that intersection of the celestial equator and ecliptic which is called the *first point* of Aries.

(50.) Great circles passing through the poles of the heavens are called *circles of declination;* and great circles passing through the poles of the ecliptic are called *circles of latitude.*

(51.) *Parallels of declination* and of *latitude* are small circles parallel respectively to the celestial equator and ecliptic.

(52.) The *declination* of a heavenly body is the arc of a circle of declination passing through its place in the celestial concave, intercepted between that place and the celestial equator.

(53.) The *right ascension* of a heavenly body is the arc of the equator, intercepted between the first point of Aries and the circle of declination passing through the place of the heavenly body in the celestial concave, measuring from the first point of Aries, eastward, from 0° to 360°.

(54.) The *latitude* of a heavenly body is the arc of a circle of latitude passing through its place in the celestial concave, intercepted between that place and the ecliptic.

(55.) The *longitude* of a heavenly body is the arc of the ecliptic intercepted between the first point of Aries and the circle of latitude passing through the place of the heavenly body in the celestial concave, measuring from the first point of Aries, eastward, from $0°$ to $360°$.

(56.) The *true altitude* of a heavenly body is the arc of a circle of altitude passing through the true place of the body intercepted between the place and the horizon.

(57.) The *azimuth*, or bearing of a heavenly body, is the arc of the horizon intercepted between the north or south points and the circle of altitude passing through the place of the body; or it is the corresponding angles at the zenith between the celestial meridian and the circle of altitude passing through the body.

(58.) The *amplitude* of a heavenly body is the distance from the east point at which it rises, or the distance from the west point at which it sets, the arcs or distances being measured on the horizon.

(59.) The *hour angle* of a heavenly body is the angle at the pole between the celestial meridian and the circle of declination passing through the place of the body.

(60.) A *solar year* is the interval between the sun's leaving the first point of Aries and returning to it again.

(61.) A *sidereal year* is the interval between the sun's leaving a fixed point, as a star, and returning to that point again.

(62.) The length of the solar years is found to differ a little from each other, on account of certain irregularities in the sun's apparent motion, and that of the first point of Aries. The *mean length* of several solar years is therefore the one made use of in the common division of time, and called the *mean solar year*.

(63.) The *sidereal day* is the interval between two successive transits of the first point of Aries over the same meridian. It begins when the first point of Aries is on the meridian.

(64.) The *apparent solar day* is the interval between two successive transits of the sun's center over the same meridian. It begins when that point is on the meridian.

(65.) The length of an apparent solar day is variable chiefly from two causes :

1st. From the variable motion of the sun in the ecliptic.

2d. From the motion of the sun being in a circle inclined to the equator.

(66.) To obtain, therefore, a proper measure of time, we proceed as follows. An imaginary, or as it is called a *mean sun*, is supposed to *move uniformly in the celestial equator with the mean velocity of the true sun*. A *mean solar day* may therefore be defined to be the interval between two successive transits of the mean sun over the same meridian. It begins when the mean sun is on the meridian.

(67.) *Sidereal* time is the angle at the pole of the heavens between the celestial meridian and a circle of declination passing through the first point of Aries, measuring from the meridian westward.

(68.) *Mean solar time* is the angle at the pole between the celestial meridian and a circle of declination passing through the mean sun, measuring from the meridian westward.

(69.) *Apparent solar time* is the angle at the pole between the celestial meridian and a circle of declination passing through the place of the sun's center, measuring from the meridian westward.

(70.) The *equation of time* is the difference in time between the places of the true and mean sun.

(71.) A *sidereal clock* is a clock adjusted so as to go 24 hours during one complete revolution of the earth; that is, during the interval of two successive transits of a fixed star: or supposing the first point of Aries to be invariable, between two successive transits of the first point of Aries.

(72.) A *mean solar clock* is a clock adjusted to go 24 hours during one complete revolution of the mean sun.

These definitions are fully explained and illustrated in *Navigation*, Part II. pp. 8 to 11, by means of constructions or diagrams similar to the following one, which is the diagram for definitions 67, 68, 69, and 70. The student should endeavour to explain each of the other definitions by a similar construction, as pointed out in Part II.

Construct a figure, and show what is meant by sidereal time, apparent solar time, mean solar time, and the equation of time.

Let NWSE represent the horizon, P the pole, AE the equator, A the first point of Aries, and AC the eclip-tic. Let x be the place of the sun in the ecliptic, and m the mean sun; through x and m draw the circles of declination PR and Pm. Then *sidereal time* is the angle QPA, or arc QA; *apparent solar time* is the angle QPR, or arc QR; and *mean solar time* is the angle QPm, or arc Qm,—these angles or arcs being always measured from the meridian NZS westward. Also the angle mPR, or arc mR, is the *equation of time*.

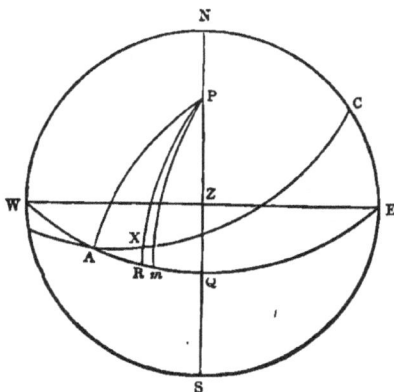

INTRODUCTORY RULES IN NAUTICAL ASTRONOMY.

Nautical day and Astronomical day.

(73.) The nautical or civil day begins at midnight and ends the next midnight. The astronomical day begins at noon and ends at noon, and is later than the civil day by 12 hours. Again, in the astronomical day the hours are reckoned throughout from 0^h to 24^h; in the nautical day there are twice 12 hours, the first 12 hours being before noon, or before the commencement of the astronomical day (denoted by A.M., *ante meridiem*); the latter are afternoon, and distinguished by the letters P.M. (*post meridiem.*)

Rule 1. *Given civil or nautical time, to reduce it to astronomical time.*

1. If the given nautical time be P.M., it will be also astronomical time; P.M. being omitted.

2. If the given nautical time be A.M., add 12^h thereto, and put the day one back, omitting the letters A.M.; thus:

(1.) April 27, at $4^h 10^m$ P.M. (civil) is April 27, at $4^h 10^m$ (astro.).
(2.) April 27, at 4 10 A.M. (civil) is April 26, at 16 10 (astro.).

EXAMPLES.

Reduce the following civil or nautical times to astronomical times.

Civil times.			Astronomical times.		
157. Sept. 10th,	$4^h 10^m$	P.M.	*Ans.* Sept. 10th,	$4^h 10^m$	
158. June 3	2 42	A.M.	„ June 2	14 42	
159. July 1	6 18	A.M.	„ June 30	18 18	
160. Dec. 10	3 42	P.M.	„ Dec. 10	3 42	

Rule 2. *Given astronomical time, to reduce it to civil or nautical time.*

1. If the astronomical time is less than 12 hours, it will also be nautical time P.M.

2. If the astronomical time be greater than 12 hours, reject 12 and put the day one forward; the result will be civil time A.M.; thus:

(1.) April 27, at $4^h 10^m$ (astro.) is April 27, at $4^h 10^m$ P.M. (civil).
(2.) April 27, at 16 10 (astro.) is April 28, at 4 10 A.M. (civil).

Reduce the following astronomical times to nautical or civil times.

Astronomical times.				Civil times.			
161. Sept. 10th, 4ʰ 32ᵐ			*Ans.*	Sept. 10th,	4ʰ	32ᵐ	P.M.
162. July	5	16 32	„	July	6	4 32	A.M.
163. July	10	18 42	„	July	11	6 42	A.M.
164. Dec.	21	23 59	„	Dec.	22	11 59	A.M.

Rule 3. *To reduce degrees into time.*

1. Divide the degrees by 15, the quotient is hours.

2. Multiply the remaining degrees, if any, by 4 ; the result is minutes in time.

3. Divide the minutes in arc by 15 ; the quotient is minutes in time.

4. Multiply the remaining minutes of arc, if any, by 4; the result is seconds of time.

5. Divide the seconds in arc by 15; the quotient is seconds in time, carried to decimals if necessary. The sum will be the arc in time.

165. Reduce 34° 44′ 34″ into time.

$$
\begin{aligned}
34° &= 2^h\ 16^m\ 0^s \\
44' &= 2\ 56 \\
34'' &= 2\cdot26 \\
\therefore 34°\ 44'\ 34'' &= \overline{2\ \ 18\ 58\cdot26}
\end{aligned}
$$

Reduce the following arcs into time.

				Ans.			
166.	84°	42′	30″		5ʰ	38ᵐ	50ˢ
167.	96	10	45	„	6	24	43
168.	108	24	22	„	7	13	37·4
169.	178	48	45	„	11	55	15
170.	140	32	10	„	9	22	8·66
171.	240	32	10	„	16	2	8·66

Rule 4. *To reduce time into degrees.*

1. Multiply the hours by 15 ; the result is degrees.

2. Divide the minutes in time by 4 ; the quotient is degrees.

3. Multiply the minutes remaining, if any, by 15; the result is minutes of arc.

4. Divide the seconds of time by 4 ; the quotient is minutes of arc.

5. Multiply the seconds (and parts of seconds) remaining, if any, by 15 ; the result is seconds of arc.

The sum will be the arc in degrees.

172. Reduce 2^h 18^m $58'\cdot26$ into degrees.

$$
\begin{array}{rcrrr}
2^h & = & 30° & 0' & 0'' \\
18^m & = & 4 & 30 & 0 \\
58'\cdot26 & = & & 14 & 33\cdot9 \\
\hline
\therefore 2^h\,18^m\,58'\cdot26 & = & 34 & 44 & 33\cdot9
\end{array}
$$

Find the arcs corresponding to the following times,

173.	3^h	52^m	4^s	*Ans.*	58°	1'	0''
174.	17	8	22	"	257	5	30
175.	8	17	15·5	"	124	18	52·5
176.	12	14	16·75	"	183	34	11·25
177.	9	13	8	"	138	17	0
178.	15	17	18·4	"	229	19	36

By means of the following Table* arcs to the nearest minute are more readily expressed in time and the converse, than by the preceding rules.

TABLE

To reduce degrees into time, and the converse.

$1' = 0^m\ 4^s$	$21' = 1^m\ 24^s$	$41' = 2^m\ 44^s$	$1° = 0^h\ 4^m$	$10° = 0^h\ 40^m$
$2' = 0\ 8$	$22' = 1\ 28$	$42' = 2\ 48$	$2° = 0\ 8$	$20° = 1\ 20$
$3' = 0\ 12$	$23' = 1\ 32$	$43' = 2\ 52$	$3° = 0\ 12$	$30° = 2\ 0$
$4' = 0\ 16$	$24' = 1\ 36$	$44' = 2\ 56$	$4° = 0\ 16$	$40° = 2\ 40$
$5' = 0\ 20$	$25' = 1\ 40$	$45' = 3\ 0$	$5° = 0\ 20$	$50° = 3\ 20$
$6' = 0\ 24$	$26' = 1\ 44$	$46' = 3\ 4$	$6° = 0\ 24$	$60° = 4\ 0$
$7' = 0\ 28$	$27' = 1\ 48$	$47' = 3\ 8$	$7° = 0\ 28$	$70° = 4\ 40$
$8' = 0\ 32$	$28' = 1\ 52$	$48' = 3\ 12$	$8° = 0\ 32$	$80° = 5\ 20$
$9' = 0\ 36$	$29' = 1\ 56$	$49' = 3\ 16$	$9° = 0\ 36$	$90° = 6\ 0$
$10' = 0\ 40$	$30' = 2\ 0$	$50' = 3\ 20$	$10° = 0\ 40$	$100° = 6\ 40$
$11' = 0\ 44$	$31' = 2\ 4$	$51' = 3\ 24$	$11° = 0\ 44$	$110° = 7\ 20$
$12' = 0\ 48$	$32' = 2\ 8$	$52' = 3\ 28$	$12° = 0\ 48$	$120° = 8\ 0$
$13' = 0\ 52$	$33' = 2\ 12$	$53' = 3\ 32$	$13° = 0\ 52$	$130° = 8\ 40$
$14' = 0\ 56$	$34' = 2\ 16$	$54' = 3\ 36$	$14° = 0\ 56$	$140° = 9\ 20$
$15' = 1\ 0$	$35' = 2\ 20$	$55' = 3\ 40$	$15° = 1\ 0$	$150° = 10\ 0$
$16' = 1\ 4$	$36' = 2\ 24$	$56' = 3\ 44$	$16° = 1\ 4$	$160° = 10\ 40$
$17' = 1\ 8$	$37' = 2\ 28$	$57' = 3\ 48$	$17° = 1\ 8$	$170° = 11\ 20$
$18' = 1\ 12$	$38' = 2\ 32$	$58' = 3\ 52$	$18° = 1\ 12$	$180° = 12\ 0$
$19' = 1\ 16$	$39' = 2\ 36$	$59' = 3\ 56$	$19° = 1\ 16$	$190° = 12\ 40$
$20' = 1\ 20$	$40' = 2\ 40$	$60' = 4\ 0$	$20° = 1\ 20$	$200° = 13\ 20$

* The table is computed to the nearest minute of arc; when seconds are to be reduced (which is seldom required) the student must proceed as pointed out in the preceding rules.

Use of the Table.

179. Reduce 34° 44′ 34″ into time.

34° 44′ 34″=34° 45′ nearly.
By Table...30°.......2ʰ 0ᵐ
4 16
45′... 3
∴ 34° 44′ 34″=2 19 nearly.

180. Reduce 2ʰ 18ᵐ 58ˢ·26 into degrees.

2ʰ 18ᵐ 58ˢ·26=2ʰ 19ᵐ nearly.
By Table...2ʰ.....30° 0′
16ᵐ 4 0
3 45
∴ 2ʰ 18ᵐ 58ˢ·26=34 45 nearly.

In some nautical tables, the angles in the log. sine table are given both in arc and in time. The reduction from degrees to hours, and the converse, is by means of such a table readily made, simply by inspection.

Given the time at Greenwich, to find the time at the same instant at any other place, and the converse.

(74.) To find the time at any place, as New York or Calcutta, corresponding to a given time at Greenwich, or the converse, we must remember that since the earth revolves through 360° in 24 hours, from west through south to east, or 15° in 1 hour; then at a place 15° to the eastward of a spectator the sun will be on the meridian 1 hour before, and at a place 15° to the westward, the sun will be on the meridian 1 hour later than at the place of the spectator: hence, when it is 10 o'clock at a given place, it will *at the same instant* be 11 o'clock at a place 15° to the eastward, and 9 o'clock at a place 15° to the westward. If, therefore, the longitude of a place is known, that is, the number of degrees it is to the east or west of Greenwich, we can readily tell what time it is at the place corresponding to a given time at Greenwich, and the converse. To find the time at Greenwich, corresponding to any given time at a place, is required in almost every nautical problem; and even if the longitude and time are only known nearly, the approximate true time at Greenwich, deduced from the estimated longitude and estimated time at the place, is an important element in nautical astronomy. The time at Greenwich, obtained in this manner, is called an approximate Greenwich time, or more frequently *the Greenwich date.*

TO FIND THE GREENWICH DATE.

Rule 5. *First method.   By estimated ship time and longitude.*

1. Express the time at the ship astronomically to the nearest minute (p. 73).

2. Reduce the longitude into time to nearest minute, and put it under ship time (p. 72).

3. If west longitude, add longitude in time to ship time; the sum, if less than 24 hours, will be the time at Greenwich, or the Greenwich date on the same day as at the ship.

But if the sum be greater than 24 hours, reject 24 hours; the result will be the Greenwich date on the day following the ship date.

4. If east longitude, subtract longitude in time from ship time, the remainder will be the Greenwich date. If the longitude in time is greater than the ship time, 24 hours must be added to the ship time before subtraction is made, and the remainder will be the Greenwich date on the day preceding the ship date.

EXAMPLES.

181. June 10, at $6^h 10^m$ P.M. mean time nearly in longitude by account 32° 42′ W.; required the Greenwich date, or the Greenwich time to the nearest minute.

Ship, June 10......$6^h 10^m$
long. in time........2 11 W.
∴ Green. June 10.....$\overline{8\ 21}$

182. July 12, at $4^h 5^m$ A.M. in long. 63° 45′ W.; required the Greenwich date.

Ship, July 11.......$16^h 5^m$
long. in time........ 4 15 W.
∴ Green. July 11......$\overline{20\ 20}$

EXAMPLES.

Required the Greenwich date in each of the following examples.

Ship times.					Answers, Greenwich dates.	
183. Mar. 7, at	$3^h 15^m$	A.M.	Long.	15° 45′ E.	. Mar. 6, at	$14^h 12^m$
184. Mar. 15 ,,	10 35	P.M.	,,	43 5 E.	. Mar. 15 ,,	7 43
185. May 12 ,,	4 30	A.M.	,,	45 50 W.	. May 11 ,,	19 33
186. May 9 ,,	5 16	P.M.	,,	90 35 E.	. May 8 ,,	23 14
187. May 5 ,,	11 30	P.M.	,,	55 47 W.	. May 5 ,,	15 13
188. May 20 ,,	10 25	A.M.	,,	150 15 W.	. May 20 ,,	8 26

The time at Greenwich, and therefore the Greenwich date, is more correctly found by means of a chronometer whose error on Greenwich mean time is known at the moment of observation as follows.

TO FIND THE GREENWICH DATE.

Rule 6. *Second method. By chronometer and its error on Greenwich mean time.*

To the time shown by chronometer, apply its error on Greenwich mean time ; adding if error is slow, and subtracting if error is fast, on Greenwich mean time ; the result is the Greenwich date in mean time. Sometimes 12 hours must be added to this result, and the day put one back. To determine when this must be done, get an approximate Greenwich date in the usual way by means of ship mean time and the estimated longitude ; if the difference between the Greenwich dates found by the two methods is nearly 12 hours, then the Greenwich date *by chronometer* must be increased by 12 hours, and the day put one day back, if necessary, so as to make the two dates more nearly agree both in the day and hour.

The following examples will remove any doubt as to putting the day one back, or not, when the 12 hours are added.

EXAMPLE 1.

189. July 10th, at 6^h 34^m P.M. mean time nearly, in longitude 60° W., a chronometer showed 10^h 42^m 3', its error on Greenwich mean time being 2^m 10' fast ; required mean time at Greenwich, or the Greenwich date.

Greenwich time by chronometer.	Greenwich date.
July 10th, chro......10^h 42^m 3'	Ship, July 10th... 6^h 34^m
Error on G. M. T..... 2 10	long. in time....... 4 0 W.
Gr. July 10th.........$\overline{10\ 39\ 53}$	Gr. July 10th......$\overline{10\ 34}$

As these two results come out near to each other, the correct Greenwich time is, July 10th, 10^h 39^m 53', and the Greenwich date is therefore July 10th, 10^h 40^m.

EXAMPLE 2.

190. Aug. 3d, at 5^h 42^m P.M. mean time nearly, in long. by account 150° 30' W., a chronometer showed 3^h 23^m 15', and its error on Greenwich mean time was 10^m $10'·4$ slow ; required the Greenwich date.

Greenwich time by chronometer.	Greenwich date.
Aug. 3d, at...... 3^h 23^m 15'	Ship, Aug. 3d... 5^h 42^m
Error............ 10 10·4	long. in time.....10 2 W.
Aug. 3d......... 3 33 25·4	Gr. Aug. 3d......$\overline{15\ 44}$
Add.............12	
Gr. Aug. 3d......$\overline{15\ 33\ 25·4}$	

In this example 12 hours must be added to the Greenwich date by chronometer, to bring the Greenwich times more nearly alike.

EXAMPLE 3.

191. March 10th, at $4^h 10^m$ A.M. mean time nearly, in longitude 20° 42′ E., a chronometer showed $2^h 2^m 50^s$, and its error on Greenwich mean time was $45^m 16^s$ slow; required the Greenwich date.

Greenwich time by chronometer.	Greenwich date.
Mar. 10th...... 2^h 2^m 50^s	Ship, Mar. 9th......16^h 10^m
Error............ 45 16	long. in time......... 1 23 E.
Mar. 10th...... $\overline{2\ 48\ 6}$	Gr. Mar. 9th.........$\overline{14\ 47}$
Add.............12	
Gr. Mar. 9th...$\overline{14\ 48\ 6}$	

In this example 12 hours must be added, and the day put one back, to bring the chronometer Greenwich time more nearly alike to the Greenwich date.

Find the Greenwich times and Greenwich dates in the following examples.

M. T. nearly.		Long.		Chro. show.		Err. of Chro.		Answers.			
192. Aug. 10	$1^h 20^m$ P.M.	35° 42′ W.		4^h 2^m 10^s		$18^m 45 \cdot 4^s$ fast		10	3^h 43^m $24 \cdot 6^s$		
193. July 13	3 42 A.M.	150 50 W.		1 30 0		10 50·6 slow		13	1 40 50·6		
194. June 10	10 42 P.M.	42 0 E.		7 44 10		8 12·0 slow		10	7 52 22		
195. June 19	6 42 A.M.	50 50 W.		10 14 15		12 3·7 fast		18	22 2 11·3		
196. Sept. 3	10 42 A.M.	19 15 E.		9 10 45		12 15·3 slow		2	21 23 0·3		
197. Dec. 30	11 45 P.M.	110 35 W.		7 10 30		9 5·0 fast		30	19 1 25		

Hence the Greenwich dates will be the above Greenwich times put down to the nearest minute : thus

Ex. 192, Gr. date is Aug. 10, $3^h 43^m$

,, 193, ,, ,, July 13, 1 41

And so on.

CHAPTER III.

(75.) The *Nautical Almanac* contains the declination, right ascension, &c., of the principal heavenly bodies, for certain fixed times at Greenwich. The declination and right ascension of the sun and planets are given for every day at $0^h\ 0^m\ 0^s$; for the moon, for every hour at Greenwich.

To obtain these quantities for any other time, we may either apply the common rules of proportion,* or—which is in most cases the simplest method—make use of certain tables computed for the purpose, called tables of *proportional logarithms*. The tables of proportional logarithms are the following :

1. The proportional logarithms (properly so called).
2. The Greenwich date proportional logarithm of the sun.
3. The Greenwich date proportional logarithm of the moon.
4. The logistic logarithms.

The construction of these tables is given in *Nav.* Part II. p. 138.

TO TAKE OUT THE SUN'S DECLINATION.

Rule 7. *First method, by using proportional logarithms.*

1. Get a Greenwich date, thus :

(*a.*) Put down the ship mean time expressed astronomically.

(*b.*) Under which put the longitude in time : add if west, subtract if east (adding or subtracting 24 hours, according to Rule 5. p. 74).

2. Take out of the *Nautical Almanac* the sun's declination for the two consecutive noons between which the Greenwich date lies.

3. Take the difference of the declinations when their names are alike (that is, both north or both south) ; but when the names of the declinations are unlike, take their sum ; the result will be the change of declination in 24 hours. Mark it + or —, or with the same or different letter, according as the declination is seen to be increasing or decreasing.

---

* The results obtained by the rules of proportion are only true when the daily or hourly change is *invariable*, and this is seldom the case in the motion or apparent motion of the heavenly bodies. When very great accuracy is required, we must apply a further correction, called *the equation of second differences.* For all the common purposes of navigation, however, a simple proportion, as directed in the following rules, will be sufficiently correct.

4. Add together Greenwich date logarithm for the sun and proportional logarithm of the change in 24 hours; the result is the proportional logarithm of change of declination for the given time, which take from the table of proportional logarithms and apply to the declination at first noon, either by subtracting or adding it, according as the declination is seen to be decreasing or increasing.

Another method of taking out the sun's declination is to make use of the hourly changes of declination given in the *Nautical Almanac*.

TO TAKE OUT THE SUN'S DECLINATION.

Rule 8. *Second method, by using hourly differences.*

1. Find a Greenwich date, as in last rule.

2. Take out of the *Nautical Almanac* the declination at noon of the Greenwich date, and put down a little to the right thereof the difference for one hour found in the same page, and close to the declination taken out. Multiply this quantity by the hours in Greenwich date, and the fractional parts of the hour if necessary; the product will be the change of declination in the time from noon; apply this, reduced to minutes and seconds, to the declination taken out, adding it if the declination is seen to be increasing, and subtracting if decreasing. The result is the declination of the sun at the time required.*

EXAMPLES.

198. March 2, at $4^h\ 23^m$ P.M. mean time nearly, in long. 32° 42' W.: required the sun's declination. (1.) By proportional logarithms. (2.) By hourly differences.

$$
\begin{array}{lll}
\text{Ship, Mar. 2.} & \ldots\ldots\ldots\ldots & 4^h\ \ 23^m \\
\text{long. in time} & \ldots\ldots\ldots\ldots & 2\ \ \ 11\ \text{W.} \\
\hline
\text{Gr. Mar. 2} & \ldots\ldots\ldots\ldots & 6\ \ \ 34
\end{array}
$$

First method.   By proportional logarithms.

$$
\begin{array}{lllll}
\text{Sun's decl. Mar. 2} & \ldots\ldots & 7° & 7' & 0''\ \text{S.} \\
\qquad\qquad\ ,,\quad 3 & \ldots\ldots & 6 & 44 & 2\ \text{S.} \\
\hline
& & 22 & 58\ \text{N.}
\end{array}
$$

$$
\begin{array}{lll}
\text{Gr. date log. } \odot \ldots & 56287 & \\
\text{pro. log. 22' 58''} . & 89417 & \\
\hline
1\cdot45704\ldots & & 6\quad 17\ \text{N.} \\
\hline
\end{array}
$$

$\therefore$ sun's decl. at $6^h\ 34^m \ldots 7\quad 0\quad 43$ S.

---

* The corrections for Greenwich date of the quantities taken out of the *Nautical Almanac*, when small, are frequently made *by inspection;* the results thus obtained are generally found sufficiently correct.

Second method. By hourly differences.

Sun's decl.

Mar. 2...7° 7' 0" S.... hourly diff....57·4" decreasing.

cor. ...   6 16·9 —   6

∴ decl.=7 0 43·1 S.

		344·4 change in 6ʰ			
30... $\frac{1}{2}$		28·7 ⎫			
3... $\frac{1}{10}$		2·9 ⎬	„	„	34ᵐ
1... $\frac{1}{3}$		0·9 ⎭			
60		376·9			

∴ cor. for 6ʰ 34ᵐ......| 6' 16·9" —

199. March 21st, at 4ʰ 23ᵐ A.M. in long. 100° 10' E.: required the sun's declination. (1.) By proportional logarithms. (2.) By hourly differences.

Ship, Mar. 20 .... 16ʰ 23ᵐ

long. in time................. 6   41 E.

Gr. Mar. 20 9   42

(1.) By proportional logarithms.

Sun's decl. Mar. 20 ....   0°   5'   32·3" S.

„   21   0   18   8·3 N.

                                  23   40·6 N.

Gr. date log. ☉ ...   ·39344

pro. log. 23' 41" ...   ·88083

                1·27427...   9   33·5 N.

∴ sun's decl. at 9ʰ 42ᵐ............   4   1·2 N.

(2.) By hourly differences.

Sun's decl.

Mar. 20...0° 5' 32·2" S....hourly diff. 59·2" decreasing.

cor. ...   9   34·1 N.   9

∴ decl.=0   4   1·9 N.

30... $\frac{1}{2}$	532·8
10... $\frac{1}{3}$	29·6
2... $\frac{1}{5}$	9·8
	1·9
60	574·1

∴ for 9ʰ 42ᵐ......| 9' 34·1" N.

200. July 30, 1845, at 3^h 20^m P.M. mean time, in long. 9° 0' W. :
required the sun's decl.   *Ans.* 18° 28' 14" N.

201. Dec. 10, 1845, at 6^h 32^m A.M. mean time, in long. 32° 30' W. :
required the sun's decl.   *Ans.* 22° 56' 4" S.

202. Aug. 1, 1845, at 4^h 52^m P.M. mean time, in long. 152° 33' E. :
required the sun's decl.   *Ans.* 18° 4' 16·5" N.

Dec. and hourly diff. from Nautical Almanac.

		Hourly diff.
July 30......18° 30' 39" N.	July 31......18° 15' 57" N.	.......36·7" dec.
Dec. 9........22 51 17 S.	Dec. 10......22 56 50 S.	.......12·7 dec.
July 31......18 15 58 N.	Aug. 1........18 0 58 N.	.......37·5 dec.

Rule 9. *To take out the* EQUATION OF TIME.

First method, by using proportional logarithms.

1. Get a Greenwich date.

2. Take out the equation of time for two consecutive noons between which the Greenwich date lies, and take their difference.

3. Add together the Greenwich date logarithm for sun and proportional logarithm of difference: the sum is the proportional logarithm of correction, which find from the table, and apply it with its proper sign to the equation of time first taken out; the result is the equation of time required.

Second method, by using hourly differences.

1. Get a Greenwich date, as in first method.

2. Take out the equation of time for the noon of Greenwich date and the hourly difference opposite thereto.

3. Multiply hourly difference by the hours of the Greenwich date, and, if great accuracy is required, by the fractional parts of hour in the Greenwich date; the result will be the correction to be applied with its proper sign to the equation of time taken out.

EXAMPLE.

203. July 12, 1853, at 5^h 8^m A.M. mean time nearly, in long. 160° W. :
required the equation of time.

$$
\begin{array}{ll}
\text{Ship, July 11}\ldots\ldots\ldots & 17^h\ \ 8^m \\
\text{long. in time}\ldots\ldots\ldots\ldots & 10\ \ 40\ \ \text{W.} \\
\text{Greenwich, July 11}\ldots & \overline{27\ \ 48} \\
\text{Greenwich, July 12}\ldots & \overline{\ 3\ \ 48} \\
\end{array}
$$

Equation of time.		Or thus, by hourly difference.	
July 12................5^m 15·7^s		Diff. for 1 hour.......0·308^s incr.	
„ 13................5 23·1			3
			‾‾‾‾‾
	7·4		0·924
Greenwich d. log. sun...0·80043		30 ½	·154
Prop. log. 7·4'............3·16419		15 ½	·077
			‾‾‾‾‾
Prop. log. cor. ....3·96462		Cor.	1·155
Cor. .... 1·2		Or,.........	1·2^s
		5^m	15·7
Equation required ......5 16·9			

Equation required 5   16·9

Find the equation of time in the following examples:

204. March 2, 1853, at 6^h 10^m P.M. mean time, in long. 38° 42′ W.
205.   „   16 „   „   5 42 A.M.   „   „   152 45 W.
206.   „   29 „   „   10 42 A.M.   „   „   87 8 E.

Equation of time from Nautical Almanac.

Diff. 1 hour.

Eq. of time, March 2, 12^m 22·1^s.........March 3, 12^m 9·3^s...0·53^s
    „   „   16 8 48·4 .... „ 17 8 30·8 ...0·72
    „   „   28 5 9·0 .... „ 29 4 50·4 ...0·77

Ans. to 204, 205, 206 : 12^m 17·4^s, 8^m 45·5^s, 4^m 56·0^s.

Rule 10. *To take out the* MOON'S SEMIDIAMETER AND HORIZONTAL PARALLAX.
The moon's semidiameter and horizontal parallax are put down in the *Nautical Almanac* for every mean noon and mean midnight at Greenwich; to find these quantities for any other time we may proceed as follows:

First. To find the moon's semidiameter.

1. Get a Greenwich date.

2. Take out of the *Nautical Almanac* the moon's semidiameter for the two times between which the Greenwich date lies, and take the difference. To the Greenwich date logarithm for moon add the proportional logarithm of the difference just found; the result will be the proportional logarithm of an arc, which being found and added to the semidiameter first taken out, or subtracted therefrom (according as the semidiameter is increasing or decreasing), will be the semidiameter at the given time.

Second. To find the moon's horizontal parallax.

Proceed in a similar manner to that pointed out above for finding the moon's semidiameter.

EXAMPLES.

207. Aug. 3, 1853, at 4^h 10^m P.M. mean time nearly, in long. 56° 15′ W.: required the moon's semidiameter and horizontal parallax.

Ship, Aug. 3.........4^h 10^m
long. in time.........3 45 W.
Greenwich, Aug. 3...7 55

G

Moon's semidiameter.		Moon's horizontal parallax.	
August 3, at noon......	15′ 6·6″	August 3, at noon......	55′ 20·6″
„ mid.......	15 10·6	„ mid.	55 35·3
	4·0		14·7
Gr. d. log. for moon...	0·18064	Gr. d. log. for moon...	0·18064
Prop. log. for 4·0″......	3·43136	Prop. log. for 14·7″....	2·86611
Prop. log. cor.	3·61200	Prop. log. cor.	3·04675
Cor.	2·6	Cor.	9·7
Required semi. 15	9·2	Required hor. par....... 55	30·3

NOTE. It is better in examples of this kind, where the differences taken out of the *Nautical Almanac* are only a few seconds, to learn to estimate the correction *at sight*: this, after a little practice, will not be difficult to do. The above method, however, by logarithms, should be adopted by beginners for the practice it gives in learning the use of the tables.

208. July 14, 1853, at $6^h 42^m$ A.M. mean time nearly, in long. 30° W.: required the moon's semidiameter and horizontal parallax.

$$\text{Ship, July 13.........}18^h 42^m$$
$$\text{long. in time .... } 2 \quad 0 \text{ W.}$$
$$\text{Greenwich, July 13..}\overline{20 \quad 42}$$

Moon's semidiameter.		Moon's horizontal parallax.	
July 13, mid.	16′ 2·7″	July 13, mid.	58′ 45·8″
„ 14, noon...............	16 7·5	„ 14, noon...............	59 3·5
	4·8		17·7
Gr. d. log. moon for $8^h 42^{m*}$	0·13966	Gr. d. log. moon for $8^h 42^m$	0·13966
Prop. log. 4·8″	3·35218	Prop. log. 17·7″	2·78545
Prop. log. cor.	3·49184	Prop. log. cor.	2·92511
Cor.	3·5	Cor.	12·8
Required semi............... 16	6·2	Required hor. par. 58	58·6

Find the moon's semidiameter and horizontal parallax in the following examples:

209. March 2, 1853.......at $6^h 42^m$ P.M.......in long. 100° 0′ W.
210. „ 14 „ ......at 3 30 A.M....... „ 120 0 W.
211. „ 24 „ ......at 10 10 P.M....... „ 60 42 E.

Moon's semi. and hor. par. from Nautical Almanac.

Moon's semidiameter.	Moon's hor. par.	Answers.
March 2, mid. 16′ 5·1″	Mid. 58′ 54·7″	Semi. 16′ 4·7″
„ 3, noon 16 2·1	Noon 58 43·6	H. P. 58 53·4
March 13, mid. 14 48·9	Mid. 54 15·7	Semi. 14 47·9
„ 14, noon 14 47·8	Noon 54 11·5	H. P. 54 11·7
March 24, noon 16 18·7	Noon 59 44·6	Semi. 16 21·2
„ 24, mid. 16 23·7	Mid. 60 1·8	H. P. 59 53·3

* When the Greenwich date exceeds 12 hours, as in this example, look out the Greenwich date logarithm moon for the excess of the Greenwich date above 12 hours.

Rule 11. *To take out the* SUN's RIGHT ASCENSION.

1. Get a Greenwich date.

2. Take out the right ascension for two consecutive noons between which the Greenwich date lies, and take their difference.

3. Add together the Greenwich date logarithm for sun and proportional logarithm of difference; the sum will be the proportional logarithm of correction to be added to the right ascension for noon of Greenwich date.

EXAMPLE.

212. July 13, 1853, at 6ʰ 31ᵐ A.M. mean time nearly, in long. 172° 10′ W.: required the sun's right ascension.

Ship, July 12.....18ʰ 31ᵐ

long. in time......11 29 W.
                   30 0

or ship, July 13... 6 0

Sun's right ascension.

July 13 ......7ʰ 30ᵐ 30ˢ

„   14 ......7 34 33
                 4 3

0·60206
1·64782
2·24988...... 1 1

∴ sun's R. A...7 31 31

This and the following examples may have been worked out by using the hourly difference, as in Rule 8. Sometimes the simplest method is to estimate the correction *by sight*, as in the above example, where we have to find the change of right ascension in 6 hours, the change in 24 hours being 4ᵐ 3ˢ, the correction is evidently 1ᵐ 1ˢ nearly.

Find the sun's right ascension in the following examples :

213. March 11, 1853...at 6ʰ 42ᵐ P.M. mean time...long. 42° 41′ W.
214.   „   21 „ ...at 10 10 A.M.   „   ... „ 100 41 E.
215.   „   21 „ ...at 0 0 A.M.   „   ... „ 142 14 W.

Sun's R. A. from Nautical Almanac.

Sun's right asc., March 11, 23ʰ 26ᵐ 26·3ˢ   March 12, 23ʰ 30ᵐ 6·6ˢ
   „   „   20 23 59 20·0   „   21 0 2 58·2
   „   „   21 0 2 58·2   „   22 0 6 36·4

Ans. to 213, 214, 215 : 23ʰ 27ᵐ 54·5ˢ, 0ʰ 1ᵐ 40·0ˢ, 0ʰ 4ᵐ 24·2ˢ.

Rule 12. *To take out the* moon's DECLINATION *and* RIGHT ASCENSION.

The moon's declination and right ascension are recorded in the *Nautical Almanac* for the beginning of every hour of mean time at Greenwich. To find them for any other time we may proceed as follows :

First. To find the moon's declination for any given time.

1. Get a Greenwich date.

2. Take out of the *Nautical Almanac* the moon's declination for two consecutive hours between which the Greenwich date lies, and take the difference.

First method.   By logistic logarithms.

1. Add together the logistic logarithm of minutes in Greenwich date and proportional logarithm of difference, the sum will be the proportional logarithm of correction, which take from the table and apply it to the declination for the hour of Greenwich date, adding or subtracting according as the declination is seen to be increasing or decreasing.   The result is the declination required.

Second method.   By 10 minutes' differences.

1. Take out "Diff. Dec. for 10ᵐ" opposite the first declination taken out.

2. Multiply the 10ᵐ diff. by the number of minutes in Greenwich date, and remove the decimal point one place to the left: the result is the correction in decl. for Greenwich date in seconds, which turn into minutes and seconds if necessary, and apply to the decl. as in first method.

To take out the MOON'S RIGHT ASCENSION.

First method.   By logistic logs.

Proceed as in the first method just given for finding the moon's declination.

Second method.   By hourly difference, or by the rule of Practice.

Multiply hourly difference, turned into seconds, by the number of minutes in Greenwich date, and divide by 60: the result will be the correction in right ascension for Greenwich date in seconds, which turn into minutes and seconds if necessary, and add to the right ascension at the hour of Greenwich date; or this correction may be obtained by the common rule of Practice.

EXAMPLES.

216.   January 24, at 5ʰ 40ᵐ P.M. mean time, in long. 60° 10′ W.: find the moon's declination and right ascension.

Ship, Jan. 24..... 5ʰ 40ᵐ
long. in time.......4   1 W.
                   ―――――――
Gr. Jan. 24.........9 41

Moon's declination.

First method.   By logistic logarithms.
24 at 9ʰ............19° 39′ 12″ N.
10 ....19 34 21 N.
                  ――――――――
                   4 51 S.

0·16537
1·56953
1·73490......   3 19 ―
∴ moon's decl. at 9ʰ 41ᵐ......19 35 53

Second method. By 10ᵐ differences.

10ᵐ diff............48·6″ dec.

$$
\begin{array}{r}
41 \\
\hline
486 \\
1944 \\
\hline
60)\overline{199\cdot26} \\
\end{array}
$$

3′ 19·3″=change in decl. as before.

Moon's right ascension.

First method. By logistic logarithms.

$$
\begin{array}{r}
24 \text{ at } 9^{h}............6^{h} \ 53^{m} \ 24\cdot1^{s} \\
10 ....6 \ 55 \ 30\cdot1 \\
\hline
2 \ 6\cdot0 \\
\end{array}
$$

$$
\begin{array}{r}
0\cdot16537 \\
1\cdot93305 \\
\hline
2\cdot09842...... \quad 1 \ 26\cdot0+ \\
\end{array}
$$

∴ moon R. A. at 9ʰ 41ᵐ......6 54 50·1

Second method. By hourly differences.   By Practice.

diff. in 60ᵐ............126ˢ	30...½ | 126·0ˢ
41	10...⅙ | 63·0
126	1...¹⁄₁₀ | 21·0
504	| 2·1
60)516·6	| 86·1
86·1	or 1ᵐ 26·1ˢ
or 1ᵐ 26·1ˢ	

The same result as before.

217. June 2, at 10ʰ 30ᵐ A.M. mean time, in long. 53° 15′ W,: find the moon's right ascension and declination.

Ans. R. A. 2ʰ 47ᵐ 29·4ˢ, decl. 17° 12′ 45″ N.

218. Sept. 7, at 4ʰ 15ᵐ A.M. mean time, in long. 56° 30′ E.: find the moon's right ascension and declination.

Ans. R. A. 14ʰ 53ᵐ 51·7ˢ, decl. 16° 51′ 39″ S.

219. July 10, at 9ʰ 30ᵐ A.M. mean time, in long. 44° 20′ W.: find the moon's right ascension and declination.

Ans. R. A. 11ʰ 27ᵐ 48·2ˢ, decl. 1° 23′ 0″ S.

Moon's R. A. and decl. from Nautical Almanac.

	Moon's right ascension.	Moon's decl.	10ᵐ diff.
June 2, at 2......	2ʰ 47ᵐ 23·1'........	17° 12' 26" N..........	63·4" incr.
,, at 3......	2 49 28·1	17 18 46 N.	
Sept. 6, at 12......14	52 44·7	16 48 19 S.	68·9" incr.
,, at 13......14	55 3·7	16 55 13 S.	
July 10, at 0......11	26 55·2	1 17 58 S.	111·7" incr.
,, at 1......11	28 52·9	1 29 9 S.	

Rule 13. *To take out the* RIGHT ASCENSION OF THE MEAN SUN (*called in the Nautical Almanac* SIDEREAL *time*).

The right ascension of the mean sun, or the sidereal time at mean noon, is given in the *Nautical Almanac* for.every day at mean noon. To find it for any other time we may proceed as in the rule for finding the right ascension of the apparent or true sun ; but as the motion of the mean sun is uniform throughout the year (the motion in every 24 hours being 3ᵐ 56·555'), the change in any given number of hours, minutes, and seconds is more easily found by means of a table. This table is given in the *Nautical Almanac*, under the title of "Time Equivalents;" it may be also found, arranged in a very convenient form, in Inman's *Nautical Tables*, p. 12*.

EXAMPLE.

220. July 23, 1853, at 2ʰ 32ᵐ P.M., in long. 80° 42' E.: required the right ascension of the mean sun.

		Right asc. mean sun.	Or thus, by table.
Ship, July 23...2ʰ 32ᵐ		July 22...8ʰ 0ᵐ 35'	July 22...8ʰ 0ᵐ 35'
long. in time.....5 23 E.		,, 23...8 4 32	cor. for 21ʰ 3 27
Green.,July 22.21 9		⎯⎯⎯⎯	,, 9ᵐ 1·5
		3 57	R.A.......8 4 3·5
		0·05490	as before.
		1·65868	
		⎯⎯⎯⎯⎯⎯	
		1·71358 3 29	

Right asc. mean sun...8 4 4

Find the right ascension of mean sun (called in the *Nautical Almanac* sidereal time) in the following examples :

221. March 2, 1853, at 10ʰ 42ᵐ P.M. mean time, in long. 48° 10' W.
222.   ,,   15   ,,   ,, 6 6 A.M.   ,,   100 0 W.
223.   ,,   21   ,,   ,, 10 10 P.M.   ,,   100 0 E.

Sid. time from Nautical Almanac, and Answers.

Sidereal time March 2, at noon, 22ʰ 40ᵐ 44·9'......*Ans.* 22ʰ 43ᵐ 2·0'
   ,,   ,,   15   ,,   23 32   0·1 ...... ,, 23 32 7·6
   ,,   ,,   21   ,,   23 55 39·4 ...... ,, 23 56 13·9

Rule 14. *To find the* LUNAR DISTANCE *for any given time at Greenwich.*

1. Get a Greenwich date.

2. Find two consecutive distances in the *Nautical Almanac* at times between which the Greenwich date lies. Take the difference of the distances. To the proportional logarithm of the excess of the Greenwich date above the first of the times taken from the *Nautical Almanac* add proportional logarithm of difference of distances; the sum will be the proportional logarithm of an arc; which arc being applied to the distance at first time with its proper sign will be the distance required.

EXAMPLE.

224. September 24, at 6^h 10^m P.M. mean time nearly, in long. 60° 15′ W.: required the distance of Aldebaran from the moon.

<table>
<tr><td></td><td></td><td colspan="2">Dist. of Aldebaran.</td></tr>
<tr><td>Ship, Sept. 24...</td><td>$6^h 10^m$</td><td>At 9^h...18°57′ 35″</td></tr>
<tr><td>long. in time ...</td><td>4   1 W.</td><td>12 ...20 23 37</td></tr>
<tr><td>Green. Sept. 24...</td><td>10 11</td><td>Proportional log....32061 ... 1 26 2</td></tr>
<tr><td></td><td></td><td>prop. log. $1^h 11^m$ ...40401</td></tr>
<tr><td></td><td></td><td>log. cor. ....72462 ... 0 33 56</td></tr>
<tr><td></td><td></td><td>∴ dist. of Aldebaran at $6^h 10^m$...19 31 31</td></tr>
</table>

Required the distance of the moon from certain stars in the following examples:

225. Jan. 24, at 4^h 30^m P.M. mean time nearly, in long. 30° 30′ E.: required the distance of Regulus from the moon.   *Ans.* 69° 33′ 6″.

226. May 20, at 6^h 20^m A.M. mean time nearly, in long. 40° 0′ E.: required the distance of α Pegasi from the moon.   *Ans.* 56° 59′ 7″.

227. June 10, at 9^h 40^m P.M. mean time nearly, in long. 32° 45′ W.: required the distance of α Aquilæ from the moon.   *Ans.* 74° 32′ 35″.

228. July 2, at 7^h 20^m A.M. mean time nearly, in long. 30° 0′ E.: required the distance of Jupiter from the moon.   *Ans.* 54° 16′ 52″.

229. Sept. 19, at 10^h 30^m A.M. mean time nearly, in long. 63° 15′ E.: required the distance of Aldebaran from the moon.   *Ans.* 72° 0′ 51″.

230. Dec. 15, at 2^h 0^m P.M. mean time nearly, in long. 19° 40′ E.: required the distance of Pollux from the moon.   *Ans.* 58° 56′ 47″.

Distances from Nautical Almanac.

Distance of Regulus	at noon	68°11′ 7″	......at	3^h	69°50′ 50″
,,	α Pegasi	,,	15^h	57 17 6	...... ,, 18	55 56 9
,,	α Aquilæ	,,	9	75 56 46	...... ,, 12	74 28 9
,,	Jupiter	,,	15	55 41 18	...... ,, 18	53 52 44
,,	Aldebaran	,,	18	72 9 14	...... ,, 21	70 40 29
,,	Pollux	,,	noon	58 32 51	...... ,, 3	60 17 54

In rule (50), for finding the longitude by lunar observations, we have to calculate the *true* distance of the moon from some heavenly body at the time of observation. If the heavenly body is one whose distance is recorded in the *Nautical Almanac* for every three hours, we may find the mean time at Greenwich corresponding to the computed true distance for the time of observation as follows :

Rule 15. *To find the* TIME AT GREENWICH *corresponding to a* GIVEN DISTANCE *of a heavenly body from the moon.*

1. Under the given distance put down the two computed distances of the same heavenly body found in the *Nautical Almanac* between which the given true distance lies.

2. Take the difference between the first and second, and also between the second and the third.

3. From the proportional logarithm of the first difference subtract the proportional logarithm of the second difference, the remainder is the proportional logarithm of the additional time to be added to the hours of the distance first taken out of the *Nautical Almanac;* the result is the mean time at Greenwich corresponding to the given distance.

EXAMPLES.

231. November 22, 1853, the true distance of Saturn from the moon was found to be 77° 52′ 45″: required Greenwich mean time when the observation was taken.

$$
\begin{array}{llr}
\text{True distance at observation} & \dots\dots 77°52' \ 45'' \\
\text{in } \textit{Naut. Almanac} \text{ dist. at } 3^h & \dots\dots 77 \ \ 14 \ \ 40 \\
6 & \dots\dots 78 \ \ 47 \ \ 24 \\
\end{array}
$$

$$
\begin{array}{lll}
\text{prop. logarithm} & \cdot 67454 \dots\dots & 38 \quad 5 \\
& \cdot 28804 \dots\dots & 1 \ \ 32 \ \ 44 \\
\hline
& \cdot 38650 \quad \text{Cor. } 1^h 13^m 55^s \\
& \text{Adding } 3 \\
\hline
\end{array}
$$

Greenwich mean time Nov. 22 ... 4 13 55

Find mean time at Greenwich from each of the following observations:

232. November 24, 1853, when true distance of Aldebaran was 93° 38′ 45″.   *Ans.* 3^h 57^m 18^s.

233. Sept. 24, 1853, when true distance of Regulus was 58° 45′ 8″.
                                          Ans. 16^h 3^m 6^s.

234. May 27, 1853, when true distance of the sun was 110° 8′ 50″.
                                          Ans. 14^h 2^m 22^s.

Distances from Nautical Almanac.

Dist. Aldebaran, Nov. 24, at 3^h ... 93° 7′ 57″......at 6^h ... 94° 44′ 42″

,,   Regulus, Sept. 24 ... ,, 15 ... 59 16 16 ......,, 18 ... 57 47 27

,,   Sun, May 27............ ,, mid....111 12 57 ......,, 15 ...109 38 38

To take out a PLANET'S RIGHT ASCENSION AND DECLINATION.

Proceed as in the similar rules for finding the sun's right ascension and declination (pp. 82, 78).

The rules above given are sufficient to enable the student to acquire a competent knowledge of the principal contents of the *Nautical Almanac.* They will be found of the greatest use in the subsequent rules for finding the latitude and longitude.

Equation of Second Differences.

We have supposed in the above examples the motion of the heavenly body to be *uniform* in the interval between the Greenwich times taken out of the *Nautical Almanac.* This is seldom the case, although in most of the questions in *Nautical Astronomy* it may be so assumed without any practical error. When, however, very accurate results are required, a correction must be used called the *equation of second differences.* The investigation of this equation belongs to the theoretical part of Navigation, and as the equation is seldom required in the common rules of Navigation, we may omit for the present examples of its application.

CHAPTER IV.

PRELIMINARY PROBLEMS AND RULES IN NAUTICAL ASTRONOMY.

CORRECTIONS FOR PARALLAX, REFRACTION, CONTRACTION OF THE MOON'S SEMI-DIAMETER, AND DIP.

RULE 16. *Given mean solar time and the equation of time: to find the apparent solar time; or*

Given apparent solar time and the equation of time: to find mean solar time.

1. Get a Greenwich date (p. 74).

2. Correct for this date the equation of time (taken out of page i. of *Nautical Almanac*, when apparent solar time is given, and out of page ii. when mean time is given) (Rule 9).

3. Apply the equation of time, with its proper sign (as shown in the *Nautical Almanac*), to the given time.

4. The result is the time required.

EXAMPLES.

235. April 27, 1846, at $9^h\ 10^m$ P.M. mean time, in long. 16° W.: required apparent solar time.

Mean time.		Equation of time (page ii. *Naut. Alm.*).		Mean time.	
Ship, April 27 .	$9^h\ 10^m$	27 . . $2^m\ 26{\cdot}9^s$ add to	Ship, April 27 .	$9^h\ 10^m\ 0^s$
long. in time . .	1 4 W.	28 . . 2 36·4 mean time.	eq. of time . . .	2 31+
Gr., April 27 . .	10 14	9·5	∴ 9 12 31
		0·37020	apparent time required.
		3·05570	
		3·42590 . . 4·1	
		∴ eq. of time . . 2 31·0+	

236. June 22, 1852, at $5^h\ 42^m$ P.M. apparent solar time, in long. 100° 30′ E.: required mean solar time.

Apparent time.		Equation of time (page i. *Naut. Alm.*). Diff. for 1 hour, 0·54^s.		Mean time.	
Ship, June 22 .	$5^h\ 42^m$ P.M.	21 . . $1^m\ 27{\cdot}2^s$ add to	Ship, June 22	$5^h\ 42^m\ 0^s$
long. in time .	6 42 E.	22 . . 1 40·2 app. time.	eq. of time .	1 39·6+
Gr., June 21 .	23 0	13·0	∴ 5 43 39·6
		23 × ·54 12·4	mean time required.
		1 39·6+	

237. July 4, 1853, at 6ʰ 10ᵐ P.M. mean time, in long. 100° W.: required apparent solar time.   *Ans.* 6ʰ 5ᵐ 53·2ˢ.

238. Dec. 10, 1853, at 4ʰ 42ᵐ P.M. apparent solar time, in long. 80° 45′ W.: required mean solar time.   *Ans.* 4ʰ 35ᵐ 18·1ˢ.

239. Feb. 23, 1848, at 10ʰ 40ᵐ A.M. apparent solar time, in long. 1° 6′ W.: required mean solar time.   *Ans.* 10ʰ 53ᵐ 43·5ˢ A.M.

Equation of time from Nautical Almanac.

July 4 ... 4ᵐ 1·1ˢ sub.	July 5 ... 4ᵐ 11·7ˢ sub.
Dec. 10 ... 6 53·5 sub.	Dec. 11 ... 6 25·8 sub.
Feb. 22 ... 13 51·0 add.	Feb. 23 ... 13 43·1 add.

Rule 17. *Given mean solar time: to find sidereal time.*

1. Get a Greenwich date (p. 74).

2. Correct the right ascension of the mean sun by the table of Time Equivalents (p. 86), or by proportional logarithms, or otherwise, for the Greenwich date.

3. Add together the corrected right ascension of mean sun and mean time at the ship.

4. The sum (rejecting 24 hours if greater than 24 hours) will be sidereal time.

EXAMPLES.

240. Feb. 24, 1848, at 10ʰ 40ᵐ 30ˢ A.M. mean time, in long. 1° 6′ W.: required sidereal time.

Right ascension mean sun.

Ship, Feb. 23	22ʰ 40ᵐ 30ˢ	Feb. 23 . .	22ʰ 10ᵐ 3·52ˢ	
long. in time .	0 4 24	cor. for 22ʰ . .	3 36·84 ⎫	by table.
Gr., Feb. 23 .	22 44 54	„ 44ᵐ . .	7·23 ⎬	
		„ 54ˢ . .	·15 ⎭	
		right asc. mean sun .	22 13 47·74	
		ship mean time . .	22 40 30·00	
		sidereal time . . .	20 54 17·74 (rejecting 24 hours).	

241. July 10, 1853, at 0ʰ 42ᵐ 10ˢ P.M. mean time, in long. 84° 42′ W.: required sidereal time.   *Ans.* 7ʰ 56ᵐ 29·7ˢ.

242. Sept. 30, 1853, at 6ʰ 42ᵐ 10ˢ A.M. mean time, in long. 100° 42′ W.: required sidereal time.   *Ans.* 7ʰ 18ᵐ 58·59ˢ.

243. Dec. 8, 1853, at 10ʰ 10ᵐ 42ˢ P.M. mean time, in long. 18° 32′ E.: required sidereal time.   *Ans.* 3ʰ 20ᵐ 47ˢ.

R. A. mean sun from Nautical Almanac.

July 10	7ʰ 13ᵐ 17·14ˢ
Sept. 30	12 36 34·64
Dec. 8	17 8 36·96

Rule 18. *Given apparent solar time : to find sidereal time.*

1. Get a Greenwich date (p. 74).

2. Correct the equation of time and also the right ascension of the mean sun for Greenwich date (pp. 80, 86).

3. Apply corrected equation of time to ship apparent time, and thus get ship mean time. Then, as in the last rule,

4. Add together ship mean time and right ascension of mean sun.

5. The sum (rejecting 24 hours if greater than 24 hours) will be sidereal time.

EXAMPLES.

244. May 24, 1853, at $6^h 8^m 40^s$ A.M. apparent solar time, in long. $20° 20'$ W. : required sidereal time.

				Equation of time.			Right ascension mean sun.			
Ship, May 23	18^h	8^m	40^s	23 . 3^m $33 \cdot 2^s$ sub. from		23d, at noon .	4^h	4^m	$2 \cdot 37^s$
long. in time .	1	21	20 W.	24 . 3 $28 \cdot 3$ app. time.		19^h . . .		3	$7 \cdot 27$
Gr., May 23 .	19	30	0	$4 \cdot 9$		30^m . . .			$4 \cdot 93$

Prop. logs.
0·09018

3·34323

3·43341 . . . . 4·0

R. A. mean sun	4	7	$14 \cdot 57$
ship M. T. . .	18	5	$10 \cdot 80$
sidereal time .	22	12	$25 \cdot 37$

eq. of time .   3   29·2 sub.
app. time .   18   8   40·0

May 23 . .   18   5   10·8 ship mean time.

245. July 4, 1853, at $3^h 42^m$ A.M. apparent solar time, in long. $84° 42'$ W. : required sidereal time.   *Ans.* $22^h 35^m 10 \cdot 53^s$.

246. Oct. 21, 1853, at $8^h 48^m$ P.M. apparent solar time, in long. $88° 8'$ E. : required sidereal time.   *Ans.* $22^h 32^m 30 \cdot 87^s$.

Eq. of time, and R. A. mean sun from Nautical Almanac.

Equation of time.						Right ascen. mean sun.			
July 3,	3^m	$50 \cdot 1^s$ add	4,	4^m	$1 \cdot 1^s$, add	......	3,	6^h	45^m	$41 \cdot 24^s$
Oct. 21, 15	$19 \cdot 1$ sub.	22, 15	$28 \cdot 1$, sub.	......	21, 13	59	$22 \cdot 26$

Rule 19. *Given mean time, or apparent time at the ship : to find what heavenly body will pass the meridian the next after that time.*

1. Get a Greenwich date (p. 74).

2. Find the right ascension of the mean sun (p. 86), and, if the Greenwich date is in apparent time, find also the equation of time (p. 80) for that date, so as to get mean time.

3. Add together ship mean time and the right ascension of mean sun.

4. The sum (rejecting 24 hours if greater than 24 hours) will be sidereal time, or the right ascension of the meridian.

5. Then that star, found in some catalogue of fixed stars, whose right ascension is the *next greater* will be the star required.

EXAMPLE.

247. Feb. 24, 1853, at 4^h 42^m P.M. mean time nearly, in long. 100° E. : find what bright star will pass the meridian the next after that date.

Right ascension mean sun.

Ship, Feb. 24 .	.	4^h	42^m	23	. . .	22^h 13^m $9\cdot0^s$	Ship, Feb. 24	.	4^h 42^m	0^s
long. in time .	.	6	40 E.	22^h .	.	3 36·8	R. A. mean sun	22	16	46
Gr., Feb. 23 .	. 22	2		2^m .	.	·3	R. A. merid.	. 2	58	46
						22 16 46·1				

Looking into the "Catalogue of the mean places of 100 principal fixed stars" (see *Nautical Almanac*), we find the star whose right ascension is next greater than 2^h 58^m is α Persei ; therefore α Persei is the bright star that will come to the meridian the next after 4^h 42^m P.M. on Feb. 24.

Sometimes it is required to find what principal stars will pass the meridian between certain convenient hours for observing their transits : as, for instance, between 8^h and 11^h P.M. To do this, we must find the right ascension of the meridian for these two times by the above rule ; then the stars whose right ascensions lie between will be the stars required.

EXAMPLES.

248. Oct. 3, 1853, in long. 90° W., find what bright stars put down in the *Nautical Almanac* will pass the meridian between the hours of 9 and 12 P.M.

Ship, Oct. 3............	9^h 0^m		Ship, Oct. 3	12^h 0^m	
long. in time	6 0 W.		long. in time	6 0 W.	
Greenwich, Oct. 3 ...	15 0		Greenwich, Oct. 3 ...	18 0	

Right ascension mean sun.			Right ascension mean sun.		
Oct 3	12^h 48^m 24^s		Oct. 3	12^h 48^m 24^s	
15^h	2 27		18^h	2 57	
	· 12 50 51			12 51 21	
ship, Oct. 3	9 0 0		ship, Oct. 3	12 0 0	
R. A. meridian ...	21 50 51		R. A. meridian ...	0 51 21	

In Catalogue p. 432, *Nautical Almanac*, the stars whose right ascensions lie between $21^h 50^m 51^s$ and $0^h 51^m 21^s$ are from α Aquarii to β Ceti inclusive.*

249. What bright stars put down in the *Nautical Almanac* will pass the meridian of a ship in long. 40° E., between 8^h and 10^h P.M. mean time, on Nov. 20, 1853?   *Ans.* From α Andromedæ to α Arietis.

250. What bright star will pass the meridian of a ship in long. 30° W. the first after $10^h 30^m$ P.M. on Oct. 10, 1853?   *Ans.* α Andromedæ.

251. What bright stars will pass the meridian of a ship in long. 56° W., between the hours of 6 and 10 P.M., on March 10, 1853?

Ans. From β Tauri to ι Argûs.

252. What bright stars put down in the *Nautical Almanac* will pass the meridian of Greenwich, between the hours of 7 and 9 P.M. mean time, on August 20, 1853?   *Ans.* From ϵ Ursæ Minoris to β Lyræ.

253. What stars named in the *Nautical Almanac* will pass the meridian of a ship in long. 86° E., on Oct. 20, 1853, between the hours of 10 P.M. and midnight?   *Ans.* From α Andromedæ to α Eridani.

254. What bright star will pass the meridian of Greenwich the first after 9^h P.M. on Sept. 12, 1853?   *Ans.* α Cygni.

R. A. mean sun from Nautical Almanac.

	R. A. mean sun.				R. A. mean sun.	
Nov. 20	15^h	57^m	39^s	Aug. 20	9^h	54^m	56^s
Oct. 10	13	16	0	Oct. 20	13	55	26
Mar. 10	23	12	17	Sep. 12	11	25	37

Rule 20. *Given sidereal time : to find mean time.*

1. Take out of the *Nautical Almanac* the right ascension of the mean sun (called in the *Nautical Almanac* sidereal time) for noon of the given day.

2. From sidereal time (increased if necessary by 24 hours) subtract the quantity just taken out; the remainder is mean time nearly.

3. Find in the table of the acceleration of sidereal on mean solar time the correction for this time, and subtract it from the mean time nearly.

4. The remainder is the mean time required.

NOTE. In strictness we ought to have entered the table with the correct mean time, instead of that used; but it is evident we may obtain a still closer approximation to the truth by repeating the work, using the last approximate value instead of the preceding one. For all practical purposes this repetition is seldom necessary.

---

* In the *Handbook for the Stars*, published by the author, there is a table of the approximate times of the transits of the principal fixed stars. This table enables the observer to find the name of the bright star that is near the meridian at any given time, and at any place, by *inspection*, and without any calculation.

255. April 27, when a sidereal clock showed 3^h 40^m 45^s: required mean time.

Sidereal time		3^h	40^m	$45\cdot00^s$
R. A. mean sun at				
mean noon		2	20	21·58
mean time nearly ...		1	20	23·42
cor. ...1^h	9·86			
„ 20^m ...	3·28			
„ 23^s	·06			13·20
∴ required mean time		1	20	10·22

256. March 2, when a sidereal clock showed 3^h 40^m 45^s: required mean time.

Sidereal time		3^h	40^m	$45\cdot00^s$
R. A. mean sun at				
mean noon		22	41	35·94
mean time nearly...		4	59	9·06
cor. ...4^h ...	39·43			
„ 59^m ..	9·69			
„ 9^s ...	·02			49·14
∴ required mean time		4	58	19·92

(76.) The clock of an observatory used for noting the transit of a heavenly body is generally adjusted to *sidereal* time. By means of the above rule we can determine the error of a chronometer or solar clock regulated to *mean* time, by comparing the chronometer with the sidereal clock at some *coincident* beat, and then, correcting the sidereal clock for its error, we can find the corresponding mean time at the instant; the difference between which and the time shown by the chronometer will be the error of the chronometer on mean time at the place.

EXAMPLE.

257. Greenwich, March 3, 1853, when a sidereal clock showed 6^h 10^m 20^s a chronometer showed 7^h 32^m 10^s: required the error of the chronometer on Greenwich mean time; the error of sidereal clock being 2^m $42\cdot5^s$ slow.

Sidereal clock		6^h	10^m	20^s	
error			2	42·5 slow.	
sidereal time		6	13	2·5	
R. A. mean sun at mean noon		22	44	41·48	
Cor. 7^h	1^m 8·99^s	7	28	21·02	
„ 28^m	4·60				
„ 21^s	·05				
	1 13·64		1	13·64	
required mean time		7	27	7·38	
chronometer showed		7	32	10·0	
error of chro. on Gr. mean time			5	2·62 fast.	

(77.) When the calculations are made for any other meridian than that of Greenwich, for which the quantities in the *Nautical Almanac* are calculated, we must take into consideration the change of the mean sun's place arising from the difference of longitude. For example, the tables of the *Connaissance des Tems* are computed for Paris, the long. of which is 9ᵐ 22ˢ to the east of Greenwich: as in that time the mean sun moves to the eastward through an arc of 1·53ˢ in time (for 24ʰ : 9ᵐ 22ˢ : : 3ᵐ 56·55ˢ : 1·53ˢ), it follows that we must add 1·53ˢ to all the right ascensions of the mean sun in the French tables to obtain those of the mean sun at mean noon at Greenwich. (See *Nav.* Part II. chap. vii.) Or thus, by Rule 20:

EXAMPLE.

258. April 27, 1841, the right ascension of the mean sun at mean noon at Paris, by the *Connaissance des Tems*, was 2ʰ 21ᵐ 10·09ˢ: required the same for Greenwich mean noon.

```
        Greenwich, April 27 ............  0ʰ  0ᵐ  0ˢ
        long. in time ....................     9  22  W.
                                             ─────────
        Paris date, April 27.............    9  22
R. A. mean sun at Paris .................  2ʰ 21ᵐ 10·09ˢ
        Cor. 9ᵐ......  1·48ˢ
         ,,  22ˢ......   ·05
                      ──────
                       1·53                      1·53
                                             ─────────
R. A. mean sun at Greenwich............    2   21  11·62
```

(78.) The longitude is usually found at sea by means of a chronometer showing Greenwich mean time at the instant the mean time at the ship is known. The mean time at the ship is deduced from the hour-angle of a heavenly body, and this hour-angle is calculated by means of the altitude of the body observed with a sextant and certain elements given in the *Nautical Almanac*.

Rules for calculating the hour-angle of a heavenly body from an observed altitude will be given hereafter. We will here suppose the hour-angle known, and proceed to show how mean time might be found from it.

Rule 21. *To find mean time at the ship, from the hour-angle of a star or the moon.*

It is proved in *Navigation*, Part II. p. 34, that

1. When the star is WEST of meridian,

Mean time=star's hour-angle + star's right ascension — right ascension of mean sun.

2. When the star is EAST of meridian,

Mean time=(24ʰ — star's hour-angle) + star's right ascension — right ascension of mean sun.

To find ship mean time, we must proceed therefore as follows :

1. Get a Greenwich date.
2. Take out the star's right ascension.
3. Take out also the right ascension of the mean sun (called in *Nautical Almanac* sidereal time), and correct it for Greenwich date.
4. When heavenly body is WEST of meridian :

To the star's hour-angle add star's right ascension, and from the sum subtract the right ascension of mean sun (adding or rejecting 24 hours if necessary); the result is ship mean time.

5. When heavenly body is EAST of meridian :

First subtract hour-angle from 24 hours, then to the remainder add star's right ascension, and from the sum subtract the right ascension of the mean sun; the result (rejecting 24 hours if necessary) is ship mean time required.

EXAMPLES.

259. Feb. 10, 1847, at $9^h 22^m$ P.M. mean time nearly, in long. 27° 15' W., the hour-angle of Regulus (α Leonis) was $3^h 15^m 17^s$ EAST of meridian: required mean time at the ship.

Ship, Feb. 10 ...	9^h	22^m		24^h	0^m	0^s
long. in time ...	1	49 W.	Star's H. A.......	3	15	17
Gr. Feb. 10	11	11		20	44	43·0
R. A. mean sun.			star's R. A.	10	0	15·3
Feb. 10	21^h	19^m	$46·0^s$			
cor...11^h		1	48·4	30	44	58·3
„ 11^m			1·8	R. A. mean sun	21	21	36·2
R. A. mean sun..	21	21	36·2	∴ ship mean time 9	23	22·1

260. Sept. 10, 1844, at $7^h 11^m$ P.M. mean time nearly, in long. 32° E., the hour-angle of Arcturus (α Bootis) was $4^h 22^m 15^s$ WEST of meridian: required mean time at the ship. *Ans.* $7^h 11^m 31·7^s$ P.M.

261. Nov. 22, 1853, at $7^h 15^m$ P.M. mean time nearly, in long. 22° 0' W., the hour-angle of Aldebaran (α Tauri) was $5^h 10^m 20^s$ EAST of meridian: required mean time at the place. *Ans.* $7^h 10^m 14^s$.

262. June 23, 1853, at $4^h 15^m$ A.M. mean time nearly, in long. 100° 40' E., the hour-angle of α Lyræ was $3^h 42^m 40^s$ WEST of meridian: required mean time at the place. *Ans.* $16^h 10^m 56^s$.

NOTE.—If the estimated ship time used for getting the Greenwich date differs several minutes from the true ship time, the R. A. mean sun, and therefore ship mean time deduced from it, may be a few seconds incorrect. To get a correct result we must use the ship mean time, found as in the above examples, instead of that first used, and thus obtain a *corrected* Greenwich

H

date, and then recalculate the R. A. mean sun for that date. It will be rarely necessary to repeat this method of approximation more than once; but the necessity for this repetition should be borne in mind in many of the subsequent rules when a *wrong Greenwich date* has been found to have been used. The following examples will show the effect of an error in the Greenwich date on the ship mean time deduced from it.

EXAMPLES.

263. August 11, 1846, at 8^h 50^m P.M. mean time nearly, in long. $90°$ W., the hour-angle of Arcturus was 3^h 56^m $55'$ WEST of meridian: required correct mean time at the place.

$$
\begin{aligned}
\text{Ship, Aug. 11} &\quad 8^h \quad 50^m \\
\text{long. in time} &\quad 6 \quad\; 0 \text{ W.} \\[4pt]
\text{Greenwich, Aug. 11} &\quad 14 \quad 50
\end{aligned}
$$

Right ascension mean sun.			Star's hour-angle ...	3^h	56^m	$55·0'$
Aug. 11	9^h 18^m	$16·51'$	,, right asc......	14	8	40·14
cor. 14^h	2	$17·99$				
,, 50^m		$8·21$		18	5	35·14
			rt. asc. mean sun...	9	20	42·71
	9 20	$42·71$				
			ship mean time ...	8	44	52·43

This result is slightly incorrect, arising from the estimated mean time, 8^h 50^m, being different from the true time. When great accuracy is required, the operation should be repeated, using mean time last found, namely 8^h 45^m, instead of the one used before ; thus,

The operation repeated.

$$
\begin{aligned}
\text{Ship, Aug. 11} &\quad 8^h \quad 45^m \\
\text{long. in time} &\quad 6 \quad\; 0 \\[4pt]
\text{Greenwich, Aug. 11} &\quad 14 \quad 45
\end{aligned}
$$

Right ascension mean sun.			Star's hour-angle...	3^h	56^m	$55·0'$
Aug. 11	9^h 18^m	$16·51'$	,, right asc. ...	14	8	40·12
cor. 14^h	2	$17·99$				
,, 45^m		$7·39$		18	5	35·12
			rt. asc. mean sun ...	9	20	41·89
	9 20	$41·89$				
			cor. ship mean time	8	44	53·23

264. June 15, 1853, at 10^h 10^m P.M., supposed mean time nearly, in long. $10°$ $42'$ W., the hour-angle of Arcturus was 2^h 2^m $30'$ EAST of meridian: required mean time at the place.

Ans. 1st approximation, 6^h 30^m; 2d approx. 6^h 30^m $35'$.

Elements from Nautical Almanac.

Right ascension mean sun.				Right ascension star.			
Sept. 10, 1853......	11ʰ	18ᵐ	28ˢ	α Bootis	14ʰ	8ᵐ	35ˢ
Nov. 22, ,, ...	16	5	32	Aldebaran......	4	27	32
June 22, ,, ...	6	2	10	α Lyræ.........	18	32	0
June 15, ,, ...	5	34	43	Arcturus	14	8	59

TO FIND SHIP MEAN TIME FROM THE HOUR-ANGLE OF THE SUN.

If the heavenly body observed be the sun, its hour-angle will also be *apparent time* at the place if P.M. at the time of observation, and what it wants of 24 hours if A.M. Therefore, to find the corresponding *mean* time, we have only to apply apparent time thus found to the equation of time, with its proper sign, as pointed out in Rule 16, p. 90.

Rule 22. *To find at what time any heavenly body will pass the meridian, and how far north or south of the zenith.*

1. Take out of the *Nautical Almanac* the right ascension of the heavenly body, and also the right ascension of the mean sun for noon of the given day.

2. From the right ascension of the heavenly body (increased if necessary by 24 hours) subtract the right ascension of the mean sun ; the remainder is mean time at the ship nearly.

3. Apply the longitude in time, and thus get a Greenwich date ; with this Greenwich date correct the right ascension of mean sun, and the right ascension of the heavenly body if necessary.

4. Then from the right ascension of the star subtract the right ascension of the mean sun thus corrected; the remainder is the mean time when the heavenly body is on the meridian.

As in the last problem, the table of acceleration for correcting the R. A. of mean sun ought to have been entered with the correct mean time; but the error in this case is inappreciable.

EXAMPLE.

265. At what time will Sirius pass the meridian of a place in long. 68° 30′ W. on Nov. 20, 1845 ?

R. A. mean sun.

Star's R. A.+24 30ʰ 38ᵐ 23ˢ	Nov. 20. .... 15ʰ 57ᵐ 26ˢ	Star's R. A. . . 30ʰ 38ᵐ 23ˢ
R. A. mean sun 15 57 26	cor. 19ʰ. . . . 3 7·3	R.A.mean sun 16 0 36
ship M.T. nearly 14 40 57	,, 15ᵐ . . . . 2·5	∴ ship M. T.. 14 37 47
long. in time .. 4 34	R. A. mean sun 16 0 35·8	
Gr. Nov. 20. . . . 19 15		

Therefore the transit of Sirius is at 14ʰ 37ᵐ 47ˢ on Nov. 20, or at 2ʰ 37ᵐ 47ˢ A.M. on Nov. 21.

To find at what time it will pass the meridian on the morning of Nov. 20, we must evidently begin one day back, and take out the right ascension of the mean sun for Nov. 19.

266. At what time will α Pegasi pass the meridian of Portsmouth, long. 1° 6′ W., on Nov. 25, 1853?   *Ans.* Nov. 25, 6ʰ 38ᵐ 58ˢ.

267. At what time will the star Regulus (α Leonis) pass the meridian of Land's End, long. 5° 42′ W., on May 30, 1845?

                                                    Ans. May 30, 5ʰ 27ᵐ 45ˢ P.M.

268. At what time will Antares pass the meridian of Portsmouth, long. 1° 6′ W., on Aug. 20, 1845?   *Ans.* Aug. 20, 6ʰ 24ᵐ 11ˢ.

269. At what time will α Leonis pass the meridian of Lisbon, long. 9° 8′ W., on June 4, 1846?   *Ans.* June 4, 5ʰ 9ᵐ 4ˢ.

270. At what time will the star Antares pass the meridian of Copenhagen, long. 12° 35′ E., on Aug. 20, 1846?   *Ans.* Aug. 20, 6ʰ 25ᵐ 21ˢ.

271. At what time will the star Fomalhaut pass the meridian of Calcutta, long. 88° 26′ E., on Nov. 20, 1846?   *Ans.* Nov. 20, 6ʰ 52ᵐ 34ˢ.

Elements from Nautical Almanac.

	Right ascension mean sun.			Right asc. of star.		
Nov. 25, 1853	16ʰ	17ᵐ	22ˢ	22ʰ	57ᵐ	26ˢ
May 30, 1845	4	31	25	10	0	9
Aug. 20, ,,	9	54	43	16	19	58
June 4, 1846	4	50	11	10	0	11
Aug. 20, ,,	9	53	45	16	20	2
Nov. 20, ,,	15	56	28	22	49	11

To find the meridian zenith distance of a heavenly body, or how far it will pass north or south of zenith.

1. Take out the declination, and correct it, if necessary, for the Greenwich date.

2. Under the latitude of the place put the declination, with their proper names N. or S.

3. If the names are *alike* (both north or both south), take the difference and mark it with the common name of the latitude and declination, if the declination be greater than the latitude, otherwise on the contrary name.

4. If the names are *unlike* (one north and one south), take the sum and mark it with the name of the declination.

5. The result will be the meridian distance of the heavenly body from the zenith N. or S., according as the result was marked N. or S.

EXAMPLE.

272. In latitude 25° N. find how far north or south of the zenith the

following heavenly bodies will pass the meridian, their declinations being 10° N., 30° N., 10° S., and 50° S. respectively :

(1.)	(2.)	(3.)	(4.)
lat. ...25° N.	lat. ...25° N.	lat. ...25° N.	lat. ...25° N.
decl. ...10 N.	decl....30 N.	decl....10 S.	decl....50 S.
diff. ...15 S.	diff. ... 5 N.	sum ...35 S.	sum ...75 S.

273. At what time will α Columbæ pass the meridian of a place in lat. 42° 20′ S. and long. 54° 40′ W. on May 10, 1856, and at what distance N. or S. of the zenith ? *Ans.* $2^h 19^m$; 8° 11′ N. of zenith.

274. At what time will Sirius pass the meridian of a place in latitude 61° N. and long. 10° W. on March 16, 1860, and at what distance N. or S. of the zenith ? *Ans.* $7^h 0^m$; 77° 32′ S. of zenith.

Elements from Nautical Almanac.

	R. A. mean sun.	Star's R. A.	Star's decl.
May 10......	$3^h 13^m 52^s$	$5^h 34^m 25^s$	34° 9′ 12″ S.
Mar. 16......	23 37 10	6 39 0	16 31 46 S.

We will conclude this chapter by giving brief explanations of some of the principal corrections required for reducing the observations used for finding the latitude, longitude, time at the ship, and variation of the compass—the subjects of the next chapter.

CORRECTIONS OF THE OBSERVED ALTITUDE OF A HEAVENLY BODY.

(79.) The altitude observed at sea by means of the sextant is called the observed or apparent altitude. To obtain the *true* altitude, or that defined in p. 68, we must apply to the observed altitude (in addition to the index error of the instrument itself) several corrections, the principal of which are the *parallax in altitude, refraction,* and *dip.*

CORRECTION FOR PARALLAX IN ALTITUDE.

(80.) Let A be the place of the spectator on the surface of the earth, c the center, z the zenith, x a heavenly body, and z m r the celestial concave.

Through x draw the two straight lines Axm_1 and cxm to the celestial concave. Then m_1 is the observed or apparent place, and m the true place of the heavenly body x.

Draw Ar, a tangent to the earth's surface, at A; draw also cR through the center parallel to Ar; then con-

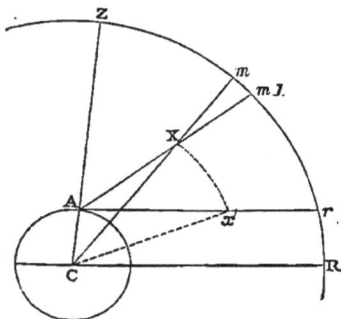

sidering the infinite distance of the points r and R from the earth, the earth's semidiameter $A C$ will subtend no angle at r or R, and $A r$ may be conceived to coincide with CR, and therefore the arc $rR = 0$. The *observed* altitude of x (without reckoning at present refraction) is measured by the arc $m_1 r$, and its true altitude by $mR = mr$. The difference $m m_1$ between the true and apparent altitudes, or the angle $A x C$, is called the *parallax in altitude.*

It appears from the figure that the effect of parallax is to *depress* bodies, so that the true altitude mR is greater on this account than the apparent altitude $m_1 R$, and that the true altitude may be obtained by *adding* the parallax in altitude to the observed altitude.

If x be the same body when in the horizon, the angle $A x C$ is called its *horizontal parallax.*

(81.) It is also evident from the figure that the parallax of a heavenly body is greatest when in the horizon, and that it diminishes to zero in the zenith; that the parallax for different bodies will differ, depending on their distance from the spectator; that the nearer the body is to the earth, the greater will be its parallax : thus the moon's parallax is the greatest of any of the heavenly bodies : the fixed stars, with perhaps a few exceptions, are at such an immense distance, that the earth dwindles to a point so indefinitely small that the radius of the earth $A C$ subtends no measurable angle at a star; hence the fixed stars are considered to have no parallax.

Since the form of the earth is an oblate spheroid, the equatorial diameter being about 26 miles longer than the polar diameter or axis, the horizontal parallax of a heavenly body, as observed from some place on the equator, will be greater than the horizontal parallax of the same heavenly body if

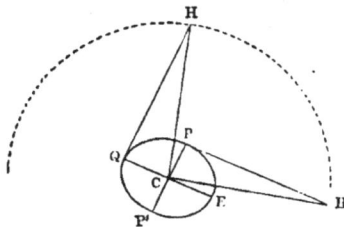

observed from the poles of the earth. For let Q be a spectator at the equator, and H a heavenly body in his horizon, then the angle H is the equatorial horizontal parallax of the body at H. Similarly to a spectator at P, the pole of the earth, the horizontal parallax of the same body would be H', which is evidently less than H, since it is subtended by a smaller radius of the earth; thus it appears from the figure that the horizontal parallax is greatest at the equator, and that it diminishes as the latitude increases. The moon's horizontal parallax put down in the *Nautical Almanac* is the *equatorial* horizontal parallax. To find the horizontal parallax for any other place a correction (see *Nav.* Part II. p. 125) must be applied, which is evidently subtractive : this correction is seldom made in the common problems of Navigation : in finding the longitude by occultations or solar eclipses, it ought not to be omitted. It is inserted in most collections of Nautical Tables.

(82.) The correction for parallax in altitude for the sun, moon, and planets has been calculated and formed into tables, so that this correction may be taken out by inspection. The tables are constructed from the following formula (*Nav.* Part II. p. 127):

Parallax in alt. = horizontal parallax × cos. altitude (corrected for refraction).

EXAMPLES.

275. Given the apparent altitude of the moon's center = 72° 42′ 15″ (cor. for ref. 18″ sub.), and horizontal parallax = 58′ 49″: find by table, and also by calculation, the parallax in altitude, and thence the true altitude.

(1.) *By Calculation.*

Obs. alt. .	72° 42′ 15″	log. cos. alt.........	9·473304	hor. par. .	58′ 49″
ref.	18−	„ hor. par. 3529″	3·547652		60
	72 41 57		3·020956		3529
par. in alt.	17 29+	∴ par. in alt.	1049″		
		or	17′ 29″		
true alt. ..	72 59 26				

(2.) *By Inman's Tables.   Cor. of moon alt., Table (w).*

Entering with hor. par. 58′	......cor. =	16′	57″
„ „ 49″	„ =		14
∴ cor. in alt., which includes refraction =		17	11+
obs. alt.......	72	42	15
∴ true alt.......	72	59	26

276. Given the apparent altitude of the sun = 13° 14′ 30″ (cor. for ref. 4′ 3″ sub.), and horizontal parallax 8·8″: find by table, and also by calculation, the parallax in alt., and thence the true altitude.

(1.) *By Calculation.*   (2.) *By Table (b).*

Obs. alt....	13° 14′ 30″	log. cos. alt.... 9·988297	Cor. by table ...8·5″
ref.........	4 3−	„ 8·8″ ...... 0·944483	
	·13 10 27	„ par. in alt. 0·932780	
par. in alt.	8·56+	∴ par. in alt....8·56″	
true alt. ...	13 10 35·6		

277. Given the apparent altitude of Mars = 14° 6′ 50″ (cor. for ref. 3′ 50″

sub.), and horizontal parallax 22″: find by table, and also by calculation, the parallax in altitude, and thence the true altitude.

	(1.) *By Calculation.*		(2.) *By Table* (b).
Obs. alt....	14° 6′ 50″	log. cos. alt.... 9·986780	Cor. by table... 21·3″
ref.	3 50—	„ 22″ ...... 1·342423	
	14 3 0	„ par. in alt. 1·329203	
par. in alt.	21·3+	∴ par. in alt....21·3″	
∴ true alt.	14 3 21·3		

CORRECTION FOR REFRACTION.

(83.) A ray of light passing obliquely from one medium to another of greater density, is found to deviate from its rectilineal course, and to bend towards a perpendicular to the surface of the denser medium. Hence to a spectator on the earth's surface, a heavenly body seen through the atmosphere appears to be raised, and its true place, on this account, is below its apparent place. Observations show that refraction is greatest when the body is in the horizon (about 34′), and that it diminishes to zero in the zenith. A table of refractions for every altitude has been computed and inserted in the Nautical Tables.

The corrections for parallax and refraction are frequently combined, so that they form one correction, called the *correction in altitude*. The two tables of the correction in altitude for the sun and moon may also be found in most collections of Nautical Tables.

(84.) The investigation of the formula for computing a table of refractions belongs more directly to a work on Optics. In any elementary book on that branch of mathematics the student will find this subject more satisfactorily explained than can be done in the brief space that could be assigned to it in the present work.

CORRECTION FOR DIP.

(85.) The altitude of a heavenly body, observed from a place above the

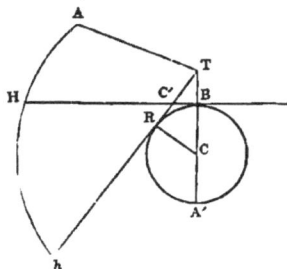

surface of the earth, as on the deck of a ship, will evidently be greater than its altitude observed from the surface, since the observer brings the image of the body down to his horizon, which is lower than the horizon seen from the surface of the sea immediately below him. The difference of altitude from this cause, expressed in minutes and seconds, is called the *dip* of the sea horizon. Let a tangent at B, the point directly beneath the spectator at T, meet the celestial concave at H; from T draw the tangent T*h*, touching the earth at ᴨ. Then, if ᴀ be the place of a heavenly body, ᴀᴨ is the altitude observed from ᴅ, the surface of the earth, and ᴀ*h* is the altitude from ᴛ. The arc ᴨ*h* is the dip (very much exaggerated in the figure) for the height ᴛᴅ of the spectator above the surface of the earth, and is evidently subtractive, to get the true altitude. This correction is found in all collections of Nautical Tables.

The table may be constructed from the following formula (*Nav.* Part II. p. 132):

$$\text{dip} = \cdot 984 \sqrt{\text{height of eye.}}$$

278. Calculate the dip for the height of the eye above the sea=110 feet.
Ans. Dip $= \cdot 984 \sqrt{110} = \cdot 984 \times 10 \cdot 488 = 10' \ 19 \cdot 2''$.

279. Find dip for 20 feet and for 30 feet. *Ans.* 4′ 24″ ; 5′ 23·4″.

(86.) The corrections just described are required in almost every example in Nautical Astronomy. Besides these, there are others of not so frequent occurrence, such as the corrections called " *The augmentation of the moon's semidiameter,*" " *The contraction of the moon's semidiameter,*" " *The correction of the moon's meridian passage;*" and in rare observations, such as *occultations*, &c., for determining the longitude, the oblate figure of the earth must be taken into consideration, and corrections called " *The correction of the moon's equatorial horizontal parallax,*" and " *Correction of the latitude for the spheroidal figure of the earth,*" must be applied to several of the terms used in the calculation. These corrections we will now very briefly describe, referring the student for fuller information to *Navigation*, Part II., where these corrections are investigated and useful practical formulæ obtained adapted to logarithms.

AUGMENTATION OF THE MOON'S HORIZONTAL SEMIDIAMETER.

(87.) When the moon is above the horizon, as at ʟ′, its distance oʟ′ from a spectator at o is less than its distance oʟ when in the horizon at ʟ. For the distance cʟ of the earth's center from the moon is about 60 times the earth's radius, therefore cʟ=60 × cɪ. But as the horizontal parallax is small, oʟ is nearly equal to cʟ, and therefore ʟɪ is less than ʟo by nearly the earth's radius. Hence if two observers were placed at o and ɪ, one would see the moon when at ʟ in his horizon, and the other in his zenith ; but to the spectator at o the moon would be a little more, and to the spectator at ɪ a little less, than 60 times its radius, and the diameter would

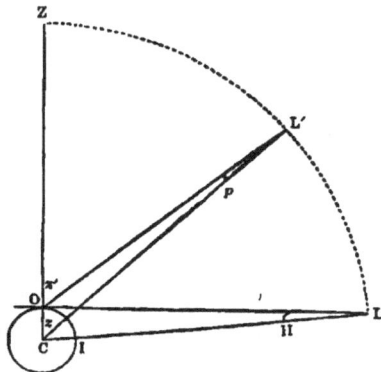

appear to the former about 30″ less than to the latter. It is evident that at any intermediate altitude, as at L′, the distance OL′ is less than OL, and therefore the moon's diameter at L′ would appear to be greater than the true or horizontal diameter at L; that is, the diameter at L′ would be *augmented*. The correction to be made to the moon's horizontal semidiameter on this account is called the *augmentation*. It has been computed for every degree of altitude, and may be found in the Nautical Tables.

In *Navigation*, Part II. p. 134, is investigated the following formula for calculating the augmentation of the moon's semidiameter.

$$\text{Aug.} = 2\,\text{R}\,.\,\text{cosec.}\,(z'-p)\,\cos.\,(z'-\tfrac{1}{2}p)\,\sin.\,\tfrac{1}{2}p$$

where R=horizontal semidiameter, z′=apparent zenith distance,
and p=parallax in altitude=hor. par. × cos. app. alt. (*Navigation*,
Part II. p. 125.)

EXAMPLE.

280. Calculate the augmentation of the moon's horizontal semidiameter when the apparent altitude of the center is 32° 42′, the horizontal parallax being 54′ 42·5″, and horizontal semi. (in *Nautical Almanac*) 14′ 56″.

$$
\begin{aligned}
\text{R} &= \quad 14'\quad 56'' = 896'' \\
z' &= 57°\ 18\quad 0 \\
p &= \text{par. in alt.} \\
&= \ 0°\ 46'\quad 3'' \\
z' &= 57\quad 18\quad 0 \\
\therefore z' - p &= \overline{56\quad 32\quad 0} \\
\tfrac{1}{2}p &= \ 0\quad 23\quad 0 \\
z' &= 57\quad 18\quad 0 \\
\therefore z' - \tfrac{1}{2}p &= \overline{56\quad 55\quad 0}
\end{aligned}
$$

1. To find par. in alt.

log. 3282·5 ....3·516205
„   cos. 32° 42′...9·925069
„   par. in alt. ....$\overline{3·441274}$
.˙. par. in alt.=2763″
=46′ 3″

2. To find augmentation.

log. 2................0·301030
„   R................2·952308
„   cosec. (z′−p)..0·081170
„   cos. (z′−½p)..9·725219
„   sin. ½p.........$\overline{7·825451}$
„   aug. ....0·885178
.˙. augmentation=7·68″

281. Calculate the augmentation of the moon's horizontal semidiameter when the apparent altitude of the center is 72° 0′, the horizontal parallax 58′ 43·4″, and horizontal semi. 16′ 0″.

Ans. By formula, 15·86″; by table, 15·8″.

CONTRACTION OF THE MOON'S SEMIDIAMETER ON ACCOUNT OF REFRACTION.

(88.) When the moon is near the horizon its disc assumes an elliptical form, as A B B', in consequence of the unequal effect of refraction at low altitudes, the lower limb being raised more than the center, and the center more than the upper limb. If, therefore, a contact is made between a distant object in the direction D and some point P on the moon's limb, the contracted semidiameter cP, to be added to the distance to obtain the distance of the centers, must be less than cA the uncontracted semidiameter. This correction has been calculated, and may be found in the Nautical Tables.

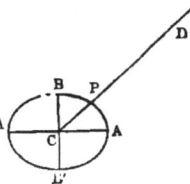

The formula investigated in *Navigation*, Part II. p. 135, for computing the contraction, is the following :

$$\text{Contraction} = c \cdot \sin{}^2\theta,$$

where c=difference of refraction for center and vertex,

θ=inclination of line joining the centers of the two bodies to the horizon.

EXAMPLE.

282. Calculate the correction for contraction of the moon's semidiameter when the altitude=4° 30', and the line joining the centers is inclined at an angle of 40°, the moon's semidiameter being 15' 30''.

Alt. of vertex...4° 45' 30''...Refraction	10'	22''	log. c...........1·447158	
„ center...4 30 0	„	10 50	„ sin. 40° ...9·808067	
∴ c=		28	„ sin. 40° ...9·808067	
			„ contr. ......1·063292	
			∴ contraction=11·57''	

283. Calculate the correction for contraction of the moon's semidiameter when the altitude=30° 0', and the line joining the centers is inclined at an angle of 36° : the moon's semi. being 16' 5''.

Ans. By formula, 0·3459''; by table, 1·0''.

CORRECTION OF MOON'S MERIDIAN PASSAGE.

(89.) The time of the transit of any heavenly body can be found by means of Rule 22, p. 99; but in the case of the moon, the following approximate method of finding the time of her passage over a given meridian may be sometimes used with advantage.

The mean time of the moon's transit for every day at Greenwich is put down in the *Nautical Almanac*. At any place to the east of Greenwich, the time of the transit, owing to the moon's proper motion to the eastward, must take place sooner (independent of that due to the difference of longitude), and to a place to the westward of Greenwich, later than the time recorded in the *Nautical Almanac*. Thus, if we suppose the moon's daily motion to be 60 minutes : to a place 90° to the east of Greenwich the transit will take place 6 hours earlier than that at Greenwich (on account of the difference of longitude) $+\frac{90}{360}$ of 60^m, or 15 minutes, due to the moon's motion, supposed equable, to the eastward in the 6 hours before she reaches the meridian of Greenwich. To a place west of the first meridian, a retardation will take place for the same reason.

The moon's daily motion in RA varies between 40^m and 60^m, so that it would not be difficult to construct a small table of the correction of the transits given in the *Nautical Almanac* for any given longitude : this has accordingly been done, and may be found in Inman's *Nautical Tables*, p. 5.

The moon's daily motion used should be that found by taking the difference between the two transits at Greenwich that happen before and after the one at the place : that is, if the place be in west longitude, the difference should be taken between the transit on the given day and the one following ; if in east longitude, that on the given day and the one preceding. By observing this rule, the error arising from the unequal motion of the moon in RA is diminished.

An example or two of finding the time at Greenwich at the transit of the moon over a given meridian will show the use of the table.

284. April 27, required Greenwich mean time nearly at the transit of the moon over the meridian of a place in longitude 50° W.

By *Nautical Almanac*, mer. pass. on 27th...11^h $46\cdot3^m$
,,   ,,   on 28th...12 $32\cdot0$
moon's motion in 24 hours... $45\cdot7$
correction from table... $6\cdot3+$
∴ time of transit at place ...11 $52\cdot6$
long. in time... 3 $20\cdot0$ W.
∴ Greenwich mean time of transit at ship...15 $12\cdot6$

285. April 27th, required Greenwich mean time nearly at the transit of the moon over the meridian of a place in longitude 50° E.

By *Nautical Almanac*, mer. pass. on 27th...11^h $46\cdot3^m$
,,   ,,   on 26th...11 $2\cdot7$
$43\cdot6$
Cor. $=\frac{50}{360}\times 43\cdot6$ correction... $6\cdot0-$
$=6\cdot06^m$ transit at place...11 $40\cdot3$
or by table $=6\cdot0$ long. in time... 3 $20\cdot0$ E.
∴ time at Greenwich... 8 $20\cdot3$

Required the mean time at the place of the moon's meridian passage on July 19 (astronomical day), in longitude 60° W., and on July 27 (astronomical day), in longitude 175° E., having given the following quantities from the *Nautical Almanac :*

Gr. mer. pass. on July 19......11ʰ 24·3ᵐ   July 27......17ʰ 30·1ᵐ
           „   20......12 19·2   „ 26......16 49·5
Ans. Mer. pass. at place on July 19 at 11ʰ 33·3ᵐ
    „   „ 27 at 17 11·1 = July 28 at 5ʰ 11·1ᵐ A.M.

The corrections for moon's *equatorial horizontal parallax* and for the *figure of the earth* are fully investigated and explained in *Navigation,* Part II. pp. 129, 123.

THE SEXTANT.

(90.) The student should begin as early as possible to learn to measure angles and take altitudes with the sextant. Before; therefore, we proceed to apply the preceding corrections to the observed altitude of a heavenly body, we will describe briefly the *construction, use,* and *principal adjustments* of this important instrument.

Construction and use of the Sextant.

(91.) The sextant is adapted for measuring angles in any plane whatever ; differing in this respect from the theodolite, which is used for observing horizontal and vertical angles only.

The construction of the sextant may be explained by means of the annexed figure.

A small piece of glass, mm', called the *movable reflector,* quicksilvered at the back, is placed at M, the center of the arc AB. It is attached to a *movable radius,* MC, by moving which the plane of its surface produced (supposed perpendicular to the plane of AB) can be made to cut the arc at any required point, C. Another piece of glass, ff', called the *fixed reflector,* also perpendicular to the plane of AB, is placed at F, the lower half only of which is quicksilvered.

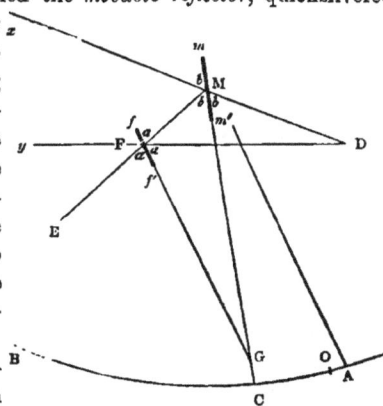

Now suppose a ray of light proceeding from the object x in the direction xD to impinge on the surface of the movable reflector M at the angle xMm ; then, by a well-known optical law, the ray will be reflected back in the direction MF, making an angle FMm', with the movable reflector equal to the angle xMm. Again, at the fixed reflector F, the ray MF will suffer another reflection in the direction F D, making, with the reflecting surface ff', the angle $\mathrm{D}\mathrm{F}f'=$ angle $\mathrm{M}\mathrm{F}f$. If we suppose an observer's eye to be placed at D, and another ray of light to proceed from the object y along the same line y F D, the two objects x and y will thus appear to come to the spectator from the same point y; the image of the object x having been transmitted to him from the quicksilvered part of F, and the direct image of y through the upper part of F, which is left transparent for that purpose. The angular distance between x and y, which is the object required to be found, is the angle D, and this angle D will be proved (see below) to be double of the angle A M C, measured by the arc A C; A M being supposed to be drawn parallel to the surface ff' of the fixed reflector F. Hence, if the arc A B, which may be supposed to be the sixth part of a circle, or to contain 60°, be so graduated that it shall contain twice that number, or 120°, then the reading off on the arc A C will be the value of the angle at D : and this is the method adopted in dividing the arc of the sextant.

To observe, therefore, the angle between any two objects, x and y, the observer at D* looks directly at the left-hand object y through the fixed reflector F : he then moves the radius M C, attached to the movable reflector M, in the plane passing through D and the two objects, until he sees the ray proceeding from x in the same direction as the object y. Then the reading off on the arc A C measures the angle at D, the angular distance between the two objects x and y; this may be proved as follows :

Proof that the arc A C measures the angle at D between the two objects x and y.

Produce M F to E and ff' to cut the line M C in G; then the angles $x\mathrm{M}m$ and D M G are equal, being vertical angles; also the angle $x\mathrm{M}m$ is equal to the reflected angle F M G ;† mark therefore these three angles at M with the same letter b; in the same manner the three angles, marked a, formed at F by the reflected ray, may be shown to be equal.

Now in the triangle M D F, the exterior angle $\mathrm{E}\mathrm{F}\mathrm{D}=\mathrm{F}\mathrm{M}\mathrm{D}+\mathrm{D}$, or $2a= 2b+\mathrm{D}$.

$$\therefore a=b+\tfrac{1}{2}\mathrm{D};$$

also in the triangle F G M, the exterior angle $a=b+\mathrm{F}\mathrm{G}\mathrm{M}$.

$$\therefore b+\tfrac{1}{2}\mathrm{D}=b+\mathrm{F}\mathrm{G}\mathrm{M}, \text{ or } \tfrac{1}{2}\mathrm{D}=\mathrm{F}\mathrm{G}\mathrm{M}=\mathrm{G}\mathrm{M}\mathrm{A},$$

since F G is parallel to M A.

But the arc A C, which measures the angle G M A, is divided into double the number of degrees due to its length, the divisions commencing at the point A; therefore the reading off on the arc A C measures the angle D, the angular distance between the two objects.

---

* The observer's eye is seldom exactly at the point D, but in some other point in the line D F; this, however, will make no appreciable difference when the objects x and y are at a considerable distance from the spectator, as the sun or moon.

† See any work on Optics for a proof of this property of light.

The index correction of the Sextant.

(92.) We have supposed above that the graduations of the arc commenced at A, the point in the arc cut by a line M A, parallel to the fixed reflector F ; but this is seldom the case, the zero point of the arc being often a little to the right or left of A.

Let us suppose the graduation to commence at o, to the left of A, then the angle D would be equal to the reading off on o c + the small arc o A. The arc o A is called the *index correction*, the value of which is usually determined by measuring a small angle, as the sun's diameter, off and on the arc, that is, to the right and left of o ; to enable us to do which, the divisions of the arc are continued a little to the right of the zero point o.

To find the index correction by measuring the sun's diameter.

Let A be the point on the arc of the sextant through which the movable radius M c (fig. p. 109) passes when its reflector M is parallel to the fixed reflector F ; then if the graduation of the arc had commenced at A, it is evident that the reading off on any arc A c (p. 109) would have measured D, the angle between the two objects x and y. But let us suppose that the commencement of the graduation on the arc, or the zero point as it is called, to be at o. Then o A is the error of the instrument or index correction to be determined.

Let P be the point on the arc through which the movable radius M c passes when there is a contact of the direct and reflected limbs of the sun *on* the arc, and Q the point through which M c passes when there is a contact of the two limbs to the right of o, and therefore called *off* the arc ; then, since the direct and reflected suns must coincide when the movable radius is at A, the arc A P = arc A Q.

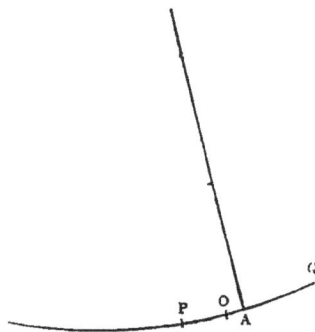

Let $a = $ o P, the measure of sun's diameter on the arc ;
$b = $ o Q, the measure of sun's diameter off the arc ;
$x = $ o A, the index correction required.

Then, since A P = A Q,
∴ $x + a = b - x$;
or, $2x = b - a$,
∴ $x = \frac{1}{2}(b - a)$;

or the index correction is equal to half the difference of the measures of the sun's diameter off and on the arc.

In this case, the index correction o A, or x, must be added to the arc o c, to get the angle between the two bodies x and y ; this is evident from the figure : hence the index correction is said to be additive when the reading on the arc is less than the reading off the arc.

In the same manner it may be shown that if the zero point o is to the right of A, the index correction $x = \frac{1}{2}(a - b)$, and is subtractive; that is, when the reading on the arc is greater than the reading off the arc, when a contact of the true and reflected limbs of the sun is made.

Line of collimation.

(93.) *The line of collimation*, or optical axis of the telescope of the sextant, is the imaginary line joining the centers of the object and eye glass. This line should be parallel to the plane of the instrument.

The visual ray coming from any point of an object, viewed through the telescope of the sextant, passes through the center of the object-glass; and the instrument must be held so that it enters the eye through the center of the eye-glass, or the middle point between the two wires at the eye end of the tube. If this ray, or line of sight, is not parallel to the plane of the instrument, the angle read off on the arc will differ from the angle between the two objects. This will be proved hereafter.

To ascertain whether the line of collimation is parallel to the plane of the arc.

Let A and B be two luminous objects, the latter of which is viewed

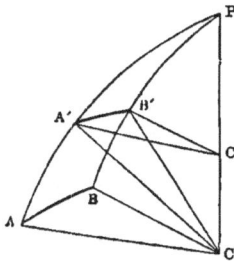

directly through the middle point between the two wires (supposed to be placed parallel to the plane of the instrument), and the reflected image of the former (A) is brought into contact with B by moving the index along the arc. Now we may ascertain if the tube is properly adjusted, by making a contact at the middle of the upper wire, and then (before any perceptible change, arising from the motion of the two bodies, takes place) bringing the same point of contact to the lower wire : if the two bodies still remain in contact, the instrument may be considered in adjustment; but if this is not the case, the difference must arise from the want of parallelism of the line of collimation. For let a contact be made between the two objects A and B, by bringing the object A, on the right hand, up to B, on the left. Then, if the instrument is properly adjusted, the angle A C B is in the plane of the instrument, and will be measured by its arc. Let now the two objects A and B be supposed to move through equal arcs, A A′ and B B′, in circles vertical to the plane of the instrument, and without moving the sextant, let the axis of the telescope be directed to B in its new position at B′, and thus inclined to the plane of the sextant. This being done, the object A will still be seen to coincide with B, while the angle they subtend at the eye, supposed to be at C, is changed; this may be shown as follows:

Let c be the eye, b and a the two objects, as the sun and moon, acb the angle between them; the instrument being supposed to be perfectly adjusted, acb will be in the plane of the instrument, and therefore will be measured by the division on the arc to which the index is set. Now the image of a being supposed in contact with b, let a and b be shifted up ap bp, through the equal arcs aa' bb', perpendicular to the plane acb. In this position of the objects the image of a' will still be seen in contact with b'. For let the eye be raised up cp, perpendicular to acb, to c', so that the plane a'c'b' is parallel to the plane acb; then the angle a'c'b' is equal to acb. Consequently, the reflectors remaining as before, and being perpendicular to a'c'b', the image of a must be transferred to a'.

But the eye is actually at c, and views a' and b' under the angle a'cb', which evidently differs from acb or a'c'b'. It follows, that when the axis of the telescope or of vision is inclined to the plane of the instrument, the image of an object as a' will be seen coincident with another object b' when the division which the index is set to differs from the angle between the two objects; and it is manifest that the difference is the same, whether the axis of vision is inclined to the same degree from or towards the plane of the sextant.

Hence is deduced the practical rule for determining whether the line of collimation is parallel to the plane of the instrument, given under the head of the third adjustment of the sextant.

Investigation of a formula for determining the error in the observed angle arising from a given error in the line of collimation.

(94.) In fig. p. 112, let a'cb' be the true angle between the objects a and b, subtended at the eye of the observer at c, and acb the instrumental angle, or the one read off on the arc, and aa' or bb' the measure of the inclination of the line of collimation of the telescope to the plane of the instrument.

$$\text{Then cos. } \mathrm{a a}' = \frac{\text{arc } \mathrm{a}'\mathrm{b}'}{\text{arc } \mathrm{a b}} \quad (\textit{Trig.} \text{ Part II. art. 69})$$

$$= \frac{\text{chord } \mathrm{a}'\mathrm{b}'}{\text{chord } \mathrm{a b}}$$

$$= \frac{2 \sin. \frac{1}{2} \mathrm{a}'\mathrm{c b}'}{2 \sin. \frac{1}{2} \mathrm{a c b}}$$

$$\therefore \sin. \tfrac{1}{2}\mathrm{a}'\mathrm{c b}' = \sin. \tfrac{1}{2}\mathrm{a c b} \,.\, \cos. \mathrm{a a}'.$$

This formula determines the angle a'cb'; the difference between which and the angle acb is the error in the angle observed.

EXAMPLES.

Required the error in the observed angles 90° and 150°, when the inclination of the line of collimation is 1°.

I

$$\sin. \tfrac{1}{2}\text{A}'\text{CB}' = \sin. \tfrac{1}{2}\text{ACB} \cdot \cos. \text{AA}'.$$

ACB$=90°$	AA$'=1°$	ACB$=150°$
(1.)		(2.)

sin. 45°.........9·849485	sin. 75°.........9·984944
cos. 1°.........9·999934	cos. 1°.........9·999934
sin. $\tfrac{1}{2}$A'O'B' . 9·849419	sin. $\tfrac{1}{2}$A'CB' . 9·984878
∴ $\tfrac{1}{2}$A'CB$=44°$ 59' 30"	∴ $\tfrac{1}{2}$ A'CB$'=74°$ 58' 0"
and $\tfrac{1}{2}$ACB$=45$	and $\tfrac{1}{2}$ACB $=75$
$\tfrac{1}{2}$ error$=0$ 0 30	$\tfrac{1}{2}$ error$=0$ 2 0

∴ The error in one case is 1' 0"; in the other 4' 0".

From this it appears that a slight inclination of the line of collimation to the plane of the instrument produces a considerable error in determining the true angle between the two objects; that this error increases as the angle increases; and that the observer should always take care, in nice observations, to make the contact as near the middle point of the field of view as possible.

Adjustments of the Sextant.

(95.) The principal adjustments are the following:

1. The movable reflector M (fig. p. 109) should be perpendicular to the plane of the instrument.

2. The fixed reflector F should be perpendicular to the plane of the instrument.

3. The line of collimation should be parallel to the plane of the instrument.

To examine the adjustments.

First adjustment.—To see if the movable reflector is perpendicular to the plane of the instrument.

Place the movable radius MC near the middle of the arc, as at c (fig. p. 109); turn the face of the instrument upwards, and look obliquely into the reflector M. Then the image of the arc BC will be seen in the reflector M; and if this image appears in one unbroken line with the arc BC itself, the reflector M is perpendicular to the plane of the instrument.

If the reflection of the line BC appears above or below the line BC itself, then the reflector M is out of adjustment, and must be adjusted by certain screws or studs at the back of the reflector. This adjustment, in good instruments, seldom requires to be made; and when it does happen, it is best to send it to the maker to be rectified.

Second adjustment.—To see if the fixed reflector is perpendicular to the plane of the instrument.

Look through the telescope and the fixed reflector F (fig. p. 109) at the sun, or any other well-defined object; hold the instrument with its face in a horizontal position; bring the index towards the commencement of the divisions, move it gently backwards and forwards until the image of the object is placed as near as possible upon the object itself; then, if the image is found to obliterate or coincide exactly with the object itself, the fixed reflector F is perpendicular to the plane of the instrument.

If any portion of the direct object is seen not coinciding with the image, then the fixed reflector F is not perpendicular to the plane of the instrument; and the adjustment is made by means of a screw, which in some instruments is under the glass, in others behind it, and in others at the side. The screw must be turned gradually till the image is made to coincide with the object. This adjustment is frequently required to be made.

Third adjustment.—To see if the optical axis of the telescope (called the line of collimation) is parallel to the plane of the instrument.

(It is usual to examine this adjustment in practice by making a contact between the sun and moon.)

The telescope being placed in the collar of the sextant, turn the eye-piece round till the two wires are parallel to the plane of the instrument. Bring the darkened image of the sun (when at a considerable distance from the moon, *i.e.* from 90° to 120°) to touch the edge of the moon at the middle point of the upper wire, and then immediately, before any perceptible change in the distance of the two bodies can take place from their own proper motion, bring the point of contact of the two bodies to the lower wire, at which, if they appear in contact, the axis of the telescope may be considered to be parallel to the plane of the instrument; if otherwise, the adjustment is made by means of two screws in the collar—by slackening one and tightening the other. In some instruments, however, these screws are wanting, the adjustment of the parallelism of the tube being supposed to be carefully made before the instrument leaves the maker's hands.

Reading off on the Sextant.

(96.) The following brief directions for reading off will be more readily understood by the student if he place a sextant before him for reference and examination.

It will be seen that the arc is divided into degrees, and (in the best instruments) into the sixths of degrees, or 10 minutes. We will suppose it is an instrument of this kind before us. The *index* lines are cut on the plate at the end of the movable radius, and therefore called the *index plate*. The index itself is the commencement of the reading off on the index plate, and is generally distinguished from the other lines on the plate by a dia-mond-shaped mark, resembling a spear-head. First let us suppose this index line to coincide exactly with some line on the arc; for example, with the second line to the left of 50°; then the reading off will be 50° 20′, since each line on the arc represents 10′. Next, let us suppose the index line not to stand exactly at any line on the arc, but somewhere between two, as, in the above example, between the second and third line from 50° ; suppose it appeared to be about halfway between the second and third lines (the reader may place it in that position), then it is evident that the reading off would be about 50° 25′. But as this is a rough and imperfect way of estimating the additional minutes and seconds beyond the second division from 50°, the exact value is found by means of the ingenious arrangement of certain lines, called the *vernier*, cut on the index plate to the left of the index line. It will be seen that the divisions of the vernier are nearer to each other than the arc divisions ; so that the line on the vernier immediately to the left of the index is somewhat nearer to the corresponding one on the arc than the small space the value of which is to be determined : and it is manifest that it must be nearer by the difference between the width of one division on the arc and one on the index plate. In like manner the second line on the vernier, reckoning from the index line, must be nearer to the corresponding line on the arc by two differences, the third by three, and so on. By carrying the eye along the vernier in this manner, it will be at length seen, by aid of the small reading-off glass, or microscope, attached to the movable radius, that a complete coincidence takes place between a line on the vernier and one on the arc.

Now it is evident, since the lines on the vernier have approached those on the arc through the small space the index is in advance of 20′, that this small space must be equal to as many times the difference of two divisions as there are lines reckoning from the index before the coincidence takes place. Hence, if we know the value of a difference, we shall know the value of the small arc to be measured ; and this may be discovered in the following manner. It will be seen, by examining the arc of the sextant before us, that 60 divisions of the vernier just cover or coincide with 59 divisions on the arc ; or the difference between a division on the arc and one on the vernier is $\frac{1}{60}$ of a division of the arc : if therefore a division on the arc is 10′, the difference in question will be $\frac{1}{60}$ of 10′, or 10″. Let us now suppose the index to stand between the second and third divisions from the 50°, and that, by carrying the eye along the vernier, we at length find the coincidence of the two lines to take place at the fourth line to the left of the line on the vernier marked 5 ; then the value of the space to be determined will be 5′ 40″, every sixth division on the vernier being distinguished by a figure indicating minutes. The magnitude of the whole angle is therefore 50°. 20′ + 5′ 40″, or 50° 25′ 40″. The sextant supposed under examination is marked to read off to the nearest 10″ ; some instruments are graduated to 15″, or 30″, &c. ; but the same method of reading off is to be followed as pointed out above.

The graduation of the arc of the sextant is usually continued to the right of 0°, or zero : this is done to enable the observer to take a small angle to the right as well as to the left of the index line, or zero ; as the measure of the sun's diameter off and on the arc to determine the index correction, &c. In this case we shall have to read off on an arc divided from left to right by means of an index, which we must suppose divided from right to left : this, however, is easily done, if we recollect that the line on the index plate marked 10′ must be considered as the commencement of the divisions; 9′ must be considered as 1′; 8′ as 2′; 7′ as 3′; &c. : thus, if the coincidence of the lines on the arc and index plate is at 6′ 40″, we must read this as 3′ 20″, and so on.

These few rules and brief observations on the adjustments and use of the sextant must be considered as introductory to other works written more expressly on the use of astronomical instruments.

Rule 23. *Given a* STAR's *observed altitude : to find its true altitude.*

The stars are at such a distance from the spectator, that (excepting probably a few) the earth's orbit subtends no angle at the star : hence a star is considered to have no parallax (p. 101) ; and the only corrections used for reducing the observed altitude to the true are the *index correction,* the *dip,* and *refraction.* Hence this rule.

1. To the observed altitude apply the index correction with its proper sign.
2. Subtract the dip (taken from table of dip of horizon).
3. Subtract the refraction (taken from table of refraction).
4. The result is the true altitude of the star.

EXAMPLE.

286. The observed altitude of Arcturus (α Bootis) was 36° 10′ 20″, index correction + 2′ 42″, and height of eye above the sea was 20 feet : required the true altitude.

Observed altitude............	36°	10′	20″
index correction		2	42 +
	36	13	2
dip		4	24 −
star's apparent altitude......	36	8	38
refraction		1	20 −
star's true altitude............	36	7	18

287. The observed altitude of Aldebaran (α Tauri) was 13° 4′ 30″, index correction −10′ 40″, and height of eye above the sea was 16 feet : required the true altitude.   *Ans.,* 12° 45′ 43″.

288. The observed altitude of γ Tauri was 62° 42′ 15″, index correction +0′ 40″, and height of eye above the sea was 20 feet : required the true altitude.   *Ans.* 62° 38′ 1″.

289. The observed altitude of α Canis Majoris (Sirius) was 32° 42′ 30″,

index correction was —3' 30", and height of eye above the sea was 12 feet: required true altitude.   *Ans.* 32° 34' 4".

Rule 24. *Given a* PLANET'S *observed altitude : to find its true altitude.*

The effect of parallax on the true altitude of a heavenly body is to diminish it (p. 101) : the correction of parallax in altitude must therefore be added to the observed, to get the true altitude. Hence this rule.

Correct the observed altitude for index correction, dip, and refraction, as in 1, 2, 3 (p. 117).

4. To the result add the parallax in altitude (taken out of the table of parallax in altitude of sun and planets).

5. The result is the true altitude of the planet.

EXAMPLE.

290. January 4, 1848, the observed altitude of Mars was 21° 41' 10", index correction +2' 42", and height of the eye above the sea 24 feet, horizontal parallax (in *Nautical Almanac*) being 10·1" : required the true altitude.

Observed altitude............	21°	41'	10"
index correction		2	42+
	21	43	52
dip		4	49 —
	21	39	3
refraction		2	26 —
	21	36	37
parallax in altitude			9 +
true altitude	21	36	46

291. Jan. 24, 1848, the observed altitude of Mars was 9° 8' 30", index correction —3' 45", and height of eye above the sea 16 feet : required the true altitude. The horizontal parallax from *Nautical Almanac* was 8·3".
                                                              Ans. 8° 55' 3".

292. Feb. 3, 1848, the observed altitude of Venus was 25° 8' 30", index correction —10' 50", and height of eye above the sea 12 feet : required the true altitude. The horizontal parallax from *Nautical Almanac* was 8·1".
                                                              Ans. 24° 52' 17".

293. Jan. 30, 1848, the observed altitude of Jupiter was 10° 20' 10", the index correction was +0' 14", and height of eye above the sea 18 feet : required the true altitude, the horizontal parallax in *Nautical Almanac* being 2·0".   *Ans.* 10° 11' 3".

Rule 25. *Given the* SUN'S *observed altitude : to find the true altitude.*

The true altitude of the sun's center is found by observing the altitude of either the upper or lower limb, and then subtracting or adding the semidiameter taken from the *Nautical Almanac;* the other corrections, namely, for index correction, dip, refraction, and parallax, being made as in the preceding rules. In some nautical tables, the two corrections for refraction and parallax of the sun are combined in one table, and called the "correction in altitude of the sun." Hence this rule.

1. Correct the observed altitude for index correction and dip, as in articles 1, 2 (p. 101).

2. To this add the sun's semidiameter, if the altitude of the lower limb is observed; but subtract if the upper limb is observed; the result is the apparent altitude of the sun's center.

3. Subtract the refraction and add the parallax taken from the proper tables; or rather take out the "correction in altitude of the sun," and subtract it.

4. The remainder is the sun's true altitude.

EXAMPLE.

294. The observed altitude of the sun's lower limb (L. L.) was 47° 32' 15", the index correction was +2' 10", and the height of the eye above the sea 15 feet: required the true altitude of the sun's center, the semidiameter in *Nautical Almanac* being 15' 49".

Observed altitude............	47°	32'	15"
index correction		2	10+
	47	34	25
dip............................		3	49−
	47	30	36
semidiameter.................		15	49+
apparent altitude	47	46	25
correction in altitude			47−
true altitude	47	45	38

295. The observed altitude of the sun's L. L. was 48° 30' 15", index correction −2' 50", and height of eye above the sea 15 feet: required the true altitude, the semidiameter being 15' 55".   *Ans.* 48° 38' 46".

296. The observed altitude of the sun's L. L. was 40° 42' 16", index correction +5' 10", and height of eye above the sea 20 feet: required the true altitude, the semidiameter being 16' 4".   *Ans.* 40° 58' 6".

297. The observed altitude of the sun's upper limb (U. L.) was 55° 57' 42", index correction −3' 40", height of eye above the sea 19 feet: required the true altitude, the semidiameter being 16' 6".   *Ans.* 55° 33' 4".

298. The observed altitude of the sun's L. L. was 39° 25' 15", index correction — 3' 15", height of eye above the sea was 15 feet: required the true altitude, the semidiameter being 16' 3".   *Ans.* 39° 33' 11".

Rule 26. *Given the* MOON's *observed altitude : to find the true altitude.*

The moon's horizontal parallax and semidiameter change so perceptibly, that they cannot be considered (as in the corresponding case of the sun) to be constant for 24 hours.   The parallax and semidiameter taken out of the *Nautical Almanac* must therefore be corrected for the Greenwich date in order to find the horizontal parallax and horizontal semidiameter at the time of the observation.   Moreover, since the moon is nearer the earth when observed than when it was in the horizon, the horizontal semidiameter must also be corrected for augmentation (p. 106).   The correction of the moon's apparent altitude for parallax and refraction is found in most of the nautical tables : it is entered with the corrected horizontal parallax at top, and the apparent altitude at the side.   Hence this rule.

1. Get a Greenwich date.

2. Correct the moon's semidiameter and horizontal parallax, taken from the *Nautical Almanac*, for the Greenwich date (p. 74).

3. Add to the semidiameter the augmentation, taken from the table of augmentation.

4. Correct the observed altitude for index correction, dip, and semidiameter, as in the preceding rules (p. 117, 119).

5. Add the moon's correction in altitude, taken out of the proper table.

6. The result is the moon's true altitude.

NOTE. The moon's correction in altitude may be found by *calculation* by the following formula (*Nav.* Part II. p. 127):

Parallax in altitude=horizontal parallax × cos. app. alt. (corrected for refraction)..

EXAMPLE.

299. April 7, 1853, at 4ʰ 47ᵐ P.M. mean time nearly, in long. 10° W., the observed altitude of the moon's L. L. was 72° 15' 0", the index correction was — 4' 20", and height of eye above the sea 15 feet: required the true altitude.

				Moon's semi.			Moon's hor. par.		
Ship, April 7 . .	4ʰ	47ᵐ	7th, at noon	.	15' 40·7"	noon	57' 32·0"		
long. in time . .	0	40 W.	„ mid.	.	15 43 8	mid.	57 50·8		
Gr. April 7 . .	5	27			5·1		18·8		
					0·34279		0·34279		
					3·32585		2·75927		
					3·66864	2·3	3·10206	8·5	
					15 43·0		57 40·5		
				aug. .	15·2				
					15 58·2		╱		

True alt. By Inman's table (w).						True alt. By Calculation.		
Obs. alt.. .	72°	15'	0"	Obs. alt. .	72° 15' 0"			
in. cor. . .		4	20—	in. cor.. .	4 20—			
	72	10	40		72 10 40	57' 40·5"		
dip . . .		3	49—	dip . . .	3 49—	60		
	72	6	51		72 6 51	3460·5=hor. par. in seconds.		
semi.. . .		15	58	semi. . .	15 58			
	72	22	49		72 22 49			
cor. for 57'.		16	57	ref.. . . .	19—			
„ 40"			12		72 22 30	cos. alt. . . . 9·481135		
∴ true alt..	72	39	58	par. in alt.	17 28+	hor. par. . . . 3·539139		
				∴ true alt.	72 39 58	3·020274		
						par. in alt. . . 1048"		
						or 17' 28"		

300. July 12, 1848, at $9^h 18^m$ P.M. mean time nearly, in long. 44° 40' W., the observed altitude of the moon's L. L. was 27° 56' 40", the index correction + 2' 20", and height of eye above the sea 20 feet: required the true altitude.   *Ans.* 28° 56' 11".

301. May 15, 1848, at $10^h 25^m$ P.M. mean time nearly, in long. 55° 40' W., the observed altitude of the moon's L. L. was 21° 14' 10", the index correction + 2' 20", and height of eye above the sea 15 feet: required the true altitude.   *Ans.* 22° 15' 15".

302. May 15, 1848, at $10^h 22^m$ P.M. mean time nearly, in long. 41° 30' W., the observed altitude of the moon's U. L. was 45° 20' 30", the index correction + 4' 10", and height of eye above the sea 20 feet: required the true altitude.   *Ans.* 45° 42' 32".

Elements from Nautical Almanac.

Moon's semidiameter.			Moon's horizontal parallax.		
July 12, mid.	14'	55·9"	.... mid.	54'	47·8"
„ 13, noon......	14	59·3	.... noon......	55	0·3
May 15, mid.	14	41·1	.... mid.	53	57·0
„ 16, noon......	14	42·3	.... noon......	53	57·7

THE ARTIFICIAL HORIZON.

(97.) When the altitude of a heavenly body is observed by means of an artificial horizon, the reading off on the instrument will be the angular distance between the heavenly body and its image in the artificial horizon, and this will be double the altitude as observed from the true horizon. This will be easily seen by the following figure. Let s A, a ray of light proceeding from the body at s, be reflected by means of an artificial horizon placed at A, in the line A E. Then, if the spec-

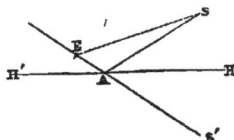

tator's eye is in the line A E, as at E, the image of the body will appear in the direction E A coming from a point s' below the horizon A. Now the observer is supposed to be placed so near A that the distance E A is inappreciable when compared with the distance A s of the heavenly body, that is, the angle observed between s and s', namely, s E s', may be considered to be = s A s', and this angle s A s' is manifestly double s A H, the altitude above the horizontal plane H H'. For by the principles of Optics it is proved that the angle s A H is equal to E A H', which is equal to the vertical or opposite angle s' A H, that is, the horizontal line A H bisects the angle observed. Hence the following rule for finding the true altitude from an observed altitude in the artificial horizon.

Rule 27. *Given the observed altitude of a heavenly body in an artificial horizon : to find the true altitude.*

1. Correct the observed altitude for index correction.
2. Half of the result will be the apparent altitude of the point observed.
3. Then proceed as in the preceding rules to find the true altitude.

303. The observed altitude of the sun's lower limb in an artificial horizon was 98° 14′ 10″, index correction − 4′ 10″ : required apparent altitude of sun's lower limb.

$$
\begin{array}{lrrr}
\text{Observed alt.} \dots\dots\dots & 98° & 14′ & 10″ \\
\text{in. cor.} \dots\dots\dots\dots\dots & & 4 & 10- \\
\hline
& 2)\,98 & 10 & 0 \\
\hline
\therefore \text{ app. alt. sun's L. L.} & 49 & 5 & 0 \\
\end{array}
$$

304. The observed altitude of moon's L. L. in an artificial horizon was 112° 32′ 15″, index correction + 3′ 25″ : required apparent altitude of moon's lower limb.

$$
\begin{array}{lrrr}
\text{Observed alt} \dots\dots\dots & 112° & 32′ & 15″ \\
\text{in. cor.} \dots\dots\dots\dots & & 3 & 25+ \\
\hline
& 2)\,112 & 35 & 40 \\
\hline
\therefore \text{ app. alt. moon's L. L.} & 56 & 17 & 50 \\
\end{array}
$$

The corrections for semidiameter and correction in altitude are then applied as in the preceding rules to obtain the true altitude.

CHAPTER V.

(98.) In the following rules for finding the latitude and longitude the earth is considered *as a sphere;* * the meridians will therefore be great circles, and the latitude may be defined to be " the arc of the celestial meridian intercepted between the zenith and the celestial equator."

Rule 28. *The* LATITUDE *by the meridian altitudes of a circumpolar star.*†

1. Correct the altitudes for index correction, height of eye, refraction, and parallax (or as many of these as are applicable to the case), and thus get the true meridian altitudes.

2. Add together the true meridian altitudes (reckoning from the same point of horizon), and half the result will be the latitude of the spectator.

3. If the heavenly body when passing the meridian above and below the pole is on different sides of the zenith, so that the altitudes are taken from opposite sides of the horizon, subtract the greater altitude from 180°, so as to reduce it to an altitude taken from the same point of the horizon as the other altitude; then proceed as directed by Rule (see Ex. 306).

EXAMPLES.

305. The meridian altitudes of α Ursæ Majoris were observed above and below the north pole to be 74° 10′ 10″ and 32° 42′ 15″ respectively (zenith south at both observations), index correction −2′ 10″, and height of eye above the sea 20 feet: required the latitude.

---

* The true figure of the earth is more nearly that of an *oblate spheroid*, that is, a solid generated by the rotation of an ellipse about its minor or shorter diameter; but as the major and minor diameters of the earth differ only about 26 miles, it is in Navigation considered, without any practical error, as a globe or sphere.

† A circumpolar star is one whose polar distance is less than the latitude of the spectator: it passes the meridian *above* the horizon, both at its superior and inferior transit.

Obs. alt. above pole	74°	10'	10"				
index cor.............		2	10 —	Obs. alt. below pole	32°	42'	15"
	74	8	0	index cor.		2	10 —
dip		4	24 —		32	40	5
	74	3	36	dip		4	24 —
ref.			17		32	35	41
true alt. above	74	3	19	ref.		1	31
„ „ below	32	34	10	...... true alt. below......	32	34	10
2)	106	37	29				
latitude...............	53	18	44·5 N.				

306. The meridian altitudes of α Aurigæ (Capella) were observed above and below the north pole to be 81° 10' 52" (zenith north of star), and 3° 42' 52" (zenith south), index correction — 3' 10", and height of eye above the sea 14 feet : required the latitude.

Obs. alt. from north } point of horizon. }	3°	42'	52"	Obs. alt. from south } point of horizon. }	81°	10'	52"
in. cor...............		3	10 —	in. cor...............		3	10 —
	3	39	42		81	7	42
dip		3	41 —	dip		3	41 —
	3	36	1		81	4	1
ref.		12	53 —	ref.			9 —
true alt.	3	23	8	true alt.	81	3	52
true alt.	98	56	8		180		
2)	102	19	16	.·. tr. alt. from north } point of horizon . }	98	56	8
.·. latitude	51	9	38				

307. The meridian altitudes of a star were observed above and below the north pole to be 69° 20' 45" and 6° 14' 30" respectively (zenith south at both observations), index correction — 1' 45", and height of eye 16 feet : required the latitude.   *Ans.* Lat. 37° 37' 35" N.

308. The meridian altitudes of a star were observed above and below the north pole to be 85° 10' 10" and 10° 10' 10" respectively (zenith south at both observations), index correction — 2' 40", and height of eye 20 feet : required the latitude.   *Ans.* Lat. 47° 30' 24" N.

309. The meridian altitudes of a star were observed above and below the north pole to be 77° 8' 10" (zenith north of star) and 3° 40' 45" (zenith south), index correction + 1' 42", and height of eye 12 feet : required the latitude.   *Ans.* Lat. 53° 10' 3" N.

310. August 12, 1850, the meridian altitudes of a star were observed above and below the south pole to be 85° 14′ 15″ (zenith south) and 4° 52′ 0″ (zenith north), index correction −8′ 14″, and height of eye above the sea was 30 feet : required the latitude. *Ans.* Lat. 49° 43′ 39″ S.

Rule 29 (*using sea horizon*). *The* LATITUDE *by the meridian altitude of the* SUN, *and its declination.*

1. Find a Greenwich date in apparent time ; namely, by adding the long. in time to 0ʰ 0ᵐ when W., and subtracting it from 24ʰ (putting the day one back) when the long. is E.

2. By means of the *Nautical Almanac* find the sun's declination for this date (p. 77). Take out also the sun's semidiameter, which is to be added to the apparent altitude when the lower limb is observed, and subtracted when the upper limb is observed.

3. Correct the observed altitude for index correction, dip, semidiameter, and correction in alt. (=refraction—parallax), and thus get the true altitude (p. 119), subtract the true altitude from 90°; the result will be the true zenith distance.

4. Mark the zenith distance N. or S. according as the zenith is north or south of the sun.

5. Add together the declination and zenith distance if they have the *same* names ; but take the difference if their names be unlike ; the result in each case will be the latitude, of the same name as the greater.

EXAMPLE.

311. April 27, 1853, in long. 87° 42′ W., the observed meridian altitude of the sun's lower limb was 48° 42′ 30″ (zenith north), the index correction was +1′ 42″, and the height of eye above the sea was 18 feet : required the latitude.

		Sun's decl. (at app. noon).							
Ship, April 27	0ʰ 0ᵐ	27	. . .	13° 43′	53″N.	Obs. alt. . .	48°	42′	36″
long. in time .	5 51 W.	28	. . .	14 2	57 N.	index cor. .		1	42+
Gr. April 27 .	5 51			19	4		48	44	12
						dip		4	11−
		0·61306					48	40	1
		0·97500				semi. . . .		15	54+
		1·58806		4	38	app. alt. center	48	55	55
		sun's decl.	13	48	31 N.	cor. in alt. .			45−
						true alt. . .	48	55	10
							90		
						true zen. dist.	41	4	50 N.
						declination .	13	48	31 N
						latitude . .	54	53	21 N.

312. January 14, 1853, in long. 72° 42′ W., the observed meridian al-

titude of the sun's L. L. was 32° 42' 10" (Z. N.), the index correction + 2' 10", and height of eye above the sea 14 feet : required the latitude.

Ans. Lat. 35° 50' 34" N.

313. March 20, 1853, in long. 72° 42' E., the observed meridian altitude of the sun's L. L. was 45° 4' 20" (Z. S.), index correction —3' 4", and height of eye above the sea 20 feet : required the latitude.

Ans. Lat. 44° 56' 54" S.

314. July 4, 1853, in long. 100° 0' W., the observed meridian altitude of the sun's L. L. was 62° 8' 7" (Z. N.), index correction — 3' 0", and height of eye above the sea 15 feet : required the latitude.

Ans. Lat. 50° 34' 59" N.

315. March 21, 1853, in long. 62° 0' W., the observed meridian altitude of the sun's U. L. was 50° 10' 5" (Z. N.), index correction + 7' 10", and height of eye 14 feet : required the latitude.   *Ans.* Lat. 40° 26' 47" N.

316. Sept. 24, 1853, in long. 33° 0' E., the observed meridian altitude of the sun's U. L. was 42° 3' 15" (Z. N.), index correction —1' 4", and height of eye above the sea 18 feet : required the latitude.

Ans. Lat. 47° 49' 39" N.

317. June 3, 1853, in long. 178° 30' W., the observed meridian altitude of the sun's U. L. was 16° 20' 0" (Z. S.), index correction +3' 30", and height of eye above the sea 20 feet : required the latitude.

Ans. Lat. 51° 35' 39" S.

Elements from Nautical Almanac.

	Sun's declination at apparent noon.				Sun's semidiameter.	
Jan.	14...21° 16' 4"S.	15...21° 5' 7"S.			14...16'	18"
March	19... 0 27 54 S.	20... 0 4 13 S.			19...16	5
July	4...22 53 8 N.	5...22 47 39 N.			4...15	46
March	21... 0 19 28 N.	22... 0 43 7 N.			21...16	5
Sept.	23... 0 8 3 S.	24... 0 31 28 S.			23...15	59
June	3...22 20 42 N.	4...22 27 50 N.			3...15	48

Meridian altitude by Artificial Horizon.

When the altitude is taken in artificial horizon, correct the observed altitude for index correction, and divide by 2.   Then correct for semi. and cor. in alt. as before to get the true alt.

EXAMPLES.

318. Oct. 21, 1853, in long. 1° 6' W., observed the meridian altitude of

sun's lower limb (in horizon) to be 56° 14′ 0″ (Z. N.), in. cor. —0′ 10″ : required the latitude.

		Sun's decl. (for app. noon).				
Ship, Oct 21 .	0ʰ 0ᵐ	21 . . 10° 35′ 22″S.		Obs. alt. . . .	56°	14′ 0″
long. in time .	4 W.	22 . . 10 56 45 S.		in. cor. . . .		0 10—
Gr. Oct. 21 .	0 4	21 23 S.		2)	56	13 50
		2·55630			28	6 55
		0·92520		semi.		16 6+
		3·48150	3	app. alt. center	28	23 1
		sun's decl. 10 35 25 S.		cor. in alt. . .		1 40—
				true alt. . . .	28	21 · 21
					90	
				true zen. dist. .	61	38 39N.
				declination . .	10	35 25 S.
				latitude . . .	51	3 14N.

319. Oct. 9, 1853, in long. 19° 20′ W., the observed meridian altitude of the sun's lower limb (in artificial horizon) was 44° 30′ 15″ (Z. S.), index correction —2′ 10″ : required the latitude.   *Ans.* Lat. 73° 53′ 28″ S.

320. June 10, 1853, in long. 23° 40′ E., the observed meridian altitude of the sun's lower limb (in quicksilver horizon) was 72° 15′ 20″ (Z. N.), index correction +4′ 5″ : required the latitude.   *Ans.* Lat. 76° 37′ 45″ N.

321. Aug. 7, 1853, in long. 62° 11′ E., the observed meridian altitude of sun's upper limb (in artificial horizon) was 83° 30′ 0″ (Z. N.), the index correction —3′ 15″ : required the latitude.   *Ans.* Lat. 65° 0′ 22″ N.

322. May 3, 1853, in long. 14° 20′ W., the observed meridian altitude of sun's upper limb (in artificial horizon) was 30° 2′ 30″ (Z. S.), index correction —1′ 15″ : required the latitude.   *Ans.* Lat. 59° 34′ 14″ S.

323. July 17, 1853, in long. 72° 30′ E., the observed meridian altitude of sun's upper limb (in artificial horizon) was 52° 30′ 0″ (Z. N.), index correction +2′ 10″ : required the latitude.   *Ans.* Lat. 85° 15′ 16″ N.

Sun's decl. and semi. from Nautical Almanac.

Oct.	9...	6°	20′	10″S.	10...	6°	42′	58″S.	semi.	16′	4″
June	9...	22	57	36 N.	10...	23	2	20 N.	,,	15	46
Aug.	6...	16	40	46 N.	7...	16	24	4 N.	,,	15	48
May	3...	15	43	50 N.	4...	16	1	18 N.	,,	15	53
July	16...	21	21	46 N.	17	21	11	43 N.	,,	15	46

Rule 30. *The* LATITUDE *by the meridian altitude of the* MOON, *and its declination, &c.*

Since the moon's declination, &c., are given in the *Nautical Almanac* for Greenwich *mean* noon, we must get a Greenwich date in *mean* time.

1. Find a Greenwich date in mean time.*

2. By means of the *Nautical Almanac* find for this date the moon's declination, moon's semidiameter, and moon's horizontal parallax, augmenting the moon's semidiameter for altitude (p. 120).

3. Correct the observed altitude for index correction, dip, semidiameter, and parallax and refraction, and thus get the true altitude; subtract the true altitude from 90°, and thus get the true zenith distance.

4. Mark the zenith distance N. or S. according as the zenith is north or south of the moon.

5. Add together the declination and zenith distance if they have the same names, but take their difference if their names be unlike; the result in each case will be the latitude—in the former of the name of either, in the latter of the name of the greater.

EXAMPLES.

324. November 12, 1853, at 2^h 20^m P.M. mean time nearly, in longitude 60° 42′ W., observed the meridian altitude of the moon's lower limb to be 30° 30′ 40″ (Z. N.), the index correction + 10′ 42″, and height of eye above the sea 16 feet: required the latitude.

Nov. 12, at ....$2^h 20^m$

long.....................4   3 +

Greenwich, Nov. 12...6   23

	Moon's declination.			Moon's semi.		Hor. par.	
Nov. 12, at 6^h...2°	44′	20″ N.	Noon...15′	6·4″......... 55′	19·7″		
„ at 7 ...2	57	38 N.	mid. ...15	2·7 55	6·4		
	13	18		3·7		13·3	
	0·41642			0·27413		0·27413	
	1·13142			3·46522		2·90957	
prop. log.	1·54784	5	6	3·73935	2·0	3·18370	7·1
decl................. 2	49	26 N.		15	4·4	55	12·6
			aug..........		7·4 +		
				15	11·8		

* When the estimated time at ship is given, the Greenwich date is found in the usual way by applying the longitude in time (Rule 5), or the Greenwich date may be found by correcting the moon's transit (see p. 108).

Moon's alt........	30°	30'	40"		Or thus; true alt. by calculation
in. cor............		10	42+		(p. 121):
	30	41	22		app. alt............ 30° 52' 38"
dip		3	56—		ref. 1 37
	30	37	26		30 51 1
semi.		15	12+		log. cos. .... 9·933747
	30	52	38		„ 3312·6" ... 3·520169
cor. in alt. ..... {		45	36		3·453916
			11		.·. par. in alt. ...... 2844"
true alt.	31	38	25		or .... 47' 24"
zenith dist.	58	21	35 N.		.·. true alt. 31 38 25
declin.............	2	49	26 N.		
latitude	61	11	1 N.		

325. January 10, 1853, at 7ʰ 40ᵐ P.M., mean time nearly, in longitude 5° 30' E., the observed meridian altitude of the moon's lower limb was 10° 20' 30" (Z. N.), the index correction —2' 20", and height of eye 14 feet: required the latitude.   *Ans.* Lat. 56° 37' 46" N.

326. February 4, 1853, at 5ʰ 40ᵐ A.M., mean time nearly, in longitude 72° 18' W., the observed meridian altitude of the moon's lower limb was 40° 20' 15" (Z. N.), index correction +3' 40", and height of eye 15 feet: required the latitude.   *Ans.* Lat. 25° 17' 10" N.

327. March 7, 1853, at 3ʰ 20ᵐ P.M., mean time nearly, in long. 19° 20'W., the observed meridian altitude of the moon's lower limb was 19° 17' 18" (Z. S.), index correction —1' 15", and height of eye 16 feet: required the latitude.   *Ans.* Lat. 88° 0' 44" S.

328. July 5, 1853, at 1ʰ 7ᵐ P.M., mean time nearly, in long. 33° 30" E., the observed meridian altitude of the moon's upper limb was 25° 42' 30" (Z. N.), the index correction +2' 15", and height of eye 20 feet: required the latitude.   *Ans.* Lat. 88° 22' 37" N.

329. August 12, 1853, at 5ʰ 4ᵐ A.M., mean time nearly, in longitude 94° 40' E., the observed meridian altitude of the moon's upper limb was 72° 20' 0" (Z. S.), the index correction +3' 40", and height of eye 22 feet: required the latitude.   *Ans.* Lat. 31° 53' 3" S.

330. December 27, 1853, at 9ʰ 12ᵐ A.M., mean time nearly, in longitude 15° 20' W., the observed meridian altitude of the moon's upper limb was 19° 50' 4" (Z. S.), the index correction —0' 30", and height of eye above the sea was 24 feet: required the latitude.   *Ans.* Lat. 87° 35' 20" S.

K

Elements from Nautical Almanac.

	Moon's declination.				Moon's semi.		Hor. par.	
Jan. 10, at	7ʰ	22°	1′	16″S.	noon 16′	0·1″	58′	36·4″
,, ,,	8	21	55	8 S.	mid. 15	54·2	58	14·9
Feb. 3, at	22	23	20	51 S.	mid. 16	8·2	59	6·1
,, ,,	23	23	24	43 S.	noon 16	6·6	59	0·3
Mar. 7, at	4	18	25	4 S.	noon 15	33·4	56	58·5
,, ,,	5	18	15	51 S.	mid. 15	29·3	56	43·7
July 4, at	22	24	33	11 N.	mid. 14	50·7	54	22·0
,, ,,	23	24	35	27 N.	noon 14	52·9	54	30·1
Aug. 11, at	10	14	4	13 S.	noon 16	6·9	59	1·1
,, ,,	11	14	16	46 S.	mid. 16	9·2	59	9·7
Dec. 26, at	22	17	55	16 S.	mid. 16	27·4	60	16·7
,, ,,	23	18	7	6 S.	noon 16	33·2	60 ·	37·6

Rule 31. *The* LATITUDE *by the meridian altitude of* A FIXED STAR, *and its declination.*

The declination of a fixed star changes so slowly, that we may, without any practical error, take it out of the *Nautical Almanac* by *inspection;* a Greenwich date will therefore be unnecessary.

1. Correct the observed altitude for index correction, dip, and refraction, and thus get the true meridian altitude; subtract this from 90° to obtain the true zenith distance.

2. Mark the same N. or S. according as the zenith is north or south of the star.

3. Take out the star's declination by inspection from the *Nautical Almanac*, and apply it to the true zenith distance in the manner pointed out in Rule 28, and thus get the latitude.

EXAMPLES.

331. Feb. 10, 1853, the observed meridian altitude of α Hydræ was 35° 50′ 40″ (zenith north of star), the index correction was +2′ 10″, and height of eye 0 feet : required the latitude.

Observed altitude	35°	50′	40″
index correction		2	10+
	35	52	50
refraction		1	21—
true altitude	35	51	29
	90		
true zenith distance	54	8	31 N.
declination	8	1	29 S. (*Naut. Alm.*)
latitude	46	7	2 N.

332. May 21, 1853, the observed meridian altitude of α Bootis was 62° 42′ 10″ (Z. N.), the index correction −4′ 4″, and height of eye 18 feet : required the latitude.   *Ans.* Lat. 47° 23′ 32″ N.

333. June 16, 1853, the observed meridian altitude of α Lyræ was 77° 1′ 50″ (Z. N.), index correction +2′ 10″, and height of eye 16 feet : required the latitude.   *Ans.* Lat. 51° 39′ 4″ N.

334. May 6, 1853, the observed meridian altitude of α Virginis was 16° 52′ 5″ (Z. N.), index correction +1′ 45″, and height of eye 20 feet : required the latitude.   *Ans.* Lat. 62° 50′ 4″ N.

335. Oct. 26, 1853, the observed meridian altitude of α Piscis Australis was 70° 10′ 0″ (Z. S.), the index correction −4′ 5″, and height of eye 10 feet : required the latitude.   *Ans.* Lat. 50° 21′ 26″ S.

336. May 10, 1853, the observed meridian altitude of α° Centauri was 10° 4′ 15″ (Z. N.), index correction −2′ 10″, and height of eye 20 feet : required the latitude.   *Ans.* Lat. 19° 54′ 9″ N.

337. August 1, 1853, the observed altitude of α Aquilæ was 50° 4′ 15″ (Z. N.), index correction −4′ 10″, and height of eye 14 feet : required the latitude.   *Ans.* Lat. 48° 33′ 32″ N.

Elements from Nautical Almanac.

May 21 ... α Bootis	Decl. 19°	56′	57″ N.
June 16 ... α Lyræ	„ 38	38	55 N.
May 6 ... α Virginis	„ 10	23	40 S.
Oct. 26 ... α Piscis Australis	„ 30	23	53 S.
May 10 ... α² Centauri	„ 60	13	31 S.
Aug. 1 ... α Aquilæ	„ 8	29	7 N.

Rule 32. *The* LATITUDE *by the meridian altitude of a* PLANET, *and its declination.*   •

1. Find a Greenwich date in mean time.

2. By means of the *Nautical Almanac* find the planet's declination for this date ; and when great accuracy is required, take out the planet's semi-diameter and horizontal parallax.

3. Correct the observed altitude for index correction, dip, refraction (and if necessary for semidiameter and parallax in altitude), and thus get the true altitude. Subtract the true altitude from 90° to get the true zenith distance.

4. Mark the zenith distance north or south according as the zenith is north or south of the planet.

5. Proceed then as in Rule 28.

EXAMPLE.

338. November 20, 1853, at 6ʰ 18ᵐ A.M., mean time nearly, in long. 62° 42′ E., observed the meridian altitude of Mars' lower limb to be 52° 10′ 45″ (Z. N.), the index correction + 4′ 0″, and height of eye above the sea 16 feet : required the latitude.

Ship, Nov. 19	18ʰ	18ᵐ		Planet's semi.		3″
long. in time	4	11 E.	,,	H. P.......		6
Greenwich, Nov. 19.........	14	7				

Planet's declination.

19	12°	55′	36″	N.
20	12	37	1	N.
		18	35	

·23048
·98615

1·21663......   10   56

Planet's decl. ...   12   44   40   N.

Obs. alt.	52°	10′	45″
in. cor.		4	0+
	52	14	45
dip		3	56
	52	10	49
semi.			3
	52	10	52
ref.			45−
	52	10	7
par. in alt.			4+
true alt.	52	10	11
	90		
true zen. dist. ...	37	49	49 N.
planet's decl. ...	12	44	40 N.
latitude....	50	34	29 N.

If the small corrections of the planet's semidiameter and parallax in altitude are neglected, the above example will be worked thus :

Ship, Nov. 19	18ʰ	18ᵐ	
long. in time	4	11 E.	
Greenwich, Nov. 19	14	7	

Planet's declination.

19	12°	55′	36″	N.
20	12	37	1	N.
		18	35	

·23048
·98615

1·21663   10   56

Planet's decl.   12   44   40   N.

Obs. alt.	52°	10′	45″
in. cor.		4	0+
	52	14	45
dip		3	56
	52	10	49
ref.			45−
true alt.	52	10	4
	90		
true zen. dist......	37	49	56 N.
decl.	12	44	40 N.
latitude...........	50	34	36 N.

339. May 4, 1853, at 2^h 45^m A.M. mean time nearly, in long. 42° 10′ W., the observed meridian altitude of Jupiter's center was 16° 42′ 10″ (Z. N.), index correction +11′ 42″, and height of eye above the sea 20 feet: required the latitude.   *Ans.* Lat. 50° 30′ 38″ N.

340. July 12, 1853, at 9^h 36^m P.M. mean time nearly, in long. 30° 30′ E., the observed meridian altitude of Jupiter's center was 10° 10′ 50″ (Z. N.), the index correction −4′ 4″, and height of eye above the sea 10 feet: required the latitude.   *Ans.* Lat. 57° 45′ 37″ N.

341. November 27, 1853, at 6^h 3^m A.M. mean time nearly, in long. 100° 0′ W., the observed meridian altitude of Mars' center was 32° 40′ 10″ (Z. S.), index correction −8′ 10″, and height of eye 16 feet: required the latitude.   *Ans.* Lat. 45° 45′ 0″ S.

342. Sept. 15, 1853, at 4^h 20^m A.M. mean time nearly, in long. 10° 6′ W., the observed meridian altitude of Saturn's center was 19° 42′ 10″ (Z. N.), index correction −6′ 45″, and height of eye 12 feet: required the latitude.   *Ans.* Lat. 88° 55′ 24″ N.

343. Jan. 12, 1853, at 7^h 9^m P.M. mean time nearly, in long. 32° 0′ W., the observed meridian altitude of Saturn's center was 62° 42′ 10″ (Z. S.), index correction −8′ 10″, and height of eye 20 feet: required the latitude.   *Ans.* Lat. 14° 36′ 41″ S.

344. June 7, 1853, at 5^h 40^m P.M. mean time nearly, in long. 72° 30′ E., the observed meridian altitude of Venus was 30° 40′ 10″ (Z. S.), index correction +4′ 20″, and height of eye 24 feet: required the latitude.   *Ans.* Lat. 35° 39′ 30″ S.

Elements from Nautical Almanac.

Jupiter, decl.	May 3...	22° 43′	11″ S.	May 4...	22° 43′	1″ S.
Jupiter,	„ July 12...	22 16	10 S.	July 13...	22 15	49 S.
Mars,	„ Nov. 27...	11 48	44 N.	Nov. 28...	11 40	44 N.
Saturn,	„ Sept. 14...	18 24	50 N.	Sept. 15...	18 24	38 N.
Saturn,	„ Jan. 12...	12 54	5 N.	Jan. 13...	12 54	24 N.
Venus,	„ June 7...	23 42	15 N.	June 8...	23 48	1 N.

Rule 33. *The* LATITUDE *by the meridian altitude of a heavenly body* BELOW *the pole, and its declination.*

1. Find the declination at the time of observation.
2. From the observed altitude get the true altitude ; then
3. Add 90° to the true altitude, and from the sum subtract the declination ; the remainder will be the latitude of the same name as the declination.

345. April 27, 1853, the meridian altitude of α Crucis below the south

pole was observed to be 14° 10' 30", the index correction was + 4' 4", and the height of eye 20 feet : required the latitude.

		°	'	"
Obs. alt.	....	14	10	30
in. cor.	....		4	4+
		14	14	34
dip	....		4	24—
		14	10	10
ref.	....		3	47—
true alt.	....	14	6	23
		90		
		104	6	23
star's decl.	....	62	17	10 S.
∴ lat.	....	41	49	13 S.

346. June 18, 1853, at apparent midnight, in long. 100° W., the observed meridian altitude of the sun's lower limb below the north pole was 8° 42' 10', the index correction — 3', and height of eye above the sea 14 feet : required the latitude.

		Sun's decl. (app. noon).										
Ship, June 18	12ʰ 0ᵐ	18	. . .	23°	25'	36" N.	Obs. alt.	. .	8°	42'	10"
long. in time	6 40 W.	19	. . .	23	26	39 N.	in. cor.	. . .		3	0—
Gr., June 18	18 40			1	3				8	39	10
							dip	. . . .		3	41—
		·10915							8	35	29
		2·23408					semi.	. . .		15	46+
		2·34323		0	49				8	51	15
		decl.	. .	23	26	25 N.	cor. in alt.	. .		5	51—
									8	45	24
									90		
									98	45	24
							sun's decl.	. .	23	26	25 N.
							∴ lat.	. . .	75	18	59 N.

347. Feb. 10, 1853, the meridian altitude of α Argûs below the pole was observed to be 6° 41' 15", index correction —2' 10", and height of eye above the sea 14 feet: required the latitude.   *Ans.* Lat. 43° 50' 18" S.

348. January 11, 1853, the observed meridian altitude of α Ursæ Majoris below the pole was 14° 14' 30", the index correction —4' 5", and height of eye 20 feet : required the latitude.   *Ans.* Lat. 41° 29' 47" N.

349. April 20, 1853, the observed meridian altitude of η Argûs below the pole was 20° 14' 15", the index correction —4' 5", and height of eye 10 feet : required the latitude.   *Ans.* Lat. 51° 9' 27" S.

350. June 1, 1853, in long. 30° 52' W., the observed meridian altitude

of the sun's lower limb below the pole was 10° 42′ 0″, the index correction +2′ 10″, and height of eye 20 feet: required the latitude.

Ans. Lat. 78° 41′ 0″ N.

351. June 10, 1853, at 2ʰ 40ᵐ A.M. mean time nearly, in long. 30° W., observed the meridian altitude of the moon's lower limb below the pole to be 14° 30′ 10″, index correction +2′ 45″, and height of eye 14 feet: required the latitude. *Ans.* Lat. 81° 32′ 31″ N.

352. July 1, 1853, at 9ʰ 30ᵐ P.M. mean time nearly, in long. 62° W., the observed meridian altitude of Mars below the pole was 10° 32′ 30″, index correction −3′ 0″, and height of eye 18 feet: required the latitude.

Ans. Lat. 79° 8′ 32″ N.

Elements from Nautical Almanac.

α Argûs	.	.	Feb. 10, decl.	.	. 52° 37′ 14″ S.	Sun's decl., June 1 .	22° 5′ 15″ N.
α Ursæ Majoris, Jan. 11,	„	. . 62 32 26 N.	„ „ June 2 .	22 13 10 N.
η Argûs	.	.	Apr. 20,	„	. . 58 54 59 S.	„ semi. . . .	15 48

	Moon's decl.	Moon's semi.	Moon's h. par.	Planet's decl.
June 9 at 16ʰ	24° 3′ 51″ N.	mid. 15′ 0·4″	54′ 57·6″	July 1 . 21° 7′ 5″ N.
„ 17	24 0 11 N.	noon 15 4·0	55 11·0	„ 2 . 21 15 9 N.

LATITUDE BY OBSERVATIONS OFF THE MERIDIAN.

In the volume of astronomical problems* by the author will be found several methods for finding the latitude depending on some particular bearing or hour-angle of the heavenly body: as when it bears due east, or when it is in the horizon, or when the hour-angle is 6 hours, &c.; but since it is difficult to determine the precise moment when the heavenly body is in any of these positions, the methods referred to are of little use in practice. Problem 131 in that volume, however, is one from which a useful rule may be derived (*Nav.* Part II. p. 54), as it depends on the declination, altitude, and hour-angle of the heavenly body; and as it requires only the common table of sines, &c., we shall select it as the second method about to be given for finding the latitude from an altitude near the meridian. The first method is deserving attention, being free from any distinction of cases; it requires, however, the tables of haversines and versines, and that the latitude should be known within a quarter of a degree of the truth, otherwise it may be necessary to repeat a part of the work perhaps more than once; but it is a useful method, and gives very accurate results. The altitude and decl. are easily obtained at sea; the hour-angle is only known accurately when the ship time is given, and this is a quantity difficult to discover independently of an observation: the ship time, however, may always be consi-

* *Problems in Astronomy, &c., and Solutions,* pp. 33, 34, &c. These Solutions of nearly 200 astronomical and nautical problems form a useful and interesting introduction to the theory of nautical astronomy.

dered to be known nearly.   To render, therefore, a rule for finding the
latitude, depending on the declination, altitude, and ship time, of practical
value, we must ascertain in what position of a heavenly body an error of a
few minutes in the ship time will produce the smallest error in the latitude
deduced from it; and this we find will be the case if the observed altitude
is taken when the body is *near the meridian* (see *Nav.* Part II. p. 57).   It
is for this reason that single altitude observations taken off the meridian for
finding the latitude are confined to bodies within half an hour of the meridian, when the time at the ship is uncertain to 3 or 4 minutes.

Another practical rule of more general application is deduced from problems 143 and 144.   Two altitudes are taken of the same or different heavenly bodies at the same or at different times, from whence the latitude may
be found.   This is called the rule by DOUBLE ALTITUDE.   In this method of
finding the latitude the heavenly bodies need not be close to the meridian,
but the effect of any error in the observations will be diminished if, in
selecting the bodies to be observed, the difference of their bearings be always
greater than the less bearing.

Rule 34. First method (using haversines).   LATITUDE *from an altitude*
of the sun NEAR THE MERIDIAN.

1. Find the Greenwich date in mean time.
2. Take out the declination and equation of time for this date, and sun's
semidiameter.
3. *To find the sun's hour-angle.*   To the Greenwich mean time found as
accurately as possible apply the longitude in time, subtracting if west, and
adding if east; the result will be ship mean time : to this apply the equation
of time with its proper sign to reduce mean time into apparent time ; the
result will be the sun's hour-angle.
4. Add together the following logarithms :

> Constant log. 6·301030.
> Log. cosine declination.
> Log. cosine estimated latitude.
> Log. haversine hour-angle.*

reject 30 in the index, and look for the result as a logarithm, and take out
its natural number.

5. Correct the observed altitude for index correction, dip, semidiameter,
correction in altitude, and thus get a zenith distance.
6. From the versine of zenith distance subtract the natural number
found as above.   The remainder will be the versine of a meridian zenith
distance, which find from the tables.

---

* Or, instead of log. haversine, take out twice the log. sine of half the hour-angle
(rejecting in this case 40 from the index).

7. Under the meridian zenith distance put the declination, and proceed to find the latitude by one of the preceding rules for finding the latitude by a meridian altitude.

NOTE. If the latitude thus found differ much from the estimated latitude used in the question, the work should be corrected by using the last latitude found, in place of the former one.

EXAMPLES.

353. August 22, 1853, A.M., in latitude by account 50° 48′ N., and long. 1° 6′ W., a chronometer showed 11^h 50^m 22ˢ, error on Greenwich mean time being 40·2ˢ fast, when the observed altitude of the sun's lower limb (in artificial horizon) was 101° 14′ 10″ (Z. N), index correction +30″: required the latitude.

Greenwich date and hour-angle.

Chr. showed......	11^h	50^m	22ˢ A.M.	Obs. alt. in hor....	101°	14′	10″
error, fast.........			40·2	in. cor.			30 +
	11	49	41·8	2)101		14	40
	12					50	37	20
Gr. date, Aug. 21 .	23	49	41·8	semi.		15	51
long. in time......		4	24·0 −			50	53	11
ship mean time...	23	45	17·8	cor. in alt.			42 −
equation of time .		2	39·1 +	sun's true alt.......	50	52	29
	23	47	56·9			90
	24			sun's zen. dist. ...	39	7	31
∴ hour-angle ...	0	12	3·1				

Sun's decl.					Equation of time.		Semi.
Aug. 21	12°	4′	57″ N.	......	Aug. 21...2^m 54ˢ add.		15′ 51″
„ 22	11	44	50 N.	......	„ 22...2 39		
		20	7		15		
·00303					·00303		
·95172					2·85733		
·95475		19	59		2·86036...	14·9	
∴ sun's decl......	11	44	58 N.		2 39·1 add.		

By First Method.

Const. log. ....6·301030		vers....39° 7′ 0224137
log. cos. decl. ......9·990803		31″ 95
„ cos. est. lat. ...9·800737		vers. zen. dist.......0224232
„ hav. hour-angle.6·839449		natural number ... 855
2·932019		vers. mer. Z. D. ∴..0223377
∴ natural number 855		39° 2′ 220
		51″ 157

mer. Z. D....39 2 51 N.
decl. ....11 44 58 N.
∴ LATITUDE 50 47 49 N.

Rule 35. Second method (using sines, &c.). LATITUDE *by altitude of sun*
NEAR THE MERIDIAN.

In *Navigation*, Part II. p. 54, it is shown that if h=hour-angle, p=polar
distance, and a=altitude of a heavenly body, then the colatitude=$y \pm x$,
where tan. x=cos. h. tan. p and cos. y=sec. p. cos. x, sin. a. From which
formulæ the colatitude, and thence the latitude, is easily found, if we attend
to the proper algebraic sign of each quantity, as pointed out in *Trigonometry*,
Part I. art. 31. We may, however, deduce from these trigonometrical expressions a direct rule, and free from the distinction of cases arising from the
use of signs, by modifying the above formulæ as follows :

$$\text{Let } z = 90 - x, \text{ and the decl.} = 90 \sim p = d,$$

Then the above formulæ become

$$\text{cot. } z = \text{cot. } d. \text{ cos. } h,$$
$$\text{cos. } y = \text{cosec. } d. \text{ sin. } z \text{ . sin. } a,$$

where the arcs z and y may be looked upon as the approximate declination,
and mer. zen. distance respectively, and marked N. or S., as in the Rule 29
for latitude by meridian altitude. Hence this direct Rule.

1. Find Greenwich date, declination, equation of time, hour-angle, and
true altitude, as in last Rule.

2. Add together log. cos. hour-angle, and log. cotangent of declination
(taking out at the same opening of the tables, and putting a little to the
right, the log. cosecant of declination).   •

3. The sum (rejecting 10 in index) of the two logarithms just added
together will be log. cotangent of arc z, which find from the tables, and
mark it N. or S., according as the declination is north or south.

4. Under log. cosecant of declination (already taken out) put log. sine of
arc z, and log. sine of altitude : the sum of these three logarithms (rejecting
20 in index) will be the log. cosine of arc y, which take out, and mark N. or
S., according as the zenith is north or south of the heavenly body.

5. Under arc z put arc y, and take their sum or difference, according as
they have the same or different names ; the result will be the latitude required, to be marked north or south, as in the rule for latitude by meridian
altitude.

354. May 10, 1853, A.M., in latitude by account 50° 50′ N., and long.
2° 10′ W., a chronometer showed 11ʰ 51ᵐ 58ˢ, error on Greenwich mean
time being 11ᵐ 31ˢ fast, when the observed altitude of the sun's lower limb
was 56° 19′ 30″ (Z. N.), index correction −3′ 20″, and height of eye 18
feet : required the latitude.

Greenwich date and hour-angle.

Chr. showed ...	11ʰ	51ᵐ	58ˢ A.M.		Obs. alt.	56°	19′	30″
error, fast		11	31		in. cor..................		3	20−
	11	40	27			56	16	10
	12				dip		4	11−
Gr. date, May 9	23	40	27			56	11	59
long. in time...		8	40−		semi.		15	52+
ship mean time	23	31	47			56	27	51
equation of time		3	49+		cor. in alt.			33−
ship app. time	23	35	36		∴ true alt.	56	27	18
	24							
hour-angle......	0	24	24					

Sun's declination				Equation of time.		Semi.
May 9 ...... 17° 24′ 37″N.				May 9 ... 3ᵐ 46·0ˢ add.		15′ 52″
„ 10 ...... 17 40 25 N.				„ 10 ... 3 49·0		
·00608	15	48			3·0	
1·05662				cor....	3·0	
1·06270	15	35			3 49·0	

∴ sun's decl. 17 40 12 N.

The Second or Direct Method.

cot. z = cot. decl. cos. hour-angle.

cos. y = cosec. decl. sin. z . sin. alt.

log. cot. decl.	0·496782	log. cosec. decl. ...	0·517773
„ cos. hour-angle.	9·997534	„ sin. z	9·484402
„ cot. z............	10·494316	„ sin. alt..........	9·920876
∴ z............17° 45′ 45″N.		„ cos. y	9·923051
y............33 6 30 N.		∴ y..................	33° 6′ 30″ N.
∴ latitude ...50 52 15 N.			

355. Nov. 14, 1853, P.M., in lat. by account 87° 41′ S., and long. 1° 0′ W., a chronometer showed 0ʰ 25ᵐ 27ˢ, error on Greenwich mean time being fast 5ᵐ 56·7ˢ, when the observed altitude of the sun's lower limb was 20° 26′ 20″ (Z. S.), index correction −2′ 20″, and height of eye 10 feet: required the latitude. *Ans.* Lat. 87° 42′ 15″ S.

356. June 30, 1853, A.M., in lat. by account 63° 20′ N., and longitude 23° 30′ W., a chronometer showed 11ʰ 30ᵐ 15ˢ, error on Greenwich mean time being 7ᵐ 32ˢ fast, when the observed altitude of the sun's upper limb was 44° 20′ 22″ (Z. N.), index correction +2′ 20″, and height of eye 14 feet: required the latitude. *Ans.* Lat. 63° 21′ N.

357. July 10, 1853, A.M., in lat. by account 57° 24' N., and longitude 3° 40' W., a chronometer showed 11ʰ 20ᵐ 15', error on Greenwich mean time being 30ᵐ 30' slow, when the observed altitude of the sun's lower limb was 54° 17' 19" (Z. N.), index correction —2' 40", and height of eye 20 feet : required the latitude.   *Ans.* Lat. 57° 25' 25" N.

358. May 20, 1853, A.M., in lat. by account 79° 48' N., and longitude 44° 30' E., a chronometer showed 11ʰ 30ᵐ 0', error on Greenwich mean time being 15ᵐ 20' slow, when the observed altitude of the sun's lower limb (in artificial horizon) was 54° 30' 20" (Z. N.), index correction — 4' 30": required the latitude.   *Ans.* Lat. 79° 48' 30" N.

359. June 16, 1853, P.M., in lat. by account 52° 25' N., and longitude 1° 6' W., a chronometer showed 1ʰ 2ᵐ 9' error on Greenwich mean time being 40ᵐ 30' fast, when the observed altitude of the sun's lower limb was 60° 37' 50" (Z. N.), index correction —2' 10", and height of eye 17 feet : required the latitude.   *Ans.* Lat. 52° 24' 15" N.

Elements from Nautical Almanac.

	Sun's declination.		Equation of time.			Sun's semi.	
Nov. 14,	18° 18'	5"S.	15ᵐ 22·9ˢ	} to be added	16' 13"		
,, 15,	18 33	30 S.	15 12·8				
June 29,	23 14	31 N.	3 3·7	} ,, subtracted	15 46		
,, 30,	23 11	3 N.	3 15·6				
July 9,	22 21	48 N.	4 50·8	} ,, subtracted	15 46		
,, 10,	22 14	22 N.	4 59·5				
May 19,	19 48	45 N.	3 47·7	} ,, added	15 50		
,, 20,	20 1	23 N.	3 44·9				
June 16,	23 22	15 N.	0 18·8	} ,, subtracted	15 46		
,, 17,	23 24	8 N.	0 31·6				

LATITUDE *by* POLE-STAR (*using Inman's Table*).

The table for correcting the altitude of the pole-star, contained in Inman's *Nautical Tables*, has recently been recalculated, and adapted to the present and several subsequent years. As this table enables us to find the latitude sufficiently near for all ordinary purposes, the practical rule (a proof of which is given in Part II.) is now inserted.

Rule 36. 1. Get a Greenwich date.

2. Take out from the *Nautical Almanac* the right ascension of the mean sun (called there sidereal time), and correct it for the Greenwich date (p. 86).

3. Add together the right ascension of mean sun so corrected to the nearest minute and ship mean time (expressed astronomically).

The result, rejecting 24ʰ if greater than 24ʰ, is the meridian right ascension, the argument of the table.

4. Correct the observed altitude of the star for index correction, dip, and refraction, and thus get the true altitude.

5. Enter the table, called the "correction of pole-star," with the meridian right ascension at the side, and with the nearest latitude to that by account at the top, and take out the correction as near as can be estimated, with its proper sign.

6. Apply the correction to the true altitude, and the result will be the latitude required.

EXAMPLE.

360. April 10, 1860, at 1ʰ 30ᵐ A.M., mean time nearly, in longitude 64° 30′ E., the observed altitude of the pole-star was 52° 30′ 40″, index correction — 1′ 20″, and height of eye above the sea 15 feet : required the latitude.

Ship, April 9 .... 13ʰ 30ᵐ

long. in time   4   18 E.

Green. April 9.........   9   12

	R. A. mean sun.		Obs. alt.		
9th	1ʰ 11ᵐ	47ˢ	52° 30′	40″	
cor. for 9ʰ 12ᵐ	1	30		1	20 —
	1 13	17	52 29	20	
ship time	13	30		3	49 —
mer. R. A.	14	43	52 25	31	
				45 —	
			52 24	46	
cor. from table......			1 19	15 +	
latitude			53 44	1 N.	

EXAMPLES.

361. June 15, 1860, at 2ʰ 20ᵐ A.M., mean time nearly, in longitude 10° 20′ W., the observed altitude of α Polaris was 46° 10′ 30″, index correction + 3′ 10″, and height of the eye 19 feet : required the latitude.

Ans. Lat. 45° 52′ N.

362. July 20, 1860, at 11ʰ 40ᵐ P.M., mean time nearly, in longitude 42° E., the observed altitude of α Polaris was 35° 30′ 40″, index correction — 4′ 10″, and height of eye above the sea 15 feet : required the latitude.

Ans. Lat. 35° 12′ N.

Elements from Nautical Almanac.

R. A. mean sun (called in *N. A.* sidereal time).

June 14th, at noon........................ 5ʰ 32ᵐ

July 20th,   „ 7 54

LATITUDE BY DOUBLE ALTITUDE.

The most general rule for finding the latitude by a double altitude of a heavenly body is the one selected as the first method; but the labour of reducing the observations is somewhat greater than in the second method, known as Ivory's Rule. The great advantage of the first method is that it may be applied to the same or different heavenly bodies, observed at the same instant or at different times, and that it is the simple application of two rules in Spherical Trigonometry.

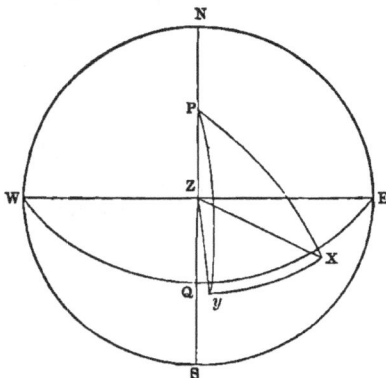

Let P be the pole, z the zenith, x and y the same heavenly body observed at different times; or different heavenly bodies observed at the same instant, or different heavenly bodies observed at different times. Let zx zy be their zenith distances. Then in the figure we know by observation zx and zy, and from the *Nautical Almanac* we can find the polar distances Px and Py; also by means of the elapsed time as measured by a watch, or from the right ascension of the bodies, or from both, we can compute the polar angle xPy; the colatitude Pz may then be computed in the following manner by the application of the common rules of Spherical Trigonometry.

1. In triangle Pyx are given two sides Px, Py and the included angle xPy, to find xy, which call arc 1.

2. In triangle Pxy are given three sides Px, Py and arc 1, to find angle Pxy, which call arc 2.

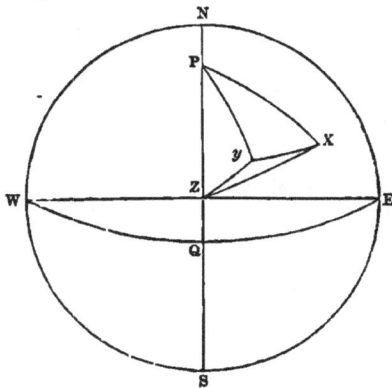

3. In triangle zxy are given three sides zx, zy and arc 1, to find angle zxy, which call arc 3.

4. Arc 2 − arc 3 = angle Pxz = arc 4. But if the arc xy drawn through x and y pass when produced between P and z the pole and the zenith, then it is evident by the annexed figure that the arc 2 + arc 3 = Pxz or arc 4. If the arc xy produced pass near z, the bodies x and y in such a position should not be observed.

Lastly. In triangle Pxz are

given the two sides PX and ZX and arc 4 (namely, the included angle PXZ), to find PZ the colatitude, and thence the latitude.

Correction for run of ship.

If the ship have moved in the interval between the observations, the second altitude will in general differ from what it would have been if both observations had been taken at the same place. On this account it is usual to apply to the first altitude a correction so as to reduce it to what it would have been if taken at the place of the second observation; this quantity is called "the correction for run of the ship," and may be calculated as follows.

When a ship describes an arc on the surface of the sea, the zenith describes a similar arc in the celestial concave : let, therefore, z be the zenith of the ship at the first observation, z' its zenith at the second observation; then arc zz' measures the distance run in the interval. Let s be the place of the heavenly body at the first observation : with center s at distance sz', describe an arc cutting sz, fig. 1, or sz produced in D, fig. 2; then the triangle zz'D being small, may be considered as a right-angled plane triangle, and zD is the correction to be applied to zs in order to get z's the distance of s from the zenith at the second observation.

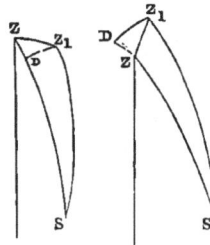

Fig. 1.   Fig. 2.

Now $zD = zz'$ cos. z' zD = distance run × cos. angle between the direction of the ship's run and the bearing of the sun at the first observation.

This correction zD may be readily found by means of the traverse table, for since (p. 43),

Diff. lat. = dist. cos. course ; if therefore in triangle zz'D the angle zz'D be considered as the course, and zz' the distance, the correction zD for run will correspond in the traverse table to the difference of latitude.

The angle z'zD is the difference between the course of the ship in the interval and the true bearing of the body, when the run of the ship has been towards the place of the body, as in fig. 1 ; and what this angle wants of 180° or 16 points when the direction of the ship's run has been from the place of the body, as in fig. 2.   In the former case it is manifest that the correction zD for run must be added to the first observed altitude, and in the second subtracted, in order to get the altitude of the body, the same as it would have been if it had been also observed at the place of the ship at the second observation.   Hence this practical Rule :

Correction for run.

1. Enter the traverse table with the distance run as a distance, and the angle (supposed less than 8 points) between the true bearing of the heavenly

body at the first observation and course of the ship, as a course, and take out the corresponding *diff. lat.*, which *add* to the first taken true altitude (the tenths in the diff. lat. being turned into seconds, by multiplying them by 60); the result will be the altitude corrected for run.

2. But if the above angle be greater than 8 points, subtract the same from 16 points, and look out the remainder as a course, and *subtract* the diff. lat. corresponding thereto from the first true altitude; the result will be the altitude corrected for run.

363. The course of the ship was N.W.$\frac{1}{2}$W. 10 miles, and bearing of the sun E. by S.: required the correction for the first altitude for run.

The angle between N.W.$\frac{1}{2}$W. and E. by S. is 13$\frac{1}{2}$ points, subtracting 13$\frac{1}{2}$ from 16 points: enter traverse table with the remainder, namely, 2$\frac{1}{2}$ as a course and 10 miles as a distance: the corresponding diff. lat. is 8·8′ = 8′ 48″ to be *subtracted* from the true altitude.

364. The course of the ship was E.N.E. 25 miles, and bearing of the sun E. by S.: required the correction of the first altitude for run.

The angle between E.N.E. and E. by S. is 3 points; entering traverse table with 3 points as a course, and 25 miles as a distance, the corresponding diff. lat. = 20·8′ = 20′ 48″ to be *added* to the true altitude.

365. The true course of the ship was S.W.$\frac{1}{2}$W. 15 miles, and the true bearing of the sun S. by E.$\frac{1}{2}$E.: required the correction of the first altitude for run.   *Ans.* +5′ 42″.

366. The true course of the ship was W.$\frac{1}{2}$N. 19 miles, and the true bearing of the sun was S. by E.$\frac{1}{2}$E.: required the correction for run.
                                        Ans. −7′ 18″.

Rule 37. DOUBLE ALTITUDE.   First method.   (1.) *The* LATITUDE *by two altitudes of the* SUN.

1. From the estimated mean time at the ship at each observation, and the longitude, get two Greenwich dates.

2. By means of the *Nautical Almanac* find the declination for each Greenwich date. Take out also from the *Almanac* the sun's semidiameter.

3. Find the polar distance at each observation by subtracting the declination from 90°, if the estimated latitude and declination are of the same name; or by adding 90° to the declination, if the estimated latitude and declination are of different names.

4. Correct the two observed altitudes for index correction, dip, semidiameter, and correction in altitude.

5. Correct also the first altitude observed for the run of the ship (p. 143).

6. Subtract the true altitudes thus obtained from 90° and thus get the zenith distances.

7. Find the polar angle or elapsed time between the observations, by subtracting the time shown by chronometer at the first observation from the time shown by chronometer (increased if necessary by 12 hours) at second observation.

NOTE. When great accuracy is required, this elapsed time should be corrected for rate of chronometer, and also for the change in the equation of time in the interval; but these corrections are seldom made.

8. *To find arc* 1 (using Inman's Tables). Add together log. sin. polar distance at greater bearing, log. sin. polar distance at lesser bearing, and log. haversine of polar angle (rejecting the tens in the index); and look out the result as a log. haversine; the arc corresponding thereto is arc 1 nearly.*

9. *To find arc* 2. Under arc 1 put polar distance at greater bearing, and take the difference, under which put polar distance at lesser bearing; take the sum and difference of the two last quantities. Add together the log. cosecants of the two first arcs put down, and halves of the log. haversines of the two last arcs put down; the sum, rejecting 10 in index, is the log. haversine of arc 2, which take from the Tables.

10. *To find arc* 3. Under arc 1 put zenith distance at greater bearing, and take the difference, under which put zenith distance at lesser bearing; take the sum and difference of the last two quantities.

Add together the log. cosecants of the two first arcs put down, and halve the log. haversines of the two last arcs put down; the sum, rejecting the tens in index, is the log. haversine of arc 3, which take from the Tables.

11. *To find arc* 4. The difference between arc 2 and arc 3 is arc 4.

NOTE. When the arc joining the places of the sun at the two observations passes, when produced, between the zenith and pole (which the observer may easily discover at the time the observation is taken), then the sum of arcs 2 and 3 is arc 4.

12. *To find arc* 5. Add together log. sin. polar dist. at greater bearing, log. sin. zenith distance at greater bearing, and log. haversine of arc 4; the sum, rejecting 10 in the index, is log. haversine of arc, which take from the Tables, and call arc 5.

Take the difference between the polar distances at the greater bearing, and the zenith distance at greater bearing.

Add together versine of arc 5 and versine of the difference just found; the sum is the versine of the colatitude, which take from the Tables, and subtract from 90°; the result is the latitude required.

NOTE. If the student have only the tables of sines, &c. he may use Rule 41, p. 156, or the Trigonometrical method given in p. 160.

* To find arc 1 correctly. To the versine of arc found as above add the versine of the difference of polar distances: the sum will be the versine of arc 1. But this is rarely necessary to be done.

L

EXAMPLE.

367. Oct. 11, 1845, in latitude by account 54° N. and long. 83° 15′ W., the following double altitude of the sun was observed :

Mean time nearly.	Chro.	Alt. sun's L. L.	Bearing.
7ʰ 45ᵐ A.M.	11ʰ 40ᵐ 15ˢ	9° 0′ 20″	E.S.E.¼E.
10 35 A.M.	2 13 20	25 3 30	S.S.E.

The run of the ship in the interval was S. by W. 15 miles, index correction +5′ 10″, and the height of eye above the sea was 18 feet : required the true latitude at the second observation.

At greater bearing.			At less bearing.		
Ship, Oct. 10	19ʰ	45ᵐ	Ship, Oct. 10.........	22ʰ	35ᵐ
long. in time	5	33 W.		5	33 W.
Oct. 10	25	18	Oct. 10.................	28	8
Gr. Oct. 11	1	18	Gr. Oct. 11............	4	8

Decl. at greater bearing.				Decl. at less bearing.			
11	7°	4′	38″ S.	11	7°	4′	38″ S.
12	7	27	15 S.	12	7	27	15 S.
		22	37			22	37

1·26627				·76391			
·90084				·90084			
2·16711 cor.		1	14	1·66475 cor.		3	54
	7	5	52 S.		7	8	32 S.
	90				90		
N. Pol. dist.......	97	5	52	N. Pol. dist.	97	8	32
	At greater bearing.				At less bearing.		

Sun's altitude at greater bearing.				Sun's altitude at less bearing.			
Obs. alt............	9°	0′	20″	Obs. alt.	25°	3′	30″
in. cor.		5	10+	in. cor..............		5	10+
	9	5	30		25	8	40
dip		4	11−	dip		4	11−
	9	1	19		25	4	29
semi.		16	3+	semi.		16	3+
	9	17	22		25	20	32
cor. in alt..........		5	33−	cor. in alt.		1	54−
	9	11	49	true alt.	25	18	38
cor. for run........		2	12+		90		
true alt.	9	14	1	Z. D.	64	41	22
	90				At less bearing.		
Z. D.	80	45	59				
	At greater bearing.						

To find arc 1.

Chro. times.					
	11ʰ	40ᵐ	15ˢ	Sin. P. D. at G. B. ...	9·996661
	14	13	20	sin. P. D. at L. B. ...	9·996617
Pol. angle	2	33	5	hav. pol. angle	9·031223
				hav. arc 1	9·024501
				arc 1 ....37° 58′ 0″	

To find arc 2.

Arc 1	37°	58′	0″	.... cosec.....	0·210982	
pol. dist. at G. B. ...	97	5	52	.... cosec.....	0·003339	
diff.	59	7	52			
pol. dist. at L. B. ...	97	8	32			
sum 156	16	24		.... ½ hav. ...	4·990618	
diff.	38	0	40	.... ½ hav. ...	4·512779	
				Hav. arc 2	9·717718	
				arc 2 ....92° 31′ 45″		

To find arc 3.

Arc 1	37°	58′	0″	.... cosec.....	0·210982
Z. D. at G. B.	80	45	59	.... cosec.....	0·005664
diff.	42	47	59		
Z. D. at L. B.	64	41	22		
sum 107	29	21		.... ½ hav. ...	4·906540
diff.	21	53	23	.... ½ hav. ...	4·278481
				Hav. arc 3	9·401667
				arc 3 ....60° 17′ 0″	

To find arc 4.

Arc 2	92°	31′	45″
arc 3	60	17	0
arc 4	32	14	45
Pol. dist. at G. B. ...	97°	5′	52″
zen. dist. at G. B. ...	80	45	59
diff.	16	19	53

To find arc 5.

Sin. pol. dist. at G. B.	9·996661
sin. zen. dist. at G. B.	9·994336
hav. arc 4	8·887148
hav. arc 5	8·878145
arc 5 ....31° 54′ 15″	

Vers.	31°	54′	15″	0151066
vers.	16	19	53	0040347
vers. colat.				0191413
∴ colat.	36°	2′	31″	
	90			
latitude..........................	53	57	29 N.	

368. June 3, 1847, in latitude by account 52° N., and long. 72° E., the following double altitude of the sun was observed :

Mean time nearly.	Chro.	Alt. sun's L. L.	Bearing.
9ʰ 50ᵐ A.M.	9ᵘ 52ᵐ 28ˢ	51° 17' 45"	S.E.b.S.
11 15 A.M.	11 14 29	59 32 15	S.b.E.

The run of the ship in the interval was W. by S. 10 miles, index correction —0' 40", and height of eye 12 feet : required the true latitude at the second observation. *Ans.* Arc 1, 18° 57' 45"; arc 2, 86° 3' 30";
arc 3, 52° 9' 15"; lat. 50° 48' N.

369. April 11, 1847, in latitude by account 56° 20' N., long. 10° 30' E., the following double altitude of the sun was observed :

Mean time nearly.	Chro.	Alt. sun's L. L.	Bearing.
11ʰ 0ᵐ A.M.	7ʰ 5ᵐ 10ˢ	40° 10' 15"	S.S.E.
2 0 P.M.	10 6 10	35 15 40	S.W.b.W.

The run of the ship in the interval was N.N.E. 29 miles, index correction +2' 10", and height of eye 18 feet : required the true latitude at the second observation. *Ans.* Arc 1, 44° 45' 45"; arc 2, 86° 38' 0";
arc 3, 66° 12' 30"; lat. 56° 55' N.

370. April 13, 1847, in latitude by account 41° 20' N., long. 156° 15' E., the following double altitude of the sun was observed :

Mean time nearly.	Chro.	Alt. sun's L. L.	Bearing.
10ʰ 45ᵐ A.M.	7ʰ 30ᵐ 20ˢ	53° 0' 20"	S.E.b.S.
2 45 P.M.	11 29 40	40 59 10	S.W.b.W.

The run of the ship in the interval was S.S.E. 25 miles, index correction —5' 20", and height of eye 14 feet : required the true latitude at second observation. *Ans.* Arc 1, 59° 3' 45"; arc 2, 85° 2' 30";
arc 3, 43° 51' 30"; lat. 41° 23' 15" N.

371. April 22, 1847, in latitude by account 50° 48' N., and longitude 148° 30' E., the following double altitude of the sun was observed :

Mean time nearly.	Chro.	Alt. sun's L. L.	Bearing.
10ʰ 0ᵐ A.M.	10ʰ 2ᵐ 25ˢ	44° 20' 0"	S.E.b.S.
11 24 A.M.	11 24 34	50 20 0	S.b.E.

The run of the ship in the interval was 0, index correction +40", and height of eye 0 : required the true latitude at second observation.
Ans. Arc 1, 20° 5' 30"; arc 2, 87° 48' 0";
arc 3, 62° 23' 30"; lat. 50° 44' 30" N.

372. Oct. 15, 1848, in latitude by account 53° N., and long. 54° E., the following double altitude of the sun was observed :

Mean time nearly.	Chro.	Alt. sun's L. L.	Bearing.
11ʰ 20ᵐ A.M.	11ʰ 15ᵐ 50ˢ	27° 31′ 50″	S.b.E.
1 20 P.M.	0 50 32	25 45 5	S.S.W.

The run of the ship in the interval was S. by W. 14 miles, index correction +2′ 55″, and height of eye above the sea 15 feet: required the true latitude at second observation.

Ans. Arc 1, 23° 24′ 15″; arc 2, 91° 43′ 15″;
arc 3, 79° 10′ 30″; lat. 53° 17′ 45″ N.

373. Oct. 24, 1849, in latitude by account 50° 40′ S., and long. 142° W., the following double altitude of the sun was observed :

Mean time nearly.	Chro.	Alt. sun's L. L.	Bearing.
10ʰ 0ᵐ A.M.	10ʰ 12ᵐ 34ˢ	44° 20′ 0″	S.E.b.S.
11 24 A.M.	11 34 34	50 20 0	S.b.E.

The run of the ship in the interval was 0, index correction 0, and height of eye above the sea 0 : required the true latitude at second observation.

Ans. Arc 1, 20° 3′ 0″; arc 2, 87° 48′ 0″;
arc 3, 62° 23′ 0″; lat. 50° 45′ 30″ S.

Elements from Nautical Almanac.

Sun's declination.								Sun's semi.	
June 2, 22°	8′	50″N.	......	June 3, 22°	16′	34″ N.	...	15′	47″
April 10, 7	49	12 N.	......	April 11, 8	11	21 N.	...	15	58
April 12, 8	33	22 N.	......	April 13, 8	55	14 N.	...	15	57
April 21, 11	44	26 N.	......	April 22, 12	4	46 N.	...	15	56
Oct. 14, 8	18	13 S.	......	Oct. 15, 8	40	28 S.	...	16	4
Oct. 24, 11	49	32 S.	......	Oct. 25, 12	10	19 S.	...	16	7

Rule 38. DOUBLE ALTITUDE. First method. (2.) *The* LATITUDE *by altitudes of* TWO STARS *taken at the* SAME INSTANT.

1. Correct the observed altitudes for index correction, dip, and refraction, and thus find the true altitudes, which subtract from 90° for the true zenith distances.

2. Take out of the *Nautical Almanac* the right ascension and declination of the two stars, and get their polar distances as in (3) p. 144.

3. *To find the polar angle.* The difference between the right ascensions of the two stars is the polar angle.

4. *To find arc* 1 (using Inman's Tables). Put down the two polar distances under each other, and take their difference. Add together the log. sin. of the polar distance at greater bearing, the log. sin. of polar distance at

less bearing, and the log. haversine of polar angle ; the result, rejecting the tens in the index, is the log. haversine of an arc, which take from the tables and call arc A.

Add together versine of arc A and versine of the difference of polar distances ; the sum will be the versine of arc 1, which find in the tables. Then proceed to find arc 2, &c., as in Rule 37, p. 145.

EXAMPLE.

374. January 1, 1846, in latitude by account 38° 10′ N., the following altitudes of the stars α Pegasi and α Aquilæ were taken at the same instant :

Obs. alt. α Pegasi.	Bearing.	Obs. alt. α Aquilæ.	Bearing.
29° 49′ 27″	E.b.S.	57° 29′ 50″	S.S.E.
In. cor. —15″		In. cor. —15″	

The height of the eye was 41 feet : required the latitude.

<table>
<tr><td colspan="4" align="center">At greater bearing.
α Pegasi.</td><td colspan="4" align="center">At less bearing.
α Aquilæ.</td></tr>
<tr><td>Observed alt. ...</td><td>29°</td><td>49′</td><td>27″</td><td>Observed alt. ...</td><td>57°</td><td>29′</td><td>50″</td></tr>
<tr><td>index correction .</td><td></td><td></td><td>15—</td><td>index correction .</td><td></td><td></td><td>15—</td></tr>
<tr><td></td><td>29</td><td>49</td><td>12</td><td></td><td>57</td><td>29</td><td>35</td></tr>
<tr><td>dip ....</td><td></td><td>6</td><td>18</td><td>dip ....</td><td></td><td>6</td><td>18—</td></tr>
<tr><td></td><td>29</td><td>42</td><td>54</td><td></td><td>57</td><td>23</td><td>17</td></tr>
<tr><td>refraction ....</td><td></td><td>1</td><td>42—</td><td>refraction.........</td><td></td><td>0</td><td>37—</td></tr>
<tr><td>true alt. ....</td><td>29</td><td>41</td><td>12</td><td>true alt. ....</td><td>57</td><td>22</td><td>40</td></tr>
<tr><td></td><td>90</td><td></td><td></td><td></td><td>90</td><td></td><td></td></tr>
<tr><td>zenith distance ...</td><td>60</td><td>18</td><td>48</td><td>zenith distance ..</td><td>32</td><td>37</td><td>20</td></tr>
<tr><td>Star's declination</td><td>14°</td><td>22′</td><td>50″ N.</td><td>Star's declination</td><td>8°</td><td>28′</td><td>2″ N.</td></tr>
<tr><td></td><td>90</td><td></td><td></td><td></td><td>90</td><td></td><td></td></tr>
<tr><td>P. D. at G. B. ...</td><td>75</td><td>37</td><td>10</td><td>pol. dist. at L. B.</td><td>81</td><td>31</td><td>58</td></tr>
<tr><td>R. A. α Pegasi ...</td><td>22ʰ</td><td>57ᵐ</td><td>6ˢ</td><td>pol. dist. at G. B.</td><td>75°</td><td>37′</td><td>10″</td></tr>
<tr><td>R. A. α Aquilæ...</td><td>19</td><td>43</td><td>15</td><td>pol. dist. at L. B.</td><td>81</td><td>31</td><td>58</td></tr>
<tr><td>polar angle ......</td><td>3</td><td>13</td><td>51</td><td>diff. pol. dists. ...</td><td>5</td><td>54</td><td>48</td></tr>
</table>

To find arc 1.*

Sin. polar distance at greater bearing	9·986177
sin. polar distance at lesser bearing............	9·995241
haversine polar angle	9·226458
haversine arc A	9·207876
arc A ....47° 22′ 30″

* Arcs 1 and 2 may be readily found by using Shadwell's Tables, from which these two arcs may be taken out by inspection, and thus materially shortening this method of finding the latitude.

vers. arc A 0322696
                                                              107
vers. difference polar distances 5319

vers. arc 1 0328122
arc 1 ....47° 47′ 16″

To find arc 2.*

Arc 1	47°	47′	16″	.... cosec.	0·130382
P. D. at G. B.	75	37	10	.... cosec.	0·013823
difference...........	27	49	54	½ hav. S. ...	4·911662
P. D. at L. B.......	81	31	58	½ hav. D....	4·654808
sum (S)	109	21	52	Haversine, arc 2	9·710675
difference (D)	53	42	4	∴ arc 2	91° 34′ 0″

To find arc 3.

Arc 1	47°	47′	16″	.... cosec.	0·130382
zen. dist. at G. B. ..	60	18	48	.. .. cosec.	0·061110
difference...........	12	31	32	½ hav. S. ...	4·584171
zen. dist. at L. B. ..	32	37	20	½ hav. D. ...	4·241725
sum (S)	45	8	52	Haversine, arc 3	9·017388
difference (D)	20	5	48	∴ arc 3.................37°	38′ 30″

To find arc 4.   To find arc 5.

Arc 2	91°	34′	0″
arc 3	37	38	30
∴ arc 4	53	55	30

Sin. pol. dist. at G. B.   9·986177
sin. Z. D. at G. B. ....   9·938890
haversine, arc 4 ......   9·312977

haversine, arc...........   9·238044
arc....................... 49° 9′ 15″
vers. arc   0345919
                                   55
... vers. difference.........   0035443
                                   28

vers. colat.   0381445
                                   363

P. D. at G. B. .... 75° 37′ 10″
Z. D. at G. B. 60 18 48
difference............ 15 18 22

                                                         82
colat. ...... 51° 47′ 22″
              90

∴ latitude ... 38 12 38 N.

* See note on previous page.

375. Sept. 17, 1844, in latitude by account 36° 45′ N., the following altitudes were observed at the same time:

Obs. alt. α Orionis.	Bearing.	Obs. alt. α Leonis.	Bearing.
55° 1′ 30″	S.E.b.E.	45° 13′ 30″	S.S.W.

The index correction was +55″, and height of eye 8 feet: required the true latitude.　　　　　　　　*Ans.* Arc 1, 62° 27′ 10″; arc 2, 88° 14′·45″; arc 3, 38° 14′ 0″; lat. 36° 44′ N.

376. Feb. 20, 1846, in latitude by account 36° 40′ N., the following altitudes were observed at the same time:

Obs. alt. Sirius.	Bearing.	Obs. alt. Spica.	Bearing.
27° 50′	S.W.	12° 56′	E.S.E.

The index correction was +1′, and height of eye above the sea 10 feet: required the true latitude.
　　　　　　　　Ans. Arc 1, 96° 10′ 30″; arc 2, 108° 5′ 15″; arc 3, 59° 39′ 15″; lat. 36° 36′ 45″ N.

377. May 1, 1845, in latitude by account 41° 20′ N., the following altitudes of stars were taken at the same instant: required the true latitude.

True alt. α Pegasi.	Bearing.	True alt. α Tauri.	Bearing.
62° 44′	S.b.E.	19° 26′ 20″	E.

　　　　　　　　Ans. Arc 1, 79° 0′ 15″; arc 2, 78° 3′ 30″; arc 3, 26° 55′ 0″; lat. 41° 22′ 30″ N.

378. March 2, 1845, in latitude by account 41° 20′ N., long. 60° E., the altitudes of the two following stars were observed at the same time: required the true latitude.

True alt. α Andromedæ.	Bearing.	True alt. α Tauri.	Bearing.
73° 14′	S.b.E.	18° 27′ 30″	E.

　　　　　　　　Ans. Arc 1, 62° 9′ 30″; arc 2, 66° 11′ 0″; arc 3, 15° 8′ 45″; lat. 41° 23′ N.

379. January 2, 1847, in latitude by account 32° 10′ N., the following altitudes of the stars α Pegasi and α Aquilæ were observed at the same instant:

Obs. alt. α Pegasi.	Bearing.	Obs. alt. α Aquilæ.	Bearing.
22° 49′ 27″	E.b.S.	57° 29′ 50″	S.S.E.

The index correction −15″, and height of eye above the sea 41 feet: required the true latitude.
　　　　　　　　Ans. Arc 1, 47° 46′ 45″; arc 2, 91° 34′ 0″; arc 3, 31° 24′ 30″; lat. 32° 43′ N.

380. Dec. 27, 1847, the following altitudes were observed at the same instant, in latitude by account 37° 10′ N. : required the true latitude.

True alt. β Orionis.	Bearing.	True alt. α Hydræ.	Bearing.
31° 5′ 11″	S.W.b.W.	39° 47′ 33″	S.E.½S.

Ans. Arc 1, 62° 30′ 0″ ; arc 2, 94° 42′ 0″ ;
arc 3, 58° 5′ 0″ ; lat. 37° 13′ N.

Elements from Nautical Almanac.

Star's right ascension.				Star's declination.			
α Leonis	10ʰ	0ᵐ	5·8ˢ	....	12°	43′	24″ N.
α Orionis	5	46	47·9	....	7	22	23 N.
Spica.........	13	17	7·4	....	10	21	30 S.
Sirius	6	38	23·8	....	16	30	52 S.
α Tauri.........	4	27	3·0	....	16	11	30 N.
α Pegasi	22	57	4·0	....	14	22	24 N.
α Andromedæ.	0	0	23·4	....	28	14	11 N.
α Pegasi	22	57	8·5	....	14	23	8 N.
α Aquilæ	19	43	18·2	....	8	28	12 N.
β Orionis	5	7	15·5	....	8	23	5 S.
α Hydræ	9	20	8·2	....	8	0	15 S.

When two heavenly bodies are observed at different times, the polar angle is to be found by the following rule :

Rule 39. 1. Subtract the time shown by the chronometer at the first observation (increased if necessary by 12 hours) from the time shown at the second observation, and thus find the elapsed time.

2. * Correct the elapsed time for rate of chronometer, if any, either by proportional logs. or by the common rule of proportion.

3. Add to the elapsed time so corrected the acceleration of sidereal on mean solar time (taken from table in *Nautical Almanac* or elsewhere). The result is the elapsed time expressed in sidereal time.

4. Add this elapsed time to the right ascension of the heavenly body first observed, and take the difference between the sum and the right ascension of the second heavenly body ; the remainder (subtracted from 24 hours if greater than 12 hours) will be the polar angle required.

EXAMPLES.

381. The altitude of α Pegasi was observed when the chronometer showed 6ʰ 42ᵐ 10ˢ, and the altitude of α Aquilæ was observed when the chronometer showed 8ʰ 32ᵐ 5ˢ : required the polar angle between the two places observed, the rate of the chronometer being 12·5ˢ gaining.

* When great accuracy is not required, and the elapsed time is small, these two corrections in 2 and 3 for rate of chronometer and acceleration may be omitted.

Times by chronometer.

At second observation	8ʰ	32ᵐ	5ˢ
at first observation	6	42	10
	1	49	55

Gr. date log. sun for 1ʰ 49ᵐ .... 1·11697
prop. log. for 12·5ˢ .... 2·93651

		4·05348			1—
			1	49	54

1ʰ	....	9·86ˢ	
49ᵐ	......	8·05	
54ˢ	......	·15	
		18·06	18+

elapsed time in sidereal time	1	50	12
right ascension α Pegasi	22	57	14
	24	47	26
α Aquilæ	19	43	25
polar angle required	5	4	1

382. The altitude of Sirius was observed when the chronometer showed 2ʰ 10ᵐ 20ˢ, and the altitude of Spica was observed when the chronometer showed 3ʰ 20ᵐ 15ˢ: required the polar angle between the two places observed, the rate of chronometer being 2·5ˢ losing.

Times by chronometer.

At second observation	3ʰ	20ᵐ	15ˢ
at first observation	2	10	20
	1	9	55
rate of chronometer			0
	1	9	55

acceleration	1ʰ	....	9·8ˢ
	9ᵐ	....	1·5
	55ˢ	....	·1
		11·4	11·4+

	1	10	6·4
right ascension Sirius	6	38	25·4
	7	48	31·8
Spica	13	17	10·9
polar angle	5	28	39·1

383. The altitude of β Orionis was observed when the chronometer showed 6ʰ 10ᵐ 25ˢ, and the altitude of α Hydræ was observed when the chronometer showed 7ʰ 17ᵐ 35ˢ: required the polar angle between the two places observed, the rate of chronometer being 6·3ˢ losing, and the right ascension of β Orionis 5ʰ 7ᵐ 15ˢ, and of α Hydræ 9ʰ 20ᵐ 8·2ˢ.

Ans. 3ʰ 5ᵐ 32ˢ.

Rule 40. DOUBLE ALTITUDE. First method. (3.) *The* LATITUDE *by altitudes of two heavenly bodies observed at* DIFFERENT *times.*

1. Proceed as in (1) and (2), p. 149, to get the zenith distances, the right ascension, and polar distances.

2. Find the polar angle, as in Rule 39, p. 153.

3. Find arc 1, as in (4), p. 149.

4. Then proceed to find arcs (2), (3), (4), &c., as in Rule 37, p. 144.

EXAMPLE.

384. Sept. 27, 1846, in latitude by account 43° 30′ N., the following altitudes of the stars α Pegasi and α Aquilæ were observed at different times :

	Observed altitude.	Time by chron.	Bearing.
α Pegasi	29° 49′ 30″	7ʰ 35ᵐ 10ˢ	S.E.
α Aquilæ	54 29 0	8 2 10	S.¼W.

The run of the ship in the interval was S. 10 miles, the index correction +1′ 10″, and height of eye above the sea 20 feet : required the true latitude at the second observation.

At greater bearing. α Pegasi.			At less bearing. α Aquilæ.		
Observed alt.	29° 49′	30″	Observed alt.	54° 29′	0″
	1	10+		1	10+
	29 50	40		54 30	10
	4	24−		4	24−
	29 46	16		54 25	46
	1	41−		0	41−
	29 44	35		54 25	5
	7	6+	zenith distance	35 34	55
	29 51	41			
zenith distance	60 8	19			

Right asc. α Pegasi, 22ʰ 57ᵐ 9ˢ.

Declination, 14° 23′ 9″ N.

Right asc. α Aquilæ, 19ʰ 43ᵐ 19ˢ.

Declination, 8° 28′ 19″ N.

To find the polar angle.

Chronom. at first observation	7ʰ	35ᵐ	10ˢ
,, second ,,	8	2	10
elapsed time	0	27	0
right ascension α Pegasi	22	57	9
	23	24	9
right ascension α Aquilæ	19	43	19
polar angle	3	40	50

To find arc 1.

Pol. dist. at G. B.	75°	36'	51"
pol. dist. at L. B.	81	31	41
diff. pol. distances	5	54	50
sin. pol. dist. at G. B.		9·986161	
sin. pol. dist. at L. B.		9·995236	
hav. pol. angle		9·331838	
hav. arc A		9·313235	
arc A	53°	56'	30"
vers. arc A		0411274	
		117	
vers. diff. pol. dists....		0005297	
		23	
vers. arc 1		0416711	
		695	
arc 1 ...... 54° 19' 4"		16	

To find arc 2.

Arc 1	54°	19'	4"
pol. dist. at G. B.	75	36	51
diff.	21	17	47
pol. dist. at L. B.	81	31	41
sum	102	49	28
diff.	60	13	54
cosec. arc 1		090309	
cosec. pol. dis. at G. B.		013839	
½ hav. sum		4·893016	
½ hav. diff.		4·700498	
hav. arc 2		9·697662	
arc 2	89°	49'	45"

To find arc 3.

Arc 1	54°	19'	4"
zenith dist. at G. B. ...	60	8	19
diff.	5	49	15
zenith dist. at L. B. ...	35	34	55
sum	41	24	10
diff.	29	45	40
cosec. arc. 1		·090309	
cosec. zen. dis. at G. B.		·061869	
½ hav. sum		4·548400	
½ hav. diff.		4·409623	
hav. arc 3		9·110201	
arc 3	42°	4'	45"
pol. dist. at G. B.	75	36	51
zen. dist. at G. B.	60	8	19
difference	15	28	32

To find arc 4.

Arc 2	89°	49'	45"
arc 3	42	4	45
arc 4....................	47	45	0

To find the latitude.

sin. pol. dis. at G. B....		9·986161	
sin. zen. dis. ,,		9·938131	
hav. arc 4		9·214358	
hav. arc 5		9·138650	
arc 5.....................	43°	33'	0"
vers. arc 5		0275227	
vers. diff. pol. dist. and } zen. dist. .... }		36214	
		41	
		0311482	
		34	
		48	
colat.	46°	29'	14"
	90		
latitude............	43	30	46 N.

Rule 41. DOUBLE ALTITUDE. Second method. *The* LATITUDE *by two alti- tudes of the* SUN. (This method requires only the common tables of sines, &c.)

1. From the time shown by the chronometer or watch at the second observation (increased if necessary by 12 hours) subtract the time shown at the first observation, divide by 2 ; the result is the half polar angle in time.

2. To the estimated mean time at the ship at the first observation add

the half polar angle ; the sum will be the ship mean time at the middle time between the observations.

3. Apply the longitude in time, and thus get a Greenwich date.

4. Take out from the *Nautical Almanac* the declination from this date, and also the sun's semidiameter in the adjacent column.

5. Correct the observed altitudes for index correction, dip, semidiameter, and parallax and refraction.

6. Correct also the first true altitude for run of ship in the interval, and thus get the true altitudes for the same place.

7. Put the first true altitude under the second true altitude, take their sum and difference, and also the half-sum and half-difference; call the half-sum S, and the half-difference D.

8. Under the log. sin. half polar angle put log. cos. declination ; at the same time take out and put a little to the right the log. sin. declination.

9. Add together the two logs. first taken out, and call the sum sin. arc 1.

10. At the same opening take out sec. arc 1, and put it under the log. sin. declination ; take out also and put down in the same horizontal line the log. cosec. arc 1, and also log. sec. arc 1.

11. Add together log. sin. declination and log. sec. arc 1 ; the sum will be log. cos. arc 2 ; the arc corresponding thereto found in the Tables will be arc 2, if the latitude and declination are of the same name ; but if the latitude and declination are of different names, subtract the arc taken out from 180° : the remainder is arc 2.

12. Under log. cosec. 1, and log. sec. 1, just taken out, put the following quantities :

Under log. cosec. 1 put log. cos. S.   Under log. cosec. 1 put log. sin. D.
„   sec. 1   „   sin. S.   „   sec. 1   „   cos. D.

Add together log. cosec. 1 and the two logs. placed beneath it ; the sum will be the log. sin. arc 3.

13. Take out the log. sec. arc 3, and put it down twice ; once under log. cos. D, and again a little to the right.

14. Add together the log. sec. 1, and the three logarithms beneath it ; the result is log. cos. arc 4, which find in the Tables.

15. Under arc 4 put arc 2, and take the difference in all cases when the line drawn through the places of the sun at the two observations will when produced *not* pass through the zenith and pole (that is, the difference must be taken, if it is seen that their sum would exceed 90°), otherwise take their sum ; the result is arc 5.

Lastly. Under log. sec. arc 3, already taken out, put log. sec. arc 5 : the sum will be the log. cosec. of the required latitude.

The arrangement on the paper of the logarithms to be taken out, as directed by the rule, will be better seen in the following blank form ; and it would also facilitate the working out questions in other rules of Navigation if blank forms, similar to the one now given, were constructed on thick drawing-paper by the student for each rule.

LATITUDE BY DOUBLE ALTITUDE OF SUN. (BLANK FORM.)

Times by Chronometer.	Mean T. at 1st observation.	Sun's declination.	1st Observed altitude.	2d Observed altitude.
1st obs. }	Ship + ½ pol. A.	At	Obs. alt.	Obs. alt. ....
Sub. from }		At	Index cor.	Index cor. ......
* 2d Obs.				
Diff or pol. A...	Ship.........	Diff	Dip (—)	Dip (—) ....
½ pol. A...	Long. in T.			
	Gr. date		Semidiameter ...	Semidiameter ...
* Add 12 to hours at 2d observation if necessary.	On		Cor. in alt. (—)	Cor. in alt. (—)
Sin. ½ pol. A. } + Cos. decl.	& Sin. decl.	Sun's decl.	Run.........	(2d) tr. alt. ......
	+		1st true alt.	(1st) tr. alt. ......
Sin. (1)	Sec. (1)	& Cosec. (1).....	& Sec. (1)	Sum.........
	Cos. (2)	Cos. (S)	& Sin. (S)	Difference
	* Arc (2)	Sin. (D)	& Cos. (D)	(S) ½ Sum
	180	Sin. (3)	& Sec. (3)	(D) ½ Difference
	Arc (2).........		Cos. (4) .... } †	& Sec. (3) } +
			Arc (4)	Sec. (5)
			Arc (5)	Cosec. ....
				Lat. ......

* If latitude and declination of different names, subtract this arc from 180° for arc (2).

† If the sum of arc (4) and (2) equal or exceed 90°, then Arc (2) — or + Arc (2) for the difference is arc (5).

EXAMPLES.

385. Oct. 11, 1845, in latitude by account 54° N., and long. 83° 15' W., the following double altitude of the sun was observed :

Mean time nearly.	Chronometer.	Alt. sun's L. L.	Bearing.
7ʰ 45ᵐ A.M.	11ʰ 40ᵐ 15ˢ	9° 0' 20"	E.S.E.
10 35 A.M.	2 13 20	25 3 30	S.S.E.

The run of the ship in the interval was S. by W. 15 miles, index correction +5' 10", and height of eye above the sea 18 feet : required the latitude at the second observation. *Ans.* 53° 59' N.

386. March 20, 1845, in latitude by account 52° 10' N., and longitude 55° 15' W., the following double altitude of the sun was taken :

Mean time nearly.	Chronometer.	Alt. sun's L. L.	Bearing.
8ʰ 35ᵐ A.M.	9ʰ 36ᵐ	20° 0' 30"	S.E.b.E.
1 45 P.M.	2 49	34 5 30	S.W.b.S.

The run of the ship in the interval was N.W. by W. 10 miles, index correction 0, and height of eye 20 feet : required the latitude at the second observation. *Ans.* 52° 27' N.

387. Dec. 11, 1845, the following double altitude of the sun was observed :

Mean time nearly.	Chronometer.	Alt. sun's L. L.	Bearing.
6ʰ 0ᵐ A.M.	6ʰ 3ᵐ 30ˢ	19° 40' 25"	E.b.S.
10 0 A.M.	10 4 25	50 20 40	N.E.

The run of the ship in the interval was E.N.E. 25 miles, index correction —1' 50", and height of eye 16 feet : required the latitude at second observation, the latitude by account being 60° S., and long. 79° 15' W. *Ans.* 56° 57' S.

388. Nov. 10, 1846, in latitude by account 35° 30' N., long. 94° 30' E., the following double altitude of the sun was observed :

Mean time nearly.	Chronometer.	Alt. sun's L. L.	Bearing.
1ʰ 15ᵐ P.M.	1ʰ 45ᵐ 15'	33° 5' 40"	S.S.W.
3 45 P.M.	4 15 17	12 55 10	S.W.b.W.

The run in the interval was S.S.E. 15 miles, index correction +4' 10", and height of eye 18 feet : required the true latitude at the second observation. *Ans.* 35° 31' N.

389. Oct. 30, 1846, in latitude by account 52° 10' N., and longitude 159° 45' E., the following double altitude of the sun was observed :

Mean time nearly.	Chronometer.	Alt. sun's L. L.	Bearing.
11ʰ 15ᵐ A.M.	11ʰ 21ᵐ 15'	25° 26' 20"	S. ¾ E.
11 30 A.M.	11 37 55	25 55 0	S. ¼ E.

The run of the ship in the interval was S. by W. 1 mile, index correction + 3' 50", and height of eye above the sea 20 feet: required the true latitude at second observation.   *Ans.* 49° 56′ N.

Elements from Nautical Almanac.

	Sun's declination.							Sun's semi.	
Oct. 11 ...	7°	4′	38″S.	...... 12 ...	7°	27′	15″S.	16′	3″
Mar. 20 ...	0	5	40 S.	...... 21 ...	0	18	1 N.	16	4
Dec. 11 ...	23	1	55 S.	...... 12 ...	23	6	54 S.	16	16
Nov. 9 ...	16	51	6 S.	...... 10 ...	17	8	9 S.	16	11
Oct. 29 ...	13	26	6 S.	...... 30 ...	13	45	56 S.	16	8

In the above rule, called Ivory's method, it is assumed that the declinations of the sun at the times of observation are the same as at the middle time between the two observations; but this is not so. The latitude deduced will therefore only be approximately true, differing perhaps 2′ or 3′ from the truth. Its principal recommendation is that it requires only the common table of log. sines, and is shorter than the first or Inman's method.

INMAN'S RULE FOR DOUBLE ALTITUDE.

Adapted to the Tables of Sines, &c.

In p. 142 it is shown that the latitude by double altitude may be found by the simple application of two rules in spherical trigonometry. These are

1. Two sides and included angle given, to find the third side.

2. Three sides given, to find an angle.

These practical rules, *adapted to the tables of sines*, are given in the author's *Trigonometry*, pp. 63, 65.

Or the several parts of the problem may be calculated by the following two formulæ, *Trig.* Part II. pp. 95, 65.

$$(\alpha)\ \sin.\ \frac{a}{2}=\frac{\sin.\ \tfrac{1}{2}\ \text{A}\sqrt{\sin.\ b\ \sin.\ c}}{\sin.\ \theta}\ \text{where tan.}\ \theta=\frac{\sin.\ \tfrac{1}{2}\ \text{A}\sqrt{\sin.\ b\ \sin.\ c}}{\sin.\ \tfrac{1}{2}\ (b\sim c)}.$$

$$(\beta)\ \sin.^c\frac{\text{A}}{2}=\text{cosec.}\ b.\ \text{cosec.}\ c.\ \sin.\ \tfrac{1}{2}\ (a+\overline{b\sim c}).\sin.\ \tfrac{1}{2}\ (a-\overline{b\sim c}).$$

Arc 1, or xy (fig. p. 142), is found by (α), where *b* and *c* represent the two polar distances, and A the polar angle. Arc 2, or angle Pxy, is found by (β), where *b* and *c* represent respectively arc 1 and polar distance at greater bearing, and *a* the polar distance at lesser bearing. Arc 3, or angle zxy, is found by (β), where *b* and *c* represent respectively arc 1 and zenith distance at greater bearing, and *a* the zenith distance at lesser bearing.

Then arc 4=arc 2∓arc 3 (p. 145).

Lastly, the colatitude Pz is found by (α), where *b* and *c* represent respectively the pol. dist. and zen. dist. at greater bearing, and A the arc 4.

These formulæ may be applied to all the cases of the double altitude; and it will be found, after a little practice, that the trigonometrical method is much easier than using the formal rules.

THERE are two methods of determining the *error of a chronometer on mean time :* the one by a single altitude of a heavenly body observed at some distance from the meridian; the other by means of equal altitudes of a heavenly body observed on both sides of the meridian.

Before going to sea, the *error* of the chronometer on Greenwich mean time at mean noon, and its *daily rate,* are supposed to have been accurately determined, either at an observatory by means of daily comparisons with an astronomical clock, or by observations taken with a sextant at a place whose longitude is known.

The *mean daily rate* of a chronometer can then be found by dividing the increase or decrease in its error by the number of days elapsed between the times when the observations were taken to determine its error; thus, suppose on April 27, at 9^h 30^m A.M., the error of a chronometer was found to be fast 10^m $10 \cdot 5^s$ on Greenwich mean time, and that on April 30th about the same hour its error was found to be 10^m $40 \cdot 5^s$ fast : then it appears that in the three days elapsed between the observations the chronometer has gained 30^s, hence its mean daily rate is 10^s gaining.

When the error and rate of a chronometer are given, we may determine what its error will be at some future time, provided the rate of the chronometer continues uniform in the interval, by the following rule.

Rule 42. Given the ERROR OF A CHRONOMETER *on Greenwich mean time at some given day and hour, and also its* MEAN DAILY RATE, *to find Greenwich mean time at some other instant.*

1. Get a Greenwich date.

2. Find the number of days and part of a day that have elapsed from the time when the error and rate were determined by the hour of the Greenwich date.

3. Multiply the rate of the chronometer by the number of days elapsed, and add thereto the proportionate part for the fraction of a day, found by

M

proportion or otherwise. The result is the accumulated error in the interval.

4. If the chronometer is gaining, subtract the accumulated error from the time shown by the chronometer; if losing, add.

5. To the result apply the original error of chronometer, adding if slow, subtracting if fast (increasing the time shown by chronometer by 24^h if necessary, and putting the day one back). The result (rejecting 24^h if greater than 24^h, and putting the day one forward) will be mean time at Greenwich at the instant of the observation.

6. If this time differs from the Greenwich date by 12 hours nearly, in that case 12 hours must be added to the Greenwich time, determined as above, and the day put one back, to get the astronomical Greenwich mean time.

390. June 13, 1851, at 10^h 52^m P.M. mean time nearly, in long. 60° W., an observation was taken when a chronometer showed 2^h 50^m 42'. On June 1, at noon, its error was known to be 3^m 10·2' fast on Greenwich mean time, and its mean daily rate was 3·5' gaining : required mean time at Greenwich when the observation was taken.

```
Ship, June 13  10ʰ 52ᵐ        Daily rate .   3·5ˢ
long. in time.  4   0 W.   Interval from June 1 to    12
                           June 13, at 14ʰ 52ᵐ is
Gr. June 13 .  14  52      12ᵈ 14ʰ 52ᵐ=12ᵈ 15ʰ   12ʰ is ½ | 42·0
                           nearly                  3 „ ¼ | 1·75
                                                           ·44
                        Accumulated rate . . .      44·2 gaining
                        chronometer showed . .  2ʰ 50ᵐ 42·0ˢ
                                                 2  49  57·8
                        original error  . . . .    3  10·2 fast
                        Greenwich, June 14 . .  2  46  47·6 A.M.
                        .·. Greenwich, June 13 .14  46  47·6
```

(12 hours are added and the day put back one, to make the time thus found agree more nearly with the Greenwich date.)

391. August 10, 1853, at 3^h 42^m A.M. mean time nearly, in long. 100° 30' W., an observation was taken when a chronometer showed 10^h 30^m 45·5'.

On August 1, its error was known to be 12^m 10·5' slow on Greenwich mean time, and its rate 11·2' gaining : required mean time at Greenwich when the observation was taken.

Ship, Aug. 9 . 15ʰ 42ᵐ
long. in time . 6 42 W.

Gr. Aug. 9 . 22 24

Interval from
Aug. 1 to Aug. 9,
at 22ʰ 24ᵐ is
8ᵈ 22ʰ 24ᵐ

Daily rate . . . 11·2ˢ
8

12ʰ is ½   89·6
8 ,, ¼   5·6
2 ,, ¼   3·7
24ᵐ,, ¼ nearly   ·9
··2

100·0

Accumulated error . .   1ᵐ 40·0 gaining
chronometer showed .   10ʰ 30 45·5

10 29 5·5
original error . . . .   12 10·5 slow

Greenwich, August 10 . 10 41 16·0 A.M.

.·. Greenwich, August 9 22 41 16·0
(adding 12 hours, as directed by rule.)

NOTE. If the Greenwich time thus determined differs considerably from the Greenwich date used, the work should be repeated, using for the Greenwich date the approximate Greenwich time first found.

EXAMPLES.

392. Nov. 20, 1851, at 6ʰ 42ᵐ P.M. mean time nearly, in long. 32° 0′ E., an observation was taken when a chronometer showed 4ʰ 30ᵐ 0ˢ.

On Oct. 9 its error was known to be 5ᵐ 52·4ˢ slow on Greenwich mean time, and its rate 2·7ˢ losing: required mean time at Greenwich when the observation was taken.   *Ans.* 4ʰ 37ᵐ 52·3ˢ.

393. Dec. 31, 1851, at 10ʰ 10ᵐ A.M. mean time nearly, in long. 150° E., an observation was taken when a chronometer showed 0ʰ 0ᵐ 22·3ˢ.

On Nov. 20 its error was known to be 3ᵐ 52·4ˢ slow on Greenwich mean time, and its rate 2·7ˢ losing: required mean time at Greenwich when the observation was taken.   *Ans.* 12ʰ 6ᵐ 4·0ˢ.

394. April 11, 1851, at 3ʰ 14ᵐ P.M. mean time nearly, in long. 56° 42′ W., an observation was taken, when a chronometer showed 7ʰ 2ᵐ 10·5ˢ.

On March 15 its error was known to be 1ᵐ 32·7ˢ fast on Greenwich mean time, and its daily rate 6·3ˢ losing: required mean time at Greenwich when the observation was taken.   *Ans.* 7ʰ 3ᵐ 29·7ˢ.

THE HOUR-ANGLE AND APPARENT TIME.

Apparent time, and thence mean time, may be found by observing the altitude of a heavenly body, and calculating its corresponding hour-angle. If the time is noted by a chronometer when the observation is taken, the error of the chronometer will be found on mean time at the place by the following rules.

Rule 43. *To find the* ERROR OF A CHRONOMETER *on mean time at the ship by a single altitude of a heavenly body.*

First. *By sun's altitude.*

1. Get a Greenwich date.

2. Correct the sun's declination and equation of time for this date. Take out of the *Nautical Almanac* the sun's semidiameter, at the same time the declination and equation of time are taken out.

3. Correct the observed altitude for index correction, dip, semidiameter, and correction in altitude, and thus get the true altitude; subtract the true altitude from 90° to obtain the zenith distance.

NOTE. The HOUR-ANGLE, and thence apparent time, may be computed either by using the table of log. haversines (by far the most simple method), or by the common table of sines, &c.

4. *To find* SHIP APPARENT TIME. First method, using log. haversines, &c.

Under the latitude put the sun's declination, and, if the names be alike, take the difference; but if unlike, take their sum. Under the result put the zenith distance, and find their sum and difference.

Add together the log. secants of the two first terms in this form, and the halves of the log. haversines of the two last; and (rejecting the tens in the index) look out the sum as a log. haversine, to be taken out at the top of the page if the sun is west of the meridian, but at the bottom of the page if the sun is east of the meridian. The result is apparent solar time at the instant of observation.

To find SHIP APPARENT TIME. Second method, using log. sines, &c.

Under the latitude put the sun's declination, and, if the names be alike, take the difference; but if unlike, take their sum. Under the result put the zenith distance, and find their sum and difference, and half-sum and half-difference.

Add together the log. secants of the two first terms in this form (rejecting the tens in index) and the log. sines of the two last, and divide the sum by 2; look out the result as a log. sine, and multiply the angle taken out by 2.

Reduce the angle thus found into time, and if the sun is west of meridian, the same will be apparent time; but if east of meridian, subtract the angle from 24 hours; the remainder will then be apparent solar time at the instant of observation.

5. *To find* SHIP MEAN TIME. To apparent solar time apply the equation

of time with its proper sign, as directed in the *Nautical Almanac;* the result is mean time at the place.

6. *To find error of chronometer on* SHIP MEAN TIME. The difference between mean time thus found and the time shown by chronometer at the observation will be the error of the chronometer on mean time at the place.

Rule 44. *To find the error of a chronometer on* MEAN TIME AT GREENWICH *by a single altitude of the* SUN.

Find mean time at the place of observation as directed in preceding Rule. See 1, 2, 3, 4, and 5.

6. To mean time at the place thus found apply long. in time, adding if west, and subtracting if east (rejecting or adding 24 hours if necessary): the result will be mean time at Greenwich at time of observation.

7. The difference between which and the time shown by chronometer will be the error of chronometer on Greenwich mean time: fast or slow, according as chronometer time is fast or slow of Greenwich time.

EXAMPLE.

395. May 10, 1842, at $8^h 44^m$ A.M. mean time nearly, in latitude $50° 48'$ N., and long. $1° 6'$ W., when a chronometer showed $8^h 26^m 59\cdot7^s$, the observed altitude of the sun's lower limb was $39° 14' 30''$, index correction $+4' 24''$, and height of eye above the sea 20 feet: required the error of the chronometer on mean time at the ship, and its error on Greenwich mean time.

Ship, May 9 .	. .	$20^h 44^m$
Long. in time . .	.	4 W.
Greenwich, May 9 .	.	20 48

Sun's semi. 15' 51"

Sun's declination.		
9th . .	17° 19' 24" N.	
10th . .	17 35 18 N.	
	15 54	
·06215	. .	
1·05388		
1·11603	13 46	
Declin. .	17 33 10 N.	

Equation of time.		
9th . .	3ᵐ 46·2ˢ sub.	
10th . .	3 49·1	
	2·9	
·06215		
3·57103		
3·63318	2·5	
	3 48·7	

Obs. alt. . .	39° 14' 30"	
in. cor. . .	4 24+	
	39 18 54	
dip . . .	4 24−	
	39 14 30	
semi. . . .	15 51+	
	39 30 21	
cor. in alt. .	1 3−	
	39 29 18	
	90	
zen. dist. .	50 30 42	

First Method.			Second Method.		
Apparent time by haversines.			Apparent time by sines, &c.		
Lat.	. 50°48′ 0″ N.	. sec. lat. . 0·199263	Lat. .	. 50° 48′ 0″ N.	. sec. . . 0·199263
decl.	. 17 33 10 N.	. sec. decl. 0·020710	decl. .	. 17 33 10 N.	. sec. . . 0·020710
diff.	. 33 14 50	½ hav. S.. 4·824491		33 14 50	sin. S₁, 9·824491
zen. dis.	50 30 42	½ hav. D. 4·176307	zen. dis.	50 30 42	sin. D₁ 9·176300
sum	. 83 45 32 (S)	hav. . . 9·220771	sum .	. 83 45 32	2) 19·220764
diff.	. 17 15 52 (D)		diff. .	. 17 15 52	9·610382

.˙. apparent time . . . . 20ʰ 47ᵐ 30ˢ   ½ sum . 41 52 46 (S₁)   24° 3′ 45″
equation of time . . . .   3 48·7−   or ½ diff. 8 37 36 (D₁)   1ʰ 36ᵐ 15ˢ
                                                                                    2
.˙. ship mean time May 9 . 20 43 41·3   3 12 30
chro. showed (adding 12ʰ) 20 26 59·7   24
    .˙. error on ship M. T.   16 41·6 slow   .˙. ship ap. T. 20 47 30
                                                              as before.

To find error of chronometer on Greenwich mean time.

Ship mean time, May 9	20ʰ	43ᵐ	41·3ˢ
longitude in time..........................		4	24·0 W.
.˙. Greenwich mean time, May 9........	20	48	5·3
chronometer showed (adding 12ʰ)	20	26	59·7
.˙. error on Greenwich mean time		21	5·6 slow.

If the computed ship mean time differ several minutes from the estimated ship mean time, it will be advisable, when great accuracy is required, to recalculate the sun's declination and the hour-angle, using the approximate ship time just found to determine the Greenwich date; the following example will show the mode of proceeding:

396. March 16, 1844, at 10ʰ 10ᵐ A.M., mean time nearly, in lat. 50° 48′ N., and long. 1° 6′ W., when a chronometer showed 10ʰ 15ᵐ 47·2ˢ, the observed altitude of the sun's lower limb was 58° 46′ 30″ (in artificial horizon), the index correction +1′ 20″; required the error of chronometer on Greenwich mean time.

			Sun's declination.			Equation of time.		
March 15 . .	22ʰ	10ᵐ	15th . .	1° 58′ 28″ S.		15th . . .	0ᵐ 1·7ˢ add	
long. in time .		4 W.	16th . .	1 34 45 S.		16th . . .	8 44·4	
Gr. March 15 .	22	14		23 43			17·3	
			·03321			·03321		
			·88022			2·79538		
Sun's semi.			———			———		
16′ 5′			·91343	21 58		2·82859	16·0	
			Declination 1 36 30 S.				8 45·7	

Obs. alt.	.	58° 46' 30"	Lat. .	. 50° 48' 0" N.	. . .	sec.	0·199263
in. cor. .	.	1 20+	decl.	. 1 36 30 S.	. . .	sec.	0·000171

2) 58 47 50	sum . . 52 24 30	¼ hav. (S). 4·920520
	zen. dist. 60 21 34	¼ hav. (D). 3·840866
29 23 55		
semi. . . 16 5	sum . . 112 46 4 (S)	hav. . . 8·960820
	diff. . . 7 57 4 (D)	
29 40 0	Apparent time 21ʰ 39ᵐ 14ˢ	
cor. in alt. . 1 34—	equation of time . . . . 8 45·7+	
29 38 26	mean time . . . . . . 21 47 59·7	
90	long. in time . . . . . 4 24·0 W.	
zen. dist. . 60 21 34	Greenwich mean time . . 21 52 23·7	
	chronometer showed . . 22 15 47·2	
	error of chronometer on } 23 23·5 Greenwich mean time . }	

The mean time at the place is found to be 21ʰ 47ᵐ 59·7ˢ, but the mean time used for computing the declination and equation of time was 22ʰ 10ᵐ. Now this has rendered the declination slightly incorrect, and therefore the mean time computed from it. When it is desirable to obtain mean time at the place as correctly as possible, we must recalculate the declination and apparent time, using the approximate mean time for finding a more correct Greenwich date; thus the mean time at the place is found above to be 21ʰ 47ᵐ 59·7ˢ; let us therefore assume the mean time to be 21ʰ 48ᵐ, and obtain in this manner a second Greenwich date, then recompute the sun's declination and hour-angle for this more correct Greenwich date as follows:

Mar. 15, M. T. 21ʰ 48ᵐ	Decl. . . 1° 36' 52" S. . . sec. . . . 0·000171	
long. in time . 4 W.	lat. . . . 50 48 0 N. . . sec. . . . 0·199263	
Gr. Mar. 15 . 21 52	52 24 52	
	zen. dist. . 60 21 34	
Sun's declination.		
15th. . . 1° 58' 28" S.	sum . . 112 46 26 (S) . . ¼ hav. (S) 4·020540	
16th. . . 1 34 45 S.	diff. . . 7 56 42 (D) . . ¼ hav. (D) 3·840630	
23 43	8·960605	
·04043	Apparent time 21ʰ 39ᵐ 17ˢ, which differs 3ˢ from the	
·88022	previous result; whence the error of chonometer is fast	
	23ᵐ 20·5ˢ on Greenwich mean time.	
·92065 21 36		
∴ sun's decl. 1 36 52 S.		

397. May 20, 1847, at 5ʰ 20ᵐ P.M., mean time nearly, in lat. 47° 20' N., and long. 94° 30' E., when a chronometer showed 11ʰ 5ᵐ 20ˢ, the observed altitude of the sun's lower limb was 20° 0' 15", the index correction — 4' 10", and height of eye above the sea 20 feet: required the error of chronometer on Greenwich mean time. *Ans.* Fast 0ᵐ 42·3ˢ.

398. Feb. 3, 1847, at 10ʰ 30ᵐ A.M., mean time nearly, in lat. 49° 30′ N., and long. 22° W., when a chronometer showed 0ʰ 2ᵐ 30′, the observed altitude of the sun's lower limb was 19° 21′ 30″, the index correction +3′ 20″, and height of eye above the sea 18 feet: required the error of chronometer on Greenwich mean time.   *Ans.* Fast 10ᵐ 7·4′.

399. March 25, 1847, at 3ʰ 20ᵐ P.M., mean time nearly, in lat. 52° 10′ N., and long. 36° 58′ 15″ W., when a chronometer showed 5ʰ 40ᵐ 58′, the observed altitude of the sun's lower limb was 25° 10′ 20″, the index correction —6′ 10″, and height of eye above the sea 20 feet : required the error of chronometer on Greenwich mean time.   *Ans.* 9ᵐ 16′ slow.

400. May 19, 1847, at 3ʰ 0ᵐ P.M., mean time nearly, in lat. 49° 50′ N., and long. 21° 4′ 45″ E., when a chronometer showed 1ʰ 23ᵐ 20′, the observed altitude of the sun's lower limb was 42° 50′ 30″, the index correction +4′ 10″, and height of eye above the sea 20 feet: required the error of chronometer on Greenwich mean time.   *Ans.* 10ᵐ 37·6′ slow.

Elements from Nautical Almanac.

Sun's declination.				Equation of time.			Semi.	
May 19......	19°	41′	37″N.	......	3ᵐ	49·6′ sub.	...... 15′	49″
„ 20......	19	54	26 N.	......	3	46·2		
Feb. 2	16	54	7 S.	......	13	58·8 add.	 16	14
„ 3	16	36	41 S.	......	14	5·6		
March 25 ...	1	40	56 N.	......	6	13·8 add.	 16	3
„ 26 ...	2	4	29 N.	......	5	55·2		
May 19......	19	41	37 N.	......	3	49·5 sub.	 15	49
„ 20......	19	54	26 N.	......	3	46·9		

Rule 45. *To find the error of chronometer by a* STAR'S *altitude.*

1. Get a Greenwich date.

2. Take out of the *Nautical Almanac* the right ascension and declination of the star, and also the right ascension of the mean sun for mean noon of the Greenwich date.

3. Correct the right ascension of mean sun for Greenwich date (p. 86).

4. Correct the observed altitude for index correction, dip, and refraction, and thus get the true altitude, which subtract from 90° for the true zenith distance.

5. *To find star's hour-angle.* First method (using haversines).

Under the latitude put the star's declination ; add if the names be unlike, subtract if like. Under the result put the true zenith distance of star, and take the sum and difference.

Add together the log. secants of the two first terms in this form (omitting the tens in each index), and halves of the log. haversines of the two last; the sum (rejecting 10 in the index) will be the log. haversine of hour-angle, to be taken out at top of page if heavenly body be west of meridian, but at bottom if east of meridian.

6. *To find star's hour-angle.* Second method (using sines, &c.).

Proceed as in the corresponding Rule (43) for finding apparent time, p. 164 and example 395.

7. To the hour-angle thus found add the star's right ascension, and from the sum (increased if necessary by 24 hours) subtract the right ascension of mean sun; the remainder is mean time at the place at the instant of observation.

8. Under mean time at place put the time shown by chronometer; the difference will be the error of chronometer on mean time at place.

To find error on Greenwich mean time.

Proceed as in the corresponding Rule for the sun, p. 165 and ex. 395.

EXAMPLE.

401. June 3, 1842, at $12^h 9^m$ P.M., mean time nearly, in lat. 50° 48′ N., and long. 1° 6′ 3″ W., observed the altitude of α Bootis (west of meridian) to be 89° 53′ 30″ in artificial horizon, when a chro. showed $0^h 14^m 22.3$; the index correction was $-10″$: required the error of the chronometer on mean time at the place, and also on Greenwich mean time.

Ship, June 3	. 12^h	9^m	Star's R. A.	. . 14^h	$8^m 30.5^s$	Obs. alt.	. 89° 53′	30″
long. in time	.	4 W.	star's decl.	. . 20°	0′ 15″ N.	in. cor.	.	10
Gr. June 3	. . 12	13	R.A. mean sun	$4^h 46^m$	7.1		2)89 53	20
			cor. for 12^h	. . . 1	58.3		44 56	40
			„ 0 13^m		2.1	ref. . . .		58
			∴ R.A. for 12 13	. 4 48	7 5		44 55	42
							90	
						zen. dist.	45 4	18

<table>
<tr><td align="center">First method.
Hour-angle by haversines.</td><td align="center">Second method.
Hour-angle by sines, &c.</td></tr>
</table>

Lat. . . 50°48′ 0″N. . sec. . . . 0·199263	Lat. . . 50°48′ 0″N. . sec. . . . 0·199263
decl. . . 20 0 15 N. . sec. . . . 0·027026	decl. . . 20 0 15 N. . sec. . . . 0·027026
30 47 45	½ hav. S. 4·788699	30 47 45	sin. S_1 . 9·788694
zen. dis. 45 4 18	½ hav. D. 4·094305	zen. dis. 45 4 18	, sin. D_1 . 9·094300
sum . . 75 52 3 (S)	hav. . . 9·109293	sum . . 75 52 3	2)19·109283
diff. . . 14 16 33 (D)	h.-ang. $2^h 48^m 8^s$	diff. . . 14 16 33	sin. . . . 9·554641
star's R. A.	14 8 30·5	½ sum . 37 56 1 (S_1)	½ h.-ang. $1^h 24^m 4^s$
	16 56 38·5	½ diff. . 7 8 16 (D_1)	2
R. A. mean sun . .	4 48 7·5	∴ hour-angle . 2 48 8
∴ ship mean time. .	12 8 31·0	as before.
chro. showed (add 12^h)	12 14 22·3	
∴ error of chro. . .	5 51·3 fast.	

To find the error of chronometer on Greenwich mean time.

Mean time at place	12^h	8^m	31·0^s
long. in time............................		4	24·2 W.
Greenwich mean time	12	12	55·2
chronometer showed....................	12	14	22·5
error of chron. on Gr. mean time......		1	27·1 fast.

402. May 4, 1847, at 4^h 40^m A.M., mean time nearly, in lat. 40° 10′ 20″ N., and long. 81° 47′ 15″ E., when a chronometer showed 11^h 13^m 50^s, the observed altitude of α Bootis (west of meridian) was 20° 45′ 45″, the index correction − 2′ 10″, and the height of eye above the sea 18 feet : required the error of the chronometer on Greenwich mean time.

Ans. 0^m 35·0^s slow.

403. Feb. 10, 1847, at 9^h 22^m P.M., mean time nearly, in lat. 28° 30′ N., and long. 27° 15′ W., a chronometer showed 11^h 17^m 20^s, when the observed altitude of α Leonis (east of meridian) was 42° 10′ 0″, the index correction − 3′ 20″, and height of eye above the sea 20 feet : required the error of the chronometer on Greenwich mean time.

Ans. 4^m 57·3^s fast.

404. April 18, 1848, at 0^h 40^m A.M., mean time nearly, in lat. 46° 32′ N., and long. 43° 36′ 15″ E., when a chronometer showed 10^h 13^m 45^s, the observed altitude of the star α Aquilæ was 14° 45′ 15″ (east of meridian), the index correction + 4′ 5″, and height of eye above the sea 18 feet : required the error of the chronometer on Greenwich mean time.

Ans. 19^m 31·7^s fast.

405. Aug. 11, 1848, at 8^h 10^m P.M., mean time nearly, in lat. 50° 20′ N., and long. 29° 53′ 15″ E., when a chronometer showed 6^h 6^m 20·0^s, the observed altitude of α Bootis (Arcturus) was 39° 5′ 10″ (west of meridian), the index correction − 2′ 10″, and height of eye above the sea 18 feet : required the error of the chronometer on Greenwich mean time.

Ans. 11^m 17·0^s slow.

Elements from Nautical Almanac.

Right ascen. mean sun.				Right ascen. and decl. of star.						
May 3 ...	2^h	43^m	3·3^s	α Bootis ...	14^h	8^m	43·4^s ...	19°	58′	46″N.
Feb. 10 ...	21	19	46·0	α Leonis...	10	0	15·3 ...	12	42	30 N.
April 17 ...	1	42	57·3	α Aquilæ...	19	43	22·7 ...	8	28	16 N.
Aug. 11 ...	9	20	17·8	α Bootis ...	14	8	45·0 ...	19	58	39 N.

EQUAL ALTITUDES.

When the sun's center is on the meridian of any place, the apparent time is then either 0^h or 24^h. To obtain mean time at the same instant, we have only to apply the equation of time with its proper sign. We thus find mean time at the instant the sun is on the meridian; and if we can also ascertain what a chronometer showed at the same instant, it is manifest that the error of the chronometer on mean time at the place is known, since it will be the difference between the two times.

To find the time shown by the chronometer at apparent noon, we have recourse to the method of equal altitudes, which consists in noting the time shown by the chronometer when the heavenly body has the same altitude on both sides of the meridian; half the interval between the observations being added to what the chronometer showed at the first observation will be the time shown by the chronometer when the heavenly body is on the meridian, if *the declination is supposed to be invariable* in the interval between the observations.

But the sun's declination is not invariable during the interval, but increases or decreases by a small number of minutes; so that the declination at the second equal altitude is not the same as at the first; and therefore half the interval between the observations being added to the time shown by chronometer at the first observation will *not* be the time the chronometer shows when the sun is on the meridian, but will differ by a *few seconds*. This difference or quantity of time is called *the equation of equal altitudes*, and is found by the following rule:

Rule 46. *To find the equation of equal altitudes and error of a chronometer on mean time at the place by* EQUAL ALTITUDES OF THE SUN.

1. Find mean time nearly of apparent noon at the place by taking out of the *Nautical Almanac* the equation of time to the nearest minute, and applying it with its proper sign to 0^h or 24^h, according as the *Nautical Almanac* directs it to be added to or subtracted from apparent time, putting the day one back in the latter case.

2. To mean time nearly thus found apply the longitude in time, adding if west, and subtracting if east; the result will be a Greenwich date.

3. Correct the equation of time for this date.

4. From the P.M. time when the second altitude was taken (increased by 12 hours) subtract the A.M. time when the first altitude was taken; the remainder is elapsed time, as shown by the chronometer. Take half the

elapsed time and subtract it from the above date (increased if necessary by 24 hours and the day put one back); the remainder is a second Greenwich date.

5. Take out the sun's declination for this date.

6. *To find the equation of equal altitudes.* Under heads (1) and (2) put down the following quantities:

Under (1) put A, taken from annexed table.
 ,, (2) put B, ,, ,,
 ,, (1) put log. cotangent latitude.
 ,, (2) put log. cotangent declination.
 ,, both (1) and (2) put proportional log. change of declination in 24 hours.

7. Add together logarithms under (1) and (2), and reject the tens in the index; look out the result as a proportional logarithm, and take out the seconds and tenths corresponding thereto.

8. Mark the quantities under (1) with the sign plus ($+$) if the declination is decreasing, and of the same name as the latitude; or if increasing and of a different name. Otherwise mark the quantity minus ($-$).

9. Mark the quantity under (2) plus ($+$) if the declination is increasing, but minus ($-$) if decreasing.

10. Take the sum or difference of these quantities, according as they have the same or different signs; the result will be the correction or *equation of equal altitudes* required.

11. Add together A.M. time and half-elapsed time, and to the same apply the correction just found, with its proper sign; the result will be the time shown by the chronometer when the sun's center is on the meridian.

12. Find mean time at the same instant by applying the equation of time to 0^h or 24^h, with the proper sign, as directed in the *Nautical Almanac*.

13. Put down under each other the results determined in (11) and (12), and take the difference, which will be the error of the chronometer on mean time *at the place*.

14. *To find error of the chronometer on Greenwich mean time.* To mean time at the place as found in (12) apply the longitude in time, and thus get mean time at Greenwich, under which put the time shown by chronometer as found in (11); the difference will be the error of the chronometer on *Greenwich* mean time.

EQUATION OF EQUAL ALTITUDES.

Elapsed time.	A	B	Elapsed time.	A	B	Elapsed time.	A	B
1 30	1·97148	1·97991	4 30	1·94886	2·02901	7 30	1·90212	2·15738
1 40	1·97082	1·98123	4 40	1·94692	2·03356	7 40	1·89876	2·16854
1 50	1·97009	1·98272	4 50	1·94490	2·03833	7 50	1·89531	2·18033
2 0	1·96930	1·98435	5 0	1·94281	2·04334	8 0	1·89177	2·19280
2 10	1·96843	1·98614	5 10	1·94064	2·04861	8 10	1·88815	2·20602
2 20	1·96750	1·98808	5 20	1·93840	2·05414	8 20	1·88444	2·22003
2 30	1·96649	1·99017	5 30	1·93608	2·05996	8 30	1·88064	2·23493
2 40	1·96541	1·99243	5 40	1·93368	2·06605	8 40	1·87676	2·25081
2 50	1·96426	1·99484	5 50	1·93122	2·07246	8 50	1·87278	2·26775
3 0	1·96305	1·99743	6 0	1·92866	2·07918	9 0	1·86870	2·28587
3 10	1·96176	2·00019	6 10	1·92604	2·08624	9 10	1·86454	2·30531
3 20	1·96040	2·00312	6 20	1·92333	2·09365	9 20	1·86029	2·32623
3 30	1·95897	2·00623	6 30	1·92054	2·10143	9 30	1·85593	2·34882
3 40	1·95747	2·00954	6 40	1·91767	2·10961	9 40	1·85148	2·37334
3 50	1·95589	2·01303	6 50	1·91473	2·11821	9 50	1·84692	2·40003
4 0	1·95424	2·01671	7 0	1·91170	2·12725	10 0	1·84427	2·42928
4 10	1·95252	2·02060	7 10	1·90859	2·13678	10 10	1·83752	2·46152
4 20	1·95073	2·02470	7 20	1·90539	2·14680	10 20	1·83267	2·49733

EXAMPLE.

406. Aug. 7, 1851, in lat. 50° 48′ N., and long. 1° 6′ W., the sun had equal altitudes at the following times by chronometer :

A.M.

9^h 25^m 42·5^s

P.M.

2^h 59^m 55·6^s

Required the error of chronometer on mean time at the place, and also on mean time at Greenwich.

August 7.................. 0^h 0^m apparent time.

equation of time .... 5 +

0 5 mean time

long. in time 4

Greenwich, August 7... 0 9 1st date for eq. of time.

½ elapsed time............ 2 47

Greenwich, August 6... 21 22 2d date for decl.

Equation of time.	Diff. for 1 hour.		
Aug. 7 ... 5ᵐ 33·03'	10ᵐ is ½	0·3' sub.	P.M. ...... 14ᵇ 59ᵐ 55·6'
·05	0·05		A.M. ...... 9 25 42·5
5 32·98+			elapsed T. 5 34 13·1
			½ elap. T. 2 47 6·55

Sun's declination.	(1).	(2).
6th ....16° 49' 12"N.	A.............1·93608	B............2·05996
7th ....16 32 38 N.	cot. lat. .....9·91147	cot. decl. ...0·52631
16 34	prop. log....1·03604	prop. log....1·03604
·05048	prop. log....2·88359	prop. log....3·62231
1·03604	14·2'+	2·55—
1·08652 14 45	2·55—	
Declination...16 34 27 N.	Eq. of equal } ——	
	altitudes.. } 11·65+	

A.M.	9ᵇ	25ᵐ	42·5'
¼ elapsed time..................................	2	47	6·55
	0	12	49·05
equation of equal altitude			11·65+
time by chronometer at apparent noon......		13	0·70
apparent time at apparent noon..............	0ᵇ	0ᵐ	0'
equation of time		5	32·98+
mean time at apparent noon	0	5	32·98
time by chronometer at apparent noon......	0	13	0·70
error of chronometer at place		7	27·72 fast.

To find error on Greenwich mean time.

Mean time at apparent noon	0ᵇ	5ᵐ	32·98'
longitude in time		4	24·00+
mean time at Greenwich	0	9	56·98
time by chronometer............................	0	13	0·70
ERROR OF CHRONOMETER on Gr. mean time..		3	3·72 fast.

407. August 7, 1851, in latitude 50° 48' N., and longitude 1° 6' W., the sun had equal altitudes at the following times by chronometer.

A.M.	P.M.
9ᵇ 3ᵐ 42·31'	3ᵇ 21ᵐ 54·22'

Required the error of the chronometer on Greenwich mean time.

For Elements from *Nautical Almanac*, see preceding example.

Ans. 3ᵐ 3·48' fast.

408. August 21, 1851, in latitude 50° 48' N., and longitude 1° 6' W., the sun had equal altitudes at the following times by chronometer,

A.M.   P.M.
10ʰ 49ᵐ 15·4ˢ   1ʰ 27ᵐ 27·6ˢ

Required the error of the chronometer on Greenwich mean time.

Ans. 1ᵐ 8·83ˢ fast.

Elements from Nautical Almanac.

Equation of time 3ᵐ 1·94ˢ + difference for 1ʰ 0·606ˢ —
Declination, 20th, 12° 34′ 59″ N.   21st, 12° 15′ 9″ N.

409. September 10, 1851, in latitude 50° 48′ N., and long. 1° 6′ W.,
the sun had equal altitudes at the following times by chronometer.

A.M.   P.M.
9ʰ 45ᵐ 55·2ˢ   2ʰ 20ᵐ 39·9ˢ

Required the error of the chronometer on Greenwich mean time.

Ans. 3ᵐ 47·28ˢ slow.

Elements from Nautical Almanac.

Equation of time 2ᵐ 58·43ˢ + difference in 1ʰ 0·866ˢ +
Declination, 9th, 5° 27′ 27″ N.   10th, 5° 4′ 45″ N.

410. May 14, 1844, in latitude 50° 48′ N., and longitude 15° 0′ W., the
sun had equal altitudes at the following times by chronometer.

A.M.   P.M.
10ʰ 46ᵐ 57·0ˢ   1ʰ 39ᵐ 42·0ˢ

Required the error of the chronometer on the mean time at the place, and
also on Greenwich mean time.

Ans. Fast on mean time at place, 17ᵐ 4·7ˢ.
Slow on Greenwich mean time, 42ᵐ 55·3ˢ.

Elements from Nautical Almanac.

Equation of time 3ᵐ 53·8ˢ — difference in 1ʰ 0·01ˢ —
Declination, 13th, 18° 28′ 49″ N.   14th, 18° 43′ 21″ N.

*To find the approximate time by chronometer when the P.M. altitudes should
be observed.*

After taking the observations in the morning, it will often be convenient
to estimate nearly at what *time by the chronometer* the observer should prepare to take the P.M. sights. To do this the error of the chronometer on
mean time *at the place* must be supposed to be known within a few minutes.
Thus suppose (as in the last example) a chronometer is known to be about
17 minutes *fast* of mean time at the place, the time of the A.M. observation
was by chronometer at 10ʰ 46ᵐ 57ˢ, equation of time 4 minutes subtractive
from apparent time. It is required to find the time the chronometer will
show in the afternoon when the sun has the same altitude.

Let $a=$ estimated error of chronometer on mean time at place (supposed fast),

$t=$ time shown by chronometer at A.M. observation.

Then $t-a=$ mean time at A.M. observation nearly.

Let E$=$ equation of time (supposed subtractive from apparent time).

$\therefore t-a+$E$=$ apparent time at A.M. observation.

$\therefore 12-(t-a+$E$)=$ apparent time from noon,

$=$ apparent time of P.M. observation.

$\therefore 12-(t-a+$E$)-$E$=$ mean time of P.M. observation.

And $12-(t-a+$E$)-$E$+a=$ mean time of P.M. observation *by chronometer*.

$\therefore$ Mean time of P.M. observation as shown by the chronometer

$$=12-(t-a+\text{E})-\text{E}+a$$
$$=12-t+2\,(a-\text{E}).$$

Thus (see Example) let $t=10^{\text{h}}\ 46^{\text{m}}\ 57^{\text{s}}$, $a=17^{\text{m}}$, E$=4^{\text{m}}$;

$\therefore$ Time by chronometer$=1^{\text{h}}\ 13^{\text{m}}\ 3^{\text{s}}+26^{\text{m}}=1^{\text{h}}\ 39^{\text{m}}$.

It appears from this that the observer need not prepare to take his P.M. sights until $1^{\text{h}}\ 30^{\text{m}}$ by chronometer.

A similar formula may be made to suit any other case.

A *blank form* for finding the error and rate of a chronometer by equal altitudes is given in page 186, Part II.

The use of blank forms diminishes very much the labour of calculation in some of the problems in Nautical Astronomy, such as the lunar, occultations, double altitudes, &c.

CHAPTER VII.

THE two principal methods for finding the longitude at sea by astronomical observations, are (1) by means of a chronometer, whose error is known on Greenwich mean time; and (2) by observing the distance of the moon from some well-known star, and calculating from thence Greenwich mean time: ship mean time is to be obtained in both methods by the same kind of observation.

To find the longitude by chronometer, an altitude of a heavenly body is to be taken—an operation requiring very little skill in the observer; but to find the longitude by lunar observations, the distance of the moon from some other heavenly body must be observed with considerable accuracy: the skill necessary to do this can only be acquired by practice; for this reason the method of finding the longitude by chronometer is the one chiefly in use, although the longitude deduced from it depends on the regular going of a timekeeper, whose rate from various causes is continually liable to change, while the other, which in fact is (within certain limits) correct and independent of all errors of chronometer, is rarely applied. Another objection usually urged against the use of the method of finding the longitude by lunar observation, is the labour required in reducing the observations; but we will endeavour to show that this ought not to deter the student; for that the work, although certainly more laborious than that required by the other method, is simple, and no ambiguity or distinction of cases need occur to distract the observer. From our own impression of the utility of lunars we feel it right to devote more than usual space to this method of finding the longitude, and we shall therefore give a series of distinct rules to suit every case that can occur.

LONGITUDE BY CHRONOMETER.

When a chronometer is taken to sea, the error on Greenwich mean time and its daily rate are supposed to accompany it: knowing then the error and rate, it is easy to determine the Greenwich mean time *at any instant* after-

N

wards by applying its original error and the accumulated loss or gain in the interval.

The corresponding mean time at the ship may be found by observing the altitude of the sun, or any other heavenly body, when it bears as nearly east or west as possible.

The difference between the two times is the longitude of ship.

Rule 47. Longitude by sun chronometer.

1. Get a Greenwich date.

2. Find Greenwich mean time at the instant of the observation, by bringing up the error of the chronometer by Rule 42, p. 161.

3. Take out of the *Nautical Almanac* both the declination of the sun and the equation of time, for the noon before and the noon after the Greenwich date; take out at the same time the sun's semidiameter.

4. Correct the declination and equation of time for the Greenwich date (or rather for the Greenwich mean time as shown by the chronometer), either by proportional logarithms or by hourly differences.

5. Correct the observed altitude for index correction, dip, semi., correction in alt., and thus get the true altitude, which subtract from 90° to obtain the zenith distance.

6. *To find ship apparent time.* First method (using log. haversines). Under the latitude put the sun's declination, and, if the names be alike, take the difference; but if unlike take the sum. Under the result put the zenith distance, and find the sum and difference. Add together the log. secants of the two first terms in this form (omitting the tens in each index) and the halves of the log. haversines of the two last, and (rejecting the ten in the index) look out the sum as a log. haversine, *to be taken out at the top of the page if the sun is west of the meridian, but at the bottom of the page if the sun is east of the meridian*; the result is apparent time at the ship at the instant of observation.

7. *To find ship apparent time.* Second method (using log. sines, &c.). Proceed as pointed out in Rule 43, second method, and Example 395.

8. *To find ship mean time.* To the apparent time just obtained, apply the equation of time, with its proper sign as directed in the *Nautical Almanac:* the result is mean time at the ship or place of observation.

9. *To find the longitude.* Under ship mean time put Greenwich mean time as known by the chronometer; the difference is the longitude in time, *west* if the Greenwich time is greater than ship time, otherwise *east.*

EXAMPLE. SUN WEST OF MERIDIAN.

411. Sept. 23, 1845, at 4^h 42^m P.M. mean time nearly, in latitude 50° 30′ N., and longitude by account 110° 0′ W., when a chronometer showed 11^h 59^m 30′, the observed altitude of the sun's lower limb was 11° 0′ 50″, the index correction — 3′ 20″, and the height of the eye above the sea 20 feet: required the longitude. On August 21 the chronometer was fast on Greenwich mean time 0^m 45·5′, and its daily rate was 5·7 losing.

		Sun's declination.	Eq. of time.	Semi.
Ship, Sept. 23	4^h 42^m	Sept. 23 . . 0° 6′ 56″ S.	Sept. 23 . 7^m 42·0ˢ sub.	15′ 58″
long. in time	7 20 W.	„ 24 . . 0 30 21 S.	„ 24 . 8 2·5	
Gr. Sept. 23 . 12	2	23 25	20·5	
		·29983	·29983	
Interval of time		·88575	2·72167	
from Aug. 21				
to Gr. date . 33ᵈ 12ʰ		1·18558 11 44	3·02150 10·2	
		∴ sun's decl. 0 18 40 S.	7 52·2	

Daily rate	5·7ˢ losing	Obs. alt.	11° 0′ 50″
	33	index cor.	3 20 —
	171		10 57 30
	171	dip	4 24 —
	188·1		10 53 6
12ʰ . . ¼	2·8	semi.	15 58
	60) 190·9		11 9 4
		cor. in alt.	4 38 —
accumulated loss . .	3ᵐ 10·9		11 4 26
chron. showed . . . 11ʰ 59	30·0		90
12	2 40·9		
original error . . .	0 45·5 fast.	∴ zenith distance . . .	78 55 34
∴ Gr. Sept. 23 . . 12	1 55·4		

Ship apparent time by First Method, using haversines.

Latitude	50° 30′ 0″ N. . .	log. sec.	0·196490
declination . . .	0 18 40 S. . .	„ sec.	0·000006
	50 48 40	„ ½ hav. S . .	4·956810
zenith distance .	78 55 34	„ ½ hav. D . .	4·385440
sum	129 44 14 (S)		9·538746
difference . . .	28 6 54 (D)		
	ship app. time	4^h 48^m 7ˢ	
	equation of time	7 52·2	
	ship mean time	4 40 14·8	
	Greenwich mean time . .	12 1 55·4	
	longitude in time	7 21 40·6	
	∴ LONGITUDE	110° 25′ 9″ W.	

Or thus : Ship apparent time, by second method, using sines, &c. only.

Latitude . . .	50°	30′	0″ N.	. . log. sec.	0·196490
declination . .	0	18	40 S.	. . ,, sec.	0·000006
	50	48	40	,, sin. S_1 . .	9·956803
zenith distance .	78	55	34	,, sin. D_1 . .	9·385445
sum.	129	44	14	2) 19·538744
difference. . .	28	6	54	,, sin.	9·769372
½ sum	64	52	7 (S_1)	36° 4′ 0″
¼ difference . .	14	3	27 (D_1)	or 2ʰ 24ᵐ 4ˢ
				2
			∴ ship app. time	4 48 8

EXAMPLE. SUN EAST OF MERIDIAN.

412. April 18, 1844, at 9ʰ 18ᵐ A.M. mean time nearly, in latitude 50° 48′ N., and longitude by account 1° 0′ W., when a chronometer showed 9ʰ 27ᵐ 48ˢ, the observed altitude of the sun's lower limb was 76° 16′ 46″ (in artificial horizon), index correction 3′ 46″ − : required the longitude. On April 1 the chronometer was fast 1ᵐ 58·7ˢ on Greenwich mean time, and its mean daily rate was 11·2ˢ gaining.

		Sun's declination.			Eq. of time.		Semi.
Ship, April 17	21ʰ 18ᵐ	April 17 .	.10° 36′ 49″ N.	17 . .	0′ 31·3″ sub.	15′ 56″
long. in time	4 W.	,, 18 .	.10 57 46 N.	18 . .	0 45·1	
Gr. April 17	21 22		20 57		13·8	
		·05048			·05048	
Interval from		·93409			2·89354	
Ap. 1 at noon		·98457 .	18 39		2·94402	12·3
to April 17, 16ᵈ 21¾ʰ	sun's decl.	. 10 55 28 N.			0 43·6	

Daily rate	11·2ˢ gaining.		Obs. alt.		76°	16′	46″
	16		index cor..			3	46—
	672				2) 76	13	0
	112				38	6	30
	179·2						
12ʰ . . ½	5·6		semi.		15	56	
6 . . ¼	2·8				38	22	26
3 . . ⅛	1·4						
¾ . . 1/32	·2		cor. in alt.		1	7—	
60	189·2		sun's true alt.		38	21	19
accum. gain . . .	3ᵐ 9·2ˢ				90		
chro. showed . . .	9 27 48·0						
	9 24 38·8		,, zenith distance . .		51	38	41
original error . . .	1 58·7 fast.						
	9 22 40·1 A.M.						
	12						
Gr, April 17 . . .	21 22 40·1						

Ship apparent time. First method, using haversines.

Latitude	50° 48'	0" N.	. . log. sec. . . .	0·199263
declination . . .	10 55 28 N.	. . „ sec. . . .	0·007943
	39 52 32	„ ½ hav. S. .	4·855173
zenith distance . .	51 38 41	„ ½ hav. D. .	4·010890
sum	91 31 13 (S)	„ hav. . . .	9·073269
difference	11 46 9 (D)		

ship apparent time 21ʰ 19ᵐ 0ˢ
equation of time . . . . .   43·6 −
ship mean time . . . . . . 21 18 16·4
Greenwich mean time . . . 21 22 40·1
longitude in time . . . . .   4 23·7
or LONGITUDE . . . . . 1° 6' W.

Or thus : Ship apparent time. Second method, using sines, &c. only.

Latitude	50° 48' 0" N.	. . log. sec. . . .	0·199263
declination . . .	10 55 28 N.	. . „ sec. . . .	0·007943
	39 52 32	„ sin. S₁ . .	9·855158
zenith distance . .	51 38 41	„ sin. D₁ . .	9·010737
sum	91 31 13	2) 19·073101
difference	11 46 9	9·536500
½ sum	45 45 36 (S₁)	20° 7' 15"
¼ difference . . .	5 53 4 (D₁)	or 1ʰ 20ᵐ 29ˢ

                                                              2
                                                    ─────────
                                                    2 40 58
                                                   24
                                          ───────────
∴ ship apparent time . . . 21 19   2

413. Sept. 25, 1845, at 4ʰ 20ᵐ P.M. mean time nearly, in latitude 59° 30' N., and longitude by account 112° 30' W., when a chronometer showed 11ʰ 44ᵐ 20ˢ, the observed altitude of the sun's lower limb was 10° 50' 10", the index correction + 6' 10", and height of eye above the sea 18 feet : required the longitude. On Sept. 20 the chronometer was fast on Greenwich mean time 0ᵐ 30·7ˢ, and its daily rate was 10·5ˢ losing.
Ans. 112° 33' W.

414. May 30, 1845, at 3ʰ 10ᵐ P.M. mean time nearly, in latitude 30° 12' 0" S., and longitude by account 156° 0' E., the observed altitude of the sun's lower limb was 21° 8' 40", when a chronometer showed 4ʰ 44ᵐ 56ˢ, the index correction − 1' 10", and the height of eye above the sea 30 feet : required the longitude. On May 19 the chronometer was fast 5ᵐ 16ˢ on Greenwich mean time, and its daily rate was 3·5ˢ gaining. *Ans.* 156° 23' E.

415. July 8, 1849, at 1ʰ 40ᵐ P.M. mean time nearly, in latitude 50° 48' N., and longitude by account 1° 1' W., the observed altitude of the sun's lower limb, taken by the artificial horizon, was 109° 54' 44", the chronometer

showed 1ʰ 44ᵐ 14', the index correction +1' 25": required the longitude. On July 1 the chronometer was slow on Greenwich mean time 8ᵐ 18·4', and its daily rate was 3·5' losing.

Ans. 1° 6' 0" W.

416. January 20, 1846, at 6ʰ 40ᵐ A.M. mean time nearly, in latitude 56° 20' S., and longitude by account 83° 10' W., when a chronometer showed 0ʰ 14ᵐ 50', the observed altitude of the sun's lower limb was 20° 20' 30", the index correction —1' 30", and the height of the eye above the sea 20 feet: required the longitude. On Jan. 2 the chronometer was fast on Greenwich mean time 5ᵐ 20', and on Jan. 6 it was fast 4ᵐ 52', from which may be found its mean daily rate.

Ans. 83° 5' 0" W.

417. Feb. 10, 1846, at 7ʰ 50ᵐ A.M. mean time nearly, in latitude 50° 48' N., and longitude by account 170° 30' E., when a chronometer showed 9ʰ 59ᵐ 25', the observed altitude of the sun's lower limb was 51° 9' 10", the index correction —3' 20", and the height of eye above the sea 16 feet: required the longitude. On Jan. 31, at Greenwich noon, the chronometer was fast 1ʰ 34ᵐ 43', and its daily rate was 20·6' losing.

Ans. 170° 34' 15" E.

418. May 14, 1859, at 7ʰ 20ᵐ A.M. mean time nearly, in lat. 50° 50' N., and longitude by account 4° 10' E., when a chronometer showed 7ʰ 0ᵐ 20', the observed altitude of the sun's lower limb was 27° 20' 10", the index correction —7' 20", and the height of the eye above the sea 19 feet: required the longitude. On April 24, at noon, the chronometer was slow on Greenwich mean time 1ᵐ 10·5', and its daily rate was 5·7' losing.

Ans. 4° 16' E.

Elements from Nautical Almanac.

	Sun's declination.	Equation of time.	Semi.
Sept. 25	0° 53' 47" S.	8ᵐ 23·0' sub.	15' 59"
„ 26	1 17 12 S.	8 43·4	
May 29	21 38 43 N.	2 56·4 sub.	15 47
„ 30	21 47 47 N.	2 48·5	
July 8	22 29 9 N.	4 40·9 add	15 45
„ 9	22 22 7 N.	4 50·1	
Jan. 20	20 8 22 S.	11 19·2 add	16 16
Feb. 9	14 41 33 S.	14 31·0 add	16 13
„ 10	14 22 11 S.	14 32·0	
May 13	18 19 29 N.	3 52·4 sub.	15 51
„ 14	18 23 13 N.	3 53·2	

Rule 48. LONGITUDE BY STAR CHRONOMETER.

1. Get a Greenwich date.

2. Find Greenwich mean time by bringing up the error of the chronometer to the instant of observation by Rule 42.

3. Take out of the *Nautical Almanac* the right ascension and declination of the star, and also the right ascension of the mean sun (called in *Nautical Almanac* sidereal time) for mean noon of the Greenwich date.

4. Correct the right ascension of the mean sun for Greenwich date (or rather for the Greenwich mean time as shown by the chronometer).

5. Correct the observed altitude for index correction, dip, and refraction, and thus get the true altitude, which subtract from 90° to obtain the zenith distance.

6. *To find the star's hour-angle.* First method, using log. haversines. Under the latitude put the star's declination; add if the names be unlike, subtract if like; under the result put star's zenith distance, and take the sum and difference. Add together the log. secants of the two first terms in this form (omitting the tens in each index), and the halves of the log. haversines of the two last; the sum, rejecting ten in the index, will be the log. haversine of star's hour-angle, to be taken out at top of page if heavenly body be west of meridian, but at bottom if east of meridian.

(7. *To find star's hour-angle.* Second method, using sines, &c. Proceed as pointed out in Rule 43, Ex. 395.)

8. *To find mean time at ship.* To the hour-angle thus found add the star's right ascension, and from the sum, increased if necessary by 24 hours, subtract the right ascension of the mean sun; the remainder is mean time at the place at the instant of observation.

9. *To find the longitude.* Under ship mean time put Greenwich mean time as known by the chronometer; the difference is the longitude in time, *west* if Greenwich time is greater than ship time, otherwise *east.*

EXAMPLE. STAR WEST OF MERIDIAN.

419. Sept. 10, 1844, at $7^h 15^m$ P.M. mean time nearly, in latitude 48° 20′ N., and longitude by account 32° E., when a timekeeper showed $5^h 1^m 28^s$, the observed altitude of α Bootis (Arcturus) W. of meridian was

31° 5' 40″, the index correction −4' 10″, and height of eye above the sea
20 feet : required the longitude.   On Aug. 25 the chronometer was slow on
Greenwich mean time 2ᵐ 40ˢ, and its daily rate was 4·3ˢ gaining.

Ship, Sep. 10 . .	7ʰ 15ᵐ	Daily rate . . 4·3ˢ gaining.		Obs. alt. . .	31°	5'	40″
long. in time . .	2 8 E.	interval 16ᵈ 5ʰ. 16		in. cor. . .		4	10−
Gr. Sep. 10 . .	5 7	258			31	1	30
		43		dip		4	24−
Star's R. A. 14ʰ 8ᵐ 34·65ˢ		68·8			30	57	6
Star's decl.. 19° 59' 44″N.		4 . . ⅙ 0·7		ref.		1	37−
R. A. mean sun.		1 . . ¼ ·2			30	55	29
10 . . . 11ʰ 18ᵐ 28·15ˢ		60 69·7			90		
cor. 5ʰ . . 49·28				zen. dist.. . 59 4 31			
3ᵐ . ·49		accum. rate 1ᵐ 9·7ˢ					
R. A. mean } 11 19 17·92		chro. show. 5 1 28·0					
sun . . }		5 0 18·3					
		orig. error . 2 40·0 slow.					
		Gr. M. T. . 5 2 58·3					

Hour-angle by First Method, using haversines.

Latitude . . .	48° 20' 0″N.	. . log. sec.. . .	0·177312
declination . .	19 59 44 N.	. . „ sec.. . .	0·027003
	28 20 16	„ ½ hav. (S)	4·839453
zenith distance .	59 4 31	„ ½ hav. (D)	4·423295
sum	87 24 47 (S)	„ hav. . .	9·467063
diff.	30 44 15 (D)	.·. hour-angle . 4ʰ 22ᵐ 15ˢ	
		star's R. A. . . 14 8 35	
		18 30 50	
		R. A. mean sun . 11 19 18	
		ship mean time . 7 11 32	
		Gr. mean time . 5 2 58	
		long. in time . . 2 8 34	
		.·. long. . . . 32° 8' 80″ E.	

Hour-angle by Second Method, using sines, &c. only.

Latitude . . .	48° 20' 0″N.	. . log. sec. . . .	0·177312
declination . .	19 59 44 N.	. . „ sec. . . .	0·027003
	28 20 16	„ sin. (S₁) .	9·839470
zenith distance .	59 4 31	„ sin. (D₁) .	9·423238
sum	87 24 47		19·467028
diff.	30 44 15	„ sin. . . .	9·733511
½ sum	43 42 23 (S₁)	32° 46' 45″	
½ diff.	15 22 7 (D₁)	or 2ʰ 11ᵐ 7ˢ	
		2	
		.·. star's hour-angle 4 22 14	

EXAMPLE.   STAR EAST OF MERIDIAN.

420. May 24, 1844, at 11^h 11^m P.M. mean time nearly, in latitude 50° 48′ N., and longitude by account 1° 0′ W., when a timekeeper showed 11^h 12^m 11·8ˢ, the observed altitude of α Lyræ (Vega) E. of meridian was 109° 29′ 18″ in artificial horizon, the index correction −3′ 46″: required the longitude.   On May 14, at Greenwich mean noon, the chronometer was slow 1^m 15·8ˢ, and its mean daily rate was 7·4ˢ losing.

Ship, May 24 . .	11^h 11^m	Daily rate . . .	7·4ˢ		Obs. alt. .	. 109° 29′ 18″
long. in time . .	4	interval, 10^d 11^h 15^m	10		index cor. .	3 46−
Gr., May 24 . .	11 15			74·0		2) 109 25 32
		8^h is ⅓	2·5			54 42 46
star's R. A.	18^h 31^m 42·2ˢ	2 ,, ¼	·6	ref. . . .		41−
star's decl. .	38° 38′ 24″N.	1 ,, ⅟₁₂	·3			54 42 5
		15^m,, ⅟₄	·1			90
Right ascension mean sun.			(60) 77·5		zen. dist. .	35 17 55
24th . . .	4^h 8^m 43·56ˢ					
11^h . .	1 48·42	accum. rate .	1^m 17·5ˢ			
14^m . .	2·30	chro. showed 11^h 12	11·8			
45ˢ . .	·13	11 13	29·3			
	4 10 34·4	original error	1 15·8			
		Gr. M. T. .	. 11 14 45·1			

Hour-angle by First Method, using haversines.

Latitude	50° 48′ 0″ N.	. . log. sec. . . .	0·199263
declination . . .	38 38 24 N. . .	,, sec. . . .	0·107300
	12 9 36	,, ½ hav. (S) .	4·604673
zenith distance . .	35 17 55	,, ½ hav. (D) .	4·302209
sum	47 27 31 (S)	,, hav. . . .	9·213446
difference	23 8 19 (D)		
Hour-angle		20^h 49^m 13ˢ	
star's right ascension . . .		18 31 42·2	
		39 20 55·2	
right ascension mean sun . .		4 10 34·4	
ship mean time		11 10 20·8	
Greenwich mean time . . .		11 14 45·1	
longitude in time		4 24·3=1° 6′ W.	

Hour-angle by Second Method, using sines, &c. only.

Latitude . . .	50° 48′ 0″ N.	. . log. sec.	0·199263
declination . .	38 38 24 N. . .	,, sec.	0·107312
	12 9 36	,, sin. (S_1) . .	9·604673
zenith distance .	35 17 55	,, sin. (D_1) . .	9·302200
sum	47 27 31		19·213458
diff.	23 8 19	,, sin.	9·606729
½ sum	23 43 45 (S_1)		23° 51′ 0″
½ diff.	11 34 9 (D_1)	or	1^h 35^m 24ˢ
			2
			3 10 48
			24
∴ star's hour-angle . . .			20 49 12

421. Aug. 20, 1845, at 0ʰ 30ᵐ A.M. mean time nearly, in lat. 50° 20′ N., and long. by account 142° 0′ E., when a chronometer showed 2ʰ 41ᵐ 12′, the observed altitude of the star α Aquilæ (Altair) was 36° 59′ 50″ west of the meridian, the index correction + 6′ 30″, and height of eye above the sea 20 feet: required the longitude. On August 1 the chronometer was slow on Greenwich mean time 17ᵐ 45·0′, and its daily rate was 4·3′ losing.

Ans. 142° 14′ 15″ E.

422. Sept. 10, 1844, at 4ʰ 21ᵐ A.M. mean time nearly, in lat. 40° 36′ N., and longitude by account 73° E., when a chronometer showed 11ʰ 21ᵐ 56′, the observed altitude of β Geminorum (Pollux) was 39° 0′ 10″ east of meridian, the index correction − 4′ 10″, and height of eye above the sea 20 feet: required the longitude. On August 20 the chronometer was slow on Greenwich mean time 3ᵐ 19·9′, and its daily rate was 9·3′ gaining.

Ans. 72° 44′ E.

423. January 16, 1845, at 8ʰ 0ᵐ P.M. mean time nearly, in latitude 49° 56′ 50″ N., and longitude by account 90° 30′ E., when a chronometer showed 2ʰ 24ᵐ 30′, the observed altitude of α Leonis (Regulus) was 8° 4′ 20″ east of meridian, the index correction − 4′ 20″, and height of eye above the sea 25 feet: required the longitude. On January 1 the chronometer was fast on Greenwich mean time 5ᵐ 30·5′, and its daily rate was 5·5′ losing.

Ans. 86° 6′ E.

424. January 20, 1846, at 8ʰ 30ᵐ P.M. mean time nearly, in latitude 50° 48′ N., and long. by account 7° 10′ W., when a chronometer showed 8ʰ 32ᵐ 50′, the observed altitude of ε Leonis was 28° 0′ 10″ east of meridian, the index correction − 6′ 20″, and height of eye above the sea 20 feet: required the longitude. On January 2 the chronometer was fast on Greenwich mean time 30ᵐ 30′, and its mean daily rate was 15·5′ losing.

Ans. 7° 19′ E.

Elements from Nautical Almanac.

	Right ascen. mean sun.			Star's right ascen.			Star's decl.		
Aug. 19	9ʰ	50ᵐ	46·5′	19ʰ	43ᵐ	17·0′	8° 28′	7″ N.
Sept. 9	11	14	31·6	7	35	48·6	28 23	40 N.
Jan. 16	19	43	7·2	10	0	8·8	12 43	6 N.
Jan. 20	19	57	55·9	9	37	8·3	24 28	33 N.

LONGITUDE BY LUNAR OBSERVATIONS.

The *time at the ship* is obtained by the same kind of observation as that for finding the longitude by chronometer, selecting of the two bodies observed that whose bearing is the greatest. The *time at Greenwich* is found by calculating the *true distance* of the moon from the sun or some other heavenly body at the moment of observation, and comparing it with the true distance of the moon from the same heavenly body as recorded in the *Nautical Almanac* for some given time at Greenwich.

TO CALCULATE THE TRUE DISTANCE.

The true distance is found by clearing the observed distance of the effects of parallax and refraction, by the following or some other similar methods.

In *Nav.* Part II. are given the investigations of several methods of clearing the distance. We will here confine ourselves to two (the first and sixth in Part II.); the former, because it is the most simple method when Inman's Tables are at hand; the latter, because it is adapted to any book of logarithms that contains tables of log. sines, &c.

The practical inconvenience of this last method arises from the necessity of taking out the log. sines, &c. to the nearest second, a work of considerable labour with the common tables of log. sines, &c., which seldom give the arcs nearer than a minute or 15″. We have also to add together the unusual number of six logarithms, and this can seldom be done with accuracy unless great attention is given to the addition. This last objection may be somewhat removed by making use of a small auxiliary table, investigated in *Nav.* Part II. (which we will reprint here and call table log. C). The correction taken out of this table is equivalent to the sum of two of the above six logarithms; but notwithstanding this simplification the method is tedious, the labour of proportioning for seconds not being got rid of. For these reasons the first method is to be preferred, being simple and direct; and as it is made to depend on the table of versines, which is computed to seconds, all the quantities can be taken out *by inspection*, thereby entirely avoiding the trouble of proportioning.

TO CLEAR THE LUNAR DISTANCE.

First method : by Inman's Tables.

In *Nav.* Part II. it is proved that

$$\text{vers. } D = \text{vers. } (z+z_1) + \text{ver. } (a+a_1+A) + \text{ver. } (a+a_1 \smallfrown A)$$
$$+ \text{ver. } (d+A) + \text{ver. } (d \smallfrown A) - 4{,}000{,}000;$$

where z and z_1 are the true zenith distances, a and a_1 are the apparent altitudes, d the apparent distance of the centers of the two heavenly bodies observed, $A = \frac{1}{2} \cdot \frac{\sin. z \sin. z_1}{\cos. a \cos. a_1}$ a quantity tabulated and called the auxiliary angle A, and D the true distance to be found. From this formula the following rule may be deduced.

Rule 49. To clear the lunar distance.

1. Under the sun's or star's true zenith distance put the moon's true zenith distance; take the sum, which mark vers.

2. Under the apparent distance of the two centers put the auxiliary angle A; take their sum and difference, against both of which mark vers.

3. Under the sun's or star's apparent altitude put the moon's apparent altitude, and take their sum; under which put the auxiliary angle A: take the sum and difference, against both of which mark vers.

4. Add together the five last figures of the versines of the quantities marked vers., rejecting all but the last five in the result, which look for in the column of versines under the apparent distance, or under the adjacent one: take out the arc corresponding thereto, which will be the true distance required.

NOTE. The auxiliary angle A is found in the Nautical Tables of Inman, Riddle, Norie, and others.

The student will be able to determine the relative value of the two methods by working an example by each.

EXAMPLE.

425. Required the true distance of the moon from the sun, having given

App. dis. of the centers	35°	47′	24″		True alt. sun	34°	20′	14″
app. alt. sun	34	21	32		true alt. moon	57	40	11
app. alt. moon	57	11	25		auxiliary angle A	60	25	16

Sun's true zen. dist. z. 55° 39′ 46″
moon's true zen. dist. z_1 32 19 50

$z + z_1$ .... 87 59 36 vers.

apparent distance d ... 35 47 24
aux. angle A 60 25 16

$d + A$ .... 96 12 40 vers.
$d \sim A$ .... 24 37 52 vers.

moon's app. alt. a 57 11 25
sun's app. alt. a_1 ...... 34 21 32

91 32 57
auxiliary angle A 60 25 16

$a + a_1 + A$ ......151 58 13 vers.
$a + a_1 \sim A$ ...... 31 7 41 vers.

Parts for seconds.

964810 174
1107999 195
90885 104
1882674 30
143883 102

4190251 605
605

4190856
4000000

vers. 190856

∴ True dist. ... 35° 59′ 15″

In practice it is not necessary to take from the table of versines more than the last five figures, rejecting also all but these last five in the sum, since the true distance will be always either in the same column with the apparent distance or the adjacent one. Thus, taking the above example, it may be worked thus :

				Vers.	Parts for seconds.
55°	39'	46″			
32	19	50			
87	59	36	vers.	64810	 174
				07999	 195
35	47	24		90885	 104
60	25	16		82674	 30
96	12	40	vers.	43883	 102
24	37	52	vers.	90251	 605
				605	
57	11	25		90856	...... 35° 59'
34	21	32		812	
91	32	57		44	15″
60	25	16		True dist. 35 59 15	
151	58	13	vers.		
31	7	41	vers.		

NOTE. The figures 90856 are looked for in the table of versines in the column under the degrees of the apparent distance, 35°, or the one adjacent.

TO CLEAR THE LUNAR DISTANCE.

Second method : by the common table of sines, &c.

In *Nav.* Part II. it is proved that

$$\text{M} = \sqrt{\cos A_1 \sec a_1 \cos A \sec a \cos \tfrac{1}{2}(a + a_1 + d) \cdot \cos \tfrac{1}{2}(a + a_1 - d)}$$

$$\tan \theta = \frac{\text{M}}{\sin \tfrac{1}{2}(\text{A} + \text{A}_1)}, \text{ and } \cos \frac{\text{D}}{2} = \frac{\text{M}}{\sin \theta},$$

where A and A_1 are the true altitudes of the moon and other heavenly body observed ; a and a_1 their apparent altitudes ; d the apparent distance of the centers of the two heavenly bodies observed, and D the true distance to be found.

From the above formulæ the following method may be deduced for clearing the lunar distance :

1. Add together the true altitudes, and divide by 2 : call the result $\tfrac{1}{2}(\text{A} + \text{A}_1)$.

2. Add together the apparent altitudes, under which put the apparent distance. Take the sum and difference, and divide each by 2, and call the result the half-sum and half-difference.

3. Add together the following six logarithms :

log. cosine of the half-sum,

" cosine of the half-difference,

" cosine of moon's true altitude,

" secant of moon's apparent altitude (rejecting 10 in index),

" * cosine of sun or star's true altitude,

" * secant of sun or star's app. altitude (rejecting 10 in index).

Divide the sum by 2, and call the result log. M.

4. Put down log. M a second time, a little to the right.

5. From log. M subtract log. sin. $\frac{1}{2}$ ($A+A_1$); the remainder will be the log. tangent of an arc, which take from the tables and call θ.

6. From log. M subtract log. sine of θ : the remainder is the log. sine of half the true distance. Take this from the tables, and multiplied by 2 will be the TRUE DISTANCE required.

TABLE LOG. C.

| FOR THE *SUN.* | | FOR A *STAR* OR *PLANET.* | |
| ARGUMENT. SUN'S APP. ALT. | | ARGUMENT. STAR'S APP. ALT. | |
App. alt.	Log. C.	App. alt.	Log. C.
90	0·000104	90	0·000123
75	0·000105	30	0·000122
65	0·000106	20	0·000121
60	0·000107	15	0·000120
55	0·000108	12	0·000119
50	0·000109	10	0·000118
45	0·000110	9	0·000117
40	0·000111	8	0·000116
35	0·000112	7	0·000114
30	0·000113	6	0·000111
25	0·000114	5	0·000107
20	0·000115	4	0·000100
10	0·000114	3	0·000090
8	0·000113		
7	0·000112		
6	0·000109		
4	0·000100		

Use of Table log. C.

Instead of taking out the last two of the above six logarithms (viz. those marked thus *), enter table log. C with the sun or star's apparent altitude, and take out the corresponding logarithm. This quantity, added to the four logarithms already taken out (the index 10 of sec. being retained), will give the same result as the sum of the six logarithms.

426. The apparent distance of centers of sun and moon, 35° 47′ 24″; sun's apparent altitude, 34° 21′ 32″; moon's apparent altitude, 57° 11′ 25″; sun's true altitude, 34° 20′ 14″; moon's true altitude, 57° 40′ 10″: required the true distance.

Apparent dist. (d) . . 35° 47′ 24″

Sun's app. alt. (a_1) .	34°	21′	32″	Sun's true alt. (A_1)	34°	20′	14″
moon's app. alt. (a) .	57	11	25	moon's true alt. (A)	57	40	10
$a+a_1$	91	32	57	$A+A_1$	92	0	24
d	35	47	24	$\therefore \frac{1}{2}(A+A_1)$	46	0	12
$a+a_1+d$ . .	127	20	21				
$a+a_1-d$ . . .	55	45	33				

$\therefore \frac{1}{2}(a+a_1+d)$ 63 40 10 . . log. cos. . 9·646939
$\frac{1}{2}(a+a_1-d)$ 27 52 46 . . „ cos. . 9·946419
„ cos. A 9·728194
„ sec. a . 0·266121
To find log. M by Table log. C. „ cos. A_1 9·916840
„ sec. a_1 0·083273
9·646939   39·587786
9·946419
9·728194 $\therefore$ log. M . . . . 19·793893 . . . . . 19·793893
10·266121 „ sin. $\frac{1}{2}$ (A+A_1) 9·856959 log. sin. θ . 9·815670
log. C . . ___112___
39·587785 „ tan. θ . . . 9·936934 „ cos.$\frac{D}{2}$. 9·078223
log. M . . 19·793893
as before. $\therefore$ θ . . . . 40° 51′ 16″ $\therefore$ $\frac{D}{2}$. 17° 59′ 36″
_____2_____
$\therefore$ TRUE DISTANCE = 35 59 12

EXAMPLES.

Find the TRUE DISTANCE in the following examples:

427. Sun's app. alt. is 30° 29′ 48″; moon's app. alt., 50° 54′ 38″; moon's true zen. dist., 38° 30′ 40″; sun's true zen. dist., 59° 31′ 44″; the auxiliary angle A, 60° 24′ 12″; and app. dist., 88° 49′ 58″.

Ans. True distance, 88° 24′ 17″.

428. Sun's app. alt. is 54° 29′ 33″; moon's app. alt., 5° 25′ 59″; moon's true zen. dist., 83° 48′ 29″; sun's true zen. dist., 35° 31′ 3″; auxiliary angle A, 60° 2′ 11″; and app. dist., 105° 5′ 47″.

Ans. True distance, 104° 26′ 18″.

429. Sun's app. alt. is 17° 39′ 31″; moon's app. alt., 24° 13′ 45″; moon's zen. dist., 64° 56′ 45″; sun's zen. dist., 72° 23′ 22″; auxiliary angle A, 60° 12′ 33″; and app. dist., 111° 20′ 45″.

Ans. True distance, 110° 56′ 0″.

430. Sun's app. alt. is 54° 47′ 4″; moon's app. alt., 21° 20′ 1″; moon's zen. dist., 67° 51′ 5″; sun's zen. dist., 35° 13′ 32″; auxiliary angle A, 60° 10′ 44″; and app. dist., 71° 16′ 44″.

Ans. True distance, 70° 38′ 5″.

431. Sun's app. alt. is 12° 19′ 30″; moon's app. alt., 20° 40′ 18″; moon's zen. dist., 68° 28′ 19″; sun's zen. dist., 77° 44′ 42″; auxiliary angle ʌ, 60° 10′ 53″; and app. dist., 124° 44′ 32″.

Ans. True distance, 124° 19′ 11″.

432. Sun's app. alt. is 57° 53′ 52″; moon's app. alt., 35° 3′ 2″; moon's zen. dist., 54° 11′ 56″; sun's zen. dist., 32° 6′ 40″; auxiliary angle ʌ, 60° 17′ 54″; and app. dist., 65° 34′ 42″.

Ans. True distance, 64° 58′ 10″.

433. Sun's app. alt. is 15° 43′ 48″; moon's app. alt., 16° 5′ 5″; moon's zen. dist., 73° 1′ 32″; sun's zen. dist., 74° 19′ 28″; auxiliary angle ʌ, 60° 8′ 36″; and app. dist., 119° 44′ 31″.

Ans. True distance, 119° 19′ 51″.

434. App. alt. of a star is 20° 13′ 26″; moon's app. alt., 31° 17′ 22″ star's zen. dist., 69° 49′ 11″; moon's zen. dist., 57° 57′ 44″; auxiliary angle ʌ, 60° 15′ 21″; and app. dist., 72° 42′ 16″.

Ans. True distance, 72° 33′ 4″.

435. App. alt. of a star is 29° 59′ 16″; moon's app. alt., 32° 30′ 10″; star's zen. dist., 60° 2′ 24″; moon's zen. dist., 56° 41′ 33″; auxiliary angle ʌ, 60° 17′ 23″; and app. dist., 58° 44′ 19″.

Ans. True distance, 58° 30′ 21″.

Rule 50. To FIND THE LONGITUDE BY LUNAR OBSERVATIONS.

Objects observed, sun and moon. Altitudes taken. Ship mean time determined from *sun's altitude.*

1. Get a Greenwich date.

2. Take from the *Nautical Almanac* and correct for Greenwich date the following quantities:

Sun's declination and semidiameter.

Equation of time (noting whether it is to be added to or subtracted from the ship apparent time).

Moon's semidiameter and horizontal parallax.

3. Correct the sun's apparent altitude for index correction, dip, semidiameter, correction in altitude, and thus get the sun's apparent and true altitudes. Subtract the true altitude from 90° for sun's zenith distance.

4. Correct the moon's observed altitude for index correction, dip, semidiameter (augmented), correction in altitude, and thus get the moon's apparent and true altitude. Subtract the true altitude from 90° for moon's zenith distance.

5. When the moon's correction in altitude is taken out of the Tables, take out also at the same opening the auxiliary angle ʌ.

6. Correct the observed distance for index correction, and to the result add the semidiameters of the sun and moon (augmented), and thus get the apparent distance of the centers.

7. *To find ship mean time*, by first method, using haversines.*

Under sun's declination put the latitude of the ship; take the *sum* if their names be *unlike*, the *difference* if the names be *alike*. Under the result put the sun's zenith distance; take the sum and difference of the last two lines put down. Add together the log. secants of the two first quantities in this form (omitting to put down the tens in the index), and half of the log. haversines of each of the two last quantities. The sum will be the log. haversine of the ship apparent time. When the sun is west of the meridian, the time corresponding to the haversine must be taken out at the top of the page; but when the sun is east it must be taken out at the bottom. The result is apparent time at the ship: to this apply the equation of time with its proper sign, and the result will be the ship mean time.

8. *To calculate the true distance.* First method, using versines.†

Add together the zenith distances of the sun and moon, and mark the sum *v*.

Add together the apparent altitudes of the sun and moon, and under the sum put the auxiliary angle A: take the sum and difference of the last two quantities, and mark each with the letter *v*.

Under the apparent distance of the centers put the auxiliary angle A, and take the sum and difference, and mark each result with the letter *v*.

Add together the *five last figures* of the versines of each of the quantities marked *v*. The five last figures in the sum being looked for in the column of versines under the apparent distance or in the adjacent column, the arc corresponding thereto will be the true distance at the time of the observation.

9. *To find Greenwich mean time corresponding to this true distance.*

Take out of the *Nautical Almanac* two distances of the sun and moon three hours apart, between which is the true distance just calculated: place the first distance taken out under the true distance, and the one three hours after under the other distance taken out. Take the difference between the first and second, and also between the second and third. From the proportional logarithm of the first difference subtract the proportional logarithm of the second difference; the remainder is the proportional logarithm of a portion of time, which take from the table, and add thereto the hours corresponding to the first distance taken out of the *Nautical Almanac*. The result is Greenwich mean time when the observation was taken.

The difference between ship mean time found above and Greenwich mean time is the longitude in time: turn it into degrees, and mark it "east if the Greenwich time is the least, and west if the Greenwich time is best."

* To find ship mean, using only the common tables of sines, &c., see second method, p. 164, ex. 395.

† To find true distance, using sines, &c., see second method, p. 189, ex. 426.

O

EXAMPLES.

436. Feb. 12, 1848, at 2^h 36^m P.M. mean time nearly, in lat. 53° 30' N., and long. by account 15° 45' E., the following lunar observation was taken :

	Obs. alt. sun's L. L.	Obs. alt. moon's L. L.	Obs. dist. N. L.
	29° 17' 26"	25° 40' 20"	99° 27' 30"
Index cor.	2 10—	1 10—	0 50—

The height of the eye above the sea was 20 feet : required the longitude.

Ship, February 12	2^h 36^m	
Longitude in time	1 3 E.	
Greenwich, February 12 . .	1 33	

Sun's declin.		Eq. of time.	Moon's semi.		Hor. par.
12th . 13° 52' 18"S. . .	. 14ᵐ 33ˢ add	12th noon . 15' 58"	58' 36"
13th . 13 32 21 S. . .	. 14 32	„ mid. . 15 54	58 23
19 57	1	4	13
1·18985	∴ cor. 0	·88885	·88885
·95533	14 33	3·43136	2·91948
2·14518 1 17		4·32021 0·4 3·80833	1·7
13 51 1 S.		15 57·6	58 34·3
sun's semi. . 16' 13"		Aug. . . 7·4	
		16 5	

Sun's alt.		Moon's alt.		Distance.	
Obs. alt. . . 29° 17' 26"	Obs. alt. . 25° 40' 20"		99° 27' 30"
index cor. . 2 10—	index cor. . 1 10—	index cor. . 0 50—
29 15 16	25 39 10	99 26 40
dip . . . 4 24—	dip . . . 4 24—	sun's semi.'. 16 13
29 10 52	25 34 46	moon's semi. 16 5
semi. . . . 16 13	semi. . . . 16 5	99 58 58
29 27 5	25 50 51	
cor. in alt. . 1 35—	cor. in alt. . 50 13	Aux. angle.
29 25 30	31	60° 13' 40"
90	26 41 35	9
sun's Z. D. . 60 34 30	90	3
	moon's Z. D. 63 18 25	60 13 52

To find ship mean time.

Sun's declination . . .	13° 51' 1" S.	. . sec. . .	0·012814
latitude	53 30 0 N.	. . sec. . .	0·225612
	67 21 1		
sun's zenith distance .	60 34 30		
sum	127 55 31	. . ½ hav. .	4·953521
difference	6 46 31	. . ½ hav. .	3·771503
		hav. . .	8·963450
Apparent time	2^h 21^m 12^s		
equation of time	14 33+		
ship mean time	2 35 45		

To find Greenwich mean time.

Sun's zenith distance .	60°	34'	30"				
moon's zenith distance	63	18	25				
	123	52	55 vers.				

Versines.

sun's app. alt. . . .	29	27	5	57262	. .	222	
moon's app. alt. . . .	25	50	51	30774	. .	210	
sum	55	17	56	03680	. .	19	
auxiliary angle A . .	60	13	52	40881	. .	83	
				31158	. .	18	
sum	115	31	48 vers.	63755	. .	552	
difference	4	55	56 vers.	552		True distance.	
apparent distance . .	99°	58	58	64307	. .	99° 27 25	
auxiliary angle A . .	60	13	52	187		98 38 0 0 hours.	
sum	160	12	50 vers.	120		100 14 7 3 hours.	
difference	39	45	6 vers.			0 49 25	
	1ʰ	32ᵐ 33ˢ				1 36 7	
	0					·56140	
Greenwich mean time	1	32	33 . .			·27247	
ship mean time . . .	2	35	45	. . .		·28893	
longitude in time . .	1	3	12				
longitude . . .	15°	48'	0" E.				

437. March 25, 1847, at 3ʰ 30ᵐ p.m. mean time nearly, in lat. 52° N., and long. by account 33° W., the following lunar was taken :

	Obs. alt. sun's L. L.	Obs. alt. moon's L. L.	Obs. dist. N. L.
	23° 10' 20"	28° 50' 10"	112° 56' 30"
Index cor.	6 10 —	5 0 +	4 20 —

The height of the eye above the sea was 20 feet : required the longitude.

· *Ans.* Hour-angle, 3ʰ 32ᵐ 9ˢ; true dist. 112° 52' 45";
longitude, 32° 59' 38" W.

438. April 20, 1847, at 2ʰ 0ᵐ p.m. mean time nearly, in lat. 50° 50' N., and long. by account 1° 40' E., the following lunar was taken :

	Obs. alt. sun's L. L.	Obs. alt. moon's L. L.	Obs. dist. N. L.
	43° 16' 30"	24° 39' 20"	69° 12' 0"
Index cor.	0 10 +	0 20 —	0 50 —

The height of the eye above the sea was 20 feet : required the longitude.

Ans. Hour-angle, 2ʰ 1ᵐ 25ˢ; true dist. 69° 11' 37";
longitude, 1° 21' E.

439. May 19, 1847, at 2ʰ 50ᵐ P.M. mean time nearly, in lat. 51° 30′ N., and long. by account 20° 40′ E., the following lunar was taken:

	Obs. alt. sun's L. L.	Obs. alt. moon's L. L.	Obs. dist. N. L.
	42° 50′ 30″	25° 10′ 20″	61° 40′ 20″
Index cor.	4 10+	6 10−	2 10+

The height of the eye above the sea was 20 feet: required the longitude.

Ans. Hour-angle, 2ʰ 57ᵐ 20ˢ; true dist. 61° 44′ 38″; longitude, 20° 41′ 26″ E.

440. Feb. 6, 1851, at 3ʰ 30ᵐ P.M. mean time nearly, in lat. 60° 20′ N., and long. by account 26° 45′ E., the following lunar was taken:

	Obs. alt. sun's L. L.	Obs. alt. moon's L. L.	Obs. dist. N. L.
	24° 20′ 0″	33° 10′ 0″	57° 30′ 10″
Index cor.	2 30+	1 20+	2 0+

The height of the eye above the sea was 11 feet: required the longitude.

Ans. Hour-angle, 3ʰ 22ᵐ 21ˢ; true dist. 57° 56′ 43″; longitude, 28° 34′ 38″ E.

441. Feb. 20, 1856, at 3ʰ 50ᵐ P.M. mean time nearly, in lat. 10° 20′ N., and long. by account 7° W., the following lunar was taken:

	Obs. alt. sun's L. L.	Obs. alt. moon's L. L.	Obs. dist. N. L.
	30° 15′ 40″	24° 10′ 10″	100° 55′ 10″
Index cor.	3 10−	1 10+	0 30+

The height of the eye above the sea was 20 feet: required the longitude.

Ans. Hour-angle, 3ʰ 44ᵐ 12ˢ; true dist. 100° 54′ 54″; longitude, 7° 3′ 45″ W.

442. Jan. 9, 1851, at 2ʰ 50ᵐ P.M. mean time nearly, in lat. 56° 10′ 20″ N., and long. by account 20 40′ E., the following lunar was taken:

	Obs. alt. sun's L. L.	Obs. alt. moon's L. L.	Obs. dist. N. L.
	19° 10′ 20″	25° 30′ 10″	77° 10′ 20″
Index cor.	2 10+	1 20−	2 20−

The height of the eye above the sea was 20 feet: required the longitude.

Ans. Hour-angle, 2ʰ 45ᵐ 40ˢ; true dist. 77° 27′ 33″; longitude, 20° 35′ E.

443. July 16, 1869, at 4ʰ 15ᵐ P.M. mean time nearly, in lat. 25° 30′ N., and longitude by account 48 18′ E., the following lunar was taken:

	Obs. alt. sun's L. L.	Obs. alt. moon's L. L.	Obs. dist. N. L.
	34° 9′ 27″	45° 42′ 40″	93° 21′ 52″
Index cor.	2 10−	2 10+	1 15+

The height of the eye above the sea was 20 feet: required the longitude.

Ans. 48° 14′ E.

Elements from Nautical Almanac.

	Sun's declin.	Eq. of time.	Moon's semi.	Hor. par.	Sun's semi.
Mar. 25,	1° 40′ 56″N. . .	6ᵐ 13·8ˢ add .	. 14′ 59″ .	. 54′ 59″	. . 16′ 3″
„ 26,	2 4 29 N. . .	5 55·2 „	. 14 55 .	. 54 44	

Distance at 3 hours, 111° 33′ 34″; at 6 hours, 112° 57′ 16″.

Apr. 20, 11	23 54 N. . .	1 3·4 sub. .	. 15 23·7 .	. 56 29·8	. . 15 56
„ 21, 11	44 26 N. . .	1 16·3 „ .	. 15 17·0 .	. 56 5·3	

Distance at noon, 68° 14′ 43″; at 3 hours, 66° 43′ 48″.

May 19, 19	41 37 N. . .	3 49·58 sub. .	. 15 11·6 .	. 55 45·5	. 15 49
„ 20, 19	54 26 N. . .	3 46·90 „ .	. 15 6·3 .	. 55 25·7	

Distance at noon, 61° 0′ 58″; at 3 hours, 62° 27′ 35″.

Feb. 6, 15	42 19 S. .	. 14 22·51 add .	. 14 55·0 .	. 54 44·3	. . 16 14
„ 7, 15	23 35 S. .	. 14 26·16 „ .	. 14 59·0 .	. 54 59·1	

Distance at 0 hours, 57° 9′ 21″; at 3 hours, 58° 32′ 36″.

Feb. 20, 10	55 53 S. .	. 14 1·7 add .	. 16 4·4 .	. 58 59·2	. . 16 11
„ 21, 10	34 16 S. .	. 13 54·7 „ .	. 16 9·6 .	. 59 18·0	

Distance at 3 hours, 100° 7′ 50″; at 6 hours, 101° 45′ 50″.

Jan. 9, 22	9 2 S. . .	7 19·4 add .	. 14 55·8 .	. 54 47·2	. . 16 17
„ 10, 22	0 23 S. . .	7 44·1 „ .	. 15 0·2 .	. 55 3·6	

Distance at 0 hours, 76° 45′ 42″; at 3 hours, 78° 8′ 46″.

July 16, 21	20 13 N. . .	5 45·0 add .	. 16 5·1 .	. 58 55·7	. . 15 46
hourly difference .	. +25″ .	. +0·24ˢ .	. 16 1·4 .	. 58 42·2	

Distance at 0 hours, 92° 49′ 59″; at 3 hours, 94° 27′ 44″.

When the sun or star is near the meridian, the ship mean time must be obtained by computing the hour-angle of the moon, and deducing from thence the ship mean time. This may be done by the following rule :

Rule 51. To FIND THE LONGITUDE BY LUNAR OBSERVATIONS.

Objects observed, moon and sun. Altitudes taken, and ship mean time obtained from *moon's altitude.*

1. Get a Greenwich date.

2. Take out of *Nautical Almanac* and correct for Greenwich date the following quantities : Sun's semidiameter and right ascension of mean sun ; right ascension and declination of moon ; semidiameter and horizontal parallax of moon.

3. Correct the sun's altitude for index correction, dip, semidiameter, and thus get the apparent altitude : from the apparent altitude subtract correction in altitude ; the result is sun's true altitude, which subtract from 90° for sun's true zenith distance.

4. Correct the moon's altitude for index correction, dip, semidiameter

(augmented); the result is the moon's apparent altitude. To the apparent altitude add the correction in altitude, the result subtract from 90° for the moon's true zenith distance.

5. When the moon's correction in altitude is taken out, take out also at the same opening of the book the auxiliary angle A.

6. Correct the observed distance for index and semidiameter.

7. *To find ship mean time from moon's altitude.* First method, using haversines.*

Under the moon's declination put the latitude of ship : take the difference if the names be alike, but their sum if the names be unlike : under the result put the moon's zenith distance, and take the sum and difference. Add together the log. secants of the two first quantities in this form (rejecting the tens in index), and the halves of the log. haversines of the two last; the sum is the log. haversine of the moon's hour-angle, to be taken out at the top of the page if the moon is west of the meridian, but at the bottom of the page if the moon is east of meridian. To the hour-angle thus found add the moon's right ascension, and from the sum (increased if necessary by 24 hours) subtract the right ascension of the mean sun ; the remainder (rejecting 24 hours if greater than 24 hours) is ship mean time at the instant of observation.

8. Then proceed to calculate the true distance and Greenwich mean time as pointed out in p. 193, arts. 8, 9.

EXAMPLES.

444. May 22, 1844, at 11^h 15^m A.M. mean time nearly, in lat. 50° 48′ N., and long. by account 1° W., the following lunar observation was taken :

	Obs. alt. moon's L. L.	
Obs. alt. sun's L. L.	E. of meridian.	Obs. dist. N. L.
57° 53′ 0″	22° 53′ 2″	56° 26′ 6″
Index cor. 0 35+	0 20−	0 35−

The height of the eye above the sea was 24 feet : required the longitude.

			R. A. mean sun.	
Ship, May 21	23^h 15^m	May 21	3^h 56^m $53·8^s$
longitude in time.	4 W.	cor. 23^h	3 46·7
Greenwich, May 21	23 19	19^m	3·1
sun's semi. . . 15′ 49″			4 0 43·6

---

* To find ship mean time using only the common tables of sines, see second method, p. 164, ex. 395.

	Moon's R. A.			Moon's decl.	
23ʰ	. . . . 7ʰ 55ᵐ 24ˢ	. . . . 17° 5′ 12″N.
0	. . . . 7 57 29	. . . . 16 57 11 N.

logistic logs.	2 5	logistic logs.	8 1
·49940		·49940	
1·93651		1·35128	
2·43591	0 40	1·85068	2 32
	7 56 4		17 2 40 N.

	Moon's semi.			Moon's hor. par.	
21, mid.	. . . . 15′ 1·3″	. . . . 55′ 7·6″
22, noon	. . . . . . 15 5·7	. . . . 55 23·7
	4·4		16·1

·02546		·02546	
3 38997		2·82930	
3·41543	. . . 4·2	2·85476	. . . . 15·1
	15 5·5		55 22·7
aug. . .	5·5		
moon's semi. . . . . . 15 11·0			

Sun's alt.	Moon's alt.	
Ob. alt. . . 57° 53′ 0″	Obs. alt. . . 22° 53′ 2″	Obs. dist. . . 56° 26′ 6″
in. cor. . . 35+	in. cor. . . 0 20−	in. cor. . . 35−
57 53 35	22 52 42	56 25 31
dip . . . . 4 49−	dip . . . . 4 49−	sun's semi. . 15 49
57 48 46	22 47 53	moon's semi. 15 11
semi. . . . 15 49	semi. . . . 15 11	M. app. dist. . 56 56 31
58 4 35	23 3 4	
cor. in alt. . 32−	cor. 48 21	Aux. ang. A.
sun's true alt. 58 4 3	21	60° 11′ 31″
90	M. true alt. . 23 51 46	5
sun's zen. dis. 31 55 57	90	3
	M. zen. dist. . 66 8 14	60 11 39

Moon's hour-angle. First method, using haversines.*

Moon's decl. . . . 17° 2′ 40″N.	. . log. sec. . . . . 0·019510
latitude . . . . . 50 48 0 N. . . „ sec. . . . . 0·199263
33 45 20	„ ½ hav. (S) . . 4·883909
moon's zen. dist. . . 66 8 14	„ ½ hav. (D) . . ′4·445373
sum . . . . . . 99 53 34 (S)	„ hav. . . . . . 9·548055
difference . . . 32 22 54 (D)	
.·. hour-angle 19ʰ 8ᵐ 17ˢ
moon's R. A. 7 56 4
27 4 21
R. A. mean sun 4 0 44
.·. ship mean time 23 3 37

* If no haversines, then find hour-angle by second method. (See ex. 395, p. 164.)

Greenwich mean time.   First method, using versines.*

Zenith distance	31°	55'	57"			
zenith distance	66	8	14			
	98	4	11 vers.			

				Versines.		
apparent altitude.	58	4	35			
apparent altitude.	23	3	4			
sum	81	7	39			
auxiliary angle A.	60	11	39	40325	. .	53
sum	141	19	18 vers.	80612	. .	55
difference	20	56	0 vers.	66003	. .	0
				56063	. .	43
apparent distance	56	56	31	01608	. .	2
auxiliary angle A.	60	11	39	44611	. .	153
sum	117	8	10 vers.	153		
difference	3	15	8 vers.	44764		
				672		
true distance	56°	16'	23"	92		
„ at 21ʰ . . .	55	15	36			
	56	41	20			

·47149 . . .	1	0	47
·32212 . . .	1	25	44
·14937			

$$2^h \quad 7^m 37^s$$
$$21$$

Greenwich mean time. . . .	23	7	37
ship mean time	23	3	37
longitude in time		4	0
or longitude	1°	0' W.	

445. May 16, 1850, at 0ʰ 50ᵐ P.M. mean time nearly, in lat. 42° 30' N., and long. by account 29° 6' W., the following lunar was taken :

	Obs. alt. moon's L. L.	
	E. of meridian.	
Obs. alt. sun's L. L.		Obs. dist. N. L.
61° 44' 30"	39° 30' 20"	63° 10' 0"
Index cor. 2 10—	1 10—	0 20+

The height of the eye above the sea was 20 feet : required the longitude.

Ans. Hour-angle, 20ʰ 32ᵐ 26ˢ; true dist. 63° 4' 0";
longitude, 29° 5' 0" W.

446. May 18, 1850, at 1ʰ 0ᵐ P.M. mean time nearly, in lat. 43° N., and long. by account 41° 36' W., the following lunar was taken:

	Obs. alt. moon's L. L.	
	E. of meridian.	
Obs. alt. sun's L. L.		Obs. dist. N. L.
64° 30' 10"	18° 10' 20"	90° 20' 10"
Index cor. 2 10+	1 10—	1 20+

The height of the eye above the sea was 15 feet : required the longitude.

Ans. Hour-angle, 18ʰ 57ᵐ 17ˢ; true dist. 90° 2' 57";
longitude, 41° 33' 45" W.

* If no versines, then find true distance by second method. (See ex. 426, p. 189.)

Elements from Nautical Almanac.

Mean sun's R.A.	Moon's semi.	Hor. par.	Moon's R.A.	Moon's decl.	S. semi.
16th, $3^h 35^m 22 \cdot 14^s$	Noon, $16' 16 \cdot 4''$	$59' 43 \cdot 0''$	$7^h 58^m 16 \cdot 79^s$	$18° 51' 56'' N.$	$15' 50''$
	Mid. 16 13·9	59 34·0	8 0 48·01	18 47 43 N.	

Distance at noon, $61° 27' 26''$; at 3 hours, $63° 7' 50''$.

| 18th, 3 43 15·25 | Noon, 16 4·3 | 58 58·8 | 9 57 4·66 | 13 17 16 N. | 15 49 |
| | Mid. 16 0·6 | 58 45·3 | 9 59 23·58 | 13 8 8 N. | |

Distance at 3 hours, $89° 31' 39''$; at 6 hours, $91° 9' 1''$.

Rule 52. To FIND THE LONGITUDE BY LUNAR OBSERVATIONS.

Objects observed, moon and star. Altitudes taken and ship mean time obtained from *star's altitude.*

1. Get a Greenwich date.

2. Take out of the *Nautical Almanac* and correct for Greenwich date the following quantities : Star's right ascension and declination ; right ascension of mean sun ; moon's semidiameter and horizontal parallax.

3. Correct the star's observed altitude for index correction and dip ; the result is the star's apparent altitude ; from this subtract refraction ; the remainder is the star's true altitude, which take from 90° to find star's true zenith distance.

4. Correct the moon's altitude for index correction, dip, semidiameter (augmented), and thus get the moon's apparent altitude ; to this add the correction in altitude : the result is the moon's true altitude. Subtract the moon's true altitude from 90° ; the remainder is the moon's true zenith distance.

5. When the correction of the moon's altitude is taken out, take out also at the same opening of the book the auxiliary angle A.

6. *To find ship mean time from star's altitude.* First method, using haversines.*

Under star's declination put latitude of ship : take the sum if the names be unlike, but if the names be like take the difference. Under the result put the star's true zenith distance ; take the sum and difference of the two last quantities put down.

Add together the log. secants of the two first quantities in this form (rejecting the tens in the index), and the halves of the log. haversines of the two last ; the sum will be the log. haversine of the star's hour-angle, to be taken out at the top of the page when the star is *west* of the meridian, and at bottom when east. To the hour-angle thus found add the star's right ascension, and from the sum (increased if necessary by 24 hours) subtract

---

* To find ship mean time, using the common tables of sines, &c., see second method, p. 164, and ex. 395.

the right ascension of the mean sun : the result (rejecting 24 hours if greater than 24 hours) will be ship mean time.

Then proceed to calculate the true distance and Greenwich mean time, as pointed out in p. 193, arts. 8, 9.

EXAMPLES.

447. June 2, 1849, at 10^h 17^m P.M. mean time nearly, in lat. 50° 50′ N., and long. by account 41° W., the following lunar observation was taken :

	Obs. alt. Regulus. W. of meridian.	Obs. alt. moon's L. L.	Dist. N. L.
	20° 21′ 40″	31° 11′ 0″	72° 36′ 30″
Index cor.	3 50—	4 10—	9 10—

The height of the eye above the sea was 20 feet : required the longitude.

Ship, June 2	10^h	17^m	
longitude in time	2	44 W.	
Greenwich, June 2	13	1	

Right asc. mean sun.				Moon's semi.	Hor. par.
2d ... 4^h 43^m $21 \cdot 2^s$	Star's R. A. 10^h 0^m 20^s	2d, mid.	14′ 49″ ...	54′ 23″
13^h 2 8·1	Star's decl. 12 42 4 N.	3d, noon	14 47 ...	54 15
	4 45 29·3		2	8
		cor.	0 ...	0
		14 49	54 23
		aug.	7	
		14 56	

	Star's altitude.		Moon's altitude.		Observed distance.	
Obs. alt....	20°21′ 40″	Obs. alt. ...	31°11′ 0″		72°36′ 30″	
in. cor. ...	3 50—	in. cor. ...	4 10—	in. cor.......	9 10—	
	20 17 50		31 6 50		72 27 20	
dip.........	4 24—	dip.........	4 24	semi.	14 56	
app. alt. ...	20 13 26		31 2 26	app. dist. ...	72 42 16	
cor. in alt.	2 37—	semi.	14 56			
	20 10 49	app. alt. ...	31 17 22		Aux. angle A.	
	90	cor. in alt..	44 34	....	60° 15′ 14″	
zen. dist....	69 49 11		20	....	7	
			32 2 16		60 15 21	
			90			
		zen. dist....	57 57 44			

To find ship mean time.

Star's declination ...	12°	42'	4" N.	...... sec.	0·010765
latitude	50	50	0 N.	...... sec.	0·199573
	38	7	56		
star's zenith dist. ...	69	49	11		
sum	107	57	7	...... ½ hav. ...	4·907820
difference	31	41	15	...... ½ hav. ...	4·436186
				Hav. ...	9·554344

Star's hour-angle.....................	4ʰ	54ᵐ	11ˢ
star's right ascension	10	0	20
	14	54	31
right ascension mean sun	4	45	29
ship mean time	10	9	2

To find Greenwich mean time.

				Versines.				
Star's zen. dist. ...	69°	49'	11"					
moon's zen. dist...	57	57	44	81360	 131			
				23453	 56			
	127	46	55 vers.	12447	 212			
				70828	 41			
star's app. alt. ...	20	13	26	11594	 24			
moon's app. alt. ..	31	17	22					
				99682	 464			
	51	30	48	464				
aux. angle ⊿......	60	15	21					
				00146	... 72°	33'	5"	
sum 111	46	9 vers.		127	72	19	35 at 12 hours.
difference 8	44	33 vers.		19	73	49	27 at 15 hours.
app. dist. 72	42	16					
aux. ⊿. 60	15	21	1·12548		13	29	
			·30167	1	29	52	
sum 132	57	37 vers.					
difference 12	26	55 vers.	·82381	... 0ʰ	27ᵐ	0ˢ	
				12			

Greenwich mean time	12	27	0
ship mean time	10	9	2
longitude in time	2	17	58
Longitude .... 34° 30' 10" W.			

448. Jan. 8, 1851, at 7ʰ 0ᵐ P.M. mean time nearly, in lat. 50° 40′ N. and long. by account 4° E., the following lunar was taken :

	Obs. alt. Aldebaran. E. of meridian.	Obs. alt. moon's L. L.	Obs. dist. F. L.
	45° 20′ 10″	30° 30′ 0″	71° 31′ 10″
Index cor.	2 20 −	2 10 +	3 30 −

The height of eye above the sea was 18 feet : required the longitude.

Ans. Hour-angle, 21ʰ 36ᵐ 57ˢ; true dist. 70° 42′ 50″; longitude, 3° 46′ 30″ E.

449. Jan. 9, 1851, at 7ʰ 50ᵐ P.M. mean time nearly, in lat. 49° 40′ N., and long. by account 10° E., the following lunar was taken :

	Obs. alt. Pollux. E. of meridian.	Obs. alt. moon's L. L.	Obs. dist. F. L.
	37° 10′ 10″	31° 50′ 10″	103° 20′ 0″
Index cor.	1 10 −	1 20 +	1 30 +

The height of eye above the sea was 18 feet: required the longitude.

Ans. Hour-angle, 19ʰ 39ᵐ 52ˢ; true dist. 102° 28′ 3″ ; longitude, 10° 20′ E.

450. April 18, 1850, at 9ʰ 40ᵐ P.M. mean time nearly, in lat. 56° 10′ N., and long. by account 23° E., the following lunar was taken :

	Obs. alt. α Virginis. E. of meridian.	Obs. alt. moon's L. L.	Obs. dist. F. L.
	19° 40′ 0″	48° 40′ 20″	91° 30′ 20″
Index cor.	1 10 +	1 20 +	0 30 −

The height of eye above the sea was 20 feet : required the longitude.

Ans. Hour-angle, 22ʰ 8ᵐ 50ˢ; true dist. 90° 55′ 16″ ; longitude, 23° 3′ 30″ E.

451. April 17, 1850, at 8ʰ 45ᵐ P.M. mean time nearly, in lat. 51° 20′ N., and long. by account 5° 10′ E., the following lunar was taken :

	Obs. alt. α Virginis. E. of meridian.	Obs. alt. moon's L. L.	Obs. dist. F. L.
	18° 15′ 30″	36° 25′ 10″	105° 33′ 28″
Index cor.	1 10 +	1 20 −	0 20 +

The height of eye above the sea was 20 feet : required the longitude.

Ans. 5° 8′ E.

452. January 11, 1868, at 5ʰ 40ᵐ P.M. mean time nearly, in latitude 50° 48′ N., and long. by account 70° W., the following lunar was taken:

	Obs. alt. α Tauri. E. of meridian.	Obs. alt. moon's L. L.	Obs. dist. N. L.
	36° 25′ 50″	30° 39′ 30″	71° 50′ 55″
Index cor.	5 30 +	2 10 −	2 15 −

The height of eye above the sea was 20 feet : required the longitude.

$$Ans.\ 69°\ 58'\ \text{W.}$$

453. March 20, 1845, at $7^h\ 50^m$ P.M. mean time nearly, in lat. 49° 50′ N., and long. by account 1° 30′ E., the following lunar was taken :

Obs. alt. Aldebaran.

W. of meridian.	Obs. alt. moon's L. L.	Obs. dist. N. L.
37° 4′ 30″	38° 17′ 20″	75° 15′ 10″
Index cor. 2 4—	1 5+	0 20+

The height of eye above the sea was 20 feet : required the longitude.

$$Ans.\ 1°\ 34'\ 30''\ \text{E.}$$

Elements from Nautical Almanac.

Right ascension mean sun.	Moon's semi.	Hor. par.	
Jan. 8, 19ʰ 9ᵐ 45·72ˢ.	. Noon, 14′ 48·8″.	. 54′ 21·8″	Star's R.A. . 4ʰ 27ᵐ 22·7ˢ
	Mid. 14 52·0 .	. 54 33·3	Star's decl. . 16° 12′ 13″ N.
Distance at 6 hours, 71° 1′ 57″; at 9 hours, 69° 32′ 16″.			
Jan. 9, 19 13 42·27 .	. Noon, 14 55·8 .	. 54 47·2	Star's R.A. . 7ʰ 36ᵐ 12ˢ
	Mid. 15 0·2 .	. 55 3·6	Star's decl. . 28° 22′ 46″ N.
Distance at 6 hours, 103° 8′ 4″; at 9 hours, 101° 37′ 51″.			
Ap. 18, 1 44 58·62 .	. Noon, 16 8·8 .	. 59 15·4	Star's R.A. . 13ʰ 17ᵐ 19ˢ
	Mid. 16 8·7 .	. 59 14·8	Star's decl. . 10° 22′ 43″S.
Distance at 6 hours, 92° 10′ 13″; at 9 hours, 90° 24′ 32″.			
Ap. 17, 1 41 2·07 .	. Noon, 16 8·2 .	. 59 13·1	Star's R.A. . 13ʰ 17ᵐ 19ˢ
	Mid. 16 8·7 .	. 59 14·9	Star's decl. . 10° 22′ 43″S.
Distance at 6 hours, 106° 16′ 11″; at 9 hours, 104° 30′ 26″.			
Jan. 11, 19 21 8·50 .	. Noon, 16 41·6 .	. 61 9·8	Star's R.A. . 4ʰ 28ᵐ 21ˢ
	Mid. 16 37·2 .	. 60 53·6	Star's decl. . 16° 14′ 21″N.
Distance at 9 hours, 70° 16′ 10″; at 12 hours, 72° 8′ 51″.			
Mar. 20, 23 51 30·11 .	. Noon, 15 12·2 .	. 55 47·6	Star's R.A. . 4ʰ 27ᵐ 3ˢ
	Mid. 15 17·4 .	. 56 6·7	Star's decl. . 16° 11′ 30″N.
Distance at 6 hours, 74° 7′ 12″; at 9 hours, 75° 42′ 9″.			

Rule 53. To FIND THE LONGITUDE BY LUNAR OBSERVATIONS.

Objects observed, moon and star. Altitudes taken and ship mean time obtained from *moon's altitude.*

This differs very little from Rule 51, p. 197.

1. Get a Greenwich date.

2. Take out of the *Nautical Almanac* and correct for Greenwich date the following quantities : Right ascension of mean sun ; moon's right ascension and declination; semidiameter, and horizontal parallax.

3. Correct the star's altitude for index correction, dip, and thus get the apparent altitude. From the star's apparent altitude subtract refraction : the result is the true altitude, which take from 90° for the star's true zenith distance.

Then proceed to calculate the moon's zenith distance, &c., ship mean time, true distance, and Greenwich mean time, as pointed out in p. 197.

Rule 54.

Objects observed, moon and planet. Altitudes taken and ship mean time from *planet's altitude*.

1. Get a Greenwich date.

2. Take out of the *Nautical Almanac* and correct for Greenwich date the following quantities: Right ascension of mean sun; planet's right ascension and declination; planet's horizontal parallax (if great accuracy is required); moon's semidiameter and horizontal parallax.

3. Correct the planet's observed altitude for index correction and dip, and thus get the apparent altitude. From the apparent altitude subtract the refraction, and add the parallax in altitude (usually neglected, being very small); the result is the planet's true altitude, which subtract from 90° to get the true zenith distance.

4. Correct the moon's altitude as in 4, p. 201.

5. Get the auxiliary angle A, as in 5, p. 201.

6. Find ship mean time, as in 6, p. 201, using planet's declination and right ascension instead of star's.

7. Then proceed to calculate the true distance and Greenwich mean time, as described in arts. 8, 9, p. 193.

EXAMPLE.

454. September 24, 1849, at $7^h 50^m$ P.M. mean time nearly, in latitude 47° 50′ N., and longitude by account 2° 30′ W., the following lunar observation was taken :

	Obs. alt. Saturn. E. of meridian.	Obs. alt. moon's L. L.	Obs. dist. F. L.
	16° 55′ 0″	15° 30′ 45″	89° 51′ 36″
Index cor.	3 5 −	3 10 −	3 0 +

The height of eye above the sea was 20 feet : required the longitude.

$$\begin{array}{ll}
\text{Ship, Sept. 24} & 7^h \quad 50^m \\
\text{long. in time} & 10 \text{ W.} \\
\hline
\text{Greenwich, Sept. 24} & 8 \quad 0
\end{array}$$

Right ascen. mean sun.				Planet's right ascen. and declination.					
24th	12ʰ	12ᵐ	48·4ˢ	24th...	0ʰ 21ᵐ 50·2ˢ	....	0° 31′ 23″ S.		
8ʰ		1	18·8	25th...	0 21 33·0	....	0 33 17 S.		
	12	14	7·2		17·2		1 54		
				Cor.	6·0	Cor.	38		
Planet's hor. par. 1·0″.					0 21 44		0 32 1 S.		

				Moon's hor. par.			Moon's semi.		
Obs. dist.	89°	51'	36"	Noon	54'	13·6"	....	14'	46·6"
index cor.		3	0+	Mid.	54	16·2	....	14	47·3
	89	54	36			2·6			0·7
semi......		14	51 —	cor.		1·7	cor. ...		0·5
app. dist.	89	39	45		54	15·3		14	47·1
							aug. ...		3·9
								14	51

Planet's altitude.				Moon's altitude.					
Obs. alt..........	16°	55'	0"	Obs. alt. ...	15°	30'	45"		
index cor.		3	5 —	index cor.		3	10 —		
	16	51	55		15	27	35		
dip		4	24 —	dip		4	24 —		
app. alt.	16	47	31		15	23	11		
refr. 3' 11" — ...				semi.......		14	51+		
par. in alt. 1" +		3	10 —		15	38	2	Angle A.	
	16	44	21	cor. in alt.		48	36	...60° 7'	29"
	90						14		2
									0
zenith dist. ...	73	15	39		16	26	52	60 7	31
					90				
				zen. dist.	73	33	8		

To find ship mean time.

Latitude	47°	50'	0"N.	Sec.	0·173090
declination........	0	32	1 S.	sec.	0·000019
				½ hav.	4·941037.
	48	22	1	⅓ hav.ʼ	4·333480
zenith distance ...	73	15	39		
				hav.	9·447626
sum	121	37	40		
difference	24	53	38		

Hour-angle	19ʰ	44ᵐ	16ˢ
planet's right ascension ...	0	21	44
	20	6	0
right ascension mean sun...	12	14	7·2
Ship mean time	7	51	52·8

To find Greenwich mean time.

				Versines.			
Zenith dist.......	73°	15'	39"				
zenith dist.	73	33	8	36764 126			
				44491 19			
	146	48	47 vers.	14471 130			
				64128 39			
App. alt.	16	47	31	29931 33			
app. alt.	15	38	2				
				89785 347			
	32	25	33	347			
aux. angle	60	7	31				
				90132 ... 89° 26' 5"			
sum	92	33	4 vers.	110	90 26 30 at 6 hours.		
difference.........	27	41	58 vers.	——	88 57 4		
				22			
App. dist.	89	39	45	·47412	1 0 25		
aux. angle	60	7	31	·30377	1 29 26		
sum	149	47	16 vers.	·17035	2ʰ 1ᵐ 36ˢ		
difference.........	29	32	14 vers.		6		

Greenwich mean time	8	1	36
ship mean time	7	51	53
longitude in time		9	43 W.
Longitude ...... 2° 25' 45" W.			

LONGITUDE BY LUNAR.—ALTITUDES CALCULATED.

To find ship mean time.

The error of the chronometer on ship mean time is found a little before or after the lunar distance is taken. For this purpose the observer selects any heavenly body whose bearing is nearly east or west, so that the error in the altitude may produce the smallest error in the resulting hour-angle (see Rule 43). Then the time being noted by the same chronometer when the distance is taken, ship mean time is known at the same instant by applying the error found by the above observation.

Rule 55. *To find longitude by lunar observations.*

Objects observed, moon and sun. Altitudes calculated.

1. Get a Greenwich date.

2. Take from the *Nautical Almanac* and correct for the Greenwich date the following quantities: Sun's declination, equation of time and semi-diameter, right ascension of mean sun. Moon's right ascension and decli-nation, moon's semidiameter and horizontal parallax.

3. *To find the sun's hour-angle.*

To the time shown by chronometer at the observation apply the error of chronometer with its proper sign, and thus get ship mean time; to this apply equation of time: the result is ship apparent time, and also the sun's hour-angle.

4. *To calculate the sun's true altitude.*

Under the latitude* put down the sun's declination; take the sum if the names be unlike, but the difference if the names be alike; call the result v; add together constant log. 6·301030, log. cos. latitude, log. cos. sun's declination, and log. haversine sun's hour-angle; reject 30 in the index, and look out the result as a logarithm, and take its natural number to the nearest unit. (If no haversines, find sun's and moon's true alt. by Rule IX., *Trig.* Part I., viz. two sides and included angle given, to find third side.)

Add together this natural number and the versine of the quantity v found above: the sum is the versine of the sun's *true zenith distance*, which find in the tables and subtract from 90°: the result is the sun's true altitude.

To find the sun's apparent altitude.

To the true altitude just found *add* correction in altitude (for parallax and refraction): the result will be the sun's apparent altitude very nearly.†   .

5. *To find the moon's hour-angle.*

To right ascension of mean sun add ship mean time, and from the sum (increased if necessary by 24 hours) subtract the moon's right ascension: the result is the moon's hour-angle (rejecting 24 hours if greater than 24 hours).

6. *To find the moon's true altitude.*

Under the latitude put the moon's declination; take the sum if the names be unlike, and the difference if the names be alike; call the result v. Add together the constant quantity 6·301030, log. cos. latitude, log. cos. moon's declination, and log. haversine of moon's hour-angle; reject 30 from the index, and look out the result as a logarithm, and take its natural number. To this natural number add the versine of the quantity v, found as above; the sum is the versine of the moon's true zenith distance, which find in the table and subtract from 90°: the result is the moon's true altitude.

---

* When great accuracy is required, the latitude and horizontal parallax should be corrected for the spheroidal figure of the earth.

† In strictness, the table for correction of altitude ought to have been entered with the *apparent* altitude, instead of the *true*, to get the correction in altitude; but in the case of the sun's altitude the above is sufficiently correct. A more exact method is followed in finding the *moon's* apparent altitude from the true (p. 210).

To find the moon's apparent altitude.

Consider the moon's true altitude just found as the apparent altitude; enter the table with it, and take out the correction in altitude thus approximately; subtract this correction from the moon's true altitude, and thus get the moon's apparent altitude nearly. Then enter the table again with this *corrected altitude*, and thus take out the correction in altitude more exactly; subtract this correction from the moon's *true* altitude, and the result will be the moon's apparent altitude very nearly.

Take out at the same opening of the table the auxiliary angle A.

Correct the moon's semidiameter for augmentation. Then proceed as in arts. 6, 8, 9, p. 193.

EXAMPLE.

455. August 19, 1843, in lat. 50° 48′ N., and long. by account 1° 6′ W., when a chronometer showed $11^h 10^m 19.8^s$ A.M., the observed distance of the nearest limb of the sun and moon was 76° 51′ 26″, the error of the chronometer on ship mean time being fast $7^m 29.3^s$, and the index correction 1′ 57″ + : required the longitude.

Time by chro....	$11^h 10^m 19.8^s$		Sun's declination.			Eq. of time.
error of chro....	7 29.3 fast		18th 14° 15′ 12″ N.			$3^m 43.2^s$ sub.
			19th 13 55 47 N.			3 30.1
ship mean time	23 2 50.5					
long. in time ...	4 24.0 W.		19 25			13.1
			.01629		.01692	
Gr. Aug. 18 ...	23 7		.96710		2.91615	
Ship mean time	23 2 50.5		.98339	18 42	2.93244	12.6
equation of time	3 30.6 sub.					
				13 56 30 N.		3 30.6
sun's hour-angle	22 59 19.9			Sun's semi.... 15′ 49″		

		Moon's right ascen.		Right ascen. mean sun.			Moon's declin.	
	18th at 23^h .	$4^h 27^m 27^s$	18th......	$9^h 44^m 48.22^s$			23° 43′ 13″ N.	
	„ 24 .	4 29 41		3 46.84			23 43 27 N.	
	.93305	2 14		1.14				
	1.42920						0 14	
				9 48 36.20	.93305			
	2.36225	16	ship M.T. 23	2 50.50	2.41018			
							0 2	
		4 27 43		32 51 26.70	3.34323			
			M.'s R.A.	4 27 43.0			23 43 15 N.	
			M.'s H.A.	4 23 43.70				

	Moon's semi.				Moon's hor. par.		
18th, at mid.........	14'	59·7"	....		55'	1·8"
19th, at noon	15	4·5	....		55	19·1

	4·8		17·8
·03321		·03321	
3·35218		2·79538	
3·38539	4·4	2·82859	16·0

	15	4·1		55	17·8
aug.		8·2			
moon's semi.........	15	12·3			

To calculate sun's altitude.	*To calculate moon's altitude.*

Latitude	50°	48'	0" N.	Latitude	50°	48'	0" N.
sun's declination ...	13	56	30 N.	moon's decl.........	23	43	15 N.
	36	51	30 v.		27	4	45 $v_{,}$

Const. log.	6·301030	Const. log.	6·301030
cos. lat.......................	9·800737	cos. lat.......................	9·800737
cos. sun's decl.	9·987014	cos. moon's decl.	9·961666
hav. sun's hour-angle ...	8·240938	hav. moon's hour-angle .	9·471439
log.	4·329719	log.	5·534872
nat. no......................	21366	nat. no......................	342667
vers. v.	199792	vers. $v_{,}$	0109522
	87		99
ver. sun's zen. dist.	0221245	vers. moon's zen. dist. ...	0452288
	209		193
	36		95

Sun's zenith distance	38°	51'	12"	Moon's zenith dist...	56°	47'	23"
	90				90		
sun's true alt.........	51	8	48	moon's true alt.	33	12	37
cor. in alt. ....·		0	42	cor. in alt.		44	0
sun's app. alt.........	51	9	30	moon's ap. alt, nearly	32	28	0

Obs. dist. . 76° 51′ 26″   Aux. angle A.   cor. in alt....... 44′ 53″
                    1 57+   60° 16′   4″   15
              ─────────   4   ─────────
              76 53 23   3   True cor. in alt.   45   8
sun's semi.   15 49   ─────────   moon's true alt. 33 12 37
moon's ,,   15 12   60 16 11 arc A.   ─────────
              ─────────   moon's app. alt. 82 27 29
app. dist... 77 24 24

To find Greenwich mean time.

Sun's zenith dist...   38° 51′ 12″   Versines.
moon's zenith dist..   56 47 23   98462 ... 169
                       ─────────   07819 ... 40
sum ....   95 38 35 v.   81784 ... 83
                                          39239 ... 128
Sun's app. alt. 51   9 30   44378 ... 13
moon's app. alt. ...   32 27 29   ─────   ───
                       ─────────   71382   433
sum 83 36 59   433
arc A. ....   60 16   5   ─────   **True dist.**
                       ─────────   71815 .... 76° 48′ 35″
sum .... 143 53 14 v.   649   77   48 59 at 21ʰ
difference...........   23 20 44 v.   ─────   76   24   3
                                            166   ─────────
App. dist. ....   77 24 24   ·47424   1   0 24
arc A. ....   60 16 15   ·32619   1 24 56
                       ─────────   ─────────
sum .... 137 40 39 v.   ·14805   2ʰ 8ᵐ 0ˢ
difference...........   17   8   9 v.   21

                    Greenwich mean time 23   8   0
                    ship mean time 23   2 50
                                                        ─────────
                    longitude in time 5 10
                         Longitude ...... 1° 17′ 30″ W.

456. September 16, 1843, in latitude 50° 48′ N., and long. by account 1° 6′ W., when a chronometer showed 9ʰ 34ᵐ 6·6ˢ A.M., the observed distance of the nearest limb of the sun and moon was 96° 26′ 18″, the index correction being +1′ 32″, and the error of chronometer on ship mean time being fast 7ᵐ 59ˢ: required the longitude.   *Ans.* 1° 35′ 30″ W.

457. October 14, 1843, in lat. 50° 48′ N., and long. by account 1° 6′ W., when a chronometer showed 9ʰ 53ᵐ 57·1ˢ A.M., the observed distance of the nearest limb of the sun and moon was 114° 58′ 22″, the error of the chronometer being slow 3ᵐ 27ˢ, and the index correction +1′ 32″: required the longitude.   *Ans.* 1° 30′ 30″ W.

458. October 16, 1843, in lat. 50° 48′ N., and long. by account 1° 6′ W., when a chronometer showed 9ʰ 58ᵐ 9·8ˢ A.M., the observed distance of the nearest limb of the sun and moon was 91° 45′ 38″, the error of the chronometer being slow 3ᵐ 26·5ˢ, and the index correction +1′ 30″: required the longitude.   *Ans.* 1° 31′ 30″ W.

459. August 17, 1843, in lat. 50° 37′ 30″ N., and long. by account 1° 6′ W., when a chronometer showed 10ʰ 42ᵐ 28·7ˢ A.M., the observed distance of the nearest limb of the sun and moon was 99° 22′ 35″, the error of the chronometer on ship mean time being fast 7ᵐ 44ˢ, and the index correction +1′ 55″: required the longitude.   *Ans.* 1° 17′ W.

460. May 25, 1843, in lat. 50° 48′ N., and long. by account 1° 6′ W., when a chronometer showed 11ʰ 19ᵐ 15·1ˢ A.M., the observed distance of the nearest limb of the sun and moon was 42° 48′ 48·3″, the error of the chronometer on ship mean time being slow 3ᵐ 29·7ˢ, and the index correction +3′ 30″: required the longitude.   *Ans.* 1° 5′ 45″ W.

461. May 25, 1843, in lat. 50° 37′ 30″ N., and long. by account 1° 6′ W., when a chronometer showed 11ʰ 4ᵐ 12·2ˢ A.M., the observed distance of the sun and moon's nearest limb was 43° 1′ 3″, the error of chronometer on ship mean time being fast 12ᵐ 53·5ˢ, and the index correction +0′ 57″: required the longitude.   *Ans.* 1° 30′ W.

Elements from Nautical Almanac.

Sun's declination.	Equation of time.	Mean sun's right ascen.
Sept. 15... 3° 11′ 23″N.	... 4ᵐ 42·5ˢ add	... 15th... 11ʰ 35ᵐ 11·72ˢ
„ 16... 2 48 15 N.	... 5 3·6	Sun's semi. 15′ 56″

Moon's right ascen.	Moon's declin.	Moon's semi.	Hor. par.
- 15th, at 21ʰ... 4ʰ 59ᵐ11·7ˢ	... 23°46′ 38·6″N....Mid.	...14′ 58·2″	...54′ 56·2″
„ 22ʰ... 5 1 25·7	... 23 47 9·0 N....Noon	...15 2·8	... 55 13·1

Distance at 21 hours, 96° 42′ 14″; at noon, 95° 17′ 50″.

Sun's declination.	Equation of time.	Mean sun's right ascen.
Oct. 13... 7° 38′ 12″S.	... 13ᵐ 36·0ˢ add	... 13th... 13ʰ 25ᵐ 35·20ˢ
„ 14... 8 0 40 S.	... 13 50·2	Sun's semi. 16′ 4″

Moon's right ascen.   Moon's declin.   Moon's semi.   Hor. par.
13th, at 22ʰ... 5ʰ 40ᵐ 1·1'... 23°21' 26"N. ... 14' 57·8"... 54' 54·7 '
  „   23ʰ... 5 42 14·6 ... 23 19 43 N.... Noon... 15   2·1 ... 55 10·3

Distance at 21 hours, 115° 27' 42"; at noon, 114° 3' 26".

Sun's declination.   Equation of time.   Mean sun's right ascen.
Oct. 15... 8° 23' 1·4"S. ... 14ᵐ 3·8' add ... 15th... 13ʰ 33ᵐ 28·31'
  „   16... 8   45 15·6 S. ... 14 16·8   Sun's semi. 16' 4"

Moon's right ascen.   Moon's declin.   Moon's semi.   Hor. par.
15th, at 22ʰ...7ʰ 26ᵐ 58·8'... 19°45' 6"N. ... Mid. ... 15' 18·0"... 56' 8·7"
  „   23ʰ...7 29 11·9 ... 19 37 45 N. ... Noon... 15 24·3 ... 56 32·1

Distance at 21 hours, 92° 27' 48"; at noon, 90° 58' 57".

Sun's declination.   Equation of time.   Mean sun's right ascen.
Aug. 16... 13° 53' 24"N. ... 4ᵐ 7·9' sub. ... 16th... 9ʰ 36ᵐ 55·12'
  „   17... 13   34 24 N. ... 3 55·8   Sun's semi. 15' 49"

Moon's right ascen.   Moon's declin.   Moon's semi.   Hor. par.
16th, at 22ʰ... 2ʰ 42ᵐ 36' ... 19°40'25"N. ... Mid. ... 14' 47·5"... 54' 16·8"
  „   23ʰ... 2 44 29 ... 19 47 38 N. ... Noon... 14 49·5 ... 54 24·3

Distance at 21 hours, 99° 59' 6"; at noon, 98° 37' 11".

Sun's declination.   Equation of time.   Mean sun's right ascen.
May 24... 20° 42' 10·8"N. ... 3ᵐ 30·9' add :.. 24th... 4ʰ 5ᵐ 44·32'
  „   25... 20   53 16·6 N. ... 3 25·6   Sun's semi. 15' 48"

Moon's right ascen.   Moon's declin.   Moon's semi.   Hor. par.
24th, at 23ʰ... 1ʰ 8ᵐ 43·11'... 12°39' 15·8"N.... Mid. ...14' 44·8"...54' 6·8"
  „   24ʰ... 1 10 36·96 ... 12 49 47·9 N.... Noon ... 14 45·7 ...54 10·3

Distance at 21 hours, 44° 0' 59"; at 0 hours, 42° 39' 15".

Rule 56. LONGITUDE BY LUNAR.

Objects observed, moon and star.   Altitudes calculated.

1. Get a Greenwich date.

2. Take from the *Nautical Almanac,* and correct for the Greenwich
date, the following quantities: Right ascension of mean sun; moon's right
ascension and moon's declination; moon's semidiameter and moon's horizontal parallax.   Take out also the star's right ascension and declination.

3. *To find star's hour-angle.*

To the right ascension of mean sun add mean time at ship,* and from the sum subtract the star's right ascension : the remainder is the hour-angle of the star.

4. *To find the moon's hour-angle.*

From the same sum (viz. right ascension of mean sun and ship mean time) subtract the moon's right ascension ; the remainder is the hour-angle of the moon.

5. *To calculate the star's true altitude.*

Proceed as in 4, p. 209, using the star's declination instead of the sun's.

To find star's apparent altitude.

To the true altitude just found add the refraction ; the result is the star's apparent altitude.

6. *To find the moon's true and apparent altitude* proceed as in 6, p. 209.

7. Proceed then as in former rule (arts. 6, 8, 9, p. 193).

EXAMPLE.

462. May 16, 1842, in lat. 50° 37' 30" N., and long. by account 1° 6' W.; when a chronometer showed 10ʰ 39ᵐ 26ˢ P.M., the observed distance of the star α Virginis from the moon's farthest limb was 64° 1' 50", index correction + 0' 40", the error of the chronometer on ship mean time being fast 4ᵐ 14ˢ : required the longitude.

In this example the lat. (50° 48' N.) and horizontal par. (59' 8" and 59' 12·6") have been corrected for the spheroidal figure of the earth (see *Nav.* Part II. pp. 122, &c.).

Time by chro. .	10ʰ	39ᵐ	26ˢ	R. A. mean sun.			Star's R. A. and decl.		
error fast		4	14	16 ... 3ʰ 35ᵐ	9·00ˢ		13ʰ	16ᵐ	55·7ˢ
ship mean time	10	35	12	1	38·6		10°	20'	25"S.
long. in time .		4	24	cor. ...	6·5				
Gr. May 16 ...	10	40		3 36	54·1				
				Ship mean time 10 35	12·0				
				14 12	6·1	....	14ʰ	12ᵐ	6·1ˢ
				star's right ascen. ... 13 16	55·7	....	9	19	40·0
				∴ star's hour-angle . 0 55	10·4	M.'s H.A.	4	52	26·1

---

* Ship mean time is usually found by an altitude of a heavenly body taken a little before or after the lunar as directed, p. 164, ex. 395.

Moon's R. A.	Moon's decl.	Moon's semi.	M.'s H. P.
At 10ʰ ... 9ˢ 18ᵐ 10·0ˢ ...	13° 25′ 13″ N. ... Noon...	16′ 7·0″	59′ 1·0″
„ 11ʰ ... 9 20 24·5 ...	13 11 47 N. ... Mid. ...	16 8·1	59 5·6

2 14·5	13 26	1·1	4·6
·17609	·17609	·05115	·05115
1·42920	·64997	3·99203	3·37067
1·60529 1 30	·82606 8 56	4·04318 1·0	3·42182 4·1
9 19 40	13 16 17 N.	16 8·0	59 5·1
		aug. 5·8	
		16 13·8	

To calculate star's altitude.   To calculate moon's altitude.

Lat. .... 50° 37′ 30″ N.	Lat. .... 50° 37′ 30″ N.
decl. .... 10 20 25 S.	decl. .... 13 16 17 N.
	60 57 55 v.		37 21 13 v,
Const. log.	6·301030	Const. log.	6·301030
cos. lat................	9·802359	cos. lat.	9·802359
cos. star's decl.	9·992887	cos. moon's decl. ..	9·988245
hav. star's H. A....	8·158830	hav. moon's H. A.	9·549884
	4·255106		5·641518
	17993		438044
ver. v................	514427	ver. v,..............	205056
	232		38
ver. zen. dist.	532652	ver. zen. dist.......	643138
	584		2990
62° 8′ 16″	68	69° 5′ 33″	148

∴ star's zen. dist.. 62° 8′ 16″	∴ moon's zen. dist. 69° 5′ 33″	
90	90	
∴ star's true alt... 27 51 44	∴ moon's true alt. 20 54 27	
cor. in alt. + ... 1 50	cor. in alt. (nearly) 52 −	
∴ star's app. alt... 27 53 34	app. alt. (nearly). 20 2 0	Aux. angle A.
obs. dist............ 64 1 50	cor. in alt. 52 48	60° 10′ 47″
index correction .. 0 40 +		5	1
64 2 30	true cor. in alt.... 52 53	0
semi. .... 16 14 −	moon's true alt.... 20 54 27	60 10 48
app. dist. 63 46 16	moon's app. alt.... 20 1 34	

To find true distance and Greenwich mean time.

Star's zen. dist. ...	62°	8'	16"	Versines.
moon's zen. dist. ...	69	5	33	58908 ... 179
				10400 ... 258
	131	13	49 *v.*	22769 ... 40
				58469 ... 16
Star's app. alt. ...	27	53	34	01955 ... 7
moon's app. alt. ...	20	1	34	
				52501
sum	47	55	8	500
arc A.	60	10	48	

True dist.

	53001	63°	26' 55"
sum .... 108 5 56 *v.*	2761	64	24 13 at 9ʰ
diff. .... 12 15 40 *v.*		62	38 17
	240		
App. dist. 63 46 16	·49712 ...	0	57 18
arc A. 60 10 48	·23024	1	45 56
sum .... 123 57 4 *v.*	·26688	1ʰ	37ᵐ 22ˢ
diff. .... 3 35 28 *v.*	9		

Gr. mean time ...	10	37	22
ship mean time...	10	35	12
long. in time ...		2	10

∴ longitude 0° 32' 30" W.

463. April 20, 1847, in lat. 50° 37' 12" N., and long. by account 1° 6' W., when a chronometer showed 8ʰ 58ᵐ 45ˢ P.M., the observed distance of the star α Leonis from the moon's farthest limb was 46° 2' 12", index correction +0' 30", the error of chronometer being fast 3ᵐ 22ˢ: required the longitude.   *Ans.* 0° 55' 45" W.

464. December 10, 1845, in lat. 50° 37' 30" N., and long. by account 1° 6' W., when a chronometer showed 9ʰ 24ᵐ 48·3ˢ P.M., the observed distance of the star Pollux from the moon's farthest limb was 65° 28' 30", index correction +0' 30", the error of the chronometer on ship mean time being fast 12ᵐ 50·8ˢ: required the longitude.   *Ans.* 1° 28' W.

465. April 19, 1847, in lat. 50° 48' N., and long. by account 1° 6' W., when a chronometer showed 8ʰ 40ᵐ 18ˢ P.M., the observed distance of the star Regulus from the moon's farthest limb was 59° 11' 1·6", index correction +30", the error of the chronometer on ship mean time being fast 9ᵐ 30·4ˢ: required the longitude.   *Ans.* 1° 7' W.

466. September 1, 1843, in lat. 50° 37' 30" N., and long. by account 1° 6' W., when a chronometer showed 8ʰ 2ᵐ 54·4ˢ P.M., the observed distance of the planet Jupiter from the moon's farthest limb was 64° 19' 57", index

correction + 1′ 50″, the error of the chronometer on ship mean time being fast 2ᵐ 2·6ˢ: required the longitude.   *Ans.* 0° 35′ 15″ W.

467. September 5, 1843, in lat. 50′ 48″ N., and long. by account 1° 6′ W., when a chronometer showed 8ʰ 52ᵐ 39ˢ P.M., the observed distance of the planet Mars from the moon's nearest limb was 45° 11′ 23·3″, index correction + 1′ 50″, the error of the chronometer being fast 4ᵐ 47·4ˢ: required the longitude.   *Ans.* 1° 19′ 45″ W.

Elements from Nautical Almanac.

Star's declin.	Star's right ascen.	Mean sun's R.A.
April 20 ... 12° 42′ 30″ N.	... 10ʰ 0ᵐ 15·35ˢ	... 1ʰ 51ᵐ 48·09ˢ

Moon's right ascen.	Moon's declin.	Moon's semi.	Hor. par.
20th, at 8ʰ...6ʰ 51ᵐ 3·8ˢ...	17°39′ 51″N. ...	Noon. 15′ 23·7″...	56′ 29·8″
„ 9ʰ...6 53 16·8 ...	17 36 40 N. ...	Mid... 15 17·0 ...	56 5·3

Distance at 6 hours, 46° 51′ 27″; at 9 hours, 45° 16′ 23″.

Star's declin.	Star's right asc.	Mean sun's right asc.
Dec. 10th ... 28° 23′ 22″N.	...... 7ʰ 35ᵐ 54·9ˢ	...... 17ʰ 16ᵐ 17·09ˢ

Moon's right ascen.	Moon's decl.	Moon's semi.	Hor. par.
10th, at 9ʰ...2ʰ 54ᵐ 19·38ˢ...	16°40′ 52″N. ...	Noon 15′ 11·5″...	55′ 45·0″
„ 10ʰ...:2 56 27·74 ...	16 47 6 N. ...	Mid.. 15 7·7 ...	55 30·9

Distance at 9 hours, 65° 12′ 56″; at mid., 63° 41′ 12″.

Star's declin.	Star's right ascen.	Mean sun's right asc.
Apr. 19th ... 12° 42′ 30″N.	...... 10ʰ 0ᵐ 15·35ˢ	...... 1ʰ 47ᵐ 51·54ˢ

Moon's right ascen.	Moon's declin.	Moon's semi.	Hor. par.
19th, at 8ʰ...5ʰ 56ᵐ41·12ˢ...	18°28′ 9·6″N....	Noon 15′ 38·4″...	57′ 23·5″
„ 9ʰ...5 58 59·68 ...	18 27 17·2 N....	Mid.. 15 30·9 ...	56 56·1

Distance at 6 hours, 59° 46′ 25″; at 9 hours, 58° 8′ 6″.

Planet's declin.	Planet's right ascen.	Mean sun's right asc.
Sept. 1st... 15° 46′ 6·8′S.	...... 21ʰ 33ᵐ 3·13ˢ...	.. 10ʰ 39ᵐ 59·98ˢ
„ 2d ... 15 48 22·9 S.	...... 21 32 35·33	

Moon's right asc.	Moon's declin.	Moon's semi.	Hor. par.
1st, at 8ʰ ...17ʰ 0′ 32·9ˢ...	23°56′ 5·7″S. ...	Noon 15′ 54·8″...	58′ 24·0″
„ 9ʰ ...17 3 2·2 ...	23 56 34·4 S. ...	Mid.. 15 49·5 ...	58 4·5

Distance at 6 hours, 65° 7′ 44″; at 9 hours, 63° 24′ 57″.

Planet's declin.	Planet's right ascen.	Mean sun's right asc.
Sept. 5th ... 26° 34′ 9·5″S.	...... 17ʰ 32ᵐ 5·8ˢ	...... 10ʰ 55ᵐ 46·19ˢ
„ 6th ... 26 34 29·5 S.	...... 17 · 34 27·8	

Moon's right ascen.	Moon's declin.	Moon's semi.	Hor. par.
5th, at 8ʰ...20ʰ 41ᵐ16·0ˢ...	15° 1′ 2″S. ...	Noon 15′ 15·7″...	56′ 0·4″
„ 9ʰ...20 43 20·7 ...	14 50 47 S. ...	Mid.. 15 11·7 ...	55 45·5

Distance at 6 hours, 44° 14′ 50″; at 9 hours, 45° 45′ 38″.

CHAPTER VIII.

THE VARIATION OF THE COMPASS.

THE deflection of the magnetic needle from the true North, or, as it is usually called, the *variation of the compass*, is found at sea either by computing the true bearing of the sun from an observed altitude, the compass bearing being noted at the time of the observation; or, without taking an altitude, determining the true bearing of the sun when in the horizon, its compass bearing being observed at the same instant. The difference between the compass bearing and true bearing thus found is *the variation of the compass.*

Sometimes it is necessary to correct the variation of the compass determined as above for the deviation of the compass itself, arising from the following local causes. The iron on board draws the needle to the east or west of the magnetic meridian; and this effect is greater or less on the needle according as the iron is distributed more or less unequally on different sides of the magnetic meridian. The deviation of the compass due to this cause is discovered, previously to the ship going to sea, by swinging her round and noting the deflection of the needle from the magnetic meridian on different points; a table is then formed similar to the one in p. 221, from which the correction of the compass for different positions of the ship's head may be readily found. The method of determining whether the variation of the compass is east or west will be best seen by means of the following examples.

EXAMPLES.

468. Suppose the true bearing of the sun was found by observation to be N. 100° 10′ E., when the compass bearing was N. 90° 42′ E.: required the variation of the compass, the ship's head being N.E.

Let N represent the true north point of the horizon, and NS the true meridian; measure (roughly) 100° 10′ from north towards east as the angle NZT; then T represents the place of the sun when the observation was taken. From T measure back towards N the compass bearing 90° 42′, as TZC; then ZC is the direction of the magnetic needle, and the angle NZC is the variation of the compass, which is evidently easterly, since the compass north is to the east of the true north: hence in this example the variation is said to be east, thus:

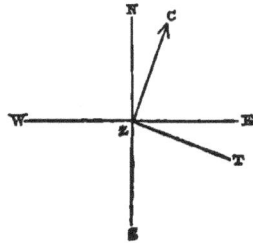

True bearing NZT.................	N. 100°	10′ E.
compass bearing CZT	N. 90	42 E.
apparent variation	0	28 E.

Now if the iron on board had no effect on the needle, this would be the true variation; but referring to the table, it appears that the needle itself is drawn or deflected 10° to the east, in consequence of the disturbing effects of the iron, when the direction of the ship's head is N.E. Placing the 10° 0′ E. under the above 9° 28′ E., and subtracting, we have the true variation of the compass corrected for local deviation, thus:

Observed variation ....   9° 28′ E.

deviation   10   0 E.

true variation   0   32 W.

469. The true bearing of the sun was found by observation to be S. 60° 42′ E., when the compass bearing was S. 50° 10′ E.: required the variation of the compass, the ship's head being N.E.

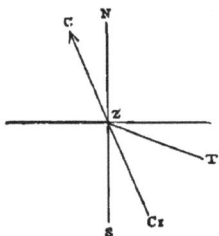

Let N z s represent as before the true meridian; draw zT, making the angle szT equal to 60° 42′ (roughly); then T represents the true place of the sun; from T measure back towards s the compass bearing Tzc′, equal 50° 10′; then zc′ is the direction of the magnetic needle, and the angle szc′ or Nzc is the observed variation of the compass, to be corrected for deviation (if any). Thus:

True bearing .... S. 60° 42′ E.

compass bearing .... S. 50   10 E.

observed variation   10   32 W.

deviation   10   0 E.

true variation   20   32 W.

For by the table it appears that the needle, by the effects of the iron, is drawn 10° to the eastward; if there had been no iron on board the needle would have been directed 10° to the westward of its observed place. Hence may be deduced the following rule to find the variation of the compass.

Rule 57. *Given the* TRUE BEARING *and* COMPASS BEARING *and* DEVIATION, *to find the* VARIATION OF THE COMPASS.

1. Reckon the compass bearing and the true bearing from the same point, north or south.

2. Take the difference of the two bearings when measured towards the same point, but the sum when measured towards different points; the result is the apparent variation of the compass; east when the true bearing is to the right of the compass bearing, west if the true bearing is to the left of the compass; the observer being supposed to be placed in the centre of the compass, and looking towards the heavenly body.

NOTE. The name of the variation, whether east or west, may also be readily found by making a figure similar to those in the preceding examples.

3. If there be no deviation to be allowed for local attraction, the above is the true variation.

4. *To correct for local deviation* (if any). Under the apparent variation just found, put the correction from the table of deviation, with the *opposite* letter to that given in the table.

5. When the names put down are alike add, putting the common letter to the result: if the names put down be unlike, subtract the less from the greater, putting to the remainder the name of the greater. The result will be the variation of the compass corrected for deviation, and therefore the true variation.

DEVIATION OF THE COMPASS OF H.M.S. ——,
(Caused by the local attraction of the Ship) for given positions of the Ship's head.

Direction of Ship's Head.	Deviation of Compass.	Direction of Ship's Head.	Deviation of Compass.
N.	2° 45′ E.	S.	3° 0′ W.
N. by E.	4 57	S. by W.	4 20
N.N.E.	7 30	S.S.W.	5 0
N.E. by N.	9 0	S.W. by S.	6 7
N.E.	10 0	S.W.	7 0
N.E. by E.	10 55	S.W. by W.	7 27
E.N.E.	10 40	W.S.W.	7 50
E. by N.	9 55	W. by S.	8 20
E.	8 50	W.	8 50
E. by S.	7 15	W. by N.	8 10
E.S.E.	5 35	W.N.W.	6 50
S.E. by E.	3 40	N.W. by W.	5 40
S.E.	1 50	N.W.	4 50
S.E. by S.	0 20 E.	N.W. by N.	3 20
S.S.E.	0 56 W.	N.N.W.	1 40 W.
S. by E.	2 20	N. by W.	1 10 E.

EXAMPLES.

470. The true bearing of the sun is N. 117° 39′ E., and compass bearing S. 71° 10′ E.: required the true variation. The ship's head being S.b.E., and therefore the deviation of the compass 2° 20′ W. (see Table). The compass bearing reckoned from the same point as the true bearing is, N. 108° 50′ E.

True bearing .... N.	117°	39′ E.
compass bearing .... N.	108	50 E.
apparent variation	8	49 E.
deviation	2	20 E.
true variation....................	11	9 E.

471. The true bearing is E. 10° N., when the compass bearing is E. 8° S.: required the true variation, the ship's head being S.W.

True bearing........................... E.	10° N.	
compass bearing .... E.	8 S.	
apparent variation	18 W.	
deviation................................	7 E.	
true variation...........................	11 W.	

472. The true bearing is S. 80° W., when the compass bearing is N. 108° W.: required the true variation, the ship's head being S.W. b. W., and therefore the deviation by Table 7½° W.

True bearing........................... S.	80° W.	
compass bearing .... S.	72 W.	
apparent variation	8 E.	
deviation	7½ E.	
true variation	15½ E.	

473. The true bearing of the sun was N. 36° E., when the compass bearing was N. 24° E., the ship's head being W. ½ N.: required the variation of the compass.   *Ans.* 20½° E.

474. The true bearing was N. 110° 42′ W., when the compass bearing was N. 90° 24′ W., the ship's head being S.S.W.: required the variation of the compass.   *Ans.* 15° 18′ W.

475. The true bearing was S. 48° 30′ W., when the compass bearing was N. 132° 33′ W., the ship's head being S.W. b. W.: required the variation of the compass.   *Ans.* 8° 30′ E.

476. The true bearing of the sun was E. 20° 20′ N., when the compass bearing was E. 32° 45′ N., the ship's head being W. : required the variation of the compass. *Ans.* 21° 15′ E.

477. The true bearing of the sun was W. 12° 32′ S., the compass bearing was W. 2° 10′ N., the ship's head being W. b. S. : required the variation of the compass. *Ans.* 6° 22′ W.

478. The true bearing of the sun was W. 30° 10′ N., the compass bearing was W. 20° 42′ N., the ship's head being N. b. E. : required the variation of the compass. *Ans.* 4° 31′ E.

479. The true bearing of the sun was W. 30° 30′ N., the compass bearing was W. 28° 15′ N., the ship's head being N.E. b. N. : required the variation of the compass. *Ans.* 6° 45′ W.

The variation of the compass is found at sea by either of the following problems :

1. Given the latitude of the ship and the sun's declination when in the horizon, to find the bearing or amplitude.

2. Given the latitude of the ship, and the altitude of the sun and declination, to find the true bearing or azimuth.

3. Given the latitude and time at the ship and the sun's declination, to find the true bearing or azimuth.

The compass bearing being observed at the time of observation, the difference of compass and true bearing—that is, the variation of the compass —is readily found by the preceding rules.

Rule 58. VARIATION OF COMPASS BY AMPLITUDE.

1. Get a Greenwich date.

2. Take out of the *Nautical Almanac* the sun's declination for this date.

3. Add together the log. sin. of the declination and log. secant of latitude; the sum, rejecting 10 in the index, is the log. sin. of amplitude, which take from the tables.

4. If the body is rising, mark it east: if setting, west : mark it also north or south, according as the declination is north or south.

5. The result is the amplitude or true bearing.

6. Under the true bearing put the compass bearing, and determine the variation of the compass by the preceding rule.

EXAMPLE.

480. September 19, 1851, at $5^h 51^m$ A.M. mean time nearly, in latitude 47° 25′ N., and longitude 72° 15′ W., the sun rose by compass E. 12° 10′ N. : required the variation, the ship's head being E. b. S.

Ship, Sept. 18............................ 17^h 51^m

longitude in time........................ 4   51   W.

Greenwich, Sept. 18.................... 22   42

Sun's declination.

18th..................	2°	0′	31″ N.	Sin. decl.		8·457103
19th..................	1	37	14 N.	sec. lat......................		0·169628
		23	17	sin. ampl.		8·626731
·02419				true bearing	E. 2°	25′ N.
·88823				comp. bearing	E. 12	10 N.
·91242		22	2	app. variation	9	45 E.
				deviation	7	15 W.
declination	1	38	29 N.	true variation	2	30 E.

481. May 6, 1846, at $5^h 30^m$ A.M. mean time nearly, in lat. 50° 48′ N., and long. 47° 12′ E., the sun rose by compass E. 2° 10′ S. : required the variation, the ship's head being S. b. W.   *Ans.* 24° 21′ 30″ W.

482. Nov. 14, 1846, at $6^h 45^m$ P.M. mean time nearly, in lat. 32° 14′ S., and long. 100° E., the sun set by compass W. 15° 40′ S. : required the variation, the ship's head being N.E.   *Ans.* 16° 1′ 30″ W.

483. Jan. 10, 1846, at $6^h 58^m$ A.M. mean time nearly, in lat. 31° 56′ N., and long. 75° 30′ W., the sun rose by compass E. 30° 10′ S. : required the variation, the ship's head being N.E. b. E.   *Ans.* 14° 55′ W.

484. March 21, 1846, at $6^h 0^m$ A.M. mean time nearly, in lat. 42° 13′ N., and long. 90° E., the sun rose by compass E. 11° 40′ S. : required the variation, the ship's head being W. b. S.   *Ans.* 3° 20′ 15″ W.

485. March 31, 1850, at $6^h 0^m$ P.M. mean time nearly, in lat. 42° 13′ N., and long. 124° W., the sun set by compass W. 11° 30′ S. : required the variation, the ship's head being N.   *Ans.* 14° 38′ 15″ E.

486. Dec. 4, 1851, at $7^h 50^m$ A.M. mean time nearly, in lat. 50° 40′ N., and long. 94° W., the sun rose by compass E. 10° 42′ S. : required the variation of the compass, the ship's head being N.E.•   *Ans.* 15° 57′ E.

Elements from Nautical Almanac.

Sun's declination.

May 5	16°	13′	22″ N.		6	16°	30′	22″ N.
Nov. 14	18	13	22 S.		15	18	28	54 S.
Jan. 10	21	58	29 S.					
March 20	0	11	37 S.		21	0	12	5 N.
March 31	4	7	44 N.		32	4	30	54 N.
Dec. 4	22	13	9 S.		5	22	21	7 S.

Rule 59. VARIATION BY AZIMUTH AND SUN'S ALTITUDE.

Given the altitude of the sun, and the compass bearing, to find the variation of the compass.

1. Get a Greenwich date.

2. Take out of the *Nautical Almanac* the sun's declination for this date, and also the sun's semidiameter.

3. Find the polar distance by adding 90° to the declination, when the latitude and declination have different names, or by subtracting the declination from 90° when the latitude and declination have the same names.

4. Correct the observed altitude for index correction, dip, semidiameter, and correction in altitude, and thus get the true altitude.

5. *To find the true bearing.* First method, by haversines.

Put down the latitude under the altitude, and take their difference; under which put the polar distance: take the sum and difference.

6. To the log. secants of the two first terms in this form (omitting the tens in the index) add the halves of the log. haversines* of the two last: the result, rejecting 10 in the index, is the log. haversine of the true bearing or azimuth, which find from the table.

7. *To find the true bearing.* Second method, by log. sines, &c.

Put down the latitude under the altitude, and take their difference; under which put the polar distance : take the sum and difference, and the half-sum and half-difference.

8. To the log. secants of the two first terms in this form (omitting the tens in the index) add the sines of the two last, and divide the sum by 2 ; the result is the log. sine of half the true bearing, which find from the tables and multiply by 2 for the true bearing or azimuth required.

9. Mark the true bearing N. or S., according as the latitude is N. or S.; mark it also E. or W., according as the heavenly body is E. or W. of the meridian.

* If the student have no table of haversines, the true bearing may be found by the second method.

Q

10. With the true bearing thus found, and the compass bearing, find the variation of the compass by Rule 57, p. 220.

by Rule 57, p. 220.

EXAMPLE.

487. June 7, 1851, at 5ʰ 50ᵐ A.M. mean time nearly, in lat. 50° 47' N., and long. 99° 45' W., when the sun bore by compass S. 92° 36' E., the observed altitude of the sun's lower limb was 18° 35' 20", index correction + 3' 10", and the height of the eye above the level of the sea was 19 feet: required the variation, the ship's head being N.E.

```
            Ship, June 6 . . . . . . . . .   17ʰ  50ᵐ
            longitude in time . . . . . . .    6   39 W.
            Greenwich, June 7 . .ˑ . . . . .    0   29
```

	Sun's declin.		Sun's semi.				Obs. alt.		
7th	22° 43' 49" N.	15' 46"					18° 35' 20"		
8th	22 49 34 N.		In. cor.				3 10+		
	5 45		dip				18 38 30		
1·69597							4 17−		
1·49560							18 34 13		
3·19157 . .	0 7		semi.				15 46		
	22 43 56 N.		app. alt.				18 49 59		
	90		cor. in alt.				2 41−		
pol. dist.	67 16 4		true alt.				18 47 18		

First method.   True bearing by haversines.

Latitude	50° 47' 0"	. . Sec. . .	0·199108
altitude	18 47 18	. . sec. . .	0·023779
	31 59 42		
polar distance	67 16 4		
sum	99 15 46 (S) .	. ¼ hav. S .	4·881893
difference	35 16 22 (D) .	. ¼ hav. D .	4·481384
			9·586164

```
True bearing . . . . . . . . . .  N. 76° 46' 30" E.
compass bearing . . . . . . . .   N. 87  24   0  E.
app. variation . . . . . .         10  37  30  W.
deviation  . . . . . . . .         10   0   0  W.
true variation . . . . . .         20  37  30  W.
```

Second method.   True bearing by sines, &c.

Latitude	50° 47' 0"	. . Sec. . .	0·199108
altitude	18 47 18	. . sec. . .	0·023779
	31 59 42	sin. S₁ . .	9·881880
polar distance	67 16 4	sin. D₁ . .	9·481434
sum	99 15 46		2)19·586201
difference	35 16 22		9·793100
¼ sum	49 37 53 (S₁)		38° 23' 15"
¼ difference	17 38 11 (D₁)		2
			76 46 30

True bearing as before.

When the ship is in harbour, or in any position where the sight of the horizon is intercepted by land, or obscured by fog, so that the altitude of the sun cannot be taken, the preceding methods are inapplicable. The following rule may then be used, in which it is supposed that the hour-angle at the ship is known, or can be found by means of the chronometer, or the time at the place.

Rule 60. Variation by Azimuth and Ship Mean Time.

Given mean time at ship and the compass bearing, to find the variation of the compass.

1. Get a Greenwich date.

2. Take out of the *Nautical Almanac* for this date the equation of time and sun's declination.

3. Find the polar distance, by adding 90° to the declination, when the latitude and declination have different names, or by subtracting the declination from 90° when the latitude and declination have the same name.

4. Under the colatitude (found by subtracting the latitude from 90°) put the polar distance, take the sum and difference, and the half sum and half difference.

5. *To find the hour-angle from ship mean time.*

Correct the time shown by chronometer when the compass bearing was observed for its error on Greenwich mean time, and thus get the mean time at Greenwich; to mean time apply the equation of time, to obtain apparent time; under this put the longitude in time, adding if east, and subtracting if west; the result will be ship apparent time, and also the hour-angle if P.M.; but if A.M. at ship, subtract the apparent time from 24 hours, the remainder will then be the hour-angle required.

6. Divide the hour-angle by 2. Then under heads (1) and (2) put down the following quantities.

7. Under both (1) and (2) put log. cotangent of half hour-angle.

Under (1) log. cosine } of half difference of polar distance and colatitude.
 (2) log. sine

 (1) log. sec. } of half sum of polar distance and colatitude.
 (2) log. cosec.

8. Add together the log. under (1) and (2) separately, and take out the angles corresponding to each as a log. tangent. Put one under the other, and take their sum, if the polar distance is greater than the colatitude, or their difference, if the polar distance is less than the colatitude: the result will be the true bearing of the sun at the time of observation.

9. Then proceed to find the variation as in Rule 57.

EXAMPLES.

488. June 23, 1847, at 10^h 58^m A.M. mean time nearly, in lat. 50° 48′ N., and long. 1° 6′ W., when a chronometer showed 11^h 3^m 37′, the bearing of the sun was observed to be N. 173° 10′ E., the error of the chronometer on Greenwich mean time being 0^m 54′ fast: required the variation.

			Equation of time.			Sun's declination.		
June 22, at . . .	22^h 58^m		22d . . .	1^m 29·72ˢ sub. . . .		23° 27′	21″ N.	
long in time . . .	4+		23d . . .	1 42·64		23 26	59 N.	
Gr. June 22 . . .	23 2			12·92		.	23	
			·01786			·01786		
colat.	39° 12′	0″	2·92283			2·69100		
pol. dist.	66 33	0	2·94069	12·40		2·70886		21
sum	105 45	0		1 42·12		23 27	0	
difference	27 21	0				90		
½ sum	52 52	30		Polar dist. . .	66	33	·0	
½ difference . . .	13 40	30						

				(1)	(2)
Time by chro. . .	11^h	3^m 37ˢ+12^h	Cot. ½ h.-ang. 10·856573	. 10·856573
error on Gr. M. T. .		0 54 fast.	cos. ½ diff. . 9·987311 Sin. ½ diff. 9·373674
Greenwich, 22d . .	23	2 43	sec. ½ sum. . 10·219283 cos. ½ sum. 10·098367
equation of time .		1 42 sub.	tan. . . . . . 11·063367 tan. . . . 10·328614
apparent time . .	23	1 1		
longitude in time .		4 24 W.	85° 3′ 30″ 64° 51′ 45″
ship app. time . .	22	56 37	64 51 45
	24		True b. N. 149 55 15 E.
hour-angle . . .	1	3 23	comp. b. N. 173 10 0 E.
½ hour-angle . . .	0	31 41	variation . 23 14 45 W.

489. April 27th, 1847, at 1^h 10^m P.M. mean time nearly, in lat. 50° 48′ N., and long. 1° 6′ W., when a chronometer showed 1^h 15^m 51′, the bearing of the sun was observed to be S. 51° 55′ W., the error of the chronometer on Greenwich mean time being 1^m 18′ fast: required the variation.

Ans. 23° 46′ W.

490. December 14, 1847, at 10^h 22^m A.M. mean time nearly, in lat. 52° 10′ N., and long. 1° 30′ W., when a chronometer showed 10^h 30^m 48′, the bearing of the sun was observed to be N., 179° 20′ E., the error of the chronometer on Greenwich mean time being 3^m 38′ fast: required the variation. *Ans.* 21° 13′ 15″ W.

491. December 14, 1847, at 1^h 55^m P.M. mean time nearly, in lat. 48° 50′ N., and long. 1° 30′ W., when a chronometer showed 1^h 59^m 55′, the bearing of the sun was observed to S. 51° 40′ W., the error of the chronometer on Greenwich mean time being 0^m 5′ fast: required the variation. *Ans.* 22° 44′ W.

492. December 14, 1848, at 11ʰ 11ᵐ A.M. mean time nearly, in lat. 39° 40′ N., and long. 0° 40′ E., when a chronometer showed 11ʰ 19ᵐ 43ˢ, the bearing of the sun was observed to be N. 167° 50′ E., the error of the chronometer on Greenwich mean time being 3ᵐ 38ˢ fast : required the variation.

Ans. 2° 50′ 13″ E.

493. March 7, 1844, at 9ʰ 59ᵐ A.M. mean time nearly, in lat. 49° 48′ N., and long. 1° 10′ E., when a chronometer showed 10ʰ 24ᵐ 8ˢ, the bearing of the sun was observed to be N. 164° 51′ 40″ E., the error of the chronometer on Greenwich mean time being fast 20ᵐ 48ˢ : required the variation.

Ans. 20° 28′ W.

494. May 26, 1851, at 9ʰ 48ᵐ A.M. mean time nearly, in lat. 50° 48′ N., and long. 1° 6′ W., when a chronometer showed 9ʰ 47ᵐ 37ˢ, the bearing of the sun was observed to be S. 31° 7′ E., the error of the chronometer on Greenwich mean time being 3ᵐ 17ˢ fast : required the variation.

Ans. 23° 35′ 15″ W.

Elements from Nautical Almanac.

Sun's declination.						Equation of time.		
April 27	13°	43′	22″N.		27	2ᵐ	24·56ˢ add
„ 28	14	2	25 N.		28	2	34·29
Dec. 13	23	8	42 S.		13	5	43·66 add
„ 14	23	12	38 S.		14	5	15·08
Dec. 14	23	12	38 S.		14	5	15·08 add
„ 15	23	16	7 S.		15	4	46·23
Dec. 13	23	11	43 S.		13	5	23·23 add
„ 14	23	15	18 S.		14	4	54·52
March 6	5	30	13 S.		6	11	25·48 sub.
„ 7	5	6	55 S.		7	11	10·81
May 25	20	53	42 N.		25	3	25·40 add
„ 26	21	4	25 N.		26	3	19·55

TO FIND THE TIME OF HIGH WATER.

The Change Tide upon which the rule is made to depend, is that tide which takes place P.M. on the day the moon changes, or is at full. The time of high water at change of the moon is given at different places in apparent time ; and indeed cannot be generally expressed in mean time. If the tide be given A.M. on that day as the *change tide*, it should be reduced to P.M. by adding 18 minutes, which may be considered as an average difference on that day.

Rule. *Given the apparent time of change tide, and the longitude of the place, to find the mean time of high water* A.M. *and* P.M.

Take out of the *Nautical Almanac* the moon's meridian passage on the given day, and also on the preceding day; also the moon's semidiameter and equation of time for the given day (roughly).

Under heads (1), (2), and (3) (see Examples), put down the following quantities :

Under (1), the time of moon's meridian passage on proposed day, as found in the *Nautical Almanac.*

,,   (3), the meridian passage on preceding day.

,,   (2), put down half the sum of these times.   (See Examples.)

Correct quantity under (1) by table (*k*), p. 5 of Inman's Tables, by entering with longitude of place at top, and difference of the times under (1) and (3) at the side : thus find the time of moon's meridian passage at the place.

Take out the correction from table (*l*), p. 5, and place it under (1). This correction is found as follows : Enter the table at top with moon's semidiameter, and at the side with meridian passage under (1), corrected by equation of time to nearest minute, so as to reduce the time of moon's meridian passage (which is given in mean time) to apparent time.

Apply the correction thus found with its proper sign, and to the result add the given apparent time of change tide.

1. When the quantity under (1) is *less* than 12 hours.

The time thus found is the mean time of high water P.M. for the proposed day.   (See Example 495.)

2. When the quantity under (1) is *greater* than 12 hours, and less than 24 hours.

Work as described above with the meridian passage under (2).   Then, if the result is greater than 12 hours, reject 12 hours; the remainder is mean time of high water on the proposed day, P.M.   (See Example 3.)   But if the result be less than 12 hours, it will be the mean time of high water A.M. on the proposed day.   (See Example 499.)

3. When the quantity under (1) is greater than 24 hours.

Work as described above, with the meridian passage under (3).   Then, if the result be greater than 24 hours, reject 24 hours; the remainder will be the mean time of high water P.M. on the proposed day.   (See Example 500.) But if the result be less than 24 hours, and greater than 12 hours, reject 12 hours; the remainder will be the mean time of high water A.M. on the proposed day.   (See Example 505.)

To find the next time of high water A.M. *or* P.M.

If the time of high water found as above is the P.M. time, subtract therefrom the difference between meridian passages under (1) and (2); the remainder will be the mean time of high water A.M. on the proposed day.

If the time of high water is the A.M. time, add thereto the difference between the meridian passages under (1) and (2), and the sum will be the mean time of high water P.M. on the proposed day.

If it be necessary to add 12 hours before this difference can be subtracted, in that case the remainder will be the mean time of high water P.M. on the preceding day; there will be no high water A.M. on the proposed day. And if in adding the difference the sum be greater than 12 hours, this sum, rejecting 12 hours, will be the mean time of high water A.M. on the following day; there will be no high water P.M. on the proposed day.

EXAMPLES.

495. Find the time of high water on January 3, 1857. Change tide 2ʰ 10ᵐ P.M. apparent time; long. 50° W.

Moon's mer. pass. Jan. 3, 6ʰ 7ᵐ   ☾ semi... 16′ 10″
    ,,   ,,   ,, 2, 5 19   Eq. of time.. 4ᵐ 54ˢ−
                         ———
                          48
                         ———
                          24
    (1)   (2) ———   (3)
  6ʰ 7ᵐ   5 43   5ʰ 19ᵐ
     7+
  ———
  6 14
  1 0−
  ———
  5 14
  2 10+
  ———
  7 24 P.M. ⎫
     24−   ⎬ January 3.
  ———   ⎭
  7 0 A.M.

496. Find the time of high water on May 28, 1857. Change tide 5ʰ 30ᵐ P.M. app. time; long. 75° E.; moon's mer. pass. on 28, 4ʰ 51ᵐ; on 27, 3ʰ 58ᵐ; ☾ semi. 15′ 41″; eq. of time 3ᵐ 1ˢ+.

Ans. 9ʰ 1ᵐ P.M.; 8ʰ 35ᵐ A.M.

497. Find the time of high water on January 12, 1857. Change tide 1ʰ 30ᵐ P.M. app. time; long. 60° E.

Moon's mer. pass. Jan. 12, 14^h 28^m ☾ semi. 15′ 22″

,,   ,,   ,, 11, 13 40   Eq. of time 8^m 42ˢ—

$$48$$

$$24$$

(1)   (2) ————   (3)
14^h 28^m 14   4 13^h 40^m
    8—   8—

14 20   13 56
 0 38—   0 29—

13 42   13 27
 1 30+   1 30+

15 12   14 57

Greater than 12 hours.   2   57 P.M. ⎫
                              24—   ⎬ January 12.
                           ——————   ⎪
                              2   33 A.M. ⎭

498. Find the time of high water on June 8, 1857. Change tide 4^h 20^m P.M. app. time; long. 40° W.; moon's mer. pass. on June 8, 13^h 4^m; on 7, 12^h 10^m; ☾ semi. 14′ 58″; eq. of time 1^m 17ˢ+.

Ans. 5^h 5^m P.M.; 4^h 38^m A.M.

499. Find the time of high water on February 5, 1857. Change tide 2^h 10^m P.M. app. time; long. 70° E.

Moon's mer. pass. Feb. 5, 9^h 37^m ☾ semi. 15′ 49″

,,   ,,   ,, 4, 8 37   Eq. of time 14^m 19ˢ—

$$60$$

$$30$$

(1)   (2) ————   (3)
9^h 37^m 9   7 8^h 37^m
   12—   12—

9 25   8 55
0 27+   0 20+

9 52   9 15
2 10+   2 10+

12   2   11 25 A.M. ⎫
                            30+   ⎬ February 5.
Greater than 12 hours.   ——————   ⎪
                    11 55 P.M. ⎭

500. Find the time of high water on September 26, 1857. Change
tide 6ʰ 40ᵐ P.M. app. time; long. 20° W.; moon's mer. pass. on Sept. 26,
6ʰ 13ᵐ; on 25, 5ʰ 20ᵐ; ☾ semi. 14′ 59″; eq. of time 8ʰ 43ˢ+.

Ans. 11ʰ 32ᵐ A.M.; 11ʰ 58ᵐ P.M.

501. Find the time of high water on January 24, 1857.   Change
tide 5ʰ 0ᵐ P.M. app. time; long. 30° W.

Moon's mer. pass. Jan. 24, 23ʰ 51ᵐ   ☾ semi. 15′ 41″
 „   „   „ 23, 22 54   Eq. of time 12ᵐ 27ˢ—

                          57

                          28

(1)   (2)   (3)
23ʰ 51ᵐ   23 23   22ʰ 54ᵐ
    5+   5+

23 56   22 59
 0 11+   0 20+

24   7   23 19
 5   0+   5   0+

29   7   28 19

Greater than 24 hours.

January 24 { 4 19 P.M. / 28— / 3 51 A.M. }

502. Find the time of high water on June 19, 1857.   Change tide
3ʰ 40ᵐ P.M. app. time; long. 120° E.; moon's mer. pass. on June 19,
22ʰ 28ᵐ; on 18, 21ʰ 27ᵐ; ☾ semi. 16′ 32″; eq. of time 0ᵐ 58ˢ—.

Ans. 1ʰ 19ᵐ P.M.; 0ʰ 49ᵐ A.M.

503. Find the time of high water on March 4, 1857.   Change tide
5ʰ 0ᵐ P.M. app. time; long. 45° W.

Moon's mer. pass. March 4, 7ʰ 32ᵐ   ☾ semi. 15′ 48″
 „   „   „ 3, 6 31   Eq. of time 11ᵐ 54ˢ—

                          61

                          30

                        7   2

(1)	(2)	(3)
7^h 32^m	7^h 2^m	6^h 31^m
$7+$	$7+$	
7 39	7 9	
0 $15-$	0 $34-$	
7 24	6 35	
5 $0+$	5 $0+$	
12 24	11 35 A.M. March 4 ; no P.M. tide.	
	$30+$	
Greater than 12 hours.	----	
	0 5 A.M. March 5.	

504. Find the time of high water on July 30, 1857. Change tide 5^h 30^m P.M. app. time; long. 90° W.; moon's mer. pass. on July 30, 7^h 6^m; on 29, 6^h 20^m; ☾ semi. 14′ 49″; eq. of time 6^m 8^s-.

Ans. 11^h 48^m A.M. July 30 ; no P.M. tide.

505. Find the time of high water on March 21, 1857. Change tide 3^h 0^m P.M. app. time; long. 100° W.

Moon's mer. pass. March 21, 21^h 11^m ☾ semi. 15′ 50″
,,   ,,   ,,   20, 20 17   Eq. of time 7^m 16$^s-$

	54	
	27	

(1)	(2)	(3)
21^h 11^m		20^h 17^m
$14+$		$14+$
21 25		20 31
0 $27+$		0 $13+$
21 52	(2) 20 44	20 44
3 $0+$		3 $0+$
24 52		March 21, A.M. 11 44 ; no P.M. tide.
		$27+$
Greater than 24 hours.		March 22, A.M. 0 11

506. Find the time of high water on November 12, 1857. Change tide 2^h 40^m P.M. app. time ; long. 70° W.; moon's mer. pass. on November 12, 21^h 24^m; on 11th, 20^h 43^m; ☾ semi. 15′ 0″; eq. of time 15^m 40^s+.

Ans. 11^h 49^m A.M. November 12; no P.M. tide.

We will conclude this treatise with a series of Examination Papers in the principal Rules of Navigation and Nautical Astronomy.

Questions.—No. I.

1. Required the course and distance from A to D.

  Lat. A...... 56° 35′ S.   Long. A...... 2° 15′ E.
  „   D...... 51   10 S.   „   D...... 3   10 W.

2. Required the course and distance from A to B.

  Lat. A...... 61° 10′ N.   Long. A...... 8° 40′ E.
  „   B...... 61   10 N.   „   B......15   10 E.

3. On May 8, at noon, a point of land in lat. 48° 10′ N., and long. 2° 2′ W., bore by compass E. by S. ½ S. distant 20 miles (variation 3¼ E.); afterwards sailed as by the following log. account : required the latitude and longitude in at noon on May 9.

Hours.	Knots.	$\frac{1}{10}$ths.	Course.	Wind.	Leeway.	—
1	3	0	N.N.W.½W.	N.E.	2¼	P.M.
2	3	2				
3	3	4				
4	2	7				
5	2	4	E. by S.¾S.	Do.	3	
6	2	4				Variation 3¼ E.
7	3	6				
8	3	2				
9	3	2				
10	4	0	E.S.E.	South.	2¾	
11	4	2				
12	4	5				
						Remarks in H.M.S. May 9, 1870.
1	5	2	N.N.E.¼E.	N.W.	1¾	A.M.
2	6	2				
3	7	2				
4	7	5				
5	8	0				
6	5	2	W.S.W.	Do.	3¼	
7	6	2				
8	6	4				
9	5	4				
10	6	2				
11	6	3				
12	7	0				

4. What bright star will pass the meridian of Canton in China the first after mean midnight on June 15, 1835, and how far N. or S. of the zenith?

5. June 15, in long. 100° 32′ E., the observed meridian altitude of the sun's lower limb was 20° 15′ 40″ (zenith S. of the sun), the index correction was + 2′ 50″, and the height of the eye above the sea was 14 feet: required the latitude.

6. April 23, at 9ʰ A.M., mean time nearly, in long. 5° 10′ W., the observed meridian altitude of the moon's lower limb was 38° 40′ 45″ (zenith N. of the moon), the index correction was − 2′ 50″, and the height of eye above the sea was 20 feet: required the latitude.

7. June 18, the observed meridian altitude of the star α Scorpii (Antares) was 20° 10′ 50″ (zenith north of the star), the index correction was + 4′ 50″, and the height of the eye above the sea was 18 feet: required the latitude.

8. June 12, the observed meridian altitude under the S. Pole of α² Crucis was 6° 40′ 10″, the index correction was + 3′ 40″, and the height of the eye above the sea was 18 feet: required the latitude.

9. December 10, at 2ʰ 10ᵐ A.M., mean time nearly, in long. 76° 12′ E., the observed altitude of α Ursæ Minoris (Polaris) was 47° 50′ 25″, the index correction was − 4′ 10″, and the height of the eye above the sea was 13 feet: required the latitude.

10. September 15, observed the following double altitude of the sun:

Mean time nearly.	Chronometer.	Obs. alt. sun's L. L.	True bearing.
10ʰ 40ᵐ A.M.	10ʰ 22ᵐ 36ˢ	40° 34′ 30″	S.b.E.¼E.
11 40 A.M.	11 22 45	42 12 0	S.½E.

The run of the ship in the interval was E.b.N. 12 miles, the index correction was + 3′ 50″, and the height of the eye above the sea was 18 feet: required the true latitude, the latitude by account being 51° N., and the longitude 50° 10′ W.

11. March 2, at 7ʰ 44ᵐ P.M., mean time nearly, in lat. 44° 25′ N., and long. by account 58° E., a chronometer showed 5ʰ 10ᵐ 42·5ˢ, and the observed altitude of α Arietis was 30° 10′ 40″ W. of meridian, the index correction was + 4′ 20″, and the height of eye above the sea was 18 feet: required the true longitude.

February 24, at Greenwich mean noon, the chronometer was fast on Greenwich mean time 1ʰ 11ᵐ 22ˢ, and its daily rate was 2·2ˢ losing.

12. Sept. 3, at 9ʰ 10ᵐ P.M., mean time nearly, in lat. 30° 10′ N., and long. by account 91° 5′ E., the following lunar observation was taken:

Obs. alt. α Arietis E. of meridian.	Obs. alt. moon's L. L.	Obs. dist. F. L.
10° 15′ 40″	36° 12′ 30″	99° 27′ 50″
+1 40	−1 10	−0 30

The height of the eye above the sea was 12 feet: required the true longitude.

13. July 5, at 7ʰ 0ᵐ P.M., mean time, in lat. 50° 53′ N., and long. 120° 10′ E., the compass bearing of the sun was W. 10° 15′ N., and the observed altitude of its lower limb was 9° 40′ 0″, the index correction was +3′ 50″, and the height of the eye above the sea was 18 feet: required the variation of the compass.

14. On Dec. 20, at 4ʰ 30ᵐ P.M., mean time nearly, in lat. 41° 12′ N., and long. 110° 45′ E., the sun set by compass S.W.: required the variation.

15. Required the time of high water at A on June 10, A.M. and P.M.

Change tide... 3ʰ 40ᵐ P.M. app. time.   Long. A... 65° W.

NOTE. In this and the following Examination Papers the compass is supposed to have no deviation arising from local attraction.

Elements from Nautical Almanac and Answers.

1. N. 30° 29′ 15″ W.   377·2 miles.

2. E. 188·1 miles.

3. Corrected courses, N.W.¾N. 20′ departure course; N.b.W.¼W. 12·3′; S. 14·8′; S.E.b.E.¼E. 12·7′; E.¾N. 34·1′; W.S.W. 42·7′. Lat. in 48° 6′ N., long. 2° 18′ W.

4. Right ascension mean sun on June 14, at Greenwich mean noon, 5ʰ 32ᵐ 11·85ˢ. γ Draconis, 28° 23′ N. of zenith.

5. Sun's declination on June 14, at Greenwich mean noon, 23° 15′ 26″ N.; on June 15, 23° 18′ 24″ N., semidiameter, 15′ 46″. Lat. 46° 14′ 16″ S.

6. Moon's declination on April 22, at 21ʰ Greenwich mean time, 11° 35′ 58″ S., at 22ʰ, 11° 23′ 53″ S.; moon's horizontal semidiameter April 22, at Greenwich mean midnight, 15′ 4·1″, April 23, at Greenwich mean noon, 15′ 0″; corresponding horizontal parallax, 55′ 17·8″ and 55′ 2·8″. Lat. 38° 57′ 50″ N.

7. Declination α Scorpii (Antares), 26° 3′ 34″ S.   Lat. 43° 47′ 34″ N.

8. Declination α² Crucis, 62° 11′ 25″ S.   Lat. 34° 20′ 28″ S.

9. Right ascension mean sun on Dec. 10, at Greenwich mean noon, 17ʰ 10ᵐ 2·21ˢ.   Lat. 47° 51′ N.

10. Sun's declination on September 15, at Greenwich mean noon,

2° 49' 54" N., on September 16, 2° 26' 42" N.; semidiameter, 15' 56". Lat. 50° 20' N.

11. Right ascension mean sun March 2, at Greenwich mean noon, 22ʰ 38ᵐ 13·55'. Right ascension α Arietis, 1ʰ 57ᵐ 51·5'; declination α Arietis, 22° 40' 42" N.; hour-angle, 4ʰ 37ᵐ 23' W. Long. 59° 12' 15" E.

12. Right ascension mean sun September 3, at Greenwich mean noon, 10ʰ 47ᵐ 36·37'. Right ascension α Arietis, 1ʰ 57ᵐ 55·3'; declination α Arietis, 22° 40' 56" N. Horizontal semidiameter moon September 3, at Greenwich mean noon, 15' 51·7", at Greenwich mean midnight, 15' 48·3"; corresponding horizontal parallax, 58' 12·5" and 57' 59·9". True distance, 98° 59' 4"; distance from *Nautical Almanac*, at iii., 99° 2' 53", at vi., 97° 22' 28"; hour-angle, 17ʰ 54ᵐ 56' W. Long. 89° 28' 30" E.

13. Sun's declination on July 4, at Greenwich mean noon, 22° 56' 37" N., on July 5, 22° 51' 24" N.; semidiameter, 15' 45"; true bearing, N. 65° 41' W. Variation, 14° 4' E.

14. Sun's declination on December 19, at Greenwich mean noon, 23° 25' 35" S., on December 20, 23° 26' 46" S.; true bearing, W. 31° 55' 30" S. Variation, 13° 4' 30" E.

15. Moon's Greenwich meridian passage June 10, 12ʰ 2ᵐ, June 9, 11ʰ 0ᵐ; moon's semidiameter, 16' 36"; equation of time, 1ᵐ sub. from apparent time. High water, 3ʰ 2ᵐ A.M. and 3ʰ 33ᵐ P.M.

NOTE. The right ascension of mean sun is found in the *Nautical Almanac* in page ii. of each month, under the heading of "Sidereal Time."

Questions.—No. II.

1. Required the course and distance from A to B.

Lat. A...... 40° 25' N.   Long. A...... 2° 10' E.
  „ B...... 35   32 N.   „ B...... 1   40 W.

2. Required the course and distance from A to B.

Lat. A...... 50° 48' N.   Long. A......... 100° E.
  „ B...... 50   48 N.   „ B......... 101 E.

3. May 10, at noon, a point of land in lat. 38° 17' N. and long. 56° 19' W. bore by compass W.b.S.¼ S. distant 17½ miles (variation of compass 2¾ E.); afterwards sailed as by the following log account: required the latitude and longitude in, May 11, at noon.

Hours.	Knots.	$\frac{1}{10}$ths.	Course.	Wind.	Leeway.	—
1	5	4	S.S.E.	E.	$2\frac{1}{4}$	P.M.
2	5	6				
3	4	9				
4	4	8				
5	4	8	S.S.W.$\frac{1}{2}$W.	W.	$2\frac{3}{4}$	
6	4	7				Variation $2\frac{3}{4}$ E.
7	5	3				
8	5	2				
9	5	1				
10	6	0	W.S.W.	S.	$2\frac{1}{2}$	
11	6	4				
12	6	8				
						H.M.S. May 11, 1870.
1	6	7				A.M.
2	5	9				
3	5	8	W.$\frac{1}{2}$N.	N.N.E.	0	
4	4	6				During the last 7
5	4	8				hours a current set
6	5	9				the ship 3 miles an
7	4	8				hour N.N.E. by
8	3	7				compass.
9	3	6	E.	S.S.E.	$2\frac{1}{2}$	
10	3	4				
11	3	5				
12	2	9				

4. What bright star will pass the meridian of Greenwich the first after 10ʰ P.M. on October 20, and how far N. or S. of the zenith?

5. October 19, in longitude 88° 49′ E., the observed meridian altitude of the sun's lower limb was 58° 37′ 56″ (zenith N. of the sun), the index correction was $+8'\ 38''$, and the height of the eye above the sea was 17 feet: required the latitude.

6. August 10, at 6ʰ 40ᵐ P.M., mean time, in long. 50° 17′ E., the observed meridian altitude of the moon's lower limb was 45° 47′ 39″ (zenith N. of the moon), the index correction was $-3'\ 18''$, and the height of the eye above the sea was 24 feet: required the latitude.

7. June 3, the observed meridian altitude of the star α Canis Majoris was 43° 29′ 47″ (zenith S. of the star), the index correction was $-3'\ 14''$, and the height of the eye above the sea was 16 feet: required the latitude.

8. February 18, the observed meridian altitude of the star α Ursæ Majoris under the North Pole was 53° 28′ 47″, the index correction was $-3'\ 49''$, and the height of the eye above the sea was 18 feet: required the latitude.

9. February 9, at 10^h 20^m P.M., mean time, in long. 85° 32' W., the observed altitude of α Ursæ Minoris (Polaris) was 50° 25' 30", the index correction was −4' 10", and the height of the eye above the sea was 15 feet: required the latitude.

10. June 9, the following double altitude of the sun was observed :

Mean time nearly.	Chronometer.	Obs. alt. sun's L. L.	True bearing.
1^h 3^m P.M.	1^h 10^m 50^s	52° 5' 40"	S.S.W.
7 6 P.M.	7 12 48	14 57 30	W.N.W.

The run of the ship in the interval was N.N.E. 5 miles, the index correction was −1' 20", and the height of the eye above the sea was 17 feet : required the true latitude at the second observation, the latitude by account being 59° N., and longitude 47° 18' E.

11. August 25, at 9^h 45^m P.M., in lat. 60° 2' N., and longitude by account 59° 15' E., when a chronometer, No. 10, showed 5^h 42^m 16^s, the observed altitude of the star α Andromedæ was 39° 32' 28" E. of the meridian, the index correction was +5' 17", and the height of the eye above the sea was 15 feet : required the true longitude.

On May 15, at Greenwich mean noon, No. 10 was *slow* on Greenwich mean time 8^m 40.5^s, and its daily rate was 7.8^s *losing*.

12. May 14, at 2^h 20^m P.M., mean time nearly, in lat. 50° 48' N., and longitude by account 60° 52' E., the following lunar observation was taken :

Obs. alt. sun's L. L.		Moon's L. L.		Obs. dist. N. L.	
	46° 48' 7"		45° 47' 38"		108° 58' 45"
Index cor.	+3 10		−1 12		+2 18

The height of the eye above the sea was 10 feet : required the true longitude.

13. May 20, at 4^h 47^m A.M., mean time nearly, in 18° 42' S. and long. 160° E., the sun rose by compass E. 21° 18' 30" N. : required the variation.

14. March 7, at 2^h 50^m P.M., mean time nearly, in lat. 51° 10' N. and long. 86° E., the compass bearing of the sun was S. 74° 42' W., and at the same time the observed altitude of the sun's lower limb was 21° 40' 45", the index correction was −2' 18", and the height of the eye above the sea was 14 feet : required the variation.

15. Required the time of high water at A on August 27, A.M. and P.M.
    Change tide at A... 5^h 18^m P.M. app. time.   Long. A... 93° E.

Elements from Nautical Almanac and Answers.

1. S. 31° 43' 30" W.   344·5'.
2. E. 37·9'.

3. Corrected courses, E. b. S. $\frac{1}{2}$ S. 17·5'; S.W. b. S. 20·7'; S.S.W. $\frac{1}{4}$ W. 25·1'; N.W. $\frac{3}{4}$ W. 31·8'; N.W. $\frac{3}{4}$ W. 29·6'; E. $\frac{1}{2}$ S. 13·4'; N.E. $\frac{3}{4}$ E. 21'. Lat. in 38° 21' N.; long. in 56° 52' W.

4. α Andromedæ 23° 17' 10" S. of zenith.

5. Sun's declination on October 18, at Greenwich mean noon, 9° 39' 16" S.; on October 19, 10° 1' 3" S.; semidiameter, 16' 5". Lat. 21° 6' 29" N.

6. Moon's declination on August 10 at 3ʰ, Greenwich mean time, 23° 15' 12" S.; on August 10 at 4ʰ, 23° 24' 37" S.; moon's horizontal semidiameter on August 10, at Greenwich mean noon, 15' 43·8"; on August 10, at Greenwich mean midnight, 15' 51·3"; corresponding horizontal parallax, 57' 43·5" and 58' 11·0". Lat. 20° 7' 1" N.

7. Declination of α Canis Majoris, 16° 29' 49" S. Lat. 63° 8' 14" S.

8. Declination of α Ursæ Majoris, 62° 37' 42" N. Lat. 80° 42' 22" N.

9. Right ascension mean sun on February 9, 1837, at Greenwich mean noon, 21ʰ 17ᵐ 28·28ˢ. Lat. 50° 34' N.

10. Sun's declination on June 8, at Greenwich mean noon, 22° 51' 58" N.; on June 9, 22° 57' 10" N.; and on June 10, 23° 1' 58" N.; semidiameter, 15' 46". Arc (1), 81° 40' 15"; Arc (2), 68° 32' 15"; Arc (3), 38° 4' 0". Lat. 60° 11' 51" N.

11. Right ascension mean sun on August 25, at Greenwich mean noon, 10ʰ 14ᵐ 9·84ˢ; declination of α Andromedæ, 28° 11' 40" N.; right ascension, 0ʰ 0ᵐ 1ˢ; hour-angle, 20ʰ 4ᵐ 23ˢ W. Long. 56° 15' 0" E.

12. Sun's declination on May 13, at Greenwich mean noon, 18° 24' 17" N.; on May 14, 18° 38' 55" N.; corresponding eq. of time, 3ᵐ 55·3ˢ S. and 3ᵐ 55·9ˢ S.; moon's horizontal semidiameter on May 13, at Greenwich mean midnight, 14' 55·2"; on May 14, at Greenwich mean noon, 14' 59"; corresponding horizontal parallax, 54' 45·1" and 54' 58·9". True distance, 108° 37' 59"; distance at xxi., 108° 4' 49"; distance on 14, at Greenwich mean noon, 109° 28' 44"; hour-angle, 2ʰ 24ᵐ 5ˢ. Long. 62° 15' E.

13. Sun's declination on May 19, at Greenwich mean noon, 19° 47' 14" N.; on May 20, 19° 59' 54" N. True bearing, E. 20° 59' 45" N. Variation, 0° 18' 45" E.

14. Sun's declination on March 6, at Greenwich mean noon, 5° 37' 27" S.; on March 7, 5° 14' 8" S.; semidiameter, 16' 8". True bearing, N. 130° 56' 30" W. Variation, 25° 38' 30" W.

15. Moon's Greenwich meridian passage on August 27, 22ʰ 8·7ᵐ; August 26, 21ʰ 19·9ᵐ; moon's semidiameter, 14' 45". Equation of time, 1ᵐ 19ˢ S. from mean time. High water, 2ʰ 18ᵐ A.M. and 2ʰ 42ᵐ P.M.

Questions.—No. III.

1. Required the course and distance from A to B.

Lat. A...... 70° 15′ S.   Long. A...... 3° 10′ W.

„ B...... 75 20 S.   „ B...... 2 15 E.

2. How many miles are there in 10° of longitude in the latitude of Portsmouth?

3. March 4, at noon, a point of land in lat. 50° 48′ N., and long. 1° 6′ W., bore by compass N.N.E. ½ E., distant 15 miles (variation 2¼ W.), afterwards sailed as by the following log account: required the latitude and longitude in, on March 5 at noon.

Hours.	Knots.	₁₀ths.	Course.	Wind.	Leeway.	
1	3	5	N.N.W. ½ W.	N.E.	1¾	P.M.
2	4	1				
3	4	3				
4	2	7	E.S.E.	Do.	2	
5	3	0				
6	3	2				Variation 2¼ W.
7	4	0				
8	5	6	S. ¾ W.	E.S.E.	2½	
9	5	2				
10	5	5				
11	4	5				
12	4	6				
						Remarks in H.M.S. Mar. 5, 1870.
1	4	7	N.E. ¼ N.	Do.	1½	A.M.
2	4	2				
3	4	4				
4	3	7				
5	3	2				
6	3	5	W. ½ N.	S.S.W.	1¼	A current set the ship N.E. the last 6 hours at the rate of 3¼ miles per hour by compass.
7	4	2				
8	3	6				
9	3	4				
10	9	5	N. by E.	South.	0	
11	10	2				
12	10	3				

4. At what time will the star α Lyræ pass the meridian of Portsmouth on May 11, and how far N. or S. of the zenith?

5. March 8, in long. 89° 48′ E., the observed meridian altitude of the sun's lower limb was 51° 49′ 30″, zenith north of the sun, the index correction was −3′ 17″, and the height of the eye above the sea 15 feet: required the latitude.

6. March 16, at 8ʰ 2ᵐ P.M. mean time nearly, in long. 110° E., the observed meridian altitude of the moon's lower limb was 48° 47′ 36″, zenith

north of the moon, the index correction was — 2' 47", and the height of the eye above the sea was 13 feet: required the latitude.

7. July 7, the observed meridian altitude of the star α Cygni was 53° 29' 38", zenith north of the star, the index correction was — 5' 12", and the height of the eye above the sea was 16 feet: required the latitude.

8. Oct. 16, the observed meridian altitude of the star α Ursæ Majoris under the North Pole was 5° 26' 10", the index correction was — 2' 10", and the height of the eye above the sea was 17 feet: required the latitude.

9. Sept. 10, 1837, at 3ʰ 42ᵐ A.M., longitude 83° 14' E., the observed altitude of α Ursæ Minoris was 39° 47' 48", the index correction was + 3' 45", and the height of the eye above the sea was 17 feet: required the latitude.

10. April 10, the following double altitude of the sun was observed:

Mean time nearly.	Chronometer.	Obs. alt. sun's L. L.	True bearing.
10ʰ 14ᵐ A.M.	10ʰ 9ᵐ 40ˢ	41° 15' 45"	S.E.
11 47 A.M.	11 43 28	46 43 12	S. by E.

The run of the ship in the interval was N.W. 6 miles, the index correction was — 4' 24", and the height of the eye above the sea was 20 feet: required the true latitude at the second observation, the latitude by account being 51° N., and the longitude 1° 6' W.

11. May 10, at 3ʰ 10ᵐ P.M. mean time nearly, in lat. 48° 12' N., and long. by account 45° 10' E., when a chronometer showed 0ʰ 10ᵐ 42ˢ, the observed altitude of the sun's lower limb was 37° 20' 10", the index correction was + 3' 10", and the height of the eye above the sea was 18 feet: required the true longitude.

On May 1, at Greenwich mean noon, the chronometer was fast on Greenwich mean time 9ᵐ 50ˢ, and its daily rate was 3·2ˢ gaining.

12. January 16, at 3ʰ 4ᵐ P.M. mean time nearly, in lat. 50° 50' N., and long. by account 65° E., the following lunar observation was taken:

	Obs. alt. sun's L. L.	Obs. alt. moon's L. L.	Obs. dist. N. L.
	8° 32' 20"	15° 42' 30"	121° 10' 30"
Index cor.	+1 10	+5 47	—5 47

The height of the eye above the sea was 16 feet: required the true longitude.

13. May 18, at 4ʰ 50ᵐ A.M. mean time nearly, in lat. 18° 45' S., and long. 99° 18' E., the sun rose by compass S. 80° 12' E.: required the variation.

14. March 7, at 9ʰ 10ᵐ A.M. mean time nearly, in lat. 51° 10' N., and long. 89° 12' E., the compass bearing of the sun was S. 74° 50' E., and at the same time the observed altitude of the sun's lower limb was 21° 40' 43", the index correction was — 2' 18", and the height of the eye above the sea was 14 feet: required the variation.

15. Required the time of high water at A on March 10, A.M. and P.M.

Change tide at A... 6ʰ 45ᵐ P.M. app. time.   Long. A... 98° E.

Elements from Nautical Almanac and Answers.

1. S. 17° 23′ E., 319·4′.

2. 379·2′.

3. Corrected courses, S. ¼ W. 15′; departure course, W. b. N. ½ N. 11·9′; E.S.E. ¼ E. 12·9′; S. ¾ W. 25·4′; N. 20·2′; W. ½ S. 14·7′; N. b. W. ¼ W. 30′; N. b. E. ¾ E. 21′. Latitude in 51° 14′ 54″ N.; longitude in 1° 35′ 24″ W.

4. At 15ʰ 12ᵐ 42ˢ : 12° 10′ 11″ S. of zenith.

5. Sun's declination on March 7, at Greenwich mean noon, 5° 14′ 8″ S.; on March 8, 4° 50′ 46″ S.; semidiameter, 16′ 7″. Lat. 33° 5′ 44″ N.

6. Moon's declination on March 16, at 0ʰ, 26° 48′ 39″ N.; at 1ʰ, 26° 44′ 20″ N., moon's horizontal semidiameter on March 16, at Greenwich mean noon, 14′ 45·1″; on March 16, at Greenwich mean midnight, 14′ 44·9″; horizontal parallax, 54′ 8·1″ and 54′ 7·4″. Lat. 67° 14′ 35″ N.

7. Declination of α Cygni, 44° 41′ 59″ N. Lat. 81° 22′ 12″ N.

8. Declination of α Ursæ Majoris, 62° 37′ 28″ N. Lat. 32° 33′ 1″ N.

9. Right ascension mean sun, on Sept. 9, at Greenwich mean noon, 11ʰ 13ᵐ 18·15ˢ. Lat. 38° 24′ N.

10. Sun's declination on April 9, at Greenwich mean noon, 7° 36′ 29″ N., on April 10, 7° 58′ 43″ N.; semidiameter, 15′ 58″. Arc (1) 23° 13′ 15″, arc (2) 88° 17′ 30″, arc (3) 65° 27′ 0″. Lat. 51° 0′ 47″ N.

11. Sun's declination on May 10, at Greenwich mean noon, 17° 38′ 34″ N.; on May 11, 17° 54′ 7″ N.; equation of time, 3ᵐ 50·2ˢ S., and 3ᵐ 52·5ˢ E.; semidiameter, 15′ 51″; hour-angle, 3ʰ 31ᵐ 21ˢ. Longitude, 51° 47′ E.

12. Sun's declination on Jan. 15, at Greenwich mean noon, 21° 6′ 42″S., on January 16, 20° 55′ 23″ S., equation of time, 9ᵐ 49·1ˢ ᴀ, and 10ᵐ 9·8ˢ ᴀ; moon's horizontal semidiameter on January 15, at Greenwich mean midnight, 15′ 2·1″, on January 16, at Greenwich mean noon, 14′ 58·0″; corresponding horizontal parallax, 55′ 10·6″ and 54′ 55·3″. True distance, 121° 20′ 29″; distance at xxi., 120° 32′ 5″; at xxiv., 121° 55′ 25″. Hour-angle, 2ʰ 54ᵐ 7ˢ. Longitude, 64° 55′ 45″ E.

13. Sun's declination on May 17, at Greenwich mean noon, 19° 20′ 53″ N., on May 18, 19° 34′ 14″ N. True bearing, E. 20° 34′ 45″ N. Variation, 30° 22′ 45″ W.

14. Sun's decl. on March 6, at Greenwich mean noon, 5° 37′ 27″ S., on March 7, 5° 14′ 8″ S.; semidiameter, 16′ 8″. True bearing, N. 131° 9′ 15″ E. Variation, 25° 59′ 15″ E.

15. Moon's Greenwich meridian passage on March 10, 3ʰ 13ᵐ mean time on March 9, 2ʰ 26ᵐ; moon's semidiameter, 15′ 37″; equation of time, 11ᵐ S. from mean time; high water 9ʰ 1ᵐ P.M. and 8ʰ 37ᵐ A.M.

Questions.—No. IV.

1. Required the course and distance from A to B.

    Lat. A ...... 60° 25′ S.   Long. A ...... 35° 22′ E.

      ,, B ...... 64 12 S.   ,, B ...... 30 10 E.

2. Sailed from Ushant due west 492·5 miles : required the latitude and longitude in.

3. May 1, at noon, a point of land in latitude 51° 10′ S., and longitude 3° 15′ E., bore by compass S.S.W. ½ W. distant 25 miles, variation 2¾ E., afterwards sailed as by the following log account : required the latitude and longitude in.

Hours.	Knots.	$\frac{1}{10}$ths.	Course.	Wind.	Leeway.	
1	3	2	S.S.E.½E.	S.W.	2¼	A.M.
2	3	4				
3	3	5				
4	4	0				
5	4	2				
6	4	3	W.N.W.	Do.	2½	Variation 2¾ E.
7	4	4				
8	4	5				
9	5	2				
10	5	6				
11	5	7				
12	6	1				
						H.M.S. May 2, 1870.
1	7	2	South.	West.	¼	P.M.
2	7	1				
3	7	3				
4	8	3				
5	8	4				
6	7	5	S.E.¾E.	S.S.W.	1½	A current set the
7	7	2				ship N.W.½ W.
8	7	1				20 miles.
9	6	7				
10	6	5				
11	6	3				
12	6	0				

4. What bright star will pass the meridian of the Land's End the first after 6ʰ 42ᵐ A.M. mean time on August 17, and how far N. or S. of the zenith ?

5. August 18, in longitude 110° 32′ E., the observed meridian altitude of the sun's lower limb was 50° 25′ 10″, zenith N. of the sun, the index correction was −2′ 50″, and the height of the eye above the sea was 15 feet : required the latitude.

6. August 18, at 8^h 0^m A.M. mean time nearly, in longitude 92° 10' W., the observed meridian altitude of the moon's lower limb was 26° 42' 10", zenith S. of the moon, the index correction was — 3' 40", and the height of the eye above the sea was 14 feet : required the latitude.

7. December 7, the observed meridian altitude of the fixed star α Arietis was 40° 25' 10", zenith N. of the star ; the index correction was — 2' 10", and the height of the eye above the sea was 18 feet : required the latitude.

8. December 7, the observed meridian altitude of α Ursæ Majoris, under the N. Pole, was 11° 10' 10" ; the index correction was + 3' 20", and the height of the eye above the sea was 19 feet : required the latitude.

9. December 7, 1835, at 1^h 20^m A.M., in long. 78° 36' E., the observed altitude of α Ursæ Minoris (Polaris) was 50° 40' 15" ; the index correction was — 5' 10", and the height of the eye above the sea was 12 feet : required the latitude.

10. July 31, observed the following double altitude of the sun :

Mean time nearly.	Chronometer.	Obs. alt. sun's L. L.	True bearing.
11^h 58^m A.M.	0^h 0^m 10^s	57° 29' 45"	S. 3° E.
0 4 P.M.	0 6 17	57 29 30	S. 3° 20' W.

The run of the ship was none, dip none, the index correction was + 0' 30" : required the true latitude, the latitude by account being 51° N., and the longitude 1° W.

11. May 14, at 9^h 30^m A.M., in latitude 50° 48' N., and long. by account 2° W., a chronometer showed 9^h 26^m 18', and the observed altitude of the sun's lower limb was 46° 48' 7", the index correction was + 3' 10", and the height of the eye above the sea was 10 feet : required the true longitude.

May 1, at Greenwich mean noon, the chronometer was slow on Greenwich mean time 4^m 2', and its daily rate was 3·5' losing.

12. September 3, at 6^h 32^m P.M. mean time nearly, in lat. 30° 10' N., and long. by account 36° 10' W., the following lunar observation was taken :

Obs. alt. α Pegasi (Markab) E. of meridian.	Obs. alt. moon's L. L.	Obs. dist. F. L.
9° 50' 40"	18° 10' 50"	55° 46' 20"
Index cor. — 1 10	+ 1 30	— 0 30

The height of the eye above the sea was 15 feet : required the true longitude.

13. August 23, at 7^h 0^m P.M. mean time nearly, in lat. 50° 48' N., and long. by account 140° 25' E., the sun set by compass W. 5° 10' S.: required the variation.

14. August 23, at 5^h 50^m A.M. mean time nearly, in lat. 51° 10' N., and long. 135° 40' W., the sun-bearing by compass was S. 92° 10' E., and the observed altitude of its lower limb was 7° 40' 50", the index correction was — 2' 50", and the height of the eye above the sea was 15 feet : required the variation.

15. Required the mean time of high water at A on Aug. 2, A.M. and P.M.
Change tide ... 3ʰ 40ᵐ app. time.   Long. A ... 70° W.

Elements from Nautical Almanac and Answers.

1. S. 32° 32′ 15″ W.   269·3′.

2. Lat. in 48° 28′ N.   Long. in 17° 25′ 48″ W.

3. Corrected courses, N.E.b.E.¼E. 25′ departure course; S.S.E. 18·3′; N.¾W. 35·8′; S.S.W.½W. 38·3′; S.E.½S. 47·3′; N.b.W.¾W. 20′.   Lat. in 51° 29′ 54″ S.   Long. in 4° 0′ 18″ E.

4. α Tauri 33° 54′ S. of zenith.

5. Sun's declination on Aug. 17, at Greenwich mean noon, 13° 35′ 52″ N., on Aug. 18, 13° 16′ 40·3″ N.; semidiameter, 15′ 49″. Latitude, 52° 48′ 51″ N.

6. Moon's declination on Aug. 18, at 2ʰ Greenwich mean time, 24° 12′ 3″ N., at 3ʰ 24° 17′ 6″ N.; moon's horizontal semidiameter on August 18, at Greenwich mean noon, 14′ 51·6″, at midn., 14′ 54·5″; corresponding horizontal parallax, 54′ 31·8″ and 54′ 42·7″. Latitude, 38° 10′ 36″ S.

7. Declination of α Arietis 22° 41′ 5″ N.   Latitude, 72° 23′ 24″ N.

8. Declination of α Ursæ Majoris, 62° 37′ 57″ N. Latitude, 38° 26′ 29″ N.

9. Right ascension mean sun on December 6, at Greenwich mean noon, 16ʰ 58ᵐ 12·52ˢ. Latitude, 50° 16′ N.

10. Sun's declination on July 31, at Greenwich mean noon, 18° 39′ 18″ N., On August 1, 18° 24′ 47″ N.; semidiameter, 15′ 47″. Latitude, 50° 53′ 30″ N.

11. Sun's declination on May 13, at Greenwich mean noon, 18° 16′ 32″ N., on May 14, 18° 31′ 19″ N.; corresponding equation of time, 3ᵐ 45·9ˢ S. and 3ᵐ 55·9ˢ S.; semidiameter, 15′ 50″; hour-angle, 21ʰ 36ᵐ 45ˢ W. Longitude, 0° 26′ 0″ E.

12. Right ascension mean sun September 3, at Greenwich mean noon, 10ʰ 47ᵐ 36·37ˢ; right ascension α Pegasi, 22ʰ 56ᵐ 35·2ˢ; declination, 14° 19′ 23″ N.; moon's horizontal semidiameter on September 3, at Greenwich mean noon, 15′ 51·7″ at midnight, 15′ 48·3″; corresponding horizontal parallax, 58′ 12·5″ and 57′ 59·9″; true distance, 55° 32′ 23″; distance from *Nautical Almanac*, at vi., 56° 49′ 7″, at ix., 55° 19′ 51″; hour-angle, 18ʰ 11ᵐ 57ˢ W. Longitude, 33° 48′ 0″ W.

13. Sun's declination on Aug. 22, at Greenwich mean noon, 11° 57′ 49″ N., on August 23, 11° 37′ 37″ N.; true bearing, W. 18° 38′ 45″ N.   Variation, 23° 48′ 45″ E.

14. Sun's declination on Aug. 23, at Greenwich mean noon, 11° 37′ 37″ N., on August 24, 11° 17′ 14″ N.; semidiameter, 15′ 51″; true bearing, N. 81° 6′ 30″ E.   Variation, 6° 43′ 30″ W.

15. Moon's Greenwich meridian passage August 2, 6ʰ 38ᵐ, August 1, 5ʰ 46ᵐ; semidiameter, 16′ 10″; equation of time, 6ᵐ S. from mean time. High water, 9ʰ 12ᵐ A.M., and 9ʰ 38ᵐ P.M.

Questions.—No. V.

1. Required the course and distance from A to B.

    Lat. A... 65° 25′S.   Long. A... 3° 28′W.

      „ B... 73   42 S.   „ B... 4   2 E.

2. Required the course and distance from C to D.

    Lat. C... 70° 15′N.   Long. C... 15° 25′ E.

      „ D... 70   15 N.   „ D... 20   25 E.

3. October 23, at noon, a point of land in latitude 34° 28″ S., and longitude 18° 28′ E., bore by compass N.W. distant 10 miles (variation of compass 2¼ W.), afterwards sailed as by the following log account: required the latitude and longitude in on October 24, at noon.

Hours.	Knots.	₁₀ths.	Course.	Wind.	Leeway.	
1	5	4	N.b.E.½E.	N.W.½W.	2¼	
2	5	2				
3	5	8				
4	6	1				
5	6	5	S.S.W.	W.N.W.	¼	
6	7	3				
7	7	0				Variation 2¼ W.
8	7	2				
9	6	8	N.W.b.W.	S.E.	0	
10	6	5				
11	6	1				
12	5	8	S.b.W.¾W.	S.E.	2¼	
						—— Oct. 24, 1870.
1	6	0			◦	
2	6	5				
3	6	8				
4	6	4	N.N.E.	N.W.	2	
5	6	0				A current set the
6	6	5				ship the last 5
7	6	8				hours N.W. 2
8	7	2	N.W.	East.	0	miles an hour by
9	7	6				compass.
10	7	9				
11	8	1				
12	8	5				

4. At what time will the star α Aquilæ (Altair) pass the meridian of the Land's End on December 8, and how far N. or S. of the zenith ?

5. December 10, in long. 55° 20′ E., the observed meridian altitude of the sun's lower limb was 25° 52′ 5″ (zenith N.), the index correction was −2′ 10″, and the height of the eye above the sea was 17 feet: required the latitude.

6. August 10, at 6ʰ 40ᵐ P.M. mean time nearly, in long. 50° 17′ E., the observed meridian altitude of the moon's lower limb was 45° 47′ 39″ (zenith N. of the moon), the index correction was −3′ 18″, and the height of the eye above the sea was 24 feet: required the latitude.

7. October 15, the observed meridian altitude of α Aquilæ was 50° 25′ 30″ (zenith N.), the index correction was −3′ 20″, and the height of the eye above the sea was 13 feet : required the latitude.

8. October 16, the observed meridian altitude of α Ursæ Majoris (Dubhe) under the North Pole was 5° 26′ 10″, the index correction was − 2′ 10″, and the height of the eye above the sea was 17 feet: required the latitude.

9. March 17, 1837, at 9ʰ 43ᵐ P.M., in long. 93° 14′ W., the observed altitude of α Ursæ Minoris (Polaris) was 32° 49′ 14″, the index correction was +7′ 49″, and the height of the eye above the sea was 12 feet: required the latitude.

10. March 14, the following double altitude of the sun was observed :

Mean time nearly.	Chronometer.	Obs. alt. sun's L. L.	True bearing.
1ʰ 5ᵐ P.M.	8ʰ 2ᵐ 25ˢ	41° 20′ 45″	S.S.W. ½ W.
5 6 P.M.	12 3 30	7 29 30	W. by S. ½ S.

The run of the ship in the interval was N.E. 18 miles, the index correction was −3′ 20″, and the height of the eye above the sea was 23 feet : required the true latitude at the second observation, the latitude by account being 45° N., and the long. 50° 20′ W.

11. February 10, at 9ʰ 20ᵐ P.M. mean time nearly, in lat. 28° 20′ N. and longitude by account 31° 2′ W., a chronometer showed 11ʰ 16ᵐ 25ˢ, and the observed altitude of the star α Leonis (Regulus) was 41° 55′ 10″ E. of the meridian, the index correction was +1′ 20″, and the height of the eye above the sea was 25 feet : required the true longitude.

On February 1, at Greenwich mean noon, the chronometer was *fast* on Greenwich mean time 5ᵐ 20·6ˢ, and its daily rate was 2·7ˢ *losing*.

12. April 27, at 2ʰ 30ᵐ A.M. mean time nearly, in lat. 45° 20′ N., and longitude by account 46° W., the following lunar observation was taken :

Obs. alt. α Virginis. W. of meridian.	Obs. alt. moon's L. L.	Obs. dist. F. L.
16° 30′ 50″	15° 38′ 56″	98° 2′ 40″
Index cor. +2 20	+5 52	+1 5

s

The height of the eye above the sea was 12 feet : required the true longitude.

13. June 15, at $8^h 10^m$ P.M. mean time, in lat. 50° 48′ N., and long. 73° 19′ E., the sun set by compass W. 30° 29′ N. : required the variation.

14. June 15, at $9^h 39^m$ A.M. mean time nearly, in lat. 50° 48′ N., and long. 99° 29′ E., the compass bearing of the sun was S. 38° 19′ 50″ E., and the observed altitude of the sun's lower limb at the time was 49° 58′ 37″, the index correction was + 10′ 43″, and the height of the eye above the sea was 12 feet : required the variation.

15. Required the time of high water at A on February 17, A.M. and P.M. Change tide at A... $11^h 42^m$ P.M. app. time.   Long. A... 2° W.

Elements from Nautical Almanac and Answers.

1. S. 17° 19′ 15″ E.   520·6′.

2. E. 101·3′.

3. Corrected courses E.b.S. ¾ S. 10′ departure course ; N.b.E. ½ E. 22·5′ ; S. ½ E. 28′ ; W. ¾ N. 19·4′ ; S.b.W. ¾ W. 25·1′ ; N.b.E. ¾ E. 25·7′ ; W.b.N. ¾ N. 39·3′ ; W.b.N. ¾ N. 10′.   Latitude in 31° 17′ 54″ S.   Longitude in 17° 32′ E.

4. At $2^h 31^m$ and 41° 37′ S. of zenith.

5. Sun's declination on Dec. 9, at Greenwich mean noon, 22° 51′ 14″ S. ; on Dec. 10, 22° 56′ 49″ S. ; semidiameter, 16° 16″.   Lat. 41° 3′ 48″ N.

6. Moon's declination on Aug. 10, at 3^h, 23° 15′ 12″ S. ; at 4^h, 23° 24′ 37″ S. ; moon's horizontal semidiameter on August 10, at Greenwich mean noon, 15′ 43·8″ ; on August 10, at Greenwich mean midnight, 15′ 51·3″ ; corresponding horizontal parallax, 57′ 43·5″ and 58′ 11·0″.   Lat. 20° 7′ 1″ N.

7. Declination of α Aquilæ 8° 26′ 42″ N.   Lat. 48° 8′ 53″ N.

8. Declination of α Ursæ Majoris 62° 37′ 28″ N.   Lat. 32° 33′ 1″ N.

9. Right ascension mean sun, on March 17, at Greenwich mean noon, $23^h 39^m 24·25^s$.   Lat. 33° 47′ N.

10. Sun's declination on March 14, at Greenwich mean noon, 2° 29′ 26″ S. ; on March 15, 2° 5′ 46″ S. ; semidiameter, 16′ 6″.   Arc (1) 60° 12′ 45″ ; arc (2) 91° 26′ 30″ ; arc (3) 46° 23′ 30″.   Lat. 43° 59′ 23″ N.

11. Sun's right ascension on February 10, at Greenwich mean noon, $21^h 21^m 24·83^s$; right ascension α Leonis, $9^h 59^m 43^s$; decl., 12° 45′ 40″ N. Hour-angle, $20^h 43^m 40^s$.   Long. 27° 50′ 45″ W.

12. Right ascension mean sun on April 26, at Greenwich mean noon, $2^h 17^m 6·41^s$; right ascension of α Virginis, $13^h 16^m 38^s$; decl., 10° 18′ 40″ S. ; moon's horizontal semidiameter on April 26, at Greenwich mean midnight,

16′ 8·3″; on April 27, at Greenwich mean noon, 16′ 8·4″; corresponding horizontal parallax, 59′ 13·5″ and 59′ 13·7″.   True distance, 97° 30′ 29″; distance at xv., 96° 0′ 34″; at xviii., 97° 46′ 47″.   Hour-angle, 3ʰ 34ᵐ 26ˢ W. Long. 45° 19′ 30″ W.

13. Sun's declination on June 15, Greenwich mean noon, 23° 19′ 50″ N.; on June 16, 23° 22′ 11″ N.   True bearing, W. 38° 48′ 45″ N.   Variation, 8° 19′ 45″ E.

14. Sun's declination on June 14, at Greenwich mean noon, 23° 17′ 5″ N.; on June 15, 23° 19′ 50″ N.; semi., 15′ 46″.   True bearing, N. 119° 53′ E. Variation, 21° 47′ 10″ W.

15. Moon's meridian passage on February 17, 10ʰ 21·9ᵐ; on February 16, 9ʰ 32·2ᵐ; semidiameter, 14′ 43″.   Equation of time, 14ᵐ S. from mean time. High water, 9ʰ 32ᵐ A.M. and 9ʰ 57ᵐ P.M.

END OF PART 1.

LONDON :
ROBSON AND SONS, PRINTERS, PANCRAS ROAD, N.W

THE THEORY

OF

NAUTICAL ASTRONOMY

AND

NAVIGATION.

PART II.

CONTAINING THE INVESTIGATIONS AND PROOFS OF THE PRINCIPAL
RULES AND CORRECTIONS.

With Practical Examples.

𝔇esigned for 𝔅eginners and advanced 𝔖tudents.

———•——

By H. W. JEANS, F.R.A.S.

LATE MATHEMATICAL MASTER AND EXAMINER AT THE ROYAL NAVAL COLLEGE, PORTSMOUTH ;

AUTHOR OF A WORK ON
" PLANE AND SPHERICAL TRIGONOMETRY;" " HANDBOOK OF THE STARS;" " PROBLEMS IN
ASTRONOMY, NAVIGATION, ETC. WITH SOLUTIONS :"

FORMERLY MATHEMATICAL MASTER IN THE ROYAL MILITARY ACADEMY, WOOLWICH ; AND AN EXAMINER OF OFFICERS
IN THE MERCHANT-SERVICE IN NAUTICAL ASTRONOMY, ETC,

THIRD EDITION.

LONDON:
LONGMANS, GREEN, READER, AND DYER.
1877.

LONDON :
ROBSON AND SONS, PRINTERS, PANCRAS ROAD, N.W.

PREFACE.

THE First Part of this Work — consisting of Practical Rules in Navigation and Nautical Astronomy, with a series of examples under each rule — was originally drawn up for the use of beginners, and as introductory to some of the larger works on the subject; but recent additions render it complete in itself, and it will now be found to contain ample directions for the guidance of the practical navigator. The rules are adapted to any of the standard Nautical Tables now in use, such as Norie's, Raper's, or Riddle's; but as the collection of Tables published by the late Dr. Inman* is almost universally adopted in the Royal Navy, these tables have in most cases been used in working out the examples. The present volume, Part II., may be considered as the scientific part of the subject, consisting of the analytical investigations and proofs of the principal rules and corrections in Navigation and Nautical Astronomy. The Author trusts that the

---

* This venerable and most useful public servant died on the 8th of February 1859, at his residence at Southsea near Portsmouth, at the advanced age of eighty-three years. The Reverend James Inman, D.D., was upwards of thirty years Professor of Mathematics at the Naval College. He was the oldest of Cambridge senior wranglers, and long possessed a just celebrity in naval circles for his application of science to navigation and ship-building. He laboured very many years unobtrusively but zealously in his country's service. He sailed round the world, having been appointed to the expedition under Flinders as astronomer; was wrecked with him; and returning to Europe in the fleet of East-India ships under the command of Commodore Dance, was present in that celebrated action in which a fleet of merchantmen repulsed the French Admiral Linois. While Professor of Mathematics at the Royal Naval College he reduced to system the previously ill-arranged methods of navigation, and published several valuable works on the subject; but he was best known by his having been the first person in this country who built ships on scientific principles. Dr. Inman's translation of Chapman, with his valuable annotations, is the text-book on which all subsequent writers have proceeded. His pupils, a long list of distinguished naval officers, will remember him as a type of the high-minded scholar, — of the loyal, the truthful, and independent man.

plan followed by him in this Part, namely, to exhibit a geometrical figure or diagram of each problem, then to give the analytical investigation from which the rule is derived, followed by a numerical example worked out to show the application of the formula, together with nearly 300 examples for practice dispersed throughout the volume, will render the book useful to beginners (for whom, in fact, it is chiefly intended), as well as deserving the attention of more advanced students, and increasing the confidence of the practical navigator. The Author's wish was to exclude from the present work every problem requiring a mathematical knowledge beyond algebra and trigonometry; he has, however, been induced to depart from this in one or two instances, at the request of several Naval Instructors, who thought that it would render the work more useful to introduce the problems for finding the longitude by an occultation, and for determining the spheroidal figure of the earth from actual measurements on its surface. These two problems belong to the more advanced part of Nautical Astronomy, which the Author contemplated at one time to publish under the title of "Nautical Astronomy, Part III." Part II. however, it is hoped, will enable the naval student to comprehend without difficulty the *principal* rules in Nautical Astronomy; and this he will more readily do, if he has previously made himself acquainted with the Author's volume of Astronomical Problems, which work may be looked upon as introductory to the more important subject of Nautical Astronomy, explained and illustrated at large in the present volume.

Langstone House, Havant,
    June 1, 1877.

CONTENTS.

PROBLEMS ON TIME.

vi   CONTENTS.

PROBLEMS FOR DETERMINING THE LATITUDE.

PROOF OF RULES IN PLANE SAILING.

SHORE OBSERVATIONS FOR TIME AND LONGITUDE.

NAUTICAL ASTRONOMY.

CHAPTER I.

ASTRONOMICAL AND NAUTICAL TERMS AND DEFINITIONS.

1. NAUTICAL Astronomy teaches the method of finding the *place* of a ship, that is, its latitude and longitude, by means of astronomical observations.

2. The following are the principal terms in Nautical Astronomy: the definitions of these terms should be thoroughly understood and carefully committed to memory. They are fully explained in pp. 8-16.

True place of a heavenly body.

Apparent place of a heavenly body.

Axis of the earth.

Terrestrial equator.

Poles of the earth.

Axis of the heavens.

Celestial equator.

Poles of the heavens.

The ecliptic.

Obliquity of the ecliptic.

True latitude of spectator.

Reduced or central latitude of spectator.

Meridians of the earth.

True zenith.

B

Reduced zenith.

Visible horizon.

Rational horizon.

Poles of the horizon.

Vertical circles, or circles of altitude.

Celestial meridian.

North and south points.

Prime vertical.

East and west points.

Circles of declination.

Circles of latitude.

Right ascension of a heavenly body.

Declination of a heavenly body.

Longitude of a heavenly body.

Latitude of a heavenly body.

Altitude of a heavenly body.

Azimuth or true bearing of a heavenly body.

Amplitude of a heavenly body.

Hour-angle of a heavenly body.

Solar year.

Sidereal year.

Mean solar year.

Sidereal day.

Apparent solar day.

Mean sun.

Mean solar day.

Sidereal time.

Apparent solar time.

Mean solar time.

Equation of time.

Sidereal clock.

Mean solar clock or chronometer.

3. By the combination of theory with astronomical observations, the motions of the sun, moon, and planets have been determined with great accuracy, so that their places may be computed beforehand. The places and relative positions of the heavenly bodies, that is, their right ascensions, declinations, &c., at certain given times, are printed every year in England in the *Nautical Almanac*. In France a similar work is published, called the *Connaissance des Tems*.

Definitions of the preceding Terms in Nautical Astronomy.

4. To a spectator on the earth the sun, moon, and stars seem to be placed on the interior surface of a hollow sphere of great but indefinite magnitude. The interior surface of this sphere is called the *celestial concave*, the center of which may be supposed to be the same as that of the earth.

5. The heavenly bodies are not in reality thus situated with respect to the spectator; for they are interspersed in infinite space at very different distances from him : the whole is an optical deception, by which an observer, wherever he is placed, is induced to imagine himself to be the center of the universe. For let us suppose the elliptical figure $p\, q\, p_1\, q_1$ to represent the earth, P Q P$_1$ Q$_1$ the celestial concave, and m a heavenly body. Then a spectator at A not being able to estimate the distance of m, would imagine it to be in the celestial concave at M.

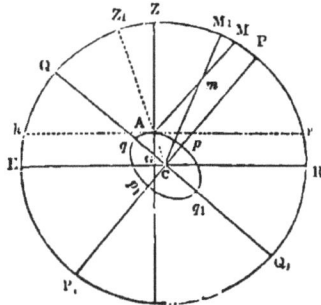

This figure will enable us to explain the terms *true* and *apparent place* of a heavenly body. The body m viewed from the surface of the earth would appear to a spectator A to be at M in the celestial concave : but if it could be seen from the center of the earth C, the point occupied by m would be M$_1$, the extremity of a line drawn from the center C of the earth through the heavenly body to the celestial concave. M is called the *apparent* place, and M$_1$ the *true* place of the heavenly body m.

6. The *axis of the earth* is that diameter about which it revolves : the *poles* of the earth are the extremities of the axis.

7. The *terrestrial equator* is that great circle on the earth that is equidistant from each pole.

8. A spectator on the earth, not being sensible of the motion by which in fact he describes daily a circle from west to east with the spot on which he stands, views in appearance the heavens moving past him in the opposite direction, or from east to west. The sphere of the fixed stars, or, as it is more usually called, the *celestial concave*, thus appears to revolve from east to west round an imaginary line which is the axis of the earth produced

to the celestial concave: this line is therefore called the *axis of the heavens.*

9. *The poles of the heavens* are the extremities of the axis of the heavens.

10. *The celestial equator* is that great circle in the celestial concave which is perpendicular to the axis of the heavens; or it may be defined to be the terrestrial equator expanded or extended to the celestial concave. The poles of the celestial equator and the poles of the heavens are therefore identical.

11. While the earth thus performs its daily revolution, it is carried with great velocity from west to east round the sun, and describes an elliptic orbit once every year. This *annual* motion of the earth round the sun causes the latter body, to a spectator on the earth, insensible of his own change of place, to appear to describe a great circle in the celestial concave

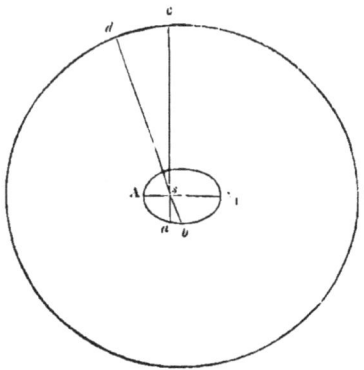

from west to east. This may be explained by a figure. Let A *b* A₁ be the earth's orbit, *s* the sun, and *c d e l* the celestial concave; then, to a spectator at *a* the sun is seen at a point *c* in the celestial concave: but when the earth has arrived at *b* the spectator (not being sensible of his motion from *a* to *b*) imagines the sun to be at *d*, and thus it would seem to have described the arc *c d* in the time the earth actually moved from *a* to *b*. It appears from this, that when the earth has arrived again at *a*, the sun will again be at *c*, having described one complete circle in the celestial concave among the fixed stars. The great circle thus described by the sun is called the *ecliptic.*

12. The axis of the earth, as it is thus carried round the sun, continues always parallel to itself, and it may be assumed without any sensible error, on account of the smallness of the earth's orbit (small, when compared with the distance of the heavenly bodies), to be always directed to the same points in the celestial concave, namely, the *poles* of the heavens.

13. From observation, the celestial equator is found to be inclined to the ecliptic at an angle of about 23° 28'. This inclination of the equator to the ecliptic is called the *obliquity of the ecliptic.* The axis of the earth, therefore, which is perpendicular to the equator, is inclined to the ecliptic, or, as it is in the same plane, to the earth's orbit, at an angle of 66° 32'.

14. In consequence of the whirling motion of the earth about its axis, the parts near the equator, which have the greatest velocity, acquire thereby a greater distance from the center than the parts near the poles. By actual measurement of a degree of latitude in different parts of the earth, it has

been computed that the equatorial diameter is longer than the axis or polar diameter by 26 miles: the former being about 7924 miles; the latter about 7898 miles, and that the form of the earth is that of an *oblate spheroid.* It is usual, however, in drawing the figure of the earth to exaggerate very much its ellipticity; this is done for the sake of drawing the lines about the figure with greater clearness; for if it were constructed according to its true dimensions, the line pp_1 (fig. art. 5) (being only about the $\frac{1}{300}$th part of itself less than qq_1) would appear to the eye of the same length as qq_1, and we should see that the figure that more nearly resembles the earth would be a sphere.

15. If A G, a perpendicular to the earth's surface, be drawn passing through A, the angle A G q formed by the line A G with the plane of the equator is the *latitude,* or true latitude of the point A.

16. If A c be a line drawn from A to c, the center of the earth, then the angle A c q is called the reduced or central latitude of the point A. The difference between the true and reduced latitude is not great: it is, however, of importance in some of the problems in Nautical Astronomy. This correction has accordingly been calculated, and forms one of the Nautical Tables.

17. Sections of the earth passing through the poles, as p A p_1, are called *meridians* of the earth. If the earth is considered as a sphere (which it is very nearly), the meridians will be circles: on this supposition, moreover, the perpendicular A G would coincide with A c, and the latitude of a place on the surface of the earth may, on this supposition, be defined to be the arc of the meridian passing through the place, intercepted between the place and the equator. If G A be produced to meet the celestial concave at z, the point z is the zenith of the spectator at A. If c A be produced to the celestial concave at z', then z' is called the *reduced* zenith of the spectator at A. The point opposite to z in the celestial concave is called the *Nadir.* In the figure the terrestrial equator $q\,q_1$ is extended to the celestial concave, and therefore Q c Q$_1$ is the plane of the *celestial* equator.

By means of this figure we may define the zenith, reduced zenith, latitude, and reduced latitude, as follows:

18. The *zenith* is that point in the celestial concave which is the extremity of the line drawn perpendicular to the place of the spectator, as z.

19. The *reduced* zenith is that point in the celestial concave which is the extremity of a straight line drawn from the center of the earth, through the place of the spectator, as z'.

20. The *latitude* of a place A on the surface of the earth, is the inclination of the perpendicular A G to the plane of the equator: thus the angle A G Q is the latitude of A. The arc z Q in the celestial concave measures the angle A G Q; hence z Q, *or the distance of the zenith from the celestial equator, is equal to the latitude of the spectator.*

21. The *reduced latitude* of the place A is the inclination of z' c or A c

to the plane of the equator: or it is the angle A C Q or arc z' Q, which measures the angle. Since the curvature of the earth diminishes from the equator to the poles, the reduced latitude z' Q must be always less than the true latitude z Q, and therefore the difference z z' must be subtracted from the true latitude to get the reduced latitude.

The formula for computing the difference between the true and reduced latitude of any place will be investigated hereafter (p. 122).

22. The *visible horizon* is that circle in the celestial concave which touches the earth where the spectator stands, as *h* A *r*; and a circle parallel to the *visible* horizon, and passing through the center of the earth, is called the *rational* horizon: thus n c R is the rational horizon. These two circles, however, form one and the same great circle in the celestial concave: thus n and r in the figure must be supposed to coincide. This may be readily conceived, when we consider that the distance of any two points on the surface of the earth will make no sensible angle at the celestial concave; therefore either of these two circles is to be understood by the word horizon. The *poles* of the horizon of any place are manifestly the zenith and nadir.

23. Great circles passing through the zenith are called *circles of altitude* or *vertical* circles. That circle of altitude which passes through the poles of the heavens is called the *celestial meridian*. The points of the horizon through which the celestial meridian passes are called the *north* and *south* points. A circle of altitude at right angles to the meridian is called the *prime vertical*. This last circle cuts the horizon in two points called the *east* and *west* points. The east and west points are manifestly the poles of the celestial meridian.

24. Since the horizon and celestial equator are both perpendicular to the celestial meridian, the points where the horizon and celestial equator intersect each other must be 90° distant from every part of the meridian (Jeans' *Trig.* P. II. art. 65); that is, the celestial equator must cut the horizon in the east and west points.

25. The ecliptic (art. 11) is divided into twelve parts, called signs, which receive their names from constellations lying near them. These divisions or signs are supposed to begin at that intersection of the celestial equator and ecliptic which is called the *first point* of Aries.

26. Great circles passing through the poles of the heavens are called *circles of declination;* and great circles passing through the poles of the ecliptic are called *circles of latitude.*

27. *Parallels of declination* and of *latitude* are small circles parallel respectively to the celestial equator and ecliptic.

28. The *declination* of a heavenly body is the arc of a circle of declination passing through its place in the celestial concave, intercepted between that place and the celestial equator.

29. The *right ascension* of a heavenly body is the arc of the equator, intercepted between the first point of Aries and the circle of declination

passing through the place of the heavenly body in the celestial concave, measuring from the first point of Aries, eastward, from 0° to 360°.

30. The *latitude* of a heavenly body is the arc of a circle of latitude passing through its place in the celestial concave, intercepted between that place and the ecliptic.

31. The *longitude* of a heavenly body is the arc of the ecliptic intercepted between the first point of Aries and the circle of latitude passing through the place of the heavenly body in the celestial concave, measuring from the first point of Aries, eastward, from 0° to 360°.

32. The *altitude* of a heavenly body is the arc of a circle of altitude passing through the place of the body intercepted between the place and the horizon.

33. The *azimuth*, or bearing of a heavenly body, is the arc of the horizon intercepted between the north or south points and the circle of altitude passing through the place of the body ; or it is the corresponding angles at the zenith between the celestial meridian and the circle of altitude passing through the body.

34. The *amplitude* of a heavenly body is the distance from the east point at which it rises, or the distance from the west point at which it sets, the arcs or distances being measured on the horizon.

35. The *hour-angle* of a heavenly body is the angle at the pole between the celestial meridian and the circle of declination passing through the place of the body.

Practical Exercises on the preceding Definitions.

In order more clearly to understand the definitions in Nautical Astronomy, the student should construct a figure or diagram *for each*, and explain the same in the manner pointed out in the answers to the following questions (see pp. 8-16).

1. Construct a figure, and show what is meant by the *true* and *apparent* places of a heavenly body.

2. Construct a figure, and show what is meant by the *axis* of the earth, the *poles* of the earth, and the *terrestrial equator*.

3. Construct a figure, and show what is meant by the *axis* of the heavens, the *poles* of the heavens, and the *celestial equator*.

4. Construct a figure, and show what is meant by the *ecliptic*.

5. Construct a figure, and show what is meant by the *true* and *reduced latitude* of a place on the earth, and also by the *true* and *reduced zenith*.

6. Construct a figure, and show what is meant by *circles of altitude*, the *prime vertical*, the *celestial meridian*, and the north, south, east, and west points of the horizon.

7. Construct a figure, and show what is meant by *circles of declination*, *circles of latitude*, and the *obliquity of the ecliptic*.

8. Construct a figure, and show how to represent on the plane of the

horizon the pole of the heavens and the celestial equator for a given latitude.

9. Construct a figure, and show what is meant by the *right ascension, declination, longitude,* and *latitude,* of a heavenly body.

10. Construct a figure, and show what is meant by the *altitude* of a heavenly body, its *polar distance, zenith distance, hour-angle,* and *azimuth* or true bearing.

11. Construct a figure, and show what is meant by the *amplitude* of a heavenly body.

ANSWERS TO THE FOREGOING QUESTIONS.

1. *The true and apparent place of a heavenly body.*

Let m be a heavenly body, and A a spectator on the surface of the earth. Through m draw the straight lines A M and C M' from the surface and center of the earth to the celestial concave at M and M'. Then M is called the *apparent* place, and M' the *true* place of the heavenly body m.

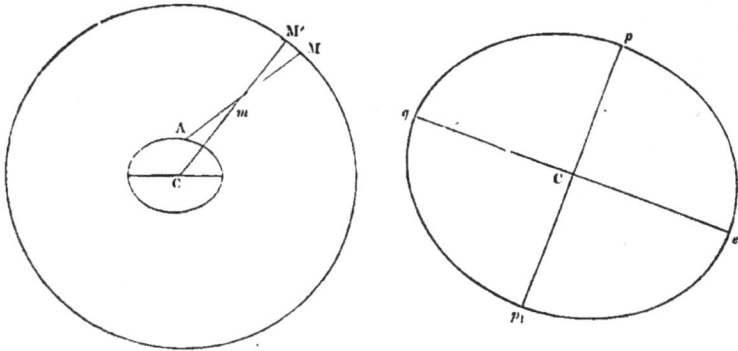

2. *The axis of the earth, the poles of the earth, and the terrestrial equator.*

Assuming the earth to be an oblate spheroid (chapter iv.), let pqp_1e represent a section of the earth passing through the center c; then if pp_1 be that diameter about which the earth revolves, the line pp_1 is called the *axis* of the earth, the extremities p and p_1 are the *poles*, and the line eq, drawn at right angles to pp_1, will be in the plane of the equator, and may be taken to represent the *terrestrial equator* itself.

3. *The axis of the heavens, the poles of the heavens, and the celestial equator.*

Let PQP₁E represent the celestial concave, pqp_1e the earth, pp_1 the axis of the earth, and eq the terrestrial equator. Expand the terrestrial equator eq to the celestial concave in Q and E, and produce pp_1 to P and P_1; then PP_1 is the *axis* of the heavens, the extremities P and P_1 are the *poles* of the heavens, and EQ is the *celestial equator*.

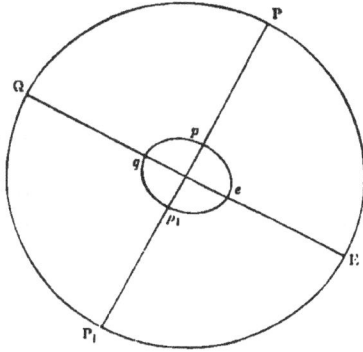

4. The ecliptic.

Let Ab_A_1 (fig. p. 4) represent the earth's orbit, s the sun, a and b the earth in two points of its orbit. Then, when the earth is at a, the sun's place in the celestial concave is c; when the earth is at b, the sun's place is d: while the earth, therefore, describes the arc ab of its orbit, the sun appears to describe an arc of a great circle cd; and when the earth has completed its revolution about the sun, the latter will appear to have described in the celestial concave a great circle cel. This circle is called the *ecliptic*.

5. The true and reduced latitude of a place on the earth; the true and reduced zenith.

Let pep_1q be a section of the earth passing through the center, which since the earth is an oblate spheroid, will be an ellipse (chapter iv.); A the given place; through A draw AG perpendicular to the tangent passing through A; join AC, and produce GA and CA to the celestial concave at z and z'; let eq be the plane of the celestial equator; then the arc zQ or angle zGQ is the *true latitude* of A, and z'Q or the angle z'CQ is the *reduced latitude* of A. The point z is the *true zenith*, and z' the *reduced zenith*, of the spectator at A.

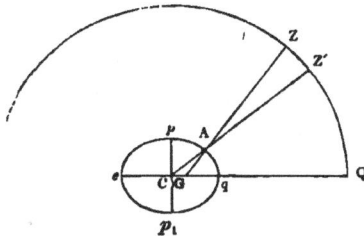

6. *Circles of altitude, the prime vertical, the celestial meridian, and the north, south, east, and west points of the horizon.*

These definitions are more easily explained by projecting the figure on the plane of the horizon. Thus, let N W S E represent the horizon of the spectator; then the point z, considered as the pole of the horizon, will be the zenith, and great circles W E, N S, Z O, will be *circles of altitude* or vertical circles. Let the circle of altitude Z N pass through the pole of the heavens P: then N Z S is the *celestial meridian* of the spectator, whose zenith is Z; and the circle of altitude W E, drawn at right angles to the celestial meridian, is the *prime vertical*. If P be the north pole, then the point of the horizon N intersected by the celestial meridian is the *north* point, and the opposite point of the horizon S is the *south* point; the points of the horizon W, E, intersected by the prime vertical, are the *west* and *east* points.

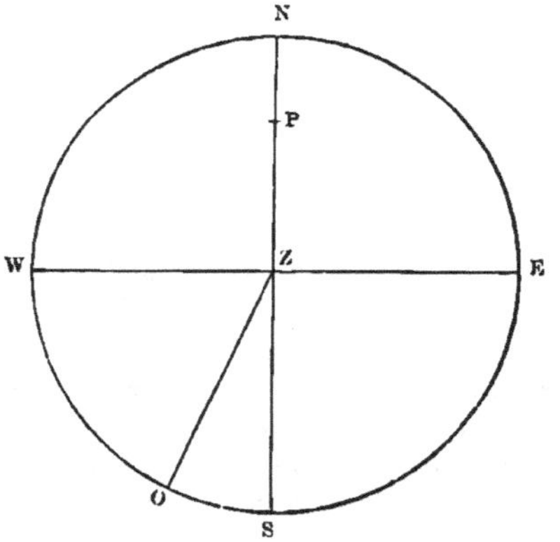

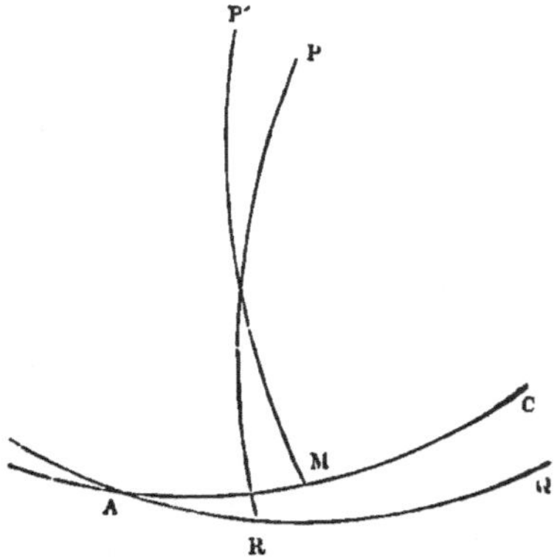

7. *Circles of declination, circles of latitude, and the obliquity of the ecliptic.*

Since the equator and ecliptic are inclined to each other at an angle of about 23° 28' (p. 4); let A Q represent the celestial equator, and A C the ecliptic. Let P be the pole of the heavens, and P' the pole of the ecliptic, and through P and P' draw P R perpendicular to A Q and P'M perpendicular to A C; then P R is a *circle of declination*, and P'M a *circle of latitude*, and the angle C A Q is the *obliquity* of the ecliptic.

8. *Represent on the plane of the horizon the pole of the heavens and the celestial equator for a given latitude.**

Let N W S E represent the horizon, N S the celestial meridian, and W E the prime vertical; then, since the distance of the zenith from the celestial

* This figure or projection will be very often used in future problems in Nautical Astronomy.

equator is equal to the latitude of the spectator (p. 5), take ZQ=the given latitude, and through the points W, Q, E draw a great circle WQE: this will represent the celestial equator, since the equator must pass through the east and west points (p. 6). From Q take QP=90°; then P will represent the pole of the heavens, since the pole of the celestial equator is the pole of the heavens (p. 4).

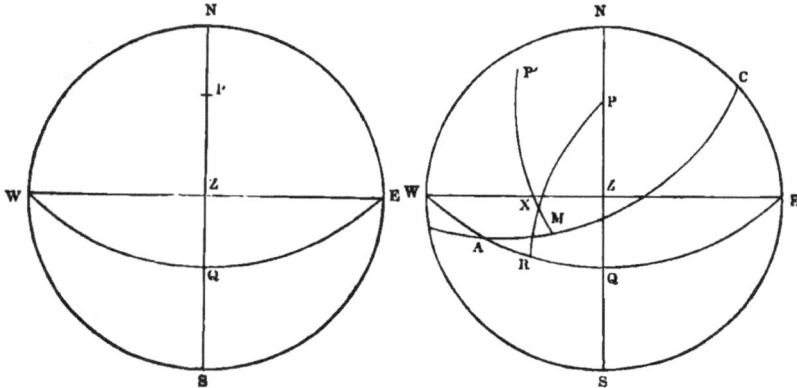

9. *The right ascension, declination, longitude and latitude, of a heavenly body.*

Let NWSE represent the horizon, P the pole, z the zenith, and x the place of a heavenly body. Draw the equator WQE and the ecliptic AC, and let P′ be the pole of the ecliptic AC and A the first point of Aries. Through x draw a circle of declination PR and a circle of latitude P′M. Then AR is the *right ascension*, XR is the *declination*, AM is the *longitude*, and XM is the *latitude* of the heavenly body x.

10. *The altitude of a heavenly body, its polar distance, zenith distance, hour-angle, and azimuth or true bearing.*

Let NWSE represent the horizon, NS the celestial meridian, P the pole, and z the zenith. Let x be the place of a heavenly body; through x draw the circle of altitude zo and circle of declination PX. Then XO is the *altitude* of the heavenly body, ZX its *zenith distance*, and PX its *polar distance*. The angle ZPX is the *hour-angle*, and the angle PZX or SZX, or the arc of the horizon NO or SO, is its *azimuth* or true bearing.

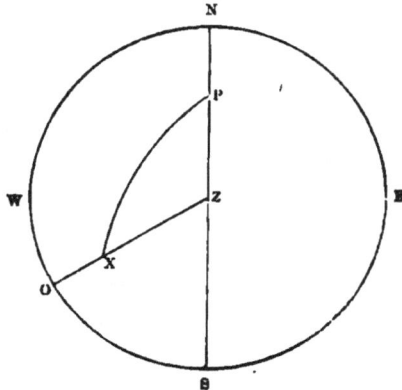

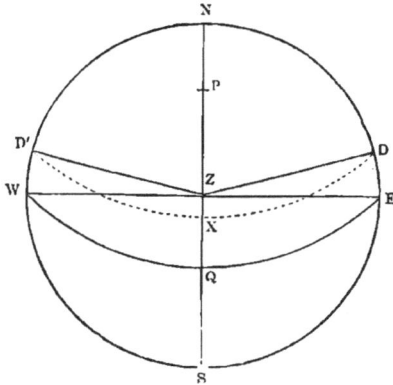

11. *The amplitude of a heavenly body.*

Let N W S E represent the horizon, and the small dotted circle D X D′ the parallel of declination described by the heavenly body x from its rising at D to setting at D′; then the angle D Z E or D′ Z W is the amplitude of x, or the arc D E or D′ W, which measures these angles.

DEFINITIONS AND PROBLEMS ON TIME.

The solar year, mean solar year, and sidereal year.

36. A *solar year* is the interval between the sun's leaving the first point of Aries and returning to it again.

37. A *sidereal year* is the interval between the sun's leaving a fixed point, as a star, and returning to that point again.

The equinoctial points have an annual motion of 50″·1, by which they are carried back to meet the sun in its apparent motion among the fixed stars from west to east.

On this account a solar year is shorter than a sidereal year by the time the sun takes to describe 50″·1.

38. The length of the solar years is found to differ a little from each other, on account of certain irregularities in the sun's apparent motion, and that of the first point of Aries. The *mean length* of several solar years is therefore the one made use of in the common division of time, and called the *mean solar year.*

To find the length of the mean solar year.

39. By comparing observations made at distant periods, it was found that the sun had described 36000° 45′ 45″ of longitude in 36525 days. Now in one solar year the sun separates from the first point of Aries 360° (taking into consideration its own apparent motion from west to east, and the actual motion of the first point of Aries in the opposite direction). Let therefore x=the length of a mean solar year ;

then, 36000° 45′ 45″ : 360° :: 36525^d : x

whence, x=365^d 5^h 48^m 51′·6=365^d·242264.[*]

* According to Bessel, the formula for determining the length of the mean solar o tropical year is

$$365^d \cdot 2422013 - \cdot 00000006686 \times t$$

where t is the number of years since 1800.

To find the length of the sidereal year.

40. Since the first point of Aries moves with a slow annual motion of about 50″·1 from east to west to meet the sun, the arc of the ecliptic described by the sun from the first point of Aries to the first point of Aries again must be 360°—50″·1=359° 59′ 9″·9, and this is the arc described by the sun in a mean solar year; but in a sidereal year the sun describes 360°; hence a sidereal is greater than a solar year by the time the sun takes to move over an arc of 50″·1. Hence the proportion,

sidereal year : mean solar year : : 360° : 360°—50″·1
or, sidereal year : 365ᵈ·242264 : : 360° : 359° 59′ 9″·9.
Therefore, sidereal year = 365ᵈ 6ʰ 9ᵐ 11ˢ·5.*

The sidereal day, the apparent solar day, and the mean solar day.

41. The *sidereal day* is the interval between two successive transits of the first point of Aries over the same meridian. It begins when the first point of Aries is on the meridian.

The *apparent solar day* is the interval between two successive transits of the sun's center over the same meridian. It begins when that point is on the meridian.

42. The length of an apparent solar day is variable chiefly from two causes :

1st. From the variable motion of the sun in the ecliptic.

2d. From the motion of the sun being in a circle inclined to the equator.

43. To explain briefly these causes of variation, let us suppose the two circles A Q, A I, to represent the celestial equator and ecliptic, and s s′ the arc of the ecliptic described by the sun in one day. The angle at r, between the two circles of declination, is measured, not by the arc s s′ described by the sun, but by the arc R R′ of the equator. Now,

1st. The velocity or motion of the sun in the ecliptic is variable, on account of the earth moving in an elliptic orbit; it sometimes describes an arc of 57′ in a day; at other times the arc described is about 61′; this is one cause of the inequality in the length of the solar days.

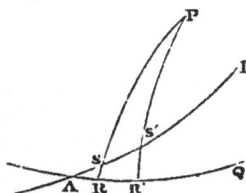
Fig. 1.
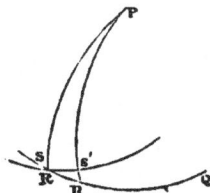
Fig. 2.
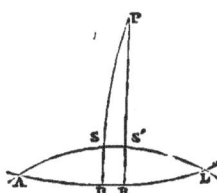
Fig. 3.

2d. But even supposing the arcs of the ecliptic described by the sun to

* The length of a sidereal year according to Bessel is
365ᵈ·256374322 = 365ᵈ 6ʰ 9ᵐ 10ˢ·7423 mean time.

be equal, yet the angles at P between the meridians, as R P R' (in the three figures) will not be so, since these angles are measured by the arc R R' of the equator, to which s s' will be differently inclined according to the place of the sun in the ecliptic. At the equinoxes, or when the sun is at R (fig. 2), the arcs s s' and R R' will be inclined to each other at an angle of about 23° 27'. At the solstices they are parallel (see fig. 3). This is the second cause of the inequality.

44. To obtain, therefore, a proper measure of time, we must proceed as follows. An imaginary, or as it is called a *mean sun*, is supposed to move *uniformly in the celestial equator with the mean velocity of the true sun*. A *mean solar day* may therefore be defined to be the interval between two successive transits of the mean sun over the same meridian. It begins when the mean sun is on the meridian.

To find the daily motion of the mean sun in the celestial equator.

45. The mean solar year, or the time the sun takes to return again to the first point of Aries, has been found to be equal to 365ᵈ·2422. Let us suppose the mean sun to describe the *equator* in this time, then we shall find its daily motion in the equator as follows : Let $x =$ daily motion,

$$365^{d}\cdot 2422 : 1^{d} : : 360° : x = 0°\cdot 9856472 = 59'\ 8''\cdot 33.$$

or, the mean sun's daily motion in the celestial equator from west to east is 59' 8''·33.

To find the arc described by a meridian of the earth, in a mean solar day.

46. Let P A M P' represent the meridian of a spectator A, drawn in some plane; as, for instance, that of the paper; M M' the terrestrial, and E Q the

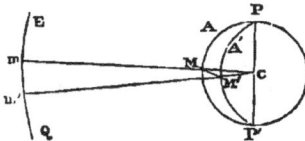

celestial equator, which must therefore be supposed at right angles to the paper. Suppose the mean sun to be at m on the meridian of A, and therefore in the plane of the paper; and let m m' be the arc of the equator described by the mean sun in one day, namely, 59' 8''·33. Join c m', cutting the terrestrial equator at M M'. Now, let the earth be supposed to revolve about P P', from west to east, until the meridian again passes through the mean sun, which has arrived at m', having moved through the arc m m' in the same time. Then the whole number of degrees described by the meridian of the spectator will evidently be one complete revolution, or 360° (by which it is again brought into the plane of the paper), together with the arc M M' = m m', or 59' 8''·33. Hence in a mean solar day a meridian describes 360° 59' 8''·33 ; and therefore a mean solar day is longer than a sidereal day in the ratio of 360° 59' 8''·33 to 360°, or 24 mean solar hours = 24ʰ 3ᵐ 56ˢ·555 in sidereal time. (See Prob. IV. p. 25.)

Sidereal time, apparent solar time, and mean solar time.

47. *Sidereal time* is the angle at the pole of the heavens between the celestial meridian and a circle of declination passing through the first point of Aries, measuring from the meridian westward.

48. *Mean solar time* is the angle at the pole between the celestial meridian and a circle of declination passing through the mean sun, measuring from the meridian westward.

49. *Apparent solar time* is the angle at the pole between the celestial meridian and a circle of declination passing through the place of the sun's center, measuring from the meridian westward.

The equation of time is the difference in time between the places of the true and mean sun.

Sidereal clock, and mean solar clock.

50. A *sidereal clock* is a clock adjusted so as to go 24 hours during one complete revolution of the earth; that is, during the interval of two successive transits of a fixed star: or supposing the first point of Aries to be invariable between two successive transits of the first point of Aries.

51. A *mean solar clock* is a clock adjusted to go 24 hours during one complete revolution of the mean sun; or while a sidereal clock is going 24ʰ 3ᵐ 56·555.

QUESTIONS ON TIME.

52. *Construct a figure, and show what is meant by sidereal time, apparent solar time, mean solar time, and the equation of time.*

Let N W S E represent the horizon, P the pole, A E the equator, A the first point of Aries, and A C the ecliptic. Let X be the place of the sun in the ecliptic, and *m* the mean sun; through X and *m* draw the circles of declination P R and P *m*. Then sidereal time is the angle QPA, or arc QA; apparent solar time is the angle QPR, or arc QR; and mean solar time is the angle QP*m*, or arc Q*m*,—these angles or arcs being always measured from the meridian N Z S westward. Also the angle *m* P R, or arc *m* R, is the equation of time.

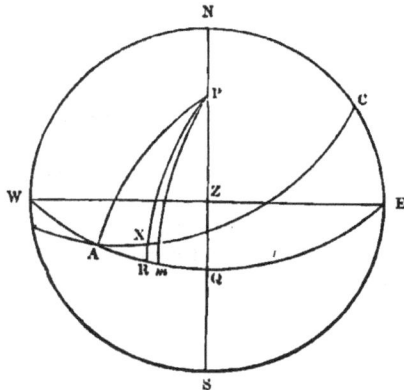

It may not be amiss sometimes for the young student to answer questions on the definitions more in detail; that is, *by referring to the definitions themselves* whilst constructing the figures: thus the last question may be answered more fully as follows.

53. *Construct a figure, and show what is meant by sidereal time, apparent solar time, mean solar time, and the equation of time.*

Let N W S E (last fig.) represent the horizon, P the pole, A E the equator, A the first point of Aries, and A C the ecliptic. Let X be the place of the sun in the ecliptic, and *m* the mean sun : through X and *m* draw the circles of declination P R and P *m*. Then sidereal time is the angle Q P A, or arc Q A ; since, by definition, sidereal time is the angle at the pole between the celestial meridian and a circle of declination passing through the first point of Aries, measuring from the meridian westward. Apparent solar time is the angle Q P R, or arc Q R ; since, by definition, apparent solar time is the angle at the pole between the celestial meridian and a circle of declination passing through the place of the sun, measuring from the meridian westward. Mean solar time is the angle Q P *m*, or arc Q *m* ; since, by definition, mean solar time is the angle at the pole between the celestial meridian and a circle of declination passing through the mean sun, measuring from the meridian westward. The equation of time is the angle *m* P R, or arc *m* R ; since, by definition, the equation of time is the difference in time between the places of the true and mean sun, or the angle at the pole between the circles of declination passing through the place of the sun and the mean sun.

In a similar manner may other questions on the definitions in pages 1 and 2 be extended or amplified when deemed necessary.

Examples on the Julian and Gregorian years.

54. If the length of the mean solar year is 365·242264 days (page 12), and we assume its length to be 365·25 days (the length of the common or Julian year) ; find the error on this account in 400 years. *Ans.* 3·0944 days.

55. To diminish the error committed by making the year to be 365·25 days = 365 days 6 hours, a correction is made in the Gregorian Calendar, or New Style, to the Julian year, as follows. Every centenary year of which the two first figures are not divisible by 4 (such as 1800, 1900, 2100, &c.), is to be considered as a common year of 365 days ; thereby deducting 3 days in every 400 years. Now supposing this correction to be made, it is required to find the number of years that must elapse before the error arising from making the year to consist of 365 days 6 hours (so corrected) shall amount to one day.   *Ans.* 4237 years.

CONSTRUCTION OF ASTRONOMICAL DIAGRAMS.

The diagrams to the examples in the following pages need not be pro·
jected by the student with very great accuracy; the quantities given, namely,
the arcs and angles, may be drawn after a little practice sufficiently near the
truth by estimating with the eye the length or magnitude of each.

56. The figures or diagrams in Nautical Astronomy are usually projected
on one of the three following planes : *the plane of the horizon, the plane of
the celestial equator, or the plane of the celestial meridian.* The first is the
one most frequently adopted, as it shows the positions of the heavenly body
during the whole of its diurnal motion. When several parts of the celestial
equator are required to be seen, then the diagram should be projected on the
plane of the celestial equator. In some cases it is indifferent which projection is used : thus, if it be required to represent by means of a figure the polar
distance, zenith distance, and hour-angle of a heavenly body, we may project
the figure on either the plane of the equator or the plane of the horizon.
If we project on the plane of the celestial equator QQ'E (first figure), then
the center P will be the pole of the heavens ; and if z be the zenith of

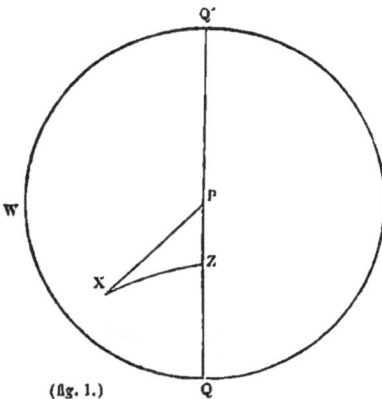
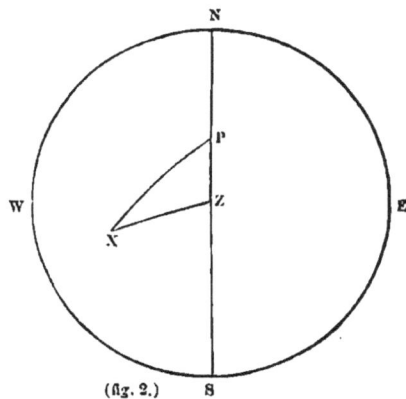

(fig. 1.)   (fig. 2.)

the spectator, and x the place of the heavenly body, then P x is the polar
distance, z x is the zenith distance, and the angle z P x is the hour-angle of
the heavenly body. If we project the figure on the plane of the horizon
N W S E (fig. 2), the center z will be the zenith ; and if P be the pole of the
heavens, and x the place of the heavenly body, P x is the polar distance, z x
is the zenith distance, and P is the hour-angle of the heavenly body. An
example of a figure projected on the plane of the celestial meridian is given
in the volume of *Solutions of Astronomical Problems,* p. 212 (published by
the author as a key to the problems in his *Trigonometry,* Part I.).

CHAPTER II.

PROBLEM I.*

GIVEN mean time and the equation of time, to find apparent time : or,
Given apparent time and the equation of time, to find mean time.

Fig. 1 is a construction on the plane of the horizon, fig. 2 on the plane of
the celestial equator. The sun's declination is supposed to be about 10° N.
The declination is given to enable the student to insert the sun's place in
the figure.

Construction.

Let P represent the north pole, WQE in fig. 1, or QRQ' in fig. 2, the
celestial equator, PQ the celestial meridian, x the place of the sun 10° N. of
the equator, and m the mean sun in the equator. Through x and m draw

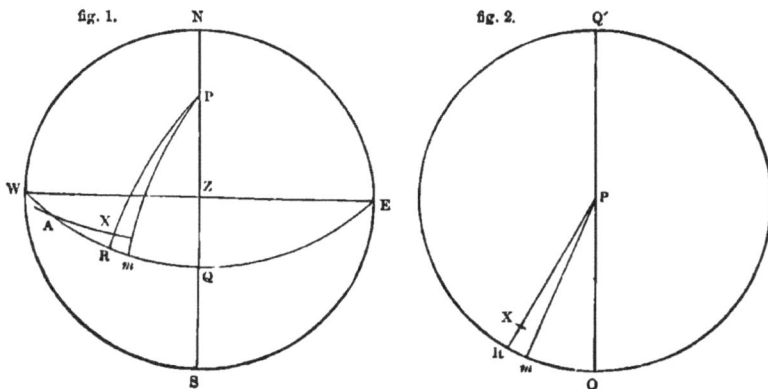

fig. 1.   fig. 2.

the circles of declination PR and Pm : then the angle QPR, or arc QR, is
apparent time ; the angle mPR, or arc mR, is the equation of time ; and the
angle QPm, or arc Qm, is mean time.

* This and some of the subsequent problems are given chiefly to afford the student
a few very easy exercises in the construction of astronomical diagrams.

By the fig., $QR = Qm + mR$,

or apparent time $=$ mean time $+$ equation of time.

From which formula, apparent time or mean time may be found when the other two quantities are given.

As the mean sun moves uniformly in the equator with the mean angular velocity of the true sun (p. 14), it is manifest that it will be sometimes in advance and sometimes behind the place of the true sun ; the above formula must therefore be modified to suit the relative positions of m and x. When the equation of time is subtractive from mean time to get apparent time, m must be on the other side of R ; when it is additive, the above figures will represent the correct positions of the bodies. The equation of time, with its proper sign, is given in the *Nautical Almanac* for every day at noon.

In the following easy examples (inserted chiefly for the sake of affording the student useful exercises in *constructing diagrams*), the positions of the true and mean suns are to be determined as nearly as possible by the eye and without the assistance of mathematical instruments.

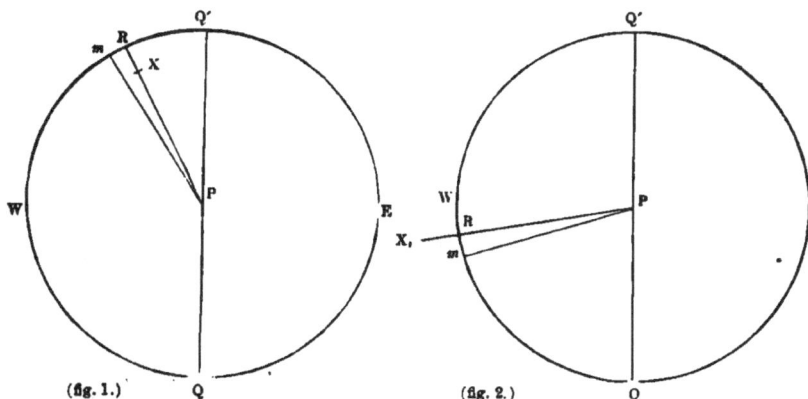

(fig. 1.)   (fig. 2.)

57. Given mean time $= 10^h\,14^m\,15^s$ P.M.; and the equation of time $= 2^m\,30^s$ *additive* to mean time ; construct a figure, and find by calculation apparent time (fig. 1).

The figure is to be constructed on the plane of the equator, and the declination is supposed to be about 10° N.

Let $Q\,w\,Q'$ represent the celestial equator, P the north pole, and $Q\,P\,Q'$ the celestial meridian. Take $Qm = 10^h\,14^m\,15^s =$ mean time ; then m is the mean sun ; and since the equation of time is additive to mean time, let mR represent the equation of time $= 2^m\,30^s$. Draw Pm, PR, circles of declination, through m and R; then the true sun will be in the circle PR. Take $Rx=10°$; then x is the place of the sun.

By the fig., $QR=Qm+mR$,
or apparent time = mean time + equation of time ;
and mean time = 10^h 14^m 15^s
equation of time =   2   30
<hr>
∴ apparent time = 10   16   45

58. Given apparent time = 5^h 10^m P.M., and the equation of time = 15^m 10^s subtractive from apparent time, the sun's declination being about 20° S ; construct a figure, and find by calculation mean time.

Let QWQ' represent the celestial equator, P the north pole, and QPQ' the celestial meridian. Take $QR=5^h$ 10^m=apparent time, and $Rm=15^m$ 10^s the equation of time. Draw the circles of declination Pm and Px through m and R, and take $Rx=20°$ to the south of the equator ; then x is the place of the sun.

By the fig., $Qm=QR-Rm$,
or mean time = apparent time − equation of time ;
apparent time = 5^h 10^m 0^s
equation of time =   15   10
<hr>
∴ mean time = 4   54   50

59. Given mean time = 10^h 20^m A.M. = 22^h 20^m P.M., and the equation of time = 3^m 13^s additive to mean time ; required apparent time, decl. 10° N. Construct the figure on the plane of the equator. Let QWQ' represent the celestial equator, P the pole and QPQ' the celestial meridian. Take $QQ'm=22^h$ 20^m to represent mean time ; and since the equation of time is additive to mean time, take $mR=3^m$ 13^s. Through m and R draw circles of declination Pm, PR, and take $Rx=10°$; then x is the place of the sun.

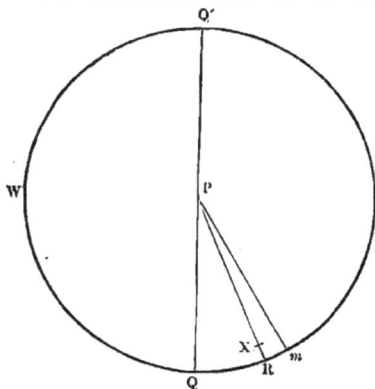

By the fig., $QQ'R=QQ'm+mR$,
or apparent time = mean time + equation of time ;
mean time = 22^h 20^m 0^s
equation of time =   3   13
<hr>
∴ apparent time = 22   23   13
or = 10   ·23   13 A.M.

60. Given apparent time $= 4^h$ 30^m P.M., and equation of time $= 3^m$ 0^s additive to apparent time. Construct a figure, and find by calculation mean time (declination 20° N.). *Ans.* Mean time $= 4^h$ 33^m 0^s.

61. Given mean time $= 2^h$ 30^m A.M., and equation of time $= 2^m$ 0^s additive to mean time. Construct a figure, and find by calculation apparent time (declination 20° N.). *Ans.* Apparent time $= 2^h$ 32^m A.M.

62. Given mean time $= 10^h$ 10^m A.M., and equation of time $= 10^m$ 0^s subtractive from mean time. Construct a figure, and find by calculation apparent time (declination 20° S.). *Ans.* Apparent time $= 10^h$ 0^m A.M.

PROBLEM II.

Given mean time, to find sidereal time.

Let $Q A Q'$ represent the celestial equator, A the first point of Aries, $Q P Q'$ the celestial meridian, and P the pole. Then, if the mean sun is west of meridian, let m (fig. 1) be the mean sun.

If the mean sun is east of the meridian, let m' (fig. 2) be the mean sun.

In the figs. (1) and (2), $Q A =$ sidereal time, since it measures the time elapsed from the first point of Aries being on the meridian (p. 15).

In fig. 1, $Q m =$ mean time.

In fig. 2, $m' Q =$ arc that must be described before the mean sun m' arrives at the meridian.

$\therefore$ $24^h - m' Q =$ mean time, $\therefore$ $m' Q = 24^h -$ mean time.

Also, $A m$, or $A m' =$ right ascension of mean sun.

Now (fig. 1) $Q A = A m + m Q$,

or sidereal time $=$ R A mean sun $+$ mean time.

In fig. 2, $Q A = A m' - m' Q$,

or sidereal time $=$ R A mean sun $- (24^h -$ mean time$)$

$= $ R A mean sun $+$ mean time $- 24^h$.

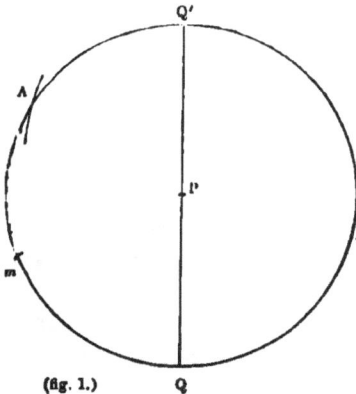

(fig. 1.)   Q

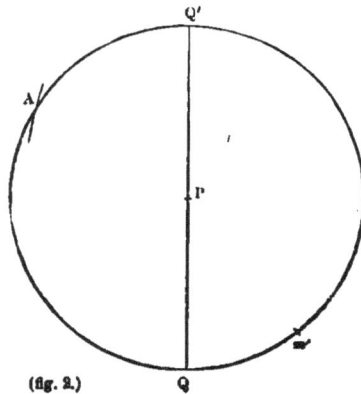

(fig. 2.)   Q

The latter expression for sidereal time is the same as the former by
adding 24 hours; and, assuming that we can always add or subtract 24
hours, the first equation may be considered true in both cases; and as we
shall get the same result for other positions of *m* and A with respect to the
meridian, therefore generally this rule is true; viz.

Sidereal time = R.A mean sun + mean time.

If apparent time is given to find sidereal time, reduce apparent time to
mean time by Problem I.

The right ascension of the mean sun is given in the *Nautical Almanac*
for every day at Greenwich mean noon, being the column marked "sidereal
time" in each month.

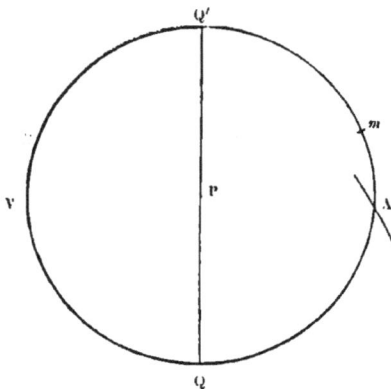

63. Given mean time = 15^h
55^m 45^s, and the right ascension
of the mean sun = 2^h 22^m 58^s.
Construct a figure, and find by
calculation sidereal time.

Let QWQ' represent the celestial
equator, P the pole, and QPQ' the
celestial meridian. From Q measure
QQ'm=15^h 55^m 45^s; then *m* is the
mean sun. From *m* take *m*A=2^h
22^m 58^s; then A is the first point
of Aries, and the arc QQ'A is sidereal time.

TO FIND SIDEREAL TIME BY FORMULA.

Sidereal time = R.A mean sun + mean time.

R.A mean sun	. .	2^h	22^m 58^s
mean time	. . .	15	55 45
∴ sidereal time =		18	18 43

EXAMPLES FOR PRACTICE.

64. Given mean time = 6^h 10^m P.M., and the right ascension of the
mean sun=2^h 22^m 41^s. Construct a figure, and find by calculation sidereal
time.   *Ans.* Sidereal time = 8^h 32^m 41^s.

65. Given mean time = 5^h 42^m 10^s A.M., and the right ascension of the
mean sun=18^h 47^m 14^s. Construct a figure, and find by calculation sidereal time.   *Ans.* Sidereal time = 12^h 29^m 24^s.

Given mean time, or apparent time at a given place; to find what heavenly body will pass the meridian the next after that time.

Let AQ be the celestial equator, PQ the celestial meridian, and x a heavenly body passing the meridian. Therefore the right ascension of the meridian will be the right ascension of the star. Then, knowing the time at the place, we can determine the position of the mean sun with respect to the meridian, and also that of the first point of Aries, since we also know the right ascension of the mean sun, from the *Nautical Almanac*.

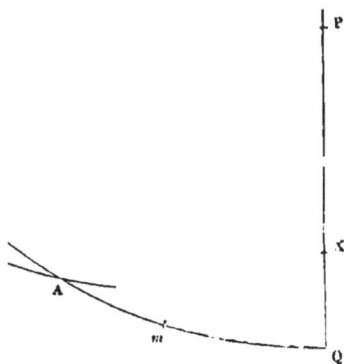

Let, therefore, m be the mean sun, A the first point of Aries.

Then $Qm =$ mean time at the place, $Am =$ right ascension of mean sun,

and $AQ =$ right ascension of the heavenly body x

$=$ right ascension of the meridian ;

and by the figure, $AQ = Am + mQ$,

or star's RA $=$ RA mean sun $+$ mean time $=$ sidereal time (Prob. II.)

∴ sidereal time $=$ right ascension of meridian.

It appears from this, that in order to find the time when a heavenly body passes the meridian, we have only to find sidereal time by the last problem, and this will be the right ascension of the meridian, and therefore of any heavenly body on the meridian. Then the star in the astronomical catalogue in the *Nautical Almanac* whose right ascension is the next greater, will evidently be the next principal star to pass the meridian, and is therefore the one required.

If it is required to know what principal stars will pass a given meridian between any two given dates, we must proceed as above to find sidereal time at each date. Then all the stars in the catalogue whose right ascensions lie between the sidereal times thus determined will be the stars required.

66. Given mean time $= 4^h 42^m$ P.M., and the right ascension of the mean sun $= 22^h 16^m 46^s$. Construct a figure, and find by calculation what bright star will pass the meridian the next after that time.

Let QWQ' represent the celestial equator, P the pole, and QPQ' the celestial meridian. From Q measure $Qm = 4^h 42^m$; then m is the mean sun. From m measure an arc $mQ'A = 22^h 16^m 46^s$; then A is the first point of Aries, and the arc AQ is the right ascension of the meridian or sidereal time, and therefore the right ascension of any star on the meridian.

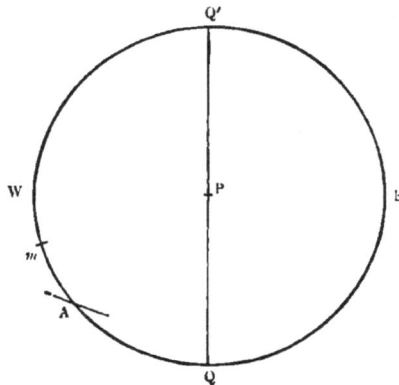

By Calculation.

RA of meridian = sidereal time

       = RA mean sun + mean time (Prob. II.).

RA mean sun . . . 22ʰ 16ᵐ 46ˢ

mean time . . . 4 42 0

∴ RA of meridian . . . 2 53 46 (rejecting 24 hours).

By catalogue it appears that the star whose right ascension is the next greater than this is α Persei, ∴ the star that passes the next after 4ʰ 42ᵐ P.M. is α Persei.

67. What bright stars will pass the meridian of the ship between the hours of 9ʰ and 12ʰ P.M., the right ascension of the mean sun at 9 o'clock being 12ʰ 49ᵐ 41ˢ, and at 12 o'clock, 12ʰ 50ᵐ 11ˢ?

By Calculation.

RA of meridian = RA mean sun + mean time.

RA mean sun at 9ʰ . . 12ʰ 49ᵐ 41ˢ   at 12ʰ . . 12ʰ 50ᵐ 11ˢ

mean time . . 9 0 0   mean time . . 12 0 0

RA of meridian = 21 49 41   RA of mer. = 0 50 11

By inspecting the catalogue, it appears that the stars whose right ascensions lie between those of α Aquarii and β Ceti will pass the meridian of the ship between 9 and 12 o'clock.

68. Given mean time=8^h 0^m P.M., and the right ascension of the mean sun=23^h 34^m. Construct a figure, and find by calculation the right ascension of the meridian and the bright star that will pass the meridian the next after that time.   *Ans.* R A meridian=7^h 34^m : Pollux.

69. Given mean time=2^h 0^m A.M., and the right ascension of the mean sun=6^h 42^m. Construct a figure, and find by calculation the right ascension of the meridian, and the bright star that will pass the meridian the next after that time.   *Ans.* R A meridian=20^h 42^m : α Cephei.

PROBLEM IV.

Given the length of a mean solar day=24 hours; to express its length in sidereal hours, and the converse.

In a mean solar day, a meridian of the earth revolves through 360° 59′ 8·33″ (art. 45) : in a sidereal day it revolves through 360° (art. 40).   A mean solar day is therefore longer than a sidereal day (or 24 sidereal hours) by the portion of sidereal time consumed in describing 59′ 8·33″. This quantity of time may be found by the following proportion :

$$360° : 59′ 8·33″ : : 24^h : x \text{ the quantity required,}$$
$$\therefore x = 3^m 56·555^s,$$
$$\therefore 24 \text{ mean solar hours} = 24^h 3^m 56·555^s \text{ sidereal hours.}$$

The quantity 3^m 56·555^s, the excess of a mean solar day over a sidereal day, is called the *acceleration of sidereal on mean solar time.*

Given the length of a sidereal day=24 hours; to express its length in mean solar time.

In a sidereal day the meridian of any place revolves through 360°; in a mean solar day the same meridian revolves through 360° 59′ 8·33″ (p. 14).

Therefore the length of a sidereal day may be found in mean solar time by this proportion :

$$24 \text{ sidereal hours} : 24 \text{ mean solar hours} : : 360° : 360° 59′ 8·33″,$$
$$\therefore 24 \text{ sidereal hours} = \frac{24 \times 360°}{360° 59′ 8·33″} = 23^h 56^m 4·0922^s \text{ mean solar hours.}$$

Formulæ for converting any portion of sidereal time into mean solar time, and the converse.

Let s=any interval of absolute time or duration expressed in sidereal time.

m=the same interval expressed in mean solar time.

Then, since the number of hours, minutes, &c., in a given interval must be inversely as the length of one of them, we have

$$\frac{\text{M}}{\text{S}} = \frac{360}{360° \ 59' \ 8\cdot3'} = \cdot9972695667,$$

$$\text{and} \ \frac{\text{S}}{\text{M}} = \frac{360° \ 59' \ 8\cdot33''}{360} = 1\cdot002737909,$$

$$\therefore \text{M} = \cdot997269 \times \text{S} = \text{S} - \cdot002731 \ \text{S} \ . \ . \ . \ (1)$$

$$\text{and} \ \text{S} = 1\cdot002738 \times \text{M} = \text{M} + \cdot002738 \ \text{M} \ . \ . \ . \ . \ (2)$$

By means of these formulæ we may convert any portion of sidereal time into mean solar time, and the converse, as in the following examples.

70. Express 12 sidereal hours in mean solar time.

By formula (1), $\text{M} = \text{S} - \cdot002731 \ \text{S}$, and $\text{S} = 12$
$$\therefore \text{M} = 12 - \cdot002731 \times 12 = 11^h \ 58^m \ 2\cdot02',$$
or 12 sidereal hours $= 11^h \ 58^m \ 2\cdot02'$ mean solar hours.

71. Express $11^h \ 58^m \ 2\cdot02'$ mean solar time in sidereal time.

By formula (2), $\text{S} = \text{M} + \cdot002738 \ \text{M}$,

$\text{M} = 11^h \ 58^m \ 2\cdot02'$ $= 11\cdot967228 + \cdot002738 \times 11\cdot967228$
$= 11\cdot967228$ $= 11\cdot999994 = 12$ hours nearly;
$\therefore 11^h \ 58^m \ 2\cdot02'$ mean solar hours $= 12$ sidereal hours.

EXAMPLES FOR PRACTICE.

72. Express 16 sidereal hours in mean solar time.
Ans. $15^h \ 57^m \ 22\cdot694'$ mean solar time.

73. Express 14 mean solar hours in sidereal time.
Ans. $14^h \ 2^m \ 17\cdot995'$ sidereal time.

(5.) Construction of " tables of time-equivalents."

By means of formulæ (1) and (2) the tables of time-equivalents given in the *Nautical Almanac* and in most collections of nautical tables may be computed.

These tables are used for readily converting intervals of sidereal time into equivalent intervals of mean solar time, and the converse; a few examples will show this.

74. Convert $8^h \ 43^m \ 51\cdot42'$ sidereal time into mean solar time. By table,

8^h (sidereal time) . . .	$7^h \ 58^m$	$41\cdot36'$	mean solar time.	
43^m	,,	. . .	$42 \quad 52\cdot96$	,,
$51'$	,,	. . .	$50\cdot86$	,,
$0\cdot42'$	,,	. . .	$\cdot42$	,,

$\therefore 8^h \ 43^m \ 51\cdot42'$ sidereal time $= 8 \quad 42 \quad 25\cdot6$,,

75. Convert 8^h 42^m $25 \cdot 6^s$ mean solar time into sidereal time. By table,

8^h (mean time)	. . . . .	8^h	1^m $18 \cdot 85^s$	sidereal time.
42^m	,,	. . . . .	42 $6 \cdot 90$	,,
25^s	,,	. . . . .	$25 \cdot 07$	,,
$0 \cdot 6$	,,	. . . . .	$\cdot 60$	,,

$\therefore$ 8^h 42^m $25 \cdot 6^s$ mean time $= 8$ 43 $51 \cdot 42$,,

EXAMPLES FOR PRACTICE.

76. Convert 10^h 10^m 30^s sidereal time into mean time.

Ans. 10^h 8^m $49 \cdot 984^s$.

77. Convert 10^h 10^m 30^s mean solar time into sidereal time.

Ans. 10^h 12^m $10 \cdot 29^s$.

PROBLEM V.

Given sidereal time at any instant, to find mean time at the same instant.

Let PQ represent the celestial meridian, AQ the celestial equator, A the first point of Aries, and m the mean sun.

Then QA=sidereal time, Qm=mean time;

and Am=right ascension of the mean sun.

By the fig., Qm=QA−Am,

or mean time=sidereal time−right ascension of mean sun

(adding 24 hours if necessary. See Prob. II.).

Hence it appears, that to find mean time we have only to subtract the right ascension of the mean sun for that in-stant of mean time from the given side-real time, and the result will be mean time at that instant. But how shall we find the right ascension of the mean sun for that instant of mean time, since in the *Nautical Almanac* the right as-cension of the mean sun is only recorded for every day at mean noon at Green-wich; and as we know not the time elapsed from mean noon at Greenwich, we cannot proportion for the change in the right ascension of the mean sun due to that elapsed time? We may proceed

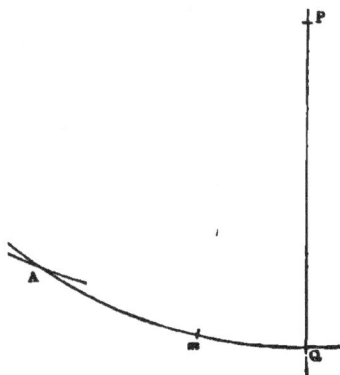

as follows. Let x=mean time required. Subtracting from sidereal time the right ascension of the mean sun at mean noon at the place, we get an approximate value of the mean time. Let this first approximation be denoted by t.

Now this time t is manifestly too great by the motion of the mean sun, expressed in time in the interval x; and this quantity is equal to $\cdot 0027379 \times x$, since the motion of the mean sun in the equator in 24 hours$=3^{m} 56 \cdot 555'=$ $\cdot 0027379$ of a day.

$$\therefore x = t - \cdot 0027379x.$$
$$\text{or } 1 \cdot 0027379x = t, \therefore x = \cdot 9972696t.$$

And this coefficient of t is the same as the factor which is used to reduce an interval of sidereal time into mean time (p. 26). Hence it appears, that to find the mean time x it will be sufficient to correct the approximate time t (obtained by using the right ascension of the mean sun at noon) as if we were about to reduce sidereal time into mean time; *i.e.* we must subtract from t the acceleration of sidereal on mean time for the interval t, which quantity may be taken out of the Table of Time Equivalents.

78. Given sidereal time$=3^{h} 40^{m} 45'$, and the right ascension of the mean sun at mean noon at the place$=12^{h} 35^{m} 24 \cdot 14'$; required mean time.

Mean time=sidereal time$-$R A mean sun.

Sidereal time	3^{h}	40^{m}	$45 \cdot 00' (+ 24^{h})$
R A mean sun at noon	12	35	$24 \cdot 14$
Mean time nearly	15	5	$20 \cdot 86 = t.$

Cor. from table, 15^{h}..	2^{m} $27 \cdot 85'$		
5^{m}	$\cdot 82$		
$21'$	$\cdot 05$	2	$28 \cdot 72$
Mean time		15	2 $52 \cdot 14$

This result, however, is not quite correct, although nearer the truth than the quantity t; for we ought to have entered the table with the *correct* mean time, instead of the approximate time, $15^{h} 5^{m} 20 \cdot 86'$. A nearer approximation, however, can now be got by repeating the work, using the last estimated mean time, $15^{h} 2^{m} 52 \cdot 14'$, instead of $15^{h} 5^{m} 20 \cdot 86'$: thus,

Cor. for 15^{h}..	2^{m} $27 \cdot 85'$
2^{m}	$\cdot 33$
$52 \cdot 14'$	$\cdot 14$
	2 $28 \cdot 32$
	15 5 $20 \cdot 86$
Mean time$=15$	2 $52 \cdot 54$

If we approximate a third time, by using this last result instead of $15^{h} 2^{m} 52 \cdot 14'$, we shall find no difference in the correction; we may therefore conclude that the correct mean time is $15^{h} 2^{m} 52 \cdot 54'$.

In almost every problem in Nautical Astronomy in which are used quantities taken from the *Nautical Almanac*, we shall find the results obtained are only approximate values of the quantities sought; and this arises from using an approximate time (called a Greenwich date) instead of the correct Greenwich time, which is seldom known. If the object of the problem is to find the correct time, we can make a second approximation, similar to the one above; but this, in the practical problems of Nautical Astronomy, is very seldom required.

EXAMPLES FOR PRACTICE.

79. Given sidereal time $=12^{h}$ 10^{m} $10'$, and the right ascension of the mean sun at mean noon at the place $=1^{h}$ 42^{m} $14.5'$; required correct mean time.   *Ans.* 10^{h} 26^{m} $12.4'$.

80. Given sidereal time $=6^{h}$ 32^{m} $40.5'$, and the right ascension of the mean sun at mean noon $=7^{h}$ 37^{m} $42.4'$; required correct mean time.   *Ans.* 22^{h} 51^{m} $12.7'$.

PROBLEM VI.

To find at what time any heavenly body will pass a given meridian.

Let x be the given heavenly body on the meridian P Q, A the first point of Aries, and m the mean sun.

Then Qm = mean time required,

     Am = right ascension of the mean sun at that time,

     A Q = star's right ascension ;

and by the figure, Qm = A Q − A m,

     or mean time = star's R A − R A mean sun.

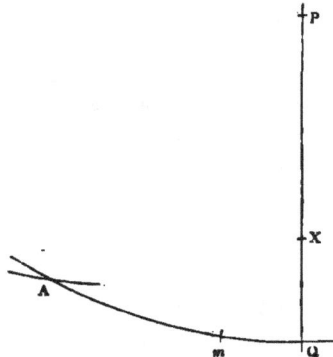

From which expression the mean time of the star's transit may be found, as in the last problem.

First. Let the given meridian be that of Greenwich, for which the quantities in the *Nautical Almanac* are calculated.

81. Find at what time Antares passed the meridian of Greenwich on October 3, 1846 ; the star's right ascension being at that time 16^h 20^m $1'$, and the right ascension of the mean sun at mean noon at Greenwich 12^h 47^m 13.8^s.

Mean time=star's R A —R A mean sun.

First approximation.				Second approximation.
R A	16^h	20^m	1.0^s	Proceed now as in the last problem, or
R A mean sun at noon .	12	47	13. 8	more simply thus :
Mean time nearly . . .	3	32	47.20	Cor. for 34.95^s . . . . 0.09^s
Cor. 3^h . . . 29.57^s				1st approximation . . . 3^h 32^m 12.25^s
32^m . . . 5.25				Cor. mean time 3 32 12.34
47^s . . . $.13$			34.95	
Mean time more nearly .	3	32	12.25	

EXAMPLES FOR PRACTICE.

82. Find at what time α Canis Majoris passed the meridian of Green-wich ; the star's R A being 6^h 38^m $52.2'$, and the right ascension of mean sun at Greenwich mean noon being 11^h 6^m $2.3'$. Construct the figure, to show the positions of the first point of Aries and the mean sun with respect to the meridian. *Ans.* 19^h 29^m $38'$ nearly.

83. Find at what time α Aquilæ passed the meridian of Greenwich, having given the right ascension of the star$=19^h$ 43^m $51.5'$, and the right ascension of the mean sun at Greenwich mean noon 0^h 6^m $40.4'$; and con-struct the figure. *Ans.* 19^h 33^m $58.3'$.

84. Find at what time α Leonis will pass the meridian of Greenwich, when its right ascension is 10^h 0^m $49.76'$, and the right ascension of the mean sun at Greenwich mean noon$=4^h$ 32^m $4.68'$; and construct the figure. *Ans.* 5^h 27^m $51.24'$.

Second. When the calculations are made for any other meridian than that of Greenwich, we must take into consideration the change of the mean sun's place corresponding to the difference of longitude between the two places. In practice, the correction of the R A of mean sun on this account is made in the Greenwich date, which is in fact the longitude in time applied to ship time.

PROBLEM VII.

Given the altitude and declination of a heavenly body, and the latitude of the place of observation, to calculate the hour-angle of the heavenly body.

Let P Z Q be the celestial meridian, P the pole, Z the zenith of spectator, and X the place of the heavenly body. Let A Q be the celestial equator ; and through X draw P X R a circle of declination, and Z X a circle of altitude.

Then, in the spherical triangle z p x, the three sides are given, to calculate P the hour-angle (*Trigonometry*, Part I., Rule VIII.)

For the polar distance P x = 90° — decl. ;
zenith distance z x = 90° — alt. ;
and colatitude P z = 90° — lat. ;

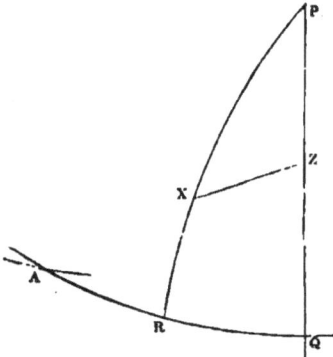

(fig. 1.)

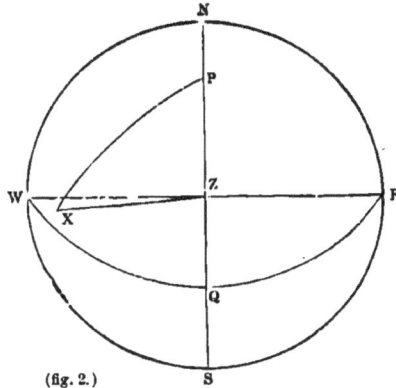

(fig. 2.)

85. Calculate the hour-angle of a heavenly body x, west of meridian, having given the latitude of observer = 48° 40′ 45″ N., the star's declination = 8° 25′ 45″ N., and star's altitude = 13° 57′ 45″ ; and construct the figure.

Construction (Fig. 2.)

Let N W S E represent the horizon, N Z S the celestial meridian, W Z E the prime vertical. Take N P = latitude; then P will represent the north pole of the heavens, since the altitude of the pole = latitude of spectator (for lat. z Q = 90 — P z and N P = 90 — P z, ∴ z Q = N P, the altitude of pole, see Ex. 8, p. 11), and P z is the colatitude. Take P Q = 90°; then Q is a point in the celestial equator : through E and w, the east and west points, and Q, draw the circle W Q E ; this will represent the celestial equator (p. 6). Let x be the place of the heavenly body at the time of the observation (estimated according to its altitude and declination) ; and through x draw x P, a circle of declination, and x z a circle of altitude. Then, in the spherical triangle z P x, are given P x, the polar distance (= 90° — decl.) = 81° 34′ 15″, z x the zenith distance (= 90° — alt.) = 76° 2′ 15″, and P z the colat. (= 90° — lat.) = 41° 19′ 15″ ; to calculate z P x, the hour-angle.

The hour-angle may be calculated either by the table of haversines, or by using only the common table of sines, &c., as follows. (See Rule VIII. in *Trigonometry*, Part I.)

BY HAVERSINES.				BY THE COMMON RULE (using only the table of log. sines, &c.).		
PX ..	81° 34′	15″	PX ..	81°	34′	15″
PZ ..	41 19	15	PZ ..	41	19	15
	40 15	0		40	15	0
XZ ..	76 2	15	ZX ..	76	2	15
S ..	116 17	15	S ..	116	17	15
D ..	35 47	15	D ..	35	47	15

BY HAVERSINES.

$$\text{log. cosec. } PX \; .. \; 0\cdot004717$$
$$,, \qquad ,, \quad PZ \; .. \; 0\cdot180275$$
$$\tfrac{1}{2}\text{ log. hav. } S \; .. \; 4\cdot929099$$
$$\tfrac{1}{2}\text{ log. hav. } D \; .. \; 4\cdot487496$$

$$\text{log. hav. } ZPX \; .. \; 9\cdot601587$$
$$\therefore ZPX = 5^h \; 13^m \; 39^s$$

BY THE COMMON RULE

$$\tfrac{1}{2} S \; .. \; 58 \quad 8 \quad 37$$
$$\tfrac{1}{2} D \; .. \; 17 \quad 53 \quad 37$$

$$\text{log. cosec. } PX \; .. \; 0\cdot004717$$
$$,, \qquad ,, \quad PZ \; .. \; 0\cdot180275$$
$$,, \quad \sin. \tfrac{1}{2} S \; .. \; 9\cdot929100$$
$$,, \quad \sin. \tfrac{1}{2} D \; .. \; 9\cdot487496$$

$$2)19\cdot601588$$

$$,, \sin. \tfrac{1}{2} ZPX \; .. \; 9\cdot800794$$
$$\tfrac{1}{2} ZPX \; .. \; 2^h \; 36^m \; 49\cdot5^s$$
$$2$$

$$\therefore \text{ HOUR-ANGLE } ZPX = 5 \quad 13 \quad 39$$

This latter method of finding the hour-angle is the one commonly adopted, since we require only such tables as Riddle's or Norie's. The concise and very superior method by means of haversines can only be used by those who have Inman's tables.

EXAMPLES FOR PRACTICE.

86. Find the hour-angle of a heavenly body, west of meridian, having given the latitude=47° 20′ N., the declination=11° 24′ 24″ N., and the altitude=42° 33′ 9″; and construct the figure.

             Ans. Hour-angle=2^h 27^m 51^s.

87. Find the hour-angle of a heavenly body, east of meridian, having given the latitude=56° 10′ N., the declination=33° 11′ 44″ N., and the altitude=59° 3′ 59″; and construct the figure.

             Ans. Hour-angle=2^h 0^m 30^s.

88. Find the hour-angle of a heavenly body, east of meridian, having given the latitude=50° 48′ 0″ N., the declination=14° 28′ 47″ S., and the altitude 3° 13′ 0″; and construct the figure.

             Ans. Hour-angle=4^h 23^m 38^s.

Given the hour-angle of the sun ; to find ship mean time.

First. Let the sun be west of meridian. Let P be the pole, z the zenith, and x the place of the sun. Then (fig. 1),

Apparent time=hour-angle zPx, or arc QR ;

and if the equation of time be *subtractive* from apparent time (and this is known from the *Nautical Almanac*), let m be the mean sun ;

Then Qm=mean time required.

By the figure, Qm=QR − Rm,

or mean time=sun's hour-angle − equation of time.

If the equation of time is additive to apparent time, then the mean sun m is in advance of the true sun x, and m should be placed in the figure on the other side of R. In this case mean time=sun's hour-angle + equation of time.

The proper sign to be used is always given in the *Nautical Almanac* with the equation of time.

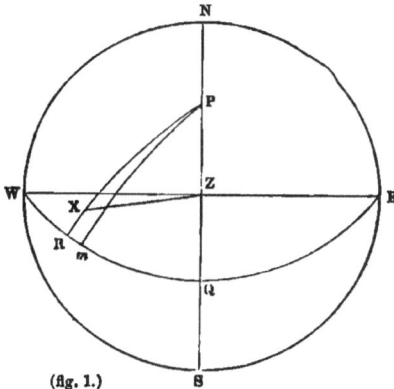

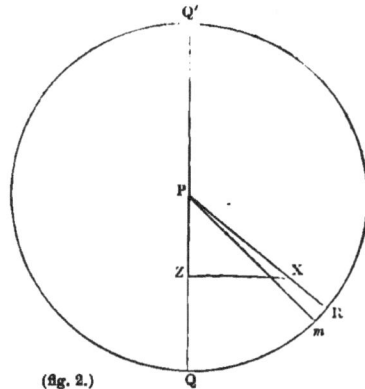

(fig. 1.)   (fig. 2.)

Next. Let the sun be east of meridian. It will be more convenient to project the figure on the plane of the celestial equator. Let therefore P be the pole, z the zenith, and x the place of the sun. Then (fig. 2),

Apparent time=arc QQ′R=24^h −hour-angle zPx ;

and if the equation of time is additive to apparent time, let m be the mean sun ; then the arc QQ′m=mean time required.

And by the figure, QQ′m=QQ′R + Rm,

or mean time=24^h −sun's hour-angle + eq. of time.

EXAMPLES FOR PRACTICE.

89. Given the sun's hour-angle=3^h 42^m 10′, west of meridian, and the equation of time=14^m 10′ additive to apparent time ; required mean time, and construct figure decl. 0°. *Ans.* Mean time=3^h 56^m 20′.

90. Given the sun's hour-angle$=6^h$ 2^m 20', east of meridian, and the equation of time$=3^m$ 42', subtractive from apparent time; required mean time. Construct figure, decl. 10° N.   *Ans.* Mean time$=5^h$ 53^m 58' A.M.

PROBLEM IX.

Given the hour-angle of any other heavenly body, as a star; to find ship mean time.

We have to prove the following rule, viz.

(1.) When the star is *west* of meridian,

Mean time=star's hour-angle + star's right ascension

—right ascension of mean sun.

(2.) When the star is *east* of meridian,

Mean time$=24^h$—star's hour-angle + star's right ascension

—right ascension of mean sun.

To do this we must show that these equations are true for all positions of the star, the mean sun, and first point of Aries, with respect to each other. The figures in this problem are projected on the plane of the equator.

First. Suppose the heavenly body to be *west* of meridian.

(*a.*) Let P be the pole, and x a heavenly body; and let A, the first point of Aries, and *m*, the mean sun, be situated with respect to x and to each other as in fig. 1.

Then QR=star's hour-angle,   A*m*=right ascen. of mean sun.

  AR=star's right ascen.   and Q*m*=mean time required.

  By the figure, Q*m*=QR+AR—A*m*,

  or mean time=hour-angle+star's RA—RA mean sun.

(fig. 1.)   (fig. 2.)

(*b.*) Let the relative positions of A, *m*, and x, be as represented in fig. 2.

Then QR=star's hour-angle, Am=24^h—RA mean sun,
AR=24^h—star's right ascen. and Qm=mean time required.

By the figure, Qm=QR—AR+Am,
or mean time=star's hour-angle —(24^h – star's RA)+24^h—RA mean sun
=hour-angle+star's RA — RA mean sun.

(c.) Let the relative positions of A, m, and x, be as represented in fig. 3.
Then QR=star's hour-angle, Am=24^h—RA mean sun,
AR=star's right ascen. and Qm=mean time required.

Then, by the figure, Qm=QR+AR+Am,
or mean time=star's hour-angle+star's RA+24^h—RA mean sun,
the same as before, by rejecting 24 hours. And the same result will be
obtained for every other position of A, m, and x, with respect to each other :
therefore, when the body is *west* of meridian,
Mean time=star's hour-angle+star's RA—RA mean sun
(rejecting, if necessary, 24 hours).

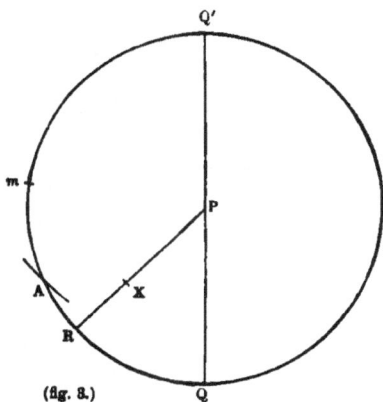 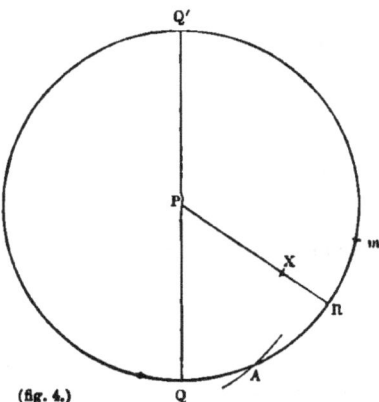

(fig. 3.) (fig. 4.)

Next. Suppose the heavenly body to be *east* of meridian.

(d.) Let the relative positions of A, m, and x, be as represented in fig. 4.
Then QR=star's hour-angle, Am=right ascen. of mean sun,
AR=star's right ascen. and Qm=24^h—mean time.

Then, by fig., Qm=QR+Am—AR,
or 24^h—mean time=hour-angle+RA mean sun—star's RA ;
∴ mean time=24^h—hour-angle+star's RA—RA mean sun.

(e.) Let the relative positions of A, m, and x, be as represented in fig. 5.

Then QR=hour-angle, AR=star's RA,

Am=RA of mean sun,

and Qm=mean time required.

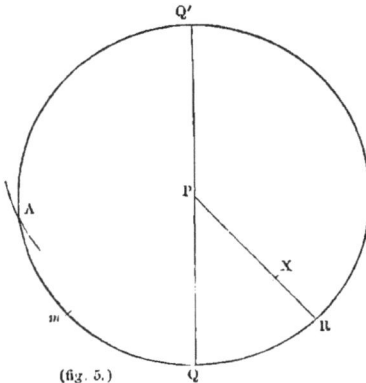

(fig. 5.)

By the figure, Qm=AR—Am—QR,
or mean time=star's RA—RA
mean sun—hour-angle.

This expression may be brought
into the required form by adding
and subtracting 24 hours; thus,

Mean time=24^h—hour angle+
star's RA—RA mean sun—24^h

the same as before, by rejecting 24
hours; and the same result may be
obtained for any other positions of
A, m, and X, with respect to each
other. Therefore, when the body is
east of meridian,

Mean time=24^h—hour-angle+star's RA—RA mean sun
(rejecting or adding 24 hours, if necessary).

This form is adopted because, if the table of haversines is used to find
the hour-angle, the quantity, 24^h—hour-angle, may be taken from the
bottom of the page *by inspection*.

We have then in all cases this convenient rule: "add the star's right
ascension to the angle taken from the table of haversines (remembering
when the heavenly body is west of meridian to take the angle from the
top, and when east, from the bottom), and from the result subtract the
right ascension of the mean sun; the remainder will be ship mean time
required (adding or rejecting 24 hours, if necessary)."

EXAMPLES FOR PRACTICE.

91. Given the hour-angle of α Aquilæ (west of meridian)=2^h 37^m 23',
star's right ascension=19^h 43^m 52', and right ascension of mean sun
=9^h 51^m 38'; find ship mean time, and construct the figure; star's declination 9° N.   *Ans.* Mean time=0^h 29^m 37' A.M.

92. Given the hour-angle of α Cygni (east of meridian)=4^h 1^m 35',
star's right ascension=20^h 36^m 36', and right ascension of mean sun=
0^h 17^m 7'; find ship mean time, and construct the figure, star's declination
being=45° N.   *Ans.* Mean time=7^h 17^m 54'.

93. Given the hour-angle of β Geminorum=1^h 15^m 57' east of meridian,
star's right ascension=7^h 36^m 35', and right ascension of mean sun=22^h
34^m 43'; find ship mean time, and construct the figure; star's declination,
28° N.   *Ans.* Mean time=7^h 45^m 55'.

PROBLEM X.

Given the altitude of a heavenly body, and the time shown by a chronometer at the instant of observation; to determine the error of the chronometer on mean time at the place.

(1.) Let the heavenly body observed be the sun.

Let P Z Q represent the celestial meridian, P the pole, Z the zenith, and X the place of the sun west of meridian, m the mean sun in the equator A Q; through m draw the circle of declination P m, and through X draw the circle of declination P R and circle of altitude Z X. Then Q m represents mean time at the place where the observation was taken; the difference between mean time and the time shown by the chronometer at the instant of observation is the error of the chronometer on mean time at the place.

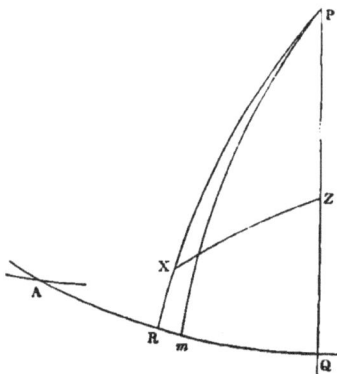

To find the value of Q m, we must calculate the hour-angle Z P X, which is in this case also apparent time, and subtract therefrom the angle X P m, the equation of time; the remainder, namely the angle Z P m, or arc Q m, will be the mean time required.

To compute the hour-angle Z P X. In the spherical triangle Z P X are given P Z = colat. of the place, P X = polar distance, or codeclination, taken out of the *Nautical Almanac*, and Z X = zenith distance, or co-altitude found from observation; to find the hour-angle Z P X = arc Q R.

The equation of time R m is taken out of the *Nautical Almanac*, as also the declination, and corrected for the time elapsed from noon by the common rule. (See Part I. p. 80, or any Practical Treatise on Navigation.)

Then Q m = Q R − R m,

or mean time = hour-angle − equation of time.

Let T = time shown by the chronometer at the instant of observation; then

Error of chronometer = mean time ∼ T.

The position of the mean sun m with respect to the true sun X, that is, whether it is in advance or behind it, is known from the sign affixed to the equation of time in the *Nautical Almanac*; in the former case, we must add the equation of time instead of subtracting it, as in the figure.

TO FIND ERROR OF CHRONOMETER BY ALTITUDE OF THE SUN
(WEST OF MERIDIAN).

94. Given the altitude of the sun = 15° 16′ 59″ (west of meridian), the latitude of observer=50° 48′ N., the sun's declination=9° 1′ 17″ S., and the equation of time=14ᵐ 25·2ˢ (additive to apparent time) ; to find the error of chronometer on mean time at the place. The chronometer showed at the instant of observation 3ʰ 7ᵐ 49·8ˢ.

Construction.

Let p be the pole, z the zenith, and x the place of the sun, west of meridian and south of the equator. Through x draw px, a circle of declina-

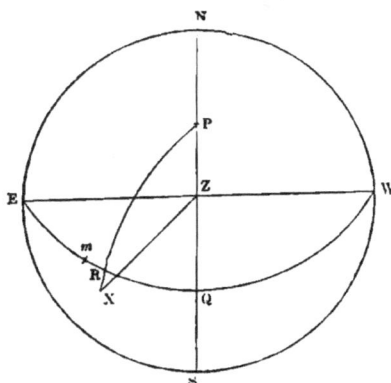

tion, and zx, a circle of altitude. Then, in the triangle zpx, the three sides are given, namely the polar distance px=99° 1′ 17″, the zenith distance zx=74° 49′ 1″, and the colatitude pz=39° 12′ 0″ ; to compute the hour-angle zpx, or arc qr, which is also apparent time (p. 16). Again, since the equation of time is additive to apparent time, let m be the position of the mean sun; then rm represents the equation of time. To arc qr add rm: the sum qm will be mean time at the instant of observation ; the difference between which and the time shown by the chronometer is the *error* of the chronometer on mean time at the place.

Calculation (1) *by haversines,* (2) *or by sines, &c.*
(Rule VIII., *Trigonometry*, Part I.)

px=99°	1′	17″...	0·005405		99°	1′	17″...	0·005405
pz=39	12	0 ...	0·199263		39	12	0 ...	0·199263
	59	49	17	4·965043		59	49	17	9·965045
zx=74	49	1	4·115578		74	49	1	9·115578
	134	38	18	9·285289		134	38	18	19·285291
	14	59	44	3ʰ 28ᵐ 25ˢ app. time,		14	59	44	
			14 25·2=rm				9·642645
					67	19	9	1 44 12·5
∴ mean time=3	42	50·2		7	29	52	2
Chronometer showed 3	7	49·8				
∴ error of chronom.=	35	0·4 slow.		∴ app. time=3 28 25

TO FIND ERROR OF CHRONOMETER BY ALTITUDE OF THE SUN
(EAST OF MERIDIAN).

95. Find the error of a chronometer on ship mean time, having given the latitude of observer=50° 48′ N., the sun's altitude=39° 29′ 18″ (east of meridian), the declination=17° 33′ 10″ N., and the equation of time =3ᵐ 48·7ˢ (to be subtracted from apparent time) ; the chronometer showing at the instant of observation 8ʰ 26ᵐ 59·7ˢ.

Construction.

Let the figure in this case be projected on the plane of the equator. Let P be the pole, z the zenith, and x the place of the sun east of the meridian QQ′. Through x draw the circle of declination PR, and circle of altitude zx. Then, in the spherical triangle zPx, are given the three sides, to calculate the hour-angle zPx, or arc QR, which measures it. Subtract this arc QR from 24 hours; the remainder is the arc QQ′R, which measures ship apparent time. Since the equation of time is subtractive from apparent time, let m be the position of the mean sun. From the arc QQ′R subtract Rm, the equation of time; the remainder, namely the arc QQ′m, represents ship mean time required: the difference between which and the time shown by the chronometer (remembering to add 12 hours to the chronometer time, if by so doing the difference between the times is made less) will be the error of the chronometer on mean time at the place.

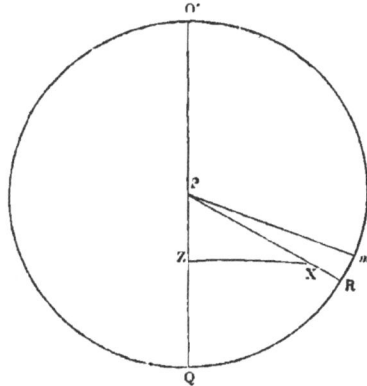

Calculation of 24ʰ—*hour-angle* (see Rule, p. 36).

P x...72°	26′	50″	......0·020705	
P z...39	12	0	......0·199263	
33	14	50	4·824491	
z x...50	30	42	4·176355	
83	45	32	9·220814	
17	15	52	20ʰ 47ᵐ 29ˢ=24ʰ—hour-angle=QQ′R,	
	3	48·7=Rm,		
20	43	40·3=QQ′m=mean time.		
Chronom. +12ʰ...20	26	59·7		

∴ error of chronom.=   16   40·6 slow.

If the common table of sines, &c., is used, and the body is east of meridian, as in this example, the hour-angle will be found as follows :

$$
\begin{array}{llll}
\text{PX}...72° & 26' & 50'' & ......0\cdot020705 \\
\text{PZ}...39 & 12 & 0 & ......0\cdot199263 \\
\hline
33 & 14 & 50 & 9\cdot824491 \\
\text{ZX}...50 & 30 & 42 & 9\cdot176355 \\
\hline
83 & 45 & 32 & 19\cdot220814 \\
17 & 15 & 52 & \overline{9\cdot610407} \\
\hline
41 & 52 & 46 & 1 \quad 36 \quad 15 \\
8 & 37 & 56 & \qquad\qquad 2 \\
\hline
\end{array}
$$

$$
\begin{array}{lll}
3 & 12 & 30 = \text{hour-angle,} \\
24 & & \\
\hline
20 & 47 & 30 = \text{QQ'R.}
\end{array}
$$

TO FIND ERROR OF CHRONOMETER BY ALTITUDE OF A STAR OR MOON (WEST OF MERIDIAN).

96. Find the error of a chronometer on ship mean time, having given the latitude of observer=50° 48′ N., the altitude of Arcturus=44° 55′ 42″ (west of meridian), its declination=20° 0′ 15″ N., right ascension=14ʰ 8ᵐ 30·5ˢ, and right ascension of mean sun=4ʰ 48ᵐ 7·5ˢ; the chronometer showing at the instant of observation 0ʰ 14ᵐ 22·3ˢ.

Before constructing the figure, it will be better to compute the hour-angle, in order to estimate more correctly the position of the heavenly body with respect to the meridian : thus (p. 32),

$$
\begin{array}{llll}
\text{Pol. dist.} \quad......... & 69° & 59' & 45''......0\cdot027026 \\
\text{Colat.} \quad............ & 39 & 12 & 0 ......0\cdot199263 \\
\hline
& 30 & 47 & 45 \quad 4\cdot788699 \\
\text{Zen. dist.}......... & 45 & 4 & 18 \quad 4\cdot094305 \\
\hline
& 75 & 52 & 3 \quad 9\cdot109293 \\
& 14 & 16 & 33 \quad \therefore 2^{\text{h}} 48^{\text{m}} 8' = \text{hour-angle.}
\end{array}
$$

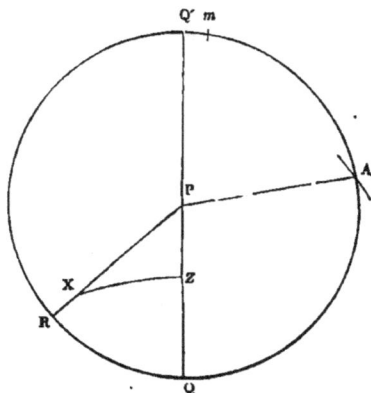

Construction.

Let P be the pole, z the zenith, and Q P Q′ the celestial meridian. On the celestial equator take QR=2ʰ 48ᵐ 8ˢ, the hour-angle of the star; then the star is on the circle of declination P R, at a point x=20° 0′ 15″ north of the equator. From R measure RQ′A=14ʰ 8ᵐ 50·5ˢ, the star's right ascension; then A is the position of the first point of Aries. Again, from A measure Am =4ʰ 48ᵐ 7·5ˢ, the right ascension of the mean sun; then m is the position of mean sun, and therefore QQ′m= mean time at the instant of observation. By Prob. IX., namely,

Mean time=hour-angle+star's RA—RA mean sun.

hour-angle	2ʰ	48ᵐ	8ˢ
star's RA	14	8	30·5
	16	56	38·5
RA mean sun	4	48	7·5
∴ ship mean time=12	8	31·0	
and chronometer+12ʰ=12	14	22·3	
∴ error of chronometer=	5	51·3	fast.

TO FIND ERROR OF CHRONOMETER BY ALTITUDE OF A STAR OR MOON
(EAST OF MERIDIAN).

97. Find the error of a chronometer on ship mean time, having given the latitude of the place=49° 57′ N., the altitude of Regulus=8° 4′ 18″ (east of meridian), its declination = 12° 39′ 49″ N., right ascension = 10ʰ 0ᵐ 46ˢ, and the right ascension of the mean sun=19ʰ 45ᵐ 8ˢ; the chronometer showing at the instant of observation 8ʰ 2ᵐ 10ˢ.

Before constructing the figure, compute the hour-angle, as in the last example, in order to estimate nearly the position of the heavenly body with respect to the meridian.

To compute hour-angle, or rather 24ʰ—hour-angle (see Rule, p. 36).

Pol. dist.	77°	20′	11″	0·010693
Colat.	40	3	0	0·191481
	37	17	11	4·935803
Zen. dist.	81	55	42	4·579546
	119	12	53	9·717523
	44	38	31	∴ 17ʰ 50ᵐ 0ˢ=24ʰ—hour-angle.

Construction.

Let P be the pole, z the zenith, and QPQ′ the celestial meridian. On the celestial equator take QQ′R=17ʰ 50ᵐ =24ʰ—the star's hour-angle; then the star is on the circle of declination PR, at a point X=12° 39′ 49″ north of the equator. From R measure RQA=10ʰ 0ᵐ 46ˢ, the star's right ascension; then A is the position of the first point of Aries. Again, from A measure ARm=19ʰ 45ᵐ 8ˢ, the right ascension of mean sun; then m is the position of the mean sun, and Qm measures mean time at the instant of observation.

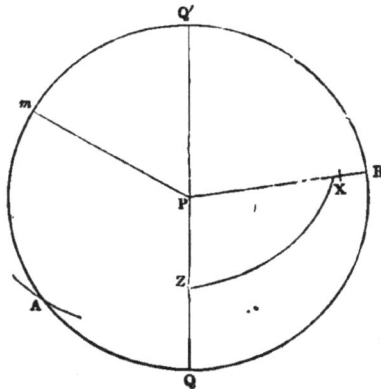

Mean time may be computed from the formula (p. 34).

Mean time=(24ʰ—star's hour-angle)+star's RA—RA mean sun.

$$
\begin{array}{lrrr}
24^{h}\text{—hour-angle}\ldots\ldots & 17^{h} & 50^{m} & 0^{s} \\
\text{star's RA} \ldots\ldots\ldots\ldots & 10 & 0 & 46 \\ \hline
 & 27 & 50 & 46 \\
\text{RA mean sun} \ldots\ldots\ldots & 19 & 45 & 8 \\ \hline
\therefore\ \text{ship mean time}= & 8 & 5 & 38 \\
\text{Chronometer}= & 8 & 2 & 10 \\ \hline
\therefore\ \text{error of chronometer}= & & 3 & 28\ \text{slow.}
\end{array}
$$

EXAMPLES FOR PRACTICE.

98. Find the error of a chronometer on ship mean time, having given the latitude of the observer=53° N., the sun's altitude=23° 17′ 20″ (west of meridian), the declination = 2° 0′ 30″ N., and the equation of time =6ᵐ 58ˢ, to be added to apparent time; the chronometer showing at the instant of observation 3ʰ 24ᵐ 46ˢ.   *Ans.* Chronometer slow, 11ᵐ 37ˢ.

99. Find the error of a chronometer on mean time at the place of observation, having given the latitude of the observer=47° 30′ N., the sun's altitude=20° 11′ 45″ (west of meridian), the declination=20° 1′ 30″ N., and the equation of time= 3ᵐ 45ˢ, to be subtracted from apparent time; the chronometer showing at the instant of observation 5ʰ 10ᵐ 32ˢ.
Ans. 0ʰ 12ᵐ 6ˢ slow.

100. Find the error of a chronometer on mean time at the place of observation, having given the latitude of the observer=50° 48′ N., the sun's altitude=3° 4′ 15″ (east of meridian), the declination=14° 35′ 20″ S., and the equation of time=14ᵐ 5ˢ, to be added to apparent time; the chronometer showing at the instant of observation 8ʰ 51ᵐ 20ˢ. *Ans.* 1ʰ 1ᵐ 16ˢ fast.

101. Find the error of a chronometer on mean time at the place of observation, having given the latitude of the observer=49° 57′ N., the altitude of α Aquilæ (Altair), 37° 0′ 30″ (west of meridian), right ascension =19ʰ 43ᵐ 51ˢ, declination=8° 29′ 43″ N., and the right ascension of mean sun=9ʰ 51ᵐ 8ˢ; the chronometer showing at the instant of observation 11ʰ 42ᵐ 17ˢ.   *Ans.* 0ʰ 49ᵐ 41ˢ slow.

102. Find the error of a chronometer on mean time at the place of observation, having given the latitude of the observer=48° 50′ N., the altitude of α Cygni=49° 33′ (east of meridian), right ascension=20ʰ 36ᵐ 36ˢ, declination=44° 46′ 22″ N., and the right ascension of the mean sun=9ʰ 15ᵐ 39ˢ; the chronometer showing at the instant of observation 5ʰ 2ᵐ 13ˢ.
Ans. 2ʰ 17ᵐ 7ˢ slow.

103. Find the error of a chronometer on mean time at the place of observation, having given the latitude of the observer=48° 50′ N., the altitude of Arcturus (east of meridian)=47° 22′ 50″, right ascension=14ʰ 9ᵐ 10ˢ, declination=19° 55′ 27″ N., and the right ascension of the mean sun = 1ʰ 50ᵐ 8ˢ; the chronometer showing at the instant of observation 10ʰ 40ᵐ 30ˢ.   *Ans.* 0ʰ 58ᵐ 24ˢ fast.

104. As it is impossible to observe at sea the altitude of a heavenly body with perfect accuracy, we will now show under what circumstances a small error committed in taking the altitude will produce the least error in the hour-angle computed from it. The problem about to be investigated will prove this very important fact, that the error in the hour-angle, and therefore in the mean time deduced from it, will be the least for a given error in the altitude *when the heavenly body is on the prime vertical*, that is, when its bearing is 8 points from the meridian. It is for this reason, namely that the unavoidable error of observation should produce the least effect on the time calculated from it, we are directed always to take our observations for time when the heavenly body bears as nearly east or west as possible. When we are able (and this may sometimes happen) to observe the altitude with very great exactness, this restriction need not be made, and we may take our observation for determining the time or hour-angle when the heavenly body is close to the meridian.

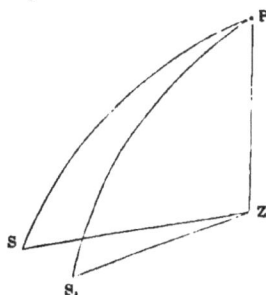

PROBLEM XI.

Given an error in the altitude, to find the corresponding error in the hour-angle.

Let P be the pole, z the zenith, and s the true place of the heavenly body. Let s_1 be the supposed place, as determined by observation.

$$\text{Let } sz \text{ the true zenith distance} = z,$$
$$zPs \text{ the true hour-angle} = h \text{ ;}$$
$$s_1z \text{ the erroneous zenith distance} = z_1,$$
$$\text{and } zPs_1 \text{ the hour-angle computed from it} = h_1.$$
$$\text{Then } h - h_1 = \text{error in the hour-angle, corresponding to}$$
$$z - z_1 = \text{the error in the altitude.}$$

Let colat. $Pz = c$, and pol. dist. Ps or $Ps_1 = p$.

In triangle sPz, $\cos. h = \dfrac{\cos. z - \cos. c \cdot \cos. p}{\sin. c \cdot \sin. p}$ (*Trig.* Art. 55)

,, s_1Pz, $\cos. h_1 = \dfrac{\cos. z_1 - \cos. c \cdot \cos. p}{\sin. c \cdot \sin. p}$,

$\therefore \cos. h - \cos. h_1 = \dfrac{\cos. z - \cos. z_1}{\sin. c \cdot \sin. p}$; or,

$\sin. \frac{1}{2} (h + h_1) \cdot \sin. \frac{1}{2} (h - h_1) = \dfrac{\sin. \frac{1}{2} (z + z_1) \cdot \sin. \frac{1}{2} (z - z_1)}{\sin. c \cdot \sin. p}.$

Now, since $h - h_1$ and $z - z_1$ are small, $\frac{1}{2} (h + h_1) = h$, and $\frac{1}{2} (z + z_1) = z$ nearly : also $\sin. \frac{1}{2} (h - h_1)$ and $\sin. \frac{1}{2} (z - z_1)$ may be replaced by $\frac{1}{2} (h - h_1)$ and $\frac{1}{2} (z - z_1)$; since the sine of an arc is nearly equal to the arc itself when the angle is small (*Trig.* p. 101). Making these substitutions, we have

$$\sin. h \,.\, \tfrac{1}{2}\,(h-h_1) = \frac{\sin. z}{\sin. c \,.\, \sin. p} \,.\, \tfrac{1}{2}\,(z-z_1) \dots (1)$$

Let $A = PZS$, the azimuth or bearing of the body; then, in triangle ZPS,

$$\frac{\sin. h}{\sin. A} = \frac{\sin. z}{\sin. p} \quad (\textit{Trigonometry, Art. } 55);$$

or $\sin. h = \dfrac{\sin. A \,.\, \sin. z}{\sin. p}$.

Substituting this value of sin. h in (1), we have

$$\frac{\sin. A \,.\, \sin. z}{\sin. p} \,.\, (h-h_1) = \frac{\sin. z}{\sin. c \,.\, \sin. p} \,.\, (z-z_1)$$

$$\therefore\ h-h_1 = \frac{1}{\sin. c \,.\, \sin. A} \,.\, (z-z_1)$$

or error in hour-angle $= \dfrac{1}{\cos. \text{lat.} \,.\, \sin. A}$. error in altitude.

From this expression, it is seen that the error in the hour-angle is the least for a given error in the altitude when the sin. azimuth, and therefore the azimuth itself, is the greatest; that is, when the heavenly body is due east or west.

The following examples will more clearly show this.

105. Find the error in the hour-angle corresponding to an error of 4 minutes in the altitude, taken at a place in latitude 50° 48′ N.

First. Supposing the bearing of the body was S. b. E.

Second. Supposing the bearing of the body was due east.

$$h - h_1 = \frac{1}{\cos. \text{lat.} \,.\, \sin. A} \,.\, (z-z_1)$$

$$= \sec. \text{lat.} \,.\, \csc. A \,.\, (z-z_1) \dots \text{ in arc}$$

$$= \tfrac{1}{15} \sec. \text{lat.} \,.\, \csc. A \,.\, (z-z_1) \dots \text{ in time.}$$

1. Bearing S. b. E.	2. Bearing East.
log. sec. 50° 48′ ....0·199263	..........................0·199263
,, cosec. 11° 15′ ......0·709764	......log. cosec. 90° ......0·000000
,, 4 ....0·602060	,, 4 ....0·602060
1·511087	0·801323
,, 15 ....1·176091	,, 15 ....1·176091
,, $h-h_1$ ....0·334996	,, $h-h_1$ ....$\overline{1}$·625232
$\therefore\ h-h_1 = 2·162^m = 2^m\ 9·7^s$	$\therefore\ h-h_1 = 0·422^m = 25·3^s$.

From these results it appears that the error in the hour-angle in one case is $2^m\ 9·7^s$, and in the other $25·3^s$, for the same error of 4 minutes in the altitude.

106. Find the error in the hour-angle corresponding to an error of two minutes in the altitude, at a place in latitude 2° 30′ N.

First. Supposing the bearing of the body was S. 15° E.

Second. Supposing the bearing of the body was S. 105° E.

Ans. Error at 1st bearing, 30·9″.

,,   2d   ,,   8·3″.

The more accurate methods of finding the error of a chronometer, by equal altitudes and by transit observations, will be given in Chapter VII.

TO FIND THE MEAN DAILY RATE OF A CHRONOMETER.

Assuming that the chronometer has gone nearly equably during the interval between two observations taken a few days apart, we may easily obtain its *mean daily rate* by taking the difference between the errors at the two observations, and dividing it by the number of intermediate days : but suppose the chronometer has (what not unseldom happens) an accelerated rate ; then the above method of obtaining its rate will give very inaccurate results. A formula for determining the error of a chronometer at any given time, having an increasing or decreasing rate, will be found investigated (among other very useful information relating to chronometers) in the valuable publications of Admiral Shadwell.

CHAPTER III

INVESTIGATION OF RULES FOR DETERMINING THE LATITUDE.

Latitude by meridian altitudes above and below pole.

PROBLEM XII.

Given the meridian altitudes of a heavenly body above and below the pole : to determine the latitude.

Let N W S E represent the horizon, P the pole, and Z the zenith ; then N Z S is the celestial meridian. Let $x x_1 d$ be a parallel of declination described by a heavenly body about the pole P, and x and x_1 its places when on the meridian ; then (fig. 1)

x N=star's meridian altitude above the pole.

x_1 N=star's meridian altitude below the pole.

Also P N=altitude of pole=latitude (p. 31).

and $Px=Px_1$=star's polar distance.

By the fig., $PN=xN-xP$,

and $PN=x_1N+x_1P$.

∴ adding, $2\overline{PN=xN+x_1N}$ ∴ since $xP=x_1P$,

or $PN=\tfrac{1}{2}(xN+x_1N)$,

∴ latitude $=\tfrac{1}{2}\{$alt. above pole+alt. below pole$\}$.

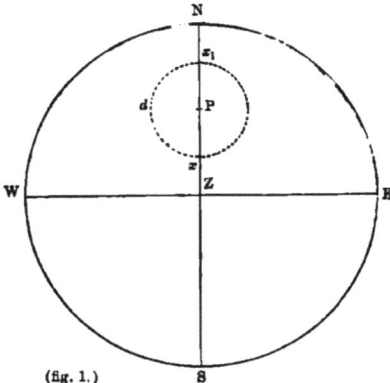 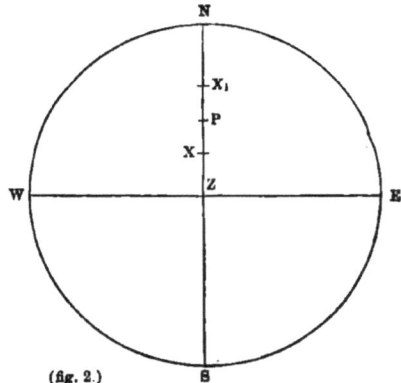

(fig. 1.)   (fig. 2.)

107. Given the meridian altitude of a heavenly body above the north pole=70° 42′, and its meridian altitude below the pole=35° 18′. Construct a figure, and find by calculation the latitude (fig. 2).

Let N W S E represent the horizon, and N Z S the celestial meridian. On N Z take $Nx = 70° 42'$, and $Nx_1 = 35° 18'$; bisect xx_1 in P, then P is the north pole and P N the altitude of the pole=latitude required.

$$\therefore \text{ by above formula, lat.} = \frac{70° 42' + 35° 18'}{2} = 53° \text{ N.}$$

EXAMPLES FOR PRACTICE.

108. The meridian altitude of a heavenly body above the north pole was 82° 10', and its meridian altitude below the pole was 40° 30'. Construct a figure, and find by calculation the latitude.　　　*Ans.* Lat. 61° 20' N.

109. The meridian altitude of a heavenly body above the south pole was 56° 42' 10", and the meridian altitude below the pole was 6° 45' 32". Construct a figure, and find by calculation the latitude.

Ans. Lat. 31° 43' 51" S.

110. The meridian altitude of a heavenly body above the south pole was 60° 0' 0", and the meridian altitude below the pole was 45° 0' 0". Construct a figure, and find by calculation the latitude.

Ans. Lat. 52° 30' 0" S.

Latitude by meridian altitude below pole.

PROBLEM XIII.

Given the meridian altitude of a heavenly body below the pole, and its declination ; to find the latitude.

Let N W S E represent the horizon, N Z S the celestial meridian, P the pole of the heavens, and x a heavenly body on the meridian below the pole (fig. 1).

Then x N=meridian alt. of heavenly body,

P x=polar distance=90° − decl.

and P N=alt. of pole=latitude.

But by the figure, P N=x N+x P,

$\therefore$ latitude=merid. alt. + 90° − decl.

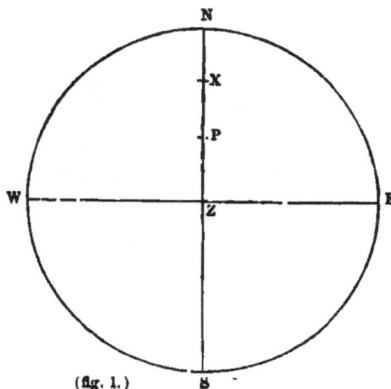
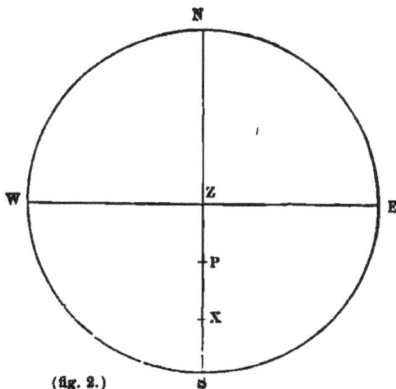

(fig. 1.)　　　　　　(fig. 2.)

111. Given the meridian altitude of α Centauri under the south pole= 27° 25' 58", and its declination=60° 14' 40" S. Construct a figure, and find by calculation the latitude (fig. 2).

Let N W S E represent the horizon, and N Z S the celestial meridian. In S Z take S X=27° 25' 58" ; then X is the place of the star under the south pole. Take X P=29° 45' 20", the star's south polar distance ; then P is the south pole, and P S=alt. of south pole=lat. required.

By formula, lat.=mer. alt.+90°−decl.

∴ lat.=27° 25' 58"+90°−60° 14' 40"

=57° 11' 18" S.

112. Given the meridian altitude of a heavenly body below the north pole=32° 42' 10", and its declination=54° 42' N. Construct a figure, and find by calculation the latitude.   *Ans.*   Lat.=68° 0' 10" N.

113. Given the meridian altitude of a heavenly body below the north pole=5° 42' 15", and its declination=45° 23' 12" N. Construct a figure, and find by calculation the latitude.   *Ans.*   Lat.=50° 19' 3" N.

114. Given the meridian altitude of a heavenly body under the south pole=10° 14' 17", and its declination=70° 41' 15" S. Construct the figure, and find by calculation the latitude.   *Ans.* 29° 33' 2" S.

Latitude by meridian altitude above pole.

PROBLEM XIV.

Given the meridian altitude of a heavenly body above the pole, its declination, and the bearing of the zenith from the body ; to find the latitude.

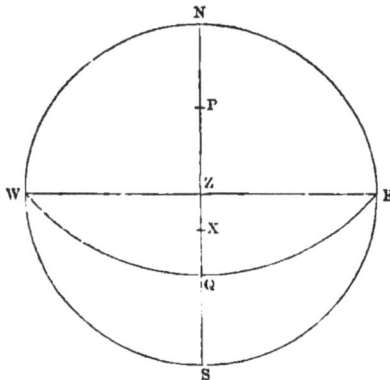

First. When the bearing of zenith from the body and declination have the same name.

Let N W S E represent the horizon, N Z S the celestial meridian, and W Z E the prime vertical. Let X be the place of a heavenly body on the meridian, zenith being north of the body, and X Q its declination north; through Q draw the great circle W Q E, then W Q E represents the celestial equator, and Z Q the latitude required.   Since X S the altitude is given, therefore X Z=90°−X S the zenith distance is known, and X Q its declination can also be found from the *Nautical Almanac.*

By the fig., Z Q=X Z+X Q, or latitude=zenith dist.+decl.

Second. When the bearing of zenith from the body and declination have different names.

Let N W S E represent the horizon, N Z S the celestial meridian, and W Z E the prime vertical. Let x be the place of a heavenly body on the meridian, zenith being north of the body, and X Q its declination south; through Q draw the great circle W Q E, then W Q E represents the celestial equator, and Z Q is the latitude required. Since X S the altitude is given, therefore $xz = 90° - xs$ the zenith distance is known, and its declination, X Q, is also given.

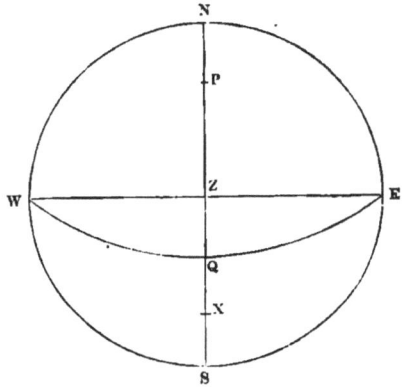

By the fig., $zq = xz - xq$, or latitude = zenith dist. − decl.

By constructing figures to suit different positions of the heavenly body with respect to the equator and zenith, we shall see that generally

latitude = mer. zen. dist. $\pm$ decl.

The sum, when bearing of zenith and the declination have the same name.

The difference, when bearing of zenith and the declination have different names.

115. Given the meridian altitude of a heavenly body = 56° 10′ 15″, zenith north of the body, and its declination = 15° 22′ 10″ N. Construct a figure, and find by calculation the latitude.

Let N W S E represent the horizon, N Z S the celestial meridian, and W Z E the prime vertical. Take S X = 56° 10′ 15″, and X Q = 15° 22′ 10″, and through Q draw the great circle W Q E; then W Q E is the celestial equator, and Z Q = latitude required.

By the figure (fig. 1, next page), $zq = xz + xq$
= merid. zen. dist. + decl.

Calculation.

90°		
56	10′	15″ = s x
33	49	45 = x z
15	22	10 = x q
∴ lat. = 49	11	55 N = z q

E

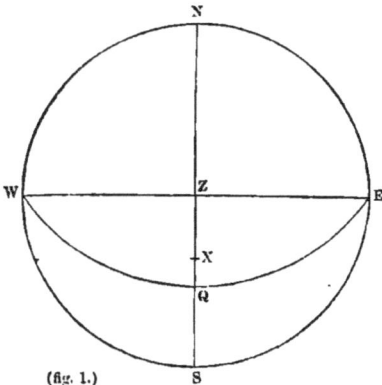
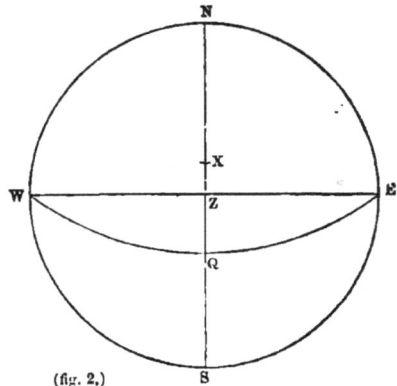

(fig. 1.)   (fig. 2.)

116. Given the meridian altitude of a heavenly body=72° 42′ 15″, zenith south of the body, and its declination=47° 32′ 14″ N. Construct a figure, and find by calculation the latitude (fig. 2).

Let NWSE represent the horizon, NZS the celestial meridian, and WZE the prime vertical. Take NX=72° 42′ 15″, and XQ=47° 32′ 14″, and through Q draw the great circle WQE; then WQE is the celestial equator, and ZQ= latitude required.

By the figure, ZQ=XQ− XZ,

or latitude=mer. zen. dist. ∼ decl.

Calculation.

$$90°$$
$$72 \quad 42′ \quad 15″$$

$$17 \quad 17 \quad 45 = \text{XZ}$$
$$47 \quad 32 \quad 14 = \text{XQ}$$

$$30 \quad 14 \quad 29 = \text{ZQ}$$

The latitude is evidently north by the figure.

EXAMPLES FOR PRACTICE.

117. Given the meridian altitude of a heavenly body=32° 42′ 15″, zenith north of the body, and its declination=10° 14′ 32″ N. Construct a figure, and find by calculation the latitude.   *Ans.* Lat.=67° 32′ 17″ N.

118. Given the meridian altitude of a heavenly body=72° 48′ 10″, zenith north of the body, and its declination=25° 36′ 42″ S. Construct a figure, and find by calculation the latitude.   *Ans.* Lat.=8° 24′ 52″ S.

119. Given the meridian altitude of a heavenly body=65° 35′ 10″, zenith south of the body, and its declination=6° 42′ 10″ S. Construct a figure, and find by calculation the latitude.   *Ans.* Lat.=31° 7′ 0″ S.

120. Given the meridian altitude of a heavenly body $=72°\ 44'\ 0''$, zenith south of the body, and its declination$=22°\ 42'\ 5''$ N. Construct a figure, and find by calculation the latitude. *Ans.* Lat. $=5°\ 26'\ 5''$ N.

Latitude by altitude near the meridian, ABOVE POLE.

FIRST METHOD, *using estimated latitude.*

PROBLEM XV.

Given the altitude of a heavenly body *near the meridian*, above the pole, its hour-angle and declination, and the estimated latitude of the observer; to find the correct latitude.

Let x be a heavenly body, P the pole, and z the zenith.

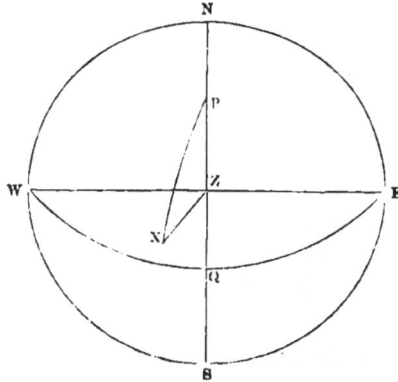

Let hour-angle $z\mathrm{P}x=h$,

latitude $z\mathrm{Q}=l$,

zenith dist. $zx=z$,

decl. of $x=d$.

Then polar distance $\mathrm{P}x=90°-d$, when the lat. and decl. have the same name, and polar distance $=90°+d$, when of different names.

In triangle $z\mathrm{P}x$, cos. $z\mathrm{P}x=\dfrac{\text{cos. } zx-\text{cos. } \mathrm{P}x\ .\ \text{cos. } \mathrm{P}z}{\text{sin. } \mathrm{P}x\ .\ \text{sin. } \mathrm{P}z}$

(*Trigonometry*, Part II. art. 55),

or cos. $h=\dfrac{\text{cos. } z-\text{sin. } l\ .\ \text{sin. } d}{\text{cos. } l\ .\ \text{cos. } d}$, when l and d have same name,

or cos. $h=\dfrac{\text{cos. } z+\text{sin. } l\ .\ \text{sin. } d}{\text{cos. } l\ .\ \text{cos. } d}$, when l and d have different names.

In the first case, cos. l . cos. d . cos. $h=$cos. $z-$sin. l . sin. d

$\therefore$ cos. $z-$sin. l . sin. $d=$cos. l . cos. d . cos. h

$=$cos. l . cos. d . $(1-$vers. $h)$

$=$cos. l . cos. $d-$cos. l . cos. d . vers. h

$\therefore$ cos. $z+$cos. l . cos. d . vers. $h=$cos. l . cos. $d+$sin. l . sin. d

$=$cos. $(l\sim d)=1-$vers. $(l\sim d)$

$\therefore$ vers. $(l\sim d)=1-$cos. $z-$cos. l . cos. d . vers. h

$=$vers. $z-$cos. l . cos. d . vers. h

In the second case, we shall find in a similar manner that

vers. $(l+d)=$vers. $z-$cos. l . cos. d . vers. h.

But $l\sim d$, or $l + d$, is the meridian zenith distance of the heavenly body at some place in the same latitude l as the spectator at the instant when its declination was d. Denote this *meridian* zenith distance by z. Then

vers. z=vers. z−cos. l . cos. d . vers. h.

The meridian zenith distance being thus found, corresponding to the declination d for the time of observation, the problem is thus reduced to finding the latitude from the meridian altitude of a heavenly body and its declination; the rest of the steps will therefore be the same as in Problem XIV.

To simplify the formula

vers. z=vers. z−cos. l . cos. d . vers. h, by adapting it to the table of versines.

Assume vers. θ=cos. l . cos. d . vers. h

=2 cos. l . cos. d . hav. h. In logarithms,

∴. log. vers. θ−6= ·301030 + log. cos. l + log. cos. d + log. hav. h−30

(*Trig.* Part I. art. 31),

∴. log. vers. θ=6·301030 + log. cos. l + log. cos. d + log. hav. h−30.

From which vers. θ may be found.

Then vers. z=vers. z− vers. θ.

From which the meridian zenith distance z is found; and thence by last problem the latitude, for

latitude=meridian zenith distance±decl.

In the above expression for finding the latitude, the latitude itself is involved. Now, as this quantity is only known approximately, it may be sometimes necessary to repeat one part of the operation, using the latitude last found, which will be nearer the truth than the given estimated latitude; and if the result thus determined differs much from the last one used, to make a further calculation until the latitude deduced does not differ from the one last found: it is seldom, however, necessary to repeat the operation more than once. This will be seen in the following example, where the approximation has been carried beyond what was requisite, in order to show the practical utility of the method; and that with an error of even one degree in the latitude a sufficiently correct result will be obtained without a second approximation.

121. Given the altitude of a heavenly body=27° 0′, zenith north of the body, its declination=11° 29′ 8″ S., and hour-angle*=0ʰ 43ᵐ 25′, the estimated latitude being 50° N. Construct a figure, and find by calculation the true latitude.

___
* The hour-angle of a star is easily deduced from the formulæ in Problem IX. p. 34, viz.

Mean time=star's hour-angle+star's RA−RA mean sun,
∴. hour-angle=mean time+RA mean sun−star's RA,

Let N W S E represent the horizon, P the pole, Z the zenith, and X the heavenly body near the meridian. Then we have given declination X R $=11°$ 29' 8" S.$=d$, zenith dist. Z X $=63°$ 0'$=z$, hour-angle Z P X $=$ 0^h 43^m 25'$=h$, and estimated latitude$=90°$ − P z $=50°=l$; to find the latitude correctly.

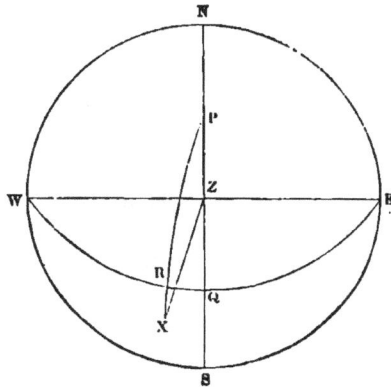

By Calculation.

vers. $\theta=2$ hav. h . cos. d . cos. l,
vers. mer. zen. dist.$=$vers. zen. dist.
$\qquad$ − vers. θ,

latitude$=$mer. zen. dist. − decl.

const. log.	....	6·301030
log. cos. d	....	9·991215
„ hav. h	....	7·951588

	2d approximation.	3d approximation.
	4·243833 .... 4·243833 .... 4·243833	
„ cos. l.................9·808067	.........9·800853	.........9·800737
„ vers. θ ....4·051900	4·044686	4·044570
∴ vers. $\theta=$ 11269	11084	11081
vers. $z=$ 546009	546009	546009
∴ vers. mer. zen. dist.$=$ 534740	534925	534928
∴ mer. zen. dist.$=62°$ 16' 23" N.	62° 17' 6" N.	62° 17' 6" N.
declination$=$11 29 8 S.	11 29 8 S.	11 29 8 S.
∴ latitude$=50$ 47 15 N.	50 47 58 N.	50 47 58 N.

As the last two results come out the same, we see that the approximation is carried far enough. When the latitude is known within a few minutes of the truth, a second approximation will seldom be necessary. Thus, in the above example, if we had supposed the estimated latitude to be 50° 28' N., which we know from the above result is 20' wrong, and had worked with this, we should have found our *first* result to be 50° 47' 40" N., and this is only 18" less than the truth ; showing in this case that an error of 20 minutes in the latitude used in the calculation will not produce any important error in the latitude deduced from it.

EXAMPLE FOR PRACTICE.

122. Given the altitude of a heavenly body$=25°$ 59' 9", zenith north of the body, its declination$=13°$ 12' 41" S., and hour-angle$=0^h$ 2^m 17', the

estimated latitude being 50° N.   Construct a figure, and find by calculation
the true latitude.   *Ans.*   1st approximation, lat.=50° 48′ 3″ N.
                                     2d   „   =50   48   3   N.

Latitude by altitude near meridian, ABOVE *pole.*

SECOND METHOD, *without using estimated latitude.*

A *direct* method of finding the latitude from an observation near the
meridian, in which the latitude by account is not required, may be obtained
from the following investigation :

PROBLEM XVI.

Given the altitude of a heavenly body off the meridian above the pole,
its hour-angle and declination ; to find the latitude.

Let N W S E represent the horizon, N Z S the celestial meridian, and X a
heavenly body near the meridian, P the pole, and Z the zenith.

Let hour-angle Z P X=h, polar distance P X=p,
zenith distance Z X=$90°-a$, a being the altitude.

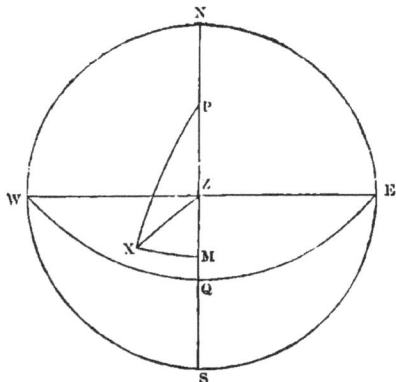

From X drop a perpendicular
X M upon the meridian.

Let P M=x, Z M=y, and X M=z.

Then the colatitude P Z=$x-y$,
when the perpendicular X M does
not fall between P and Z, as in the
figure ; and colatitude P Z=$x+y$,
when the perpendicular falls between P and Z : this will be seen by
constructing a figure. (The position of perpendicular X M with
respect to the points P and Z may
in most cases be easily determined
by the observer when the altitude
is taken.)

1. To find P M or x.   In right-angled triangle P M X,
   cos. h=cot. p . tan x, ∴ tan. x=tan. p . cos. h . . . . (1)

2. To find Z M or y.   In right-angled triangle X P M,
$$\cos. p=\cos. z . \cos. x,$$
Z X M, cos. Y Z or sin. a=cos. y . cos. z ;

dividing, to eliminate cos. z, we have $\dfrac{\cos. p}{\sin. a}=\dfrac{\cos. x}{\cos. y}$,

∴ cos. y . cos. p=cos. x . sin. a,

or cos. y=sec. p . cos. x . sin. a . . (2)

Formulæ (1) and (2) will determine x and y ; and thence the colatitude $=x\pm y$ is known.

In finding the latitude by formulæ (1) and (2), it will be necessary to attend to the magnitude of the polar distance ; for tan. x will be positive or negative according as the polar distance is less or greater than 90° ; but the value of tan. x, and also that of cos. y, are readily found in the usual manner by putting the proper signs, + or —, over the given quantities, as directed in the Rule, Art. 31, *Trigonometry*, Part I.*

123. Given the altitude of a heavenly body $=27°$ 0', zenith north of the body, its decl. $=11°$ 29' 8" S., and hour-angle $=0^h$ 43^m 25'. Construct a figure, and find by calculation the latitude.

Let N W S E represent the horizon, N Z S the celestial meridian, and X O the altitude of the heavenly body $=27°$ 0'. Take X P $=101°$ 29' 8", the polar distance; then P is the north pole. Let P Q $=90°$, and through Q draw the celestial equator W Q E ; then P Z $=$ colatitude to be determined. Let fall the perpendicular X M upon the meridian.

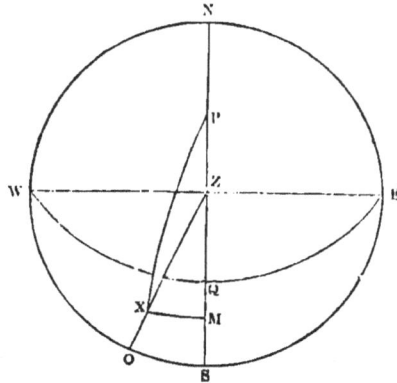

By Formulæ (1) and (2).

$$\overset{-}{\text{tan.}} \; x = \overset{+}{\text{cos.}} \; h \, . \; \overset{-}{\text{tan.}} \; p, \qquad \overset{+}{\text{cos.}} \; y = \overset{-}{\text{sec.}} \; p \; . \; \overset{-}{\text{cos.}} \; x \; . \; \overset{+}{\text{sin.}} \; a,$$

where $x=$ PM, $y=$ ZM, and $\therefore$ colat. P Z $=x-y$.

To find x.		To find y.	
log. cos. h	9·992160		
„ tan. p	0·692098	log. sec. p	0·700883
$\therefore$ „ tan. x	10·684258	„ cos. x	9·306634
Arc	78° 18' 40"	„ sin. a	9·657047
	180	„ cos. y	9·664564
$\therefore x=$	101 41 20	$\therefore y=62°$ 29' 20"	

$$
\begin{array}{rrrr}
x= & 101° & 41' & 20'' \\
y= & 62 & 29 & 20 \\
\hline
x-y= & 39 & 12 & 0 \\
& 90 & & \\
\hline
\therefore \text{lat.} = & 50 & 48 & 0 \text{ N.}
\end{array}
$$

* In *Nav.* Part I., p. 138, a NEW RULE is given deduced from formulæ (1) and (2), and which is free from any distinction of cases.

EXAMPLES FOR PRACTICE.

124. Given the altitude of a heavenly body $=50°$ $52'$ $29''$, zenith north of the body, its decl. $=11°$ $44'$ $58''$ N., and hour-angle $=0^h$ 12^m 4^s. Construct a figure, and find by calculation the latitude.   *Ans.* 50° 47′ 45″ N.

125. Given the altitude of a heavenly body $=46°$ $29'$ $15''$, zenith north of the body, its decl. $=7°$ $51'$ $53''$ N., and hour-angle $=0^h$ 33^m 51^s. Construct a figure, and find by calculation the latitude.   *Ans.* 50° 48′ 30″ N.

126. Given the altitude of a heavenly body $=45°$ $42'$ $30''$, zenith south of the body, its declination $=7°$ $51'$ $0''$ S., and hour-angle $=0^h$ 22^m 14^s. Construct a figure, and find by calculation the latitude.   *Ans.* 51° 55′ 0″ S.

Since ship mean time, from which the hour-angle has to be deduced (p. 52, note), is generally only approximately known at sea, let us now inquire under what circumstances an error in the ship time will produce the least error in the latitude determined from it. The investigation will be similar to the one in p. 44; and we shall be able to show that an error in the ship time will produce a less error in the latitude the smaller the bearing of the body is from the meridian. It is for this reason that we have assumed in the preceding problems that the altitude for determining the latitude is taken *near* noon. It is evident, however, that when the time is known very nearly, that is, within a few seconds of the truth, we need not be restricted to taking the altitudes near the meridian; but the above rule may be extended to observations taken when the bearing of the heavenly body is 3 or 4 or even more points from the meridian.

PROBLEM XVII.

Given a small error in the hour-angle of a heavenly body or time at the ship; to find the corresponding error in the latitude deduced from it (fig. next page).

$$\text{Let } \text{s} \, \text{p} \, \text{z} = h, \text{ the true hour-angle,}$$
$$\text{s}_1 \text{p} \, \text{z} = h_1, \text{ the erroneous hour-angle,}$$
$$\text{s} \, \text{z} = z, \text{ the true zenith distance,}$$
$$\text{s}_1 \text{z} = z_1, \text{ the erroneous zenith distance,}$$
$$\text{p} \text{s} = \text{p} \, \text{s}_1 = p, \text{ the polar distance,}$$
$$\text{p} \, \text{z} = c, \text{ the colatitude.}$$

In triangle s p z, $\cos. \ h = \dfrac{\cos. \ z - \cos. \ c \ . \ \cos. \ p}{\sin. \ c \ . \ \sin. \ p}$,

,,   s$_1$p z, $\cos. \ h_1 = \dfrac{\cos. \ z_1 - \cos. \ c \ . \ \cos. \ p}{\sin. \ c \ . \ \sin. \ p}$,

$\therefore \cos. \ h - \cos. \ h_1 = \dfrac{\cos. \ z - \cos. \ z_1}{\sin. \ c \ . \ \sin. \ p}$,

or $\sin. \ \tfrac{1}{2} \ (h + h_1) \ . \ \sin. \ \tfrac{1}{2} \ (h - h_1) = \dfrac{\sin. \ \tfrac{1}{2} \ (z + z_1) \ . \ \sin. \ \tfrac{1}{2} \ (z - z_1)}{\sin. \ c \ . \ \sin. \ p}.$

But since $h=h_1$ nearly, and $z=z_1$ nearly, we may assume $h=\frac{1}{2}(h+h_1)$, $z=\frac{1}{2}(z+z_1)$, $\frac{1}{2}(h-h_1)=\sin.\ \frac{1}{2}(h-h_1)$, and $\frac{1}{2}(z-z_1)=\sin.\ \frac{1}{2}(z-z_1)$.

Making these substitutions and cancelling,

$$\therefore\ \sin.\ h\ .\ (h-h_1)=\frac{\sin.\ z}{\sin.\ c\ .\ \sin.\ p}\ .\ (z-z_1).$$

Let the azimuth P Z S be denoted by A; then, in triangle Z P S, $\dfrac{\sin.\ h}{\sin.\ A}=\dfrac{\sin.\ z}{\sin.\ p}$; $\therefore\ \sin.\ h=\dfrac{\sin.\ A\ .\ \sin.\ z}{\sin.\ p}.$

Substituting this value of $\sin. h$, we have

$$\frac{\sin.\ A\ .\ \sin.\ z}{\sin.\ p}\ .\ (h-h_1)=\frac{\sin.\ z}{\sin.\ c\ .\ \sin.\ p}\ .\ (z-z_1);$$

$$\therefore\ z-z_1=\sin.\ c\ .\ \sin.\ A\ .\ (h-h_1),$$

or error in the zenith distance=cos. lat. . sin. azimuth . error in hour-angle.

But if we assume the heavenly body to be near the meridian, the zenith distance observed is nearly equal to the meridian zenith distance (=lat.$\pm$ decl.). Therefore an error in the observed zenith distance will produce the same error (nearly) in the latitude, or $z-z_1=$error in latitude nearly.

Hence error in lat.=cos. lat. sin. azimuth . error in hour-angle nearly.

From this formula it appears that the error in the latitude produced by a small error in the time or hour-angle will increase as the azimuth or bearing of the heavenly body increases: hence the rule always directs the observation to be taken as near' to the meridian as possible when the time at the ship is uncertain within a minute or two. The necessity for this restriction will be clearly shown by means of an example.

127. Find the error in the latitude corresponding to an error of 3 minutes (=45' in arc) in the hour-angle at a place in lat. 50° 48' N.

First. When the bearing of body is 1 point.

Second. When the bearing of body is 1°.

Error=cos. l . sin. bearing $(h-h_1)$.

1.	2.
log. cos. l..............9·800737	log. cos. l..............9·800737
„ sin. 11° 15'......9·290236	„ sin. 1° ....8·241855
„ 45 ....1·653213	„ 45 ....1·653213
„ error..............0·744186	„ error...............1·695805
$\therefore$ error=5·5'=5' 30''	$\therefore$ error=0·5'=30''

We thus see that in this case the error in the latitude is eleven times greater in one observation than in the other for the same error in the ship time. But if the time is known within a second or two of the truth, the observation may be taken at a considerable distance from the meridian without producing any error in the latitude on that account. The following altitudes,

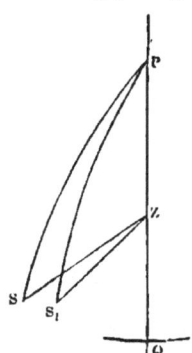

taken at Portsmouth (lat. 50° 48′ 3″ N.), will show this, the time of the observations being known to be correct within a few seconds.

EXAMPLES FOR PRACTICE.

128. Given the altitude of a heavenly body=45° 31′ 26″, zenith north of the body, its decl.=8° 14′ 1″ N., and hour-angle=1ʰ 2ᵐ 8ˢ. Construct a figure, and find by calculation the latitude. *Ans.* Lat.=50° 48′ 15″ N.

129. Given the altitude of a heavenly body=30° 23′ 29″, zenith north of the body, its decl.=2° 27′ 35″ S., and hour-angle=2ʰ 5ᵐ 28ˢ. Construct a figure, and find by calculation the latitude. *Ans.* Lat.=50° 46′ 59″ N.

130. Given the altitude of a heavenly body=35° 4′ 7″, zenith north of the body, its decl.=10° 54′ 26″ N., and hour-angle=3ʰ 5ᵐ 36ˢ. Construct a figure, and find by calculation the latitude. *Ans.* Lat.=50° 48′ 30″ N.

Latitude by altitude near the meridian, BELOW pole.

PROBLEM XVIII.

Given the altitude of a heavenly body near the meridian *below the pole*, its hour-angle and declination, and the estimated latitude of the observer; to find the true latitude.

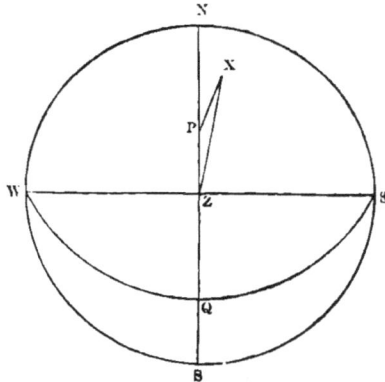

Let N W S E represent the horizon, N Z S the celestial meridian, P the pole, Z the zenith, and X a heavenly body near the meridian below the pole.

Let hour-angle ZPX=*h*,   zenith distance XZ=*z*,
     latitude PN=*l*,   declination of X=*d*.

In the spherical triangle ZPX, cos. $h=\dfrac{\cos. z - \sin. l . \sin. d}{\cos. l . \cos. d}$,

∴ cos. *d* . cos. *l* . cos. *h*=cos. *z* − sin. *l* . sin. *d*,

or—cos. *d* . cos. *l* . cos. (12ʰ−*h*)=cos. *z* − sin. *l* . sin. *d*,

$\therefore$ cos. $z-$sin. l . sin. $d=-$cos. d . cos. l . $\{1-$vers. $(12\sim h)\}$

$\qquad\qquad=-$cos. d . cos. $l+$cos. d . cos. l . vers. $(12\sim h)$,

$\therefore$ cos. $z+$cos. d . cos. $l-$sin. d . sin. $l=$cos. d . cos. l . vers. $(12-h)$,

or cos. $z+$cos. $(d+l)=$cos. d . cos. l . vers. $(12-h)$,

or$-$cos. $(180°-z)+$cos. $(d+l)=$cos. d . cos. l . vers. $(12-h)$,

$\qquad\therefore$ cos. $(d+l)=$cos. $(180°-z)+$cos. d . cos. l . vers. $(12\sim h)$ .. (1)

But $l+d=90°-$colat.$+90°-$pol. dist.$=180°-($colat.$+$pol. dist.$)$

$\qquad\qquad=180°-$a mer. zen. dist.$=180°-(90°-$a mer. alt.$)$

$\qquad\qquad=90°+$a mer. alt.

and $180°-z=180°-(90°-$alt.$)=90°+$alt.

Making these substitutions in (1), we have

cos. $(90°+$mer. alt.$)=$cos. $(90°+$alt.$)+$cos. d . cos. l . vers. $(12\sim h)$.

or vers. $(90°+$mer. alt.$)=$vers. $(90°+$alt.$)-$cos. d . cos. l . vers. $(12\sim h)$.

From which formula the meridian altitude, increased by 90°, of the heavenly body at some place in latitude l at the instant when its declination was d, may be found; and the observation is thus reduced to that of finding the latitude by a *meridian* altitude under the pole.

To adapt to logarithms the formula

vers. $(90°+$mer. alt.$)=$vers. $(90°+$alt.$)-$cos. d . cos. l . vers. $(12\sim h)$,

Let vers. $\theta =$ cos. l . cos. d . vers. $(12\sim h)$,

$\therefore$ vers. $\theta = 2$ cos. l . cos. d . hav. $(12\sim h)$,

or log. vers. $\theta = 6\cdot301030+$log. cos. $l+$log. cos. $d+$log. hav. $(12\sim h)-30$,

from which expression vers. θ may be found.

Then vers. $(90°+$mer. alt.$)=$vers. $(90°+$alt.$)-$vers. θ, which determines the value of $90°+$mer. alt.

From this we find the latitude as in the rule for finding the latitude under the pole, viz. $\qquad$ Lat.$=90°+$mer. alt.$-$decl. . . (p. 47).

131. Given the altitude of α Ursæ Majoris near the meridian below the pole $= 26°$ 10′ 0″ at 9^h 2^m P.M. mean time at the place, the estimated latitude being 53° 0′ N.; also the following data from the *Nautical Almanac:* star's right ascension $= 10^h$ 54^m 55^s, star's declination $= 62°$ 30′ 40″ N., right ascension of mean sun $= 14^h$ 54^m 21^s. Construct a figure, and find by calculation the true latitude.

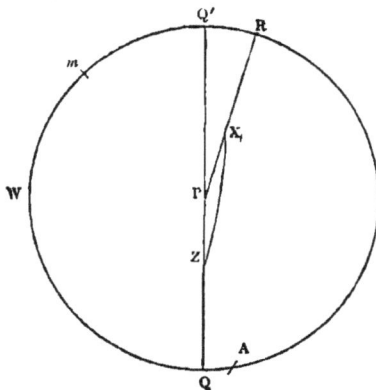

Construction.

Let Q W Q′ represent the celestial equator, Q r Q′ the celestial meridian, and P the pole. Take Qm = 9ʰ 2ᵐ; then m is the place of the mean sun. From m measure mQ′A = 14ʰ 54ᵐ 21ˢ; then A is the place of the first point of Aries. Again, from A take AR = 10ʰ 54ᵐ 55ˢ; and draw P R, a circle of declination passing through the place of the heavenly body. Let R X = 62° 30′ 40″ = star's declination, and Q z = 53° 0′ = estimated latitude; and through z draw z X, a circle of altitude: then in the spherical triangle zPX are given PX the polar distance, zX the zenith distance, and zPX the hour-angle, with zQ the estimated latitude; to find the correct latitude.

By Formulæ.

Hour-angle = mean time + R A mean sun − star's R A (p. 52, note)

vers. 0 = cos. l . cos. d . hav. $(12 \sim h)$ ....(1)

vers. (90° + mer. alt.) = vers. (90° + alt.) − vers. θ ....(2)

and latitude = 90° + alt. − decl. ....(3)

To find $12 \sim h$.

Mean time.........	9ʰ	2ⁿ	0ˢ	6·301030
R A mean sun......	14	54	21	cos. l ....9·779463
R A meridian	23	56	21	cos. d ....9·664243
Star's R A	10	54	55	hav. $(12 \sim h)$ ....8·251781
Hour-angle	13	1	26	3·996517
∴ $12 \sim h$ =	1	1	26	

vers. θ................. 9920

90° + alt. = 116° 10′ vers... 1440984

vers. (90° + mer. alt.)... 1431064

∴ 90° + mer. alt. = 115° 32′ 6″

decl. = 62 30 40

∴ latitude = 53   1 26 N.

Latitude by altitude of POLE STAR.

PROBLEM XIX.

Given the altitude of the pole star off the meridian, and the tabular correction; to find the latitude.

The bearing of α Ursæ Minoris (the pole star), which lies within 2° of the north pole, is very small in any latitude below 70° N., so that an error of a few minutes in the hour-angle is of little consequence (p. 57) when the latitude is to be determined approximately.

The rules deduced from the preceding problems for finding the latitude by an altitude near the meridian, should be applied when the latitude is required to a greater degree of accuracy than is necessary for the common purposes of navigation; but the latitude can be obtained sufficiently near for practical purposes, and with little calculation, by applying to the altitude of the pole star a *correction*, which has been formed into a table, and called the " correction of the pole star." (See Inman's Tables.)

This correction may be calculated in the following manner.

Investigation of the tabular " correction of the pole star."

Let P be the pole, x the place of the pole star, z the zenith of a place, and z' the zenith of another place a few degrees from the former. The difference of the zenith distance of the star and the colatitude will be nearly the same at both places; that is, $PZ - ZX = PZ' - Z'X$ nearly, since the polar distance PX is small. Call this difference d.

Then $d = PZ - ZX = \text{colat.} - \text{zen. dist.}$
$= 90° - \text{lat.} - 90° + \text{alt.} = \text{alt.} - \text{lat.}$
∴ latitude $= \text{alt.} - d$.

The correction d is subtractive in the present case : but it is seen by the figure that d is additive when the position of x is such that the zenith distance is greater than the colatitude, therefore, generally,

Latitude $= d \pm \text{altitude of pole star} \pm \text{tab. corr. } d$.

If, therefore, we calculate the value of d for certain latitudes, as for 20°, 30°, &c., and for certain times, as for every 10 minutes of the right ascension of the meridian,—a quantity that depends for its value on ship mean time (since right ascension of meridian = ship mean time + right ascension of mean sun [p. 24]),—and tabulate the results, then the latitude of a place near any of the assumed latitudes recorded in the table is readily found by simply applying the correction d, with its proper sign, to the altitude of the heavenly body.

The value of d for any assumed latitude and time may be computed as follows :

Let QAQ' represent the celestial equator, QQ' the celestial meridian, P the pole, and z the zenith of a place in a given latitude ZQ. Let Qm = given mean time, and suppose x the place of the star at that instant, and A the first point of Aries. Through x draw the circle of declination PR, and circle of altitude ZX. Then $d = PZ - ZX$, and ZX is unknown. We have therefore to compute ZX for the assumed latitude ZQ and time Qm.

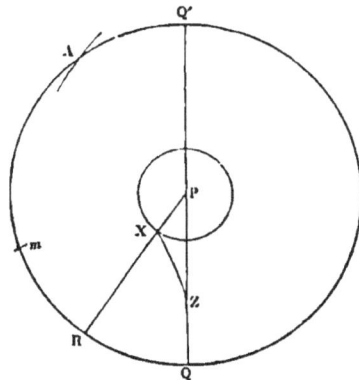

In the spherical triangle ZPX, are given PX, the star's polar distance, PZ the colatitude, and the hour-angle ZPX (since hour-angle = mean time + RA

mean sun—star's RA [note, p. 52]); to find z x (*Spherical Trigonometry,* Rule IX.) : then d is known, since $d = $ P z $-$ z x.

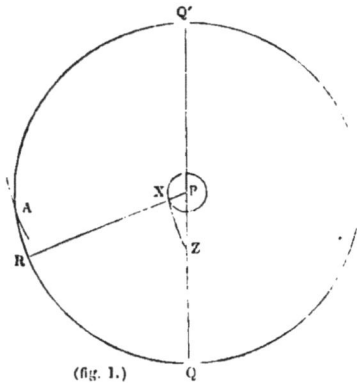

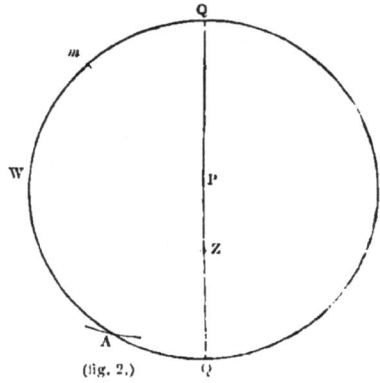

(fig. 1.)   (fig. 2.)

Given the right ascension of the meridian $= 6^h\ 10^m$; to compute the correction, the assumed latitude being 50° N., the star's RA $1^h\ 7^m\ 7^s$, and declination 88° 32′ 50″...(fig. 1).

Star's hour-angle $=$ RA mer. $-$ star's RA.

RA mer.   Q A $= 6^h\ 10^m\ 0^s$......const. log.   ....6·301030

Star's RA,   A R $= 1\quad 7\quad 7$......log. sin. colat. 9 808067

∴ star's hour-angle z P X $= 5\quad 2\quad 23$,, sin. pol. dist....8·404360

P z $= 40°\ 0′\ 0″$,, hav. H A........9·575969

P X $= 1\quad 27\quad 10$ 4·089426

38   32   50 vers.   12287

217904

230191 vers.

∴ z x $= 39°\ 39′\ 49″$

and P z $= 40\quad 0\quad 0$

∴ $d = \quad\quad 20\quad 11$

∴ the correction for the pole star, when the right ascension of meridian is $6^h\ 10^m$, is 20′ 11″ subtractive.

In the same manner may be found the corrections for right ascension of meridian 10^m, 20^m, up to $24^h\ 0^m$, and for different assumed latitudes. These results constitute the table in Inman's nautical tables.

132. Dec. 4, 1857, at $9^h\ 40^m$ P.M., mean time nearly, the true altitude of the pole star was 54° 38′ 25″. Construct a figure, and find by pole-star table the latitude ; by *Nautical Almanac* corrected RA mean sun being $16^h\ 55^m$.

Construction (fig. 2).

Let Q W Q′ represent the celestial equator, QQ′ the celestial meridian, and p the pole. Take $Qm = 9^h\ 40^m$; then m is the mean sun. From m measure $mQ'_A = 16^h\ 55^m$; then A is the first point of Aries, and AQ = right ascension of meridian, the argument of the table; or R A mer. may be found by formula (p. 24), thus,

R A mer. = mean time + R A mean sun.

R A mean sun	16^h	55^m		Tr. alt.	$54°\ 38'$	$25''$
Mean time	9	40		Cor.	1 20	36—
R A mer.	2	35		∴ Latitude = 53	17	49 N.

Entering the table with $2^h\ 35^m$ at side and $50°$ at top, the required correction is $1°\ 20'\ 36''$—.

EXAMPLES.

133. Feb. 10, 1857, at $11^h\ 40^m$ P.M. mean time, given the altitude of the pole star = $60°\ 3'\ 19''$. Construct the figure, and find by table the latitude. By *Nautical Almanac* corrected R A mean sun = $21^h\ 24^m\ 34'$.

Ans. $60°\ 51'$ N.

134. January 20, 1857, at $9^h\ 40^m$ P.M. mean time, given the altitude of the pole star = $31°\ 33'\ 19''$. Construct the figure, and find by table the latitude. By *Nautical Almanac* corrected R A mean sun = $20^h\ 0^m\ 29'$.

Ans. $31°\ 1'$ N.

135. October 4, 1857, at $11^h\ 30^m$ P.M. mean time, given the altitude of the pole star = $62°\ 2'\ 15''$. Construct a figure, and find by table the latitude. By *Nautical Almanac* corrected R A mean sun = $12^h\ 53^m\ 49'$.

Ans. $60°\ 32'$ N.

136. May 10, 1857, at $11^h\ 20^m$ P.M. mean time, given the altitude of the pole star = $52°\ 17'\ 54''$. Construct a figure, and find by table the latitude. By *Nautical Almanac* corrected R A mean sun = $3^h\ 14^m\ 18'$.

Ans. $53°\ 39'\ 0''$ N.

Problems XVI. and XVIII. applied to finding the latitude by an altitude of the pole star.

The last problem enables us to find a near approximate latitude from an altitude of the pole star, the reduction to the meridian being made by means of a table; but it is evident that if we consider the pole star as an ordinary star, we may make use of Problems XV. or XVI. to find the latitude from its altitude near the meridian above the pole, and Problem XVIII. to find the latitude from its altitude near the meridian below the pole; by which problems the latitude can be calculated from the altitude of the pole star without the aid of a table, and with a greater degree of accuracy.

In the two examples following, let the latitude be found from the altitude of the pole star by means of Problems XV. or XVI. and XVIII.

137. Given the altitude of pole star near the meridian above the pole = $54°\ 38'\ 25''$, zenith south of the body, at $9^h\ 40^m$ P.M. mean time at the place,

in estimated latitude 53° 18′ N. Construct a figure, and find by Problem XV. or XVI. the correct latitude.

Data from *Nautical Almanac:* RA mean sun=16ʰ 55ᵐ 7ʹ, star's RA= 1ʰ 7ᵐ 7ʹ, star's decl.=88° 32′ 50″ N.   *Ans.* Latitude=53° 17′ 32″ N.

138. Given the altitude of pole star near the meridian under the pole= 38° 17′ 11″, at 10ʰ 20ᵐ P.M. mean time at the ship, in latitude by account 39° 27′ N. Construct a figure, and find by Problem XVIII. the correct latitude.

Data from *Nautical Almanac :* RA mean sun=5ʰ 16ᵐ 8ʹ, star's RA= 1ʰ 7ᵐ 7ʹ, star's decl.=88° 32′ 50″ N.   *Ans.* Latitude=39° 26′ 53″ N.

Latitude by altitudes taken within a few minutes of each other.

PROBLEM XX.

Given the altitudes of a heavenly body, taken within a few minutes of each other; to find the latitude.

When the altitudes can be taken *very accurately*, the following method

of determining the latitude appears to give results sufficiently near for practical purposes; and as the observations are taken within a short time of each other, it possesses on this account an advantage over the common rule for double altitude, which requires a considerable interval to elapse between the two observations.

Let the change in the altitudes, and the corresponding change in the time between the observations, be both very accurately noted. Then, from these contemporary increments of time and altitude, the angle of position P X Z may be calculated very nearly by means of the formula about to be investigated. We shall then have given in the triangle P X Z

the polar distance P X, the zenith distance Z X (calculated for the mean or middle time between the two observations), and the included angle P X Z; to find the colatitude P Z, and this is readily done by the common rule in Trigonometry (*Trig.*, Part I. Rule IX.).

Formula for computing the angle of position PXZ from the contemporary increments of time and altitude.

Let s and s_1 (fig. p. 65) be the places of the heavenly body when their altitudes were taken.

h=hour-angle S P Z, z=zenith distance Z S,

h_1=hour-angle s_1 P Z, z_1=zenith distance Z s_1,

l=latitude and d=declination for the middle time between the observations.

Then $h - h_1$=difference of observed times,

$z - z_1$=corresponding difference of altitudes.

In triangle P S Z, $\cos. h = \dfrac{\cos. z - \sin. d \,.\, \sin. l}{\cos. d \,.\, \cos. l}$,

„　　$\text{P S}_1\text{Z}$, $\cos. h_1 = \dfrac{\cos. z_1 - \sin. d \,.\, \sin. l}{\cos. d \,.\, \cos. l}$,

$\therefore \cos. h - \cos. h_1 = \dfrac{\cos. z - \cos. z_1}{\cos. d \,.\, \cos. l}$;

or $2 \sin. \tfrac{1}{2} (h + h_1) \,.\, \sin. \tfrac{1}{2} (h - h_1)$

$$= \frac{2 \sin. \tfrac{1}{2} (z + z_1) \,.\, \sin. \tfrac{1}{2} (z - z_1)}{\cos. d \,.\, \cos. l}$$

or $\sin. h \,.\, (h - h_1) = \dfrac{\sin. z}{\cos. d \,.\, \cos. l} \,.\, (z - z_1)$

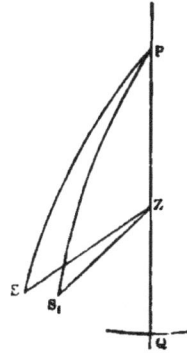

(by substituting h for $\tfrac{1}{2} (h + h_1)$, z for $\tfrac{1}{2} (z + z_1)$, and $\tfrac{1}{2} (h - h_1)$ and $\tfrac{1}{2} (z - z_1)$ for their sines, and cancelling) : in which expression z is the zenith distance and d the declination corresponding to the middle time h between the observations (fig. p. 64).

$\therefore \dfrac{\sin. h \,.\, \cos. l}{\sin. z} = \dfrac{z - z_1}{h - h_1} \,.\, \dfrac{1}{\cos. d}$　　But $\dfrac{\sin. \text{P X Z}}{\sin. h} = \dfrac{\cos. l}{\sin. z}$

$\therefore \sin. \text{P X Z} = \dfrac{\sin. h \,.\, \cos. l}{\sin. z}$; whence $\sin. \text{P X Z} = \dfrac{z - z_1}{h - h_1} \,.\, \dfrac{1}{\cos. d}$ in time,

or $\sin. \text{P X Z} = \dfrac{1}{15} \,.\, \dfrac{z - z_1}{h - h_1} \,.\, \dfrac{1}{\sin. \text{pol. dist.}}$ in arc,

$\therefore$ cosec. angle of position $\text{P X Z} = \dfrac{15 (h - h_1)}{z - z_1} \,.\, \sin. \text{pol. dist.}$

From this expression the angle P X Z may be computed, if the contemporary quantities $h - h_1$ and $z - z_1$ have been correctly noted. Knowing, then, the polar distance, zenith distance, and included angle X, in the triangle P X Z, we can compute the colatitude P Z by the common rule in *Spherical Trigonometry* (Part I. Rule IX.) for finding the third side, having given two sides and the included angle.

139. The following observations were taken at the Royal Naval College (lat. 50° 48′ 3″ N.). At $1^\text{h}\ 42^\text{m}\ 53 \cdot 7^\text{s}$ the true zenith distance of the sun's center was found to be 45° 7′ 37″, at $1^\text{h}\ 52^\text{m}\ 8 \cdot 6^\text{s}$ it was 45° 55′ 4″, the polar distance at the middle time was 80° 35′ 19″ ; to find the latitude.

Formula adapted for calculation.

cosec. $\text{P X Z} = \dfrac{15 (h - h_1)}{z - z_1} \,.\, \sin. p$,

vers. colat. = vers. $(p - z) + 2 \sin. p \,.\, \sin. z \, \text{hav. P X Z}$

$= $ vers. $(p - z) + $ vers. H.

Assuming vers. $\text{H} = 2 \sin. p \,.\, \sin. z \,.\, \text{hav. P X Z}$

(see *Trig.* Part II., the proof of Rule IX.),

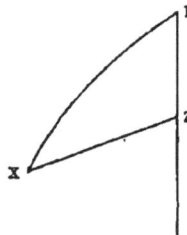

F

where $p=$pol. dist. P X, $z=$zen. dist. Z X, at the middle time between the two observations.

h ......1^h	52^m	8·6'	z ......45°	55'	4"	......z ...45°	55'	4"		
h_1......1	42	53·7	z_1......45	7	37	......z_1...45	7	37		
	9	14·9		47	27		91	2	41	
	60			60		$z=$45	31	20		
$h-h_1=$	554·9		$z-z_1=$	2847		$p=$80	35	19		
						$p-z=$35	3	59		

To find angle of position P X Z.

log. 15 ....1·176091		const. log. ....6·301030	
„ $h-h_1$.........2·744215		log. sin. p ....9·994113	
„ sin. p.........9·994113		„ sin. z ....9·853397	
13·914419		„ hav. P X Z ......8·491433	
„ $z-z_1$ ....3·454387		„ vers. H ....4·639973	
„ cosec. P X Z . 10·460032		∴ vers. H= 43649	
∴ P X Z=20° 17'		vers. ($p\sim z$) = 181516	
		vers. colat. 225165	
		∴ colat.=39° 12'	
		and latitude=50 48 N.	

EXAMPLES FOR PRACTICE.

140. August 13, 1858, A.M., the following observations were taken at the Royal Naval College (lat 50° 48' N.).

At 9^h 58^m 45·6' A.M., the true zenith distance of the sun's center was 44° 35' 59", at 10^h 9^m 44·0' it was 43° 20' 57"; required the latitude (corrected polar distance being 75° 16' 15").   *Ans.* 50° 50' N.

141. August 13, 1858, A.M., at 10^h 2^m 19·9' A.M., the true zenith distance of the sun's center was 44° 10' 56", at 10^h 17^m 26·2' it was 42° 30' 55"; required the latitude (corrected polar distance being 75° 16' 24").

     Ans. 50° 51' N.

142. August 13, 1858, A.M., at 10^h 21^m 37·5' A.M., the true zenith distance of the sun's center was 42° 3' 24", at 10^h 31^m 41·0' it was 41° 3' 23"; required the latitude (corrected polar distance being 75° 16' 33").   *Ans.* 50° 48' N.

From the great difficulty of observing at sea the altitude of a heavenly body correctly, the above method of determining the latitude is probably of little practical use. It can, however, be adopted with considerable advantage *on shore*, where the altitudes may be taken with great accuracy by means of an artificial horizon.

THE DOUBLE ALTITUDE.

The important rule for finding the latitude from two altitudes of the same or two heavenly bodies, observed at the same or at different times, may be more clearly explained if we deal with each case separately.

PROBLEM XXI.

Given the altitudes of the sun, or any other heavenly body, observed at two different times, and the elapsed time between the two observations; to find the latitude.

Let N W S E represent the horizon, N Z S the celestial meridian, P the pole, and Z the zenith of the spectator. Let X and Y be the places of the heavenly body when its altitudes were observed. Draw the circles of altitude Z O and Z O', and circles of declination P X and P Y, through X and Y. Draw also X Y, the arc of a great circle.

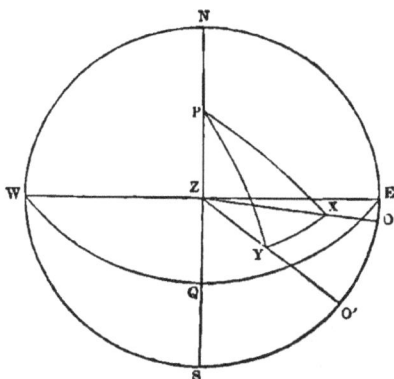

First. Suppose the arc X Y, drawn through the two places of the heavenly body, not to pass, when produced, between the zenith and pole, as in the figure.

In the two spherical triangles P X Y and Z X Y, there are given the two polar distances P X and P Y (from the *Nautical Almanac*), the angle X P Y (the interval between the observations, which will be given sufficiently near if noted by a chronometer whose rate is known), and the two zenith distances Z X and Z Y (from the observed altitudes); to find the colatitude P Z, and thence the latitude.

(1.) In the spherical triangle P X Y, are given the polar distances P X and P Y, and included angle X P Y; to calculate X Y. Call this arc 1.

(2.) In the spherical triangle P X Y, are given the polar distances P X and P Y, and arc 1, just found; to calculate the angle P X Y, the angle at the place whose bearing is the greater. Call this angle P X Y arc 2.

(3.) In the spherical triangle Z X Y, are given the zenith distances Z X and Z Y, and arc 1; to find the angle Z X Y, the angle at the greater bearing. Call this angle Z X Y arc 3.

(4.) Subtract arc 3 from arc 2; the remainder is angle P X Z, or arc 4.

(5.) *Lastly.* In the triangle P X Z, are given P X and Z X, the polar dist. and zenith dist. at the greater bearing, and the included angle P X Z, or arc 4, to calculate P Z, the colatitude; and thence Z Q, the latitude, is known.

This rule for finding the latitude, although long, is direct, and free from any ambiguity of cases, and has also the great advantage of being a simple application of two of the common rules of Spherical Trigonometry.

143. In latitude by account 50° N., at 8ʰ 58ᵐ 28ˢ A.M., the sun's altitude was 25° 3' 47", and declination 2° 58' 35" S.; and at 11ʰ 11ᵐ 43ˢ A.M. the sun's altitude was 35° 39' 28", and declination 2° 56' 30" S. Construct a figure, and find by calculation the true latitude.

Construction.

Let N W S E represent the horizon, N Z S the celestial meridian; then, the latitude being about 50° N., take N P=50°, and P will be the north pole of the heavens: take P Q= 90°, and through W, Q, E, draw the celestial equator W Q E. Since the sun has south declination, and is on the east of meridian, let x be the place of the sun at the first observation, and Y its place at the second; draw the circles of declination P X and P Y, and circles of altitude Z X and Z Y, and through X and Y draw the great circle X Y: then the colati-tude P Z may be found as follows.

(1.) In the spherical triangle P X Y, are given P X=90°+decl.=92° 58' 35", P Y=90°+decl.=92° 56' 30", and included angle X P Y=2ʰ 13ᵐ 15ˢ; to find X Y=33° 16' 0".

(2.) In the spherical triangle P X Y, are given P X=92° 58' 35", P Y=92° 56' 30", and arc X Y=33° 16' 0"; to find angle P X Y=90° 49' 45".

(3.) In the spherical triangle Z X Y, are given Z X=90°−alt.=64° 56' 13", X Y=33° 16' 0", and Z Y=90°−alt.=54° 20' 32"; to find the angle Z X Y=62° 35' 15". Hence the angle P X Z (the *difference* between P X Y and Z X Y)=28° 14' 30".

(4.) In the spherical triangle P X Z, are given P X=92° 58' 35", Z X =64° 56' 13," and included angle P X Z=28° 14' 30"; to find the colatitude P Z= 39° 12' 2". Hence the latitude Z Q=50° 47' 58" N.

Calculation.

(1.) To find arc X Y (*Trig.* Part I. Rule IX.).

P X......92° 58' 35"......6·301030
P Y......92 56   30 ......9·999414
Diff. ...   2   5   9·999427
                            8·914644

log. vers. 5·214515 ∴ vers.=163875
                        vers. diff...   000
∴ X Y=33° 16' 0" vers. X Y...163875

(2.) To find angle P X Y.

92°	58'	35"	......0·000586
33	16	0	......0·260795
59	42	35	4·987511
92	56	30	4·456316
152	39	5	9·705208
33	13	55	∴ P X Y=90° 49' 45".

(3.) To find angle Z X Y.

64°	56'	13"	......0·042946
33	16	0	......0·260795
31	40	13	4·833834
54	20	32	4·293478
86	0	45	9·431053
22	40	19	∴ Z X Y=62° 35' 15".

To find angle P X Z.

P X Y.......................90° 49' 45"
Z X Y.......................62 35 15

∴ P X Z=28 14 30

(4.) To find arc P Z, the colatitude.

	92°	58'	35"	......6·301030
	64	56	13	......9·999414
Diff....28	2	22		9·957054
				8·774664

5·032162 ∴ vers.=107686
vers. diff....117376

vers. colat....225062
∴ colatitude...39° 12' 2"

and latitude=50 47 58 N.

144. In latitude by account 50° N., at 11ʰ 23ᵐ 22ˢ A.M., as shown by chronometer, the sun's altitude was 32° 41' 45", and declination 5° 23' 9" S.; and at 3ʰ 19ᵐ 51ˢ P.M., as shown by chronometer, the sun's altitude was 21° 25' 22", and declination 5° 19' 17" S. Construct a figure, and find by calculation the true latitude.

Ans. Latitude=50° 47' 48" N.

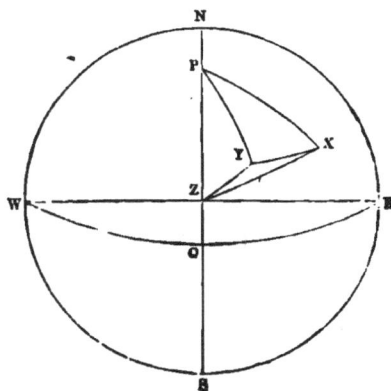

Second. Suppose the arc X Y, drawn through the two places of the same or two heavenly bodies, to pass (produced if necessary) *between* the zenith and the pole, as in the annexed figure.

Let x and y be the two places of the heavenly body when the altitudes are taken. Through x and y draw xy, the arc of a great circle, which, if produced, will evidently pass between p and z, the pole and zenith.

Then it is evident by the figure that we must *add* together the angles pxy and zxy, instead of subtracting them, as in the first case, to find the angle pxz, or arc 4.

All the other steps are the same as in the preceding example.

145. In estimated latitude 13° N., at 9ʰ 54ᵐ 25' A.M., the sun's altitude was 58° 48' 30", and declination 22° 35' 30" N.; and at 1ʰ 0ᵐ 5' P.M. the sun's altitude was 73° 0' 0", and declination 22° 34' 40" N. Construct a figure, and find by calculation the true latitude.

Construction.

Let NWSE represent the horizon, NZS the celestial meridian; then, the latitude being about 13° N., take NP=13°, and P will be the north pole of the heavens: take PQ=90°, and through W, Q, E, draw the celestial equator WQE. Let x be the place of the sun at the A.M. observation, and y its place at the P.M. observation (estimated as near as possible by means of the altitudes and declination). Through x and y draw the circles of declination PX and PY, and circles of altitude zx and zy, and join xy. It is evident by the figure that the arc xy passes between the zenith and the pole. Then the colatitude PZ may be found as follows:

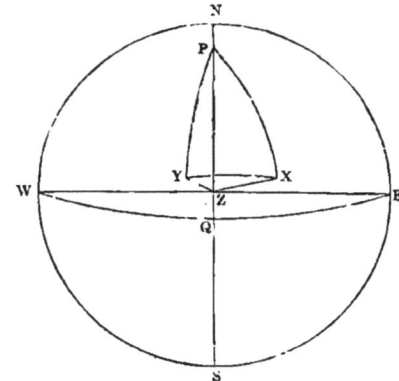

(1.) In the spherical triangle PXY, are given PX=90°−decl.=67° 24' 30", PY=90°−decl.=67° 25' 20", and angle xPY (=sum of hour-angles)= 3ʰ 5ᵐ 40'; to find xy=42° 40' 26".

(2.) In the spherical triangle PXY, are given PX=67° 24' 30", PY=67° 25' 20", and arc xy=42° 40' 26"; to find angle PXY=80° 40'.

(3.) In the spherical triangle zxy, are given zx=90°−alt.=31° 11' 30", zy=90°−alt.=17° 0' 0", and xy=42° 40' 26"; to find the angle zxy= 21° 9' 45".

Hence the angle zxP (the *sum* of PXY and zxy)=101° 49' 45".

(4.) In the spherical triangle PXZ, are given PX=67° 24' 30", zx= 31° 11' 30", and included angle zxP=101° 49' 45"; to find the colatitude zP=76° 40' 2". Hence the latitude=13° 19' 58" N.

EXAMPLE FOR PRACTICE.

146. In latitude by account 9° 19′ N., at 9ʰ 45ᵐ 12ˢ A.M., the sun's altitude was 67° 14′ 10″, and decl. 22° 13′ 14″ N. ; and at 1ʰ 10ᵐ 17ˢ P.M., the sun's altitude was 79° 8′ 30″, and decl. 22° 14′ 0″ N. Construct a figure, and find by calculation the true latitude. *Ans.* Lat. = 9° 20′ N.

PROBLEM XXII.

Given the altitudes of two heavenly bodies observed at the same time ; to find the latitude.

Let N W S E represent the horizon, N Z S the celestial meridian, P the pole, z the zenith, W Q E the celestial equator, and A the first point of Aries. Let X and Y be the places of the two heavenly bodies whose altitudes are observed. Draw the circles of altitude Z X and Z Y, and circles of declination P X and P Y, through X and Y ; draw also X Y, the arc of a great circle connecting the two bodies.

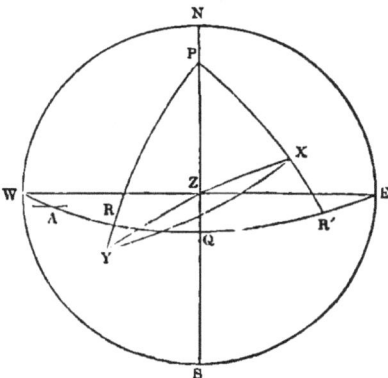

In the two spherical triangles P X Y and Z X Y, there are given the two polar distances P X and P Y (from the *Nautical Almanac*), the angle X P Y, the difference of the right ascensions of X and Y (for R R′ measures the angle X P Y, and R R′ = A R′ — A R = R A of X — R A of Y), and the two zenith distances Z X and Z Y (from the observed altitudes) ; to find the colatitude P Z, and thence the latitude.

(1.) In the spherical triangle P X Y, are given the polar distances P X and P Y, and the included angle X P Y ; to calculate X Y, or arc 1.

(2.) In the spherical triangle P X Y, are given the polar distances P X, P Y, and arc X Y or arc 1 ; to calculate the angle P X Y, the angle at the heavenly body whose bearing is the greater. Call this angle arc 2.

(3.) In the spherical triangle Z X Y, are given the zenith distances Z X and Z Y, and arc 1 or X Y ; to find the angle Z X R, the angle at the greater bearing. Call this angle arc 3.

(4.) When the arc X Y does not pass between the zenith and the pole, subtract arc 3 from arc 2, the remainder is angle P X Z. Call this arc 4. When the arc X Y does pass between the zenith and the pole, add arc 3 to arc 2 ; the sum will be angle P X Z, or arc 4.

(5.) *Lastly.* In the spherical triangle P X Z, are given P X and X Z, the polar distance and zenith distance at the greater bearing, and the included angle P x z, or arc 4, to calculate P z the colatitude ; and thence z Q, the latitude, is known.

147. In latitude by account 14° 30′ S., the altitude of Canopus was 47°·30′, bearing about S.E. b. S.; and at the same time the altitude of Achernar was 33° 40′ 20″, bearing about S.S.W. : declination of Canopus =52° 36′ 40″ S., RA=6ʰ 20ᵐ 47ˢ; declination of Achernar=57° 57′ 20″ S., RA=1ʰ 32ᵐ 37ˢ. Required the true latitude.

NOTE. It was seen, when the altitudes were taken, that a line joining the two stars would pass *between* the zenith and pole.

Construction.

Let N W S E represent the horizon, N z s the celestial meridian ; then, the

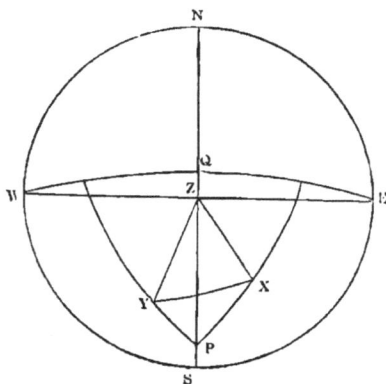

latitude being about 14° S., take s P=14°, and P will be the south pole of the heavens : take P Q= 90°, and through W, Q, E, draw the celestial equator W Q E. Since the declination of Canopus is about 50° S., and its bearing S.E. b. S., let x be the place of Canopus ; and since the declination of Achernar is about 57° S, and its bearing S.S.W., let Y be the place of Achernar. Draw the circles of declination P x and P Y, and the circles of altitude z x and z Y, and through the stars x and Y draw the great circle x Y ; then the colatitude P z may be found as follows :

(1.) In the spherical triangle P x Y, are given P x=90°−decl.=37° 23′ 20″, P Y=90°−decl.=32° 2′ 40″, and included angle x P Y (=diff. of right ascension of stars)=4ʰ 48ᵐ 20ˢ; to find x Y=39° 24′ 48″.

(2.) In the spherical triangle P x Y, are given P x=37° 23′ 20″, P Y= 32° 2′ 40″, and arc x Y=39° 24′ 48″ ; to find angle P x Y=52° 40′ 30″.

(3.) In the spherical triangle z x Y, are given z x=90°−alt.=42° 30′, z Y=90°−alt.=56° 19′ 40″, and x Y=39° 24′ 48″ ; to find the angle z x Y =92° 1′ 30″.

Hence the angle z x P (the *sum* of P x Y and z x Y)=144° 42′ 0″.

(4.) In the spherical triangle P x z, are given P x=37° 23′ 20″, z x= 42° 30′ 0″, and included angle z x P=144° 42′ 0″; to find the colatitude z P=75° 27′ 46″.

Hence the latitude z Q=14° 32′ 14″ S.

EXAMPLES FOR PRACTICE.

148. In latitude by account 41° N., the altitude of α Andromedæ was 73° 14′, bearing about S. b. E.; and at the same time the altitude of α Tauri was 18° 27′ 30″, bearing about East:

Decl. of α Andromedæ...28° 18′ 24″ N.......R.A...0ʰ 1ᵐ 2ˢ
  ,,   α Tauri ....16 13 21 N.......R.A...4 27 47
Required the true latitude.   *Ans.* Latitude=41° 26′ N.

149. In latitude by account 36° 30′ N., the altitude of Sirius was 27° 46′ 3″, bearing about S.W.; and at the same time the altitude of Spica was 12° 49′ 44″, bearing about E.S.E.

Decl. of Spica.........10° 25′ 26″ S...... R.A...13ʰ 17ᵐ 45ˢ
  ,,   Sirius ......16 31 29 S...... R.A... 6 38 53
Required the true latitude.   *Ans.* Latitude=36° 34′ N.

PROBLEM XXIII.

Given the altitudes of two heavenly bodies observed at different times, and the elapsed time; to find the latitude.

Let N W S E represent the horizon, N Z S the celestial meridian, P the pole, and Z the zenith. Let X and Y be the two places of the heavenly bodies when their altitudes X O and Y O′ were observed at different times.

First. Suppose the altitude of X to be first taken, and that the altitude of Y was taken after an interval, during which the body X had moved from X to X′; then X′PY is the difference of the right ascension of the two bodies, X P X′ is the elapsed time between the observations, and X P Y is the polar

angle between the two places of the heavenly bodies when their altitudes were taken. This polar angle is the one to be used to determine the several arcs, 1, 2, &c., and finally the latitude, as in the preceding cases. The polar angle X P Y is found by the following practical rule: "Add elapsed time (expressed in sidereal time) to the right ascension of the heavenly body first observed, and take the difference between the sum and the right ascension of the heavenly body last observed; the remainder will be the polar angle required (subtracted from 24 hours, if greater than 12 hours)." This rule may be proved as follows:

In the fig., let ♈ be the first point of Aries; then ♈ R=right ascension of x', AR'=right ascension of Y, and XPX' or arc rR=elapsed time. Also n'r=polar angle XPY required.

· To find the polar angle. R'r=Rr+AR—AR',
or polar angle=elapsed time+RA of body first observed—RA of body last observed ; which is the rule.

Second. Suppose the altitude of Y (fig. 1) first taken, and that the altitude of x was not taken till after an interval, during which Y had moved to Y'; then XPY is the polar angle between the two places of the heavenly bodies when their altitudes were taken.

To find the polar angle XPY. Rr=AR—(n'r+AR'),
or polar angle=RA of body last observed—(elapsed time+RA of body first observed) ; which is the rule.

(fig. 1.)   (fig. 2.)

150. In latitude by account 47° 54′ S., the following observations of the two stars Regulus and Antares were made at different times ; to determine the true latitude (fig. 2) :

	True alt.	Decl.	RA.	Bearing.
Regulus ...	27° 3′ 23″	12° 50′ 32″ N....	9ʰ 58ᵐ 46·6ˢ...	N. 30° W.
Antares ...	37 58 51	26 1 18 S....	16 18 23·1 ...	S. 80 E.

The altitude of Antares was taken 1ʰ 1ᵐ 19·5ˢ after that of Regulus, as measured by a chronometer.

Project the figure on the plane of the celestial equator, in order to exhibit more clearly the place of the first point of Aries. Let P' be the south pole and z the zenith of the observer, x the place of Regulus when its altitude was taken, and Y the place of Antares at that time, and Y the place of Antares when its altitude was taken ; the positions of x and Y with respect to the meridian PZ being determined from their estimated bearings.

Draw the circles of declination P′X, P′Y, and circles of altitude Z X and Z Y, and join X Y.

(1.) In the triangle P′X Y, are given P′X=102° 50′ 32″, P′Y=63° 58′ 42″, and angle X P′Y=A P′X−(A P′Y′+Y P′Y′)=5ʰ 18ᵐ 17·0ˢ; to find X Y=86° 29′ 50″.

(2.) In the triangle P′X Y, are given P′X=102° 50′ 32″, P′Y=63° 58′ 42″, and X Y=86° 29′ 50″; to find P′Y X=106° 7′ 20″.

(3.) In the triangle Z X Y, are given Z X=62° 56′ 37″, Z Y=52° 1′ 9″, and X Y=86° 29′ 50″; to find Z Y X=57° 58′ 0″.

Hence the angle P Y Z=48° 9′ 20″.

(4.) In triangle P′Z Y, are given P′Y, Z Y, and P′Y Z; to find P′Z=42° 3′ 16″, the colatitude.

Hence the latitude=47° 56′ 44″ S.

151. In latitude by account 14° 14′ N., the following observations of the two stars Dubhe and Antares were made at different times : to determine the true latitude :

	Alt.			Decl.			RA.			Bearing.
Dubhe......41°	1′	2″	...62°	43′	10″ N.	...10ʰ	52ᵐ	31·6ˢ	...N. 10° W.
Antares ...15	52	39	...26	1	18 · S.	...16	18	23·1	...S. 48 E.

The altitude of Antares was taken at 30ᵐ 6·4ˢ after that of Dubhe, as measured by a chronometer.

Construction.

The figure may be projected on the plane of the horizon or the plane of the equator, as may be seen in the annexed diagrams.

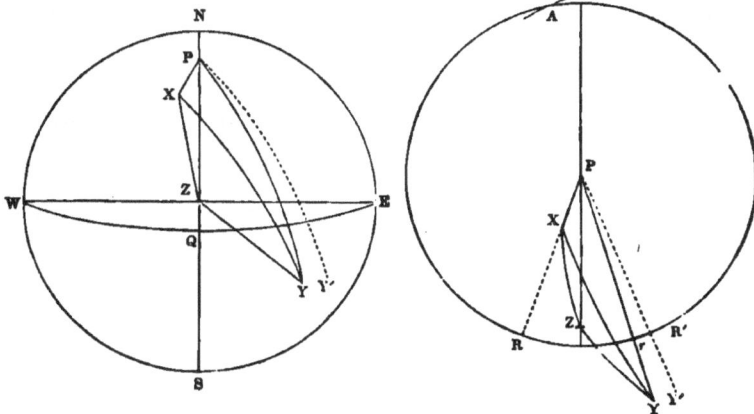

Let P be the north pole, Z the zenith, and X the place of Dubhe when its altitude was observed, and Y′ the place of Antares at that time, and Y the place of Antares when its altitude was taken ; the positions of X and Y with respect to the meridian P z being determined from their estimated bearings.   Complete the figure as in the last example.

Then the polar angle X P Y $=$ A R′ $-$ (A R $+$ r R′) $=4^h$ 55^m 45·1′.

(1.) In triangle P X Y, find   X Y $=106°$ 3′   32″
(2.)   „   P X Y,   „   P Y X $=$ 27   17   30
(3.)   „   Z X Y,   „   Z Y X $=$ 37   38   15
          Hence the angle P Y Z $=$ 64   55   45
(4.) In triangle P Z Y, find   P Z $=$ 75   44   42
          ∴ lat.$=14°$ 15′ 18″ N.

152. In latitude by account 51° N., the following observations of the two stars α Bootis and α Aquilæ were made at different times :

	Alt.			Decl.			RA.		Bearing.
α Bootis......	22°	49′	19″...	19°	55′	16″ N...	14^h	9^m 12ʻ...	W.N.W.
α Aquilæ ...	47	27	59 ...	8	29	55 N...	19	43 54 ...	S b. W.

The altitude of α Aquilæ was taken 10^m 52ʻ after that of α Bootis Construct a figure, and calculate the true latitude.

Ans. Lat.$=50°$ 57′ 30″ N,

Sumner's rule for finding the latitude.

A method of finding the latitude, that deserves the attention of the seaman, is the one called Sumner's method.   When one altitude only can be taken, the place of the observer may be assumed to be on a line, which is found as follows.   With the estimated latitude of the ship, the altitude and declination, and by means of the chronometer, showing Greenwich mean time, calculate the longitude (see Probl. XXIV. XXV.).   Mark the spot on the chart corresponding to this latitude and longitude.   Assume another latitude a few degrees different from the former, and find the longitude as before.   Mark the spot on the chart corresponding to this latitude and longitude.   Join the two spots thus found, and the place of the ship will be on or very near to the line, or the line produced.   If another altitude could be taken an hour or two afterwards, and two spots determined in like manner, the line joining them will intersect the other line on the chart in or near to the true place of the ship, and thus the position of the ship will be found very nearly.   The estimated latitudes should, if possible, be taken one greater and one less than the true, and within a few degrees of each other ; a small diagram or chart bounded by the two parallels of latitude passing through the estimated latitudes may be easily constructed, and the place of the ship indicated thereon by the intersection of the two lines found as above.   This method is more fully described in Mr. John Riddle's *Navigation,*

CHAPTER IV.

INVESTIGATION OF RULES FOR DETERMINING THE LONGITUDE, VARIATION OF THE COMPASS, ETC.

Longitude by chronometer.

153. THE rate of a chronometer, and its original error on Greenwich mean time, being given, the mean time at Greenwich at any moment is readily found, and therefore can be known at that instant when the altitude of a heavenly body is taken for finding mean time at the ship (see Problems VIII. and IX.). The altitude for determining ship mean time should be taken when the heavenly body bears as nearly east or west as possible, since in that position of the body an error in the observation will produce the least error in the hour-angle, and therefore in the mean time deduced from it (see Problem XI., p. 43).

Ship mean time being thus found, the longitude is known, since

longitude in time=Greenwich mean time∼ship mean time.

154. The principal operation, therefore, in finding the longitude by chronometer, is to calculate the ship mean time; and this may be done from the altitude of the sun, or of any other heavenly body, observed when to the east or west of the meridian. We will consider these cases separately, and give an example and construction for each.

SUN CHRONOMETER.

PROBLEM XXIV.

To find the longitude from the altitude of the sun.

The Greenwich mean time is supposed to be known by means of the chronometer, the error and rate of which are given.

To find ship mean time.

First. When the sun is *west* of meridian (fig. 1, next page).

Let x be the place of the sun (west of meridian), m the place of the mean sun; and complete the figure as before. In the spherical triangle z P x, the three sides are given, namely z x the zenith distance, found by observation, P x the polar distance, from the *Nautical Almanac*, and P z the colatitude of the ship; from which the hour-angle zPx may be computed, which is also in this case ship apparent time. Also the equation of time, namely the angle mPx, is known from the *Nautical Almanac.* Hence ship mean time, namely the angle zPm, is found by applying ship apparent time zPx to the equation of time mPx.

Then the difference between the time so deduced and the Greenwich mean time, as shown by chronometer (corrected for error and rate), is the longitude in time.

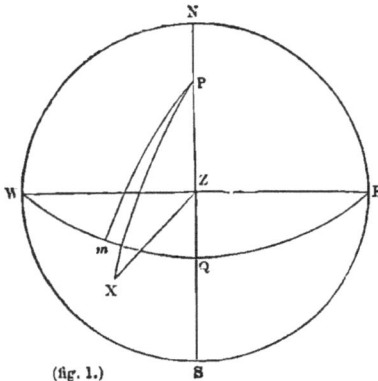 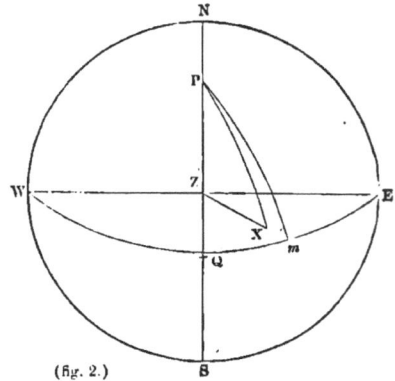

(fig. 1.)   (fig. 2.)

Second. When the sun is *east* of meridian (fig. 2).

Let x be the place of the sun (east of meridian), and m the mean sun. The hour-angle zPx, as in the former case, may be computed, and thence the ship apparent time, namely $24^h - zPx$, obtained.* Then, by applying the equation of time mPx to apparent time, ship mean time, namely the angle $24^h - zPm$, is obtained.

The difference between the time so deduced and Greenwich mean time, found from its error and rate, is the longitude in time.

STAR CHRONOMETER.

PROBLEM XXV.

To find the longitude from the altitude of any other heavenly body, as a star or the moon.

To find ship mean time.

Let x be the place of the heavenly body, AQ the equator, A the first point of Aries, and m the mean sun.

---

* This quantity, 24^h—hour-angle, may be readily found, if the table of haversines is used, by taking the quantity out from the bottom of the page instead of the top. Hence this practical rule: "West of meridian, take hour-angle from top; east of meridian, take hour-angle from bottom of table of haversines."

In the fig., Q*m* is the ship mean time required,
A*m* is the right ascension of mean sun,
A R is the star's right ascension,
and R Q measures the angle X P Z, the star's hour-angle.

Now Q*m*=R Q+A R—A*m*,
or ship mean time=star's hour-angle
+star's R A—R A mean sun.

The star's hour-angle Z P X is computed from knowing in the spherical triangle Z P X the colatitude P Z, the polar distance P X, and the zenith distance Z X, and thence ship mean time deduced.*

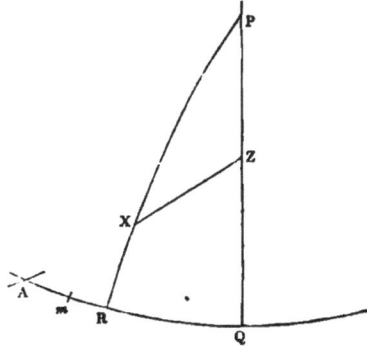

Star's R A, and R A of mean sun, are found in the *Nautical Almanac.*

Greenwich mean time, expressed astronomically, is found from the chronometer, corrected for error and rate.

Longitude by sun's altitude. West of Meridian.

155. April 21, about 2^h 30^m P.M., in latitude 46° 50′ N., given the true altitude of sun=42° 15′ 13″ (west of meridian), declination= 11° 44′ 28″ N., equation of time =1^m 4', subtractive from apparent time, and the Greenwich mean time=8^h 35^m 16·8' A.M. (=20^h 35^m 16·8', on April 20th); to find the longitude of ship.

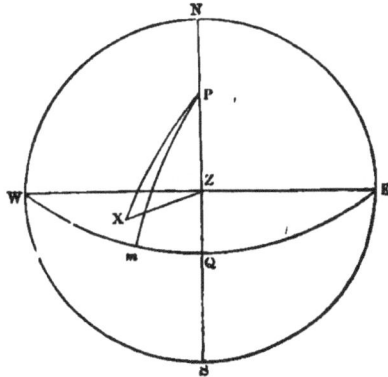

Let x be the place of the sun, P the pole, Z the zenith, and *m* the mean sun. Then, in the spherical triangle Z P X, are given Z X =47° 44′ 47″, P X=78° 15′ 32″, and P Z=43° 10′ 0″; to find the angle Z P X, or ship apparent time.

* See Problem IX., where ship mean time is found from the star's hour-angle, &c., and all the cases considered.

Calculation of apparent time z P x. .

43° 10′   0″..............0·164866
78   15   32 ....0·009184
35   5   32   4·820567
47   44   47   4·042198
82   56   19   9·036815
12   39   15   ∴ z P x =   2ʰ 34ᵐ   7ˢ
         and x P m =   1   4
∴ ship mean time, April 21 = 2   33   3 = m Q
or ship mean time, April 20 = 26   33   3
    Greenwich, April 20 = 20   35   17 (deduced from chronometer)
    ∴ longitude in time = 5   57   46 = 89° 26′ 30″ E.

NOTE. Instead of taking the difference between the polar distance and colatitude, the algebraic difference between the latitude and declination is generally used, and the secants taken out instead of the cosecants of P x and P z ; the logarithmic result will manifestly be the same.

156. May 8, about 4ʰ 0ᵐ P.M., in latitude 40° N., given the altitude of the sun = 32° 25′ 27″ (west of meridian), declination = 17° 9′ 18″ N., equation of time = 3ᵐ 42·6ˢ, subtractive from apparent time and Greenwich mean time = 7ʰ 2ᵐ 3ˢ (deduced from chronometer). Construct a figure, and find by calculation the longitude.   *Ans.* Longitude = 44° 42′ W.

Longitude by sun's altitude.   East of meridian.

157. June 9, about 9ʰ 40ᵐ A.M., in latitude 10° S., given the altitude of

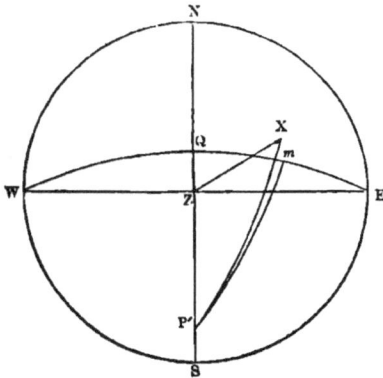

the sun = 42° 47′ 33″ (east of meridian), declination = 22° 55′ 36″ N., the equation of time = 1ᵐ 12ˢ, subtractive from apparent time, and Greenwich mean time = 7ʰ 33ᵐ 2ˢ A.M. (= 19ʰ 33ᵐ 2ˢ on June 8); to find the longitude.

Let x be the place of the sun, P′ the south pole, z the zenith, and m the mean sun. Then, in the spherical triangle z P′ x, the three sides are known, namely P′z = 90° − latitude, and z x = 90° − altitude, and P′x = 90° + declination ; from which may be computed the hour-angle z P′ x, or 24ʰ − ship apparent time.

Calculation of apparent time.

10°	0'	0" S.	....	0·006649
22	55	26 N.	....	0·035733
32	55	26		4·808669
47	12	27		4·094552
80	7	53		8·945603
14	17	1		

∴ 24^h − z p x = 21^h 41^m 46^s

and x p m = 1 12

∴ 24^h − z p m = 21 40 34 = ship mean time on 8th.

Chronometer = 19 33 2 = Greenwich mean time on 8th.

∴ long. in time = 2 7 32 = 31° 53′ E.

158. April 18, about 9^h 27^m A.M., in latitude 50° 48′ N., given the altitude of the sun = 38° 21′ 19″ (east of meridian), declination = 10° 55′ 29″ N., the equation of time = 0^m 43·6^s, subtractive from apparent time, and the Greenwich mean time = 9^h 24^m 41·8^s (A.M. at Greenwich). Construct a figure, and find by calculation the longitude.

Ans. Longitude = 4^m 25·4^s W.

Longitude by star's altitude.   West of meridian.

159. In latitude 30° 10′ N., given the altitude of Sirius = 30° 32′ 7″ (west of meridian), declination = 16° 31′ 19″ S., RA = 6^h 38^m 43^s, RA mean sun = 23^h 40^m 53^s, and the Greenwich mean time (deduced from chronometer) = 11^h 54^m 53^s; to find the longitude.

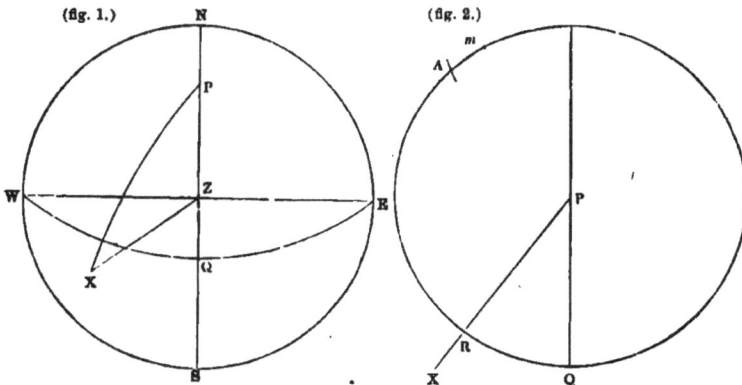

(fig. 1.)   (fig. 2.)

Fig. 1.   Let x be the place of the star (west of meridian), and complete the figure; then in the spherical triangle ᴢᴘх, the three sides are known, from which the star's hour-angle ᴢᴘх may be computed=2^h 32^m 55^s. (See below.)

Fig. 2. Take Qn=star's hour-angle=2^h 32^m 55^s, ʀᴀ=star's right ascension=6^h 38^m 43^s, and ᴀQm=right ascension of mean sun=23^h 40^m 53^s. Then Qm=ship mean time to be found, as follows:

Calculation of hour-angle and ship mean time.

30°	10′	0″ N	....	0·063201
16	31	19 S	....	0·018310
46	41	19		4·902752
59	27	13		4·046026
106	8	32		9·030289
12	45	54		

∴ ᴢᴘх, or arc Qʀ= 2^h 32^m 55^s
ᴀʀ= 6 38 43
ᴀQ= 9 11 38 +24^h
ᴀQm=23 40 53
∴ Qm= 9 30 45=ship mean time.
Chronometer=11 54 53=Greenwich mean time.
∴ longitude in time= 2 24 8=36° 2′ W.

160. In latitude 48° 30′ N., given the altitude of Arcturus=38° 37′ 13″ (west of meridian), declination=19° 56′ 42″ N., ʀᴀ=14^h 9^m 1^s, ʀᴀ mean sun=7^h 13^m $42\cdot1^s$, and the Greenwich mean time=8^h 20^m $10\cdot4^s$. Construct the figures, and find by calculation the longitude.

Ans. Longitude=32° 29′ 30″ E.

161. In latitude 30° 10′ N., given the altitude of Sirius=30° 32′ 47″ (west of meridian), declination=16° 31′ 19″ S., ʀᴀ=6^h 38^m 43^s; ʀᴀ mean sun=23^h 40^m 55^s, and the Greenwich mean time=11^h 54^m $52\cdot7^s$. Construct the figures, and find by calculation the longitude.

Ans. Longitude=36° 2′ 30″ W.

Longitude by star's altitude.   East of meridian.

162. In latitude 41° 20′ N., given the altitude of β Canis Majoris= 42° 17′ 27″ (east of meridian), declination=5° 35′ 43″ N., ʀᴀ=7^h 31^m 40^s, ʀᴀ of mean sun=21^h 21^m 30^s, and the Greenwich mean time=3^h 31^m 26^s; to find the longitude.

Fig. 1 (page 83). Let x be the place of the star (east of meridian), and complete the figure; then, in the triangle ᴢᴘх, the three sides are given to compute the hour-angle ᴢᴘх=2^h 21^m 56^s; hence 24^h−ᴢᴘх=21^h 38^m 4^s.

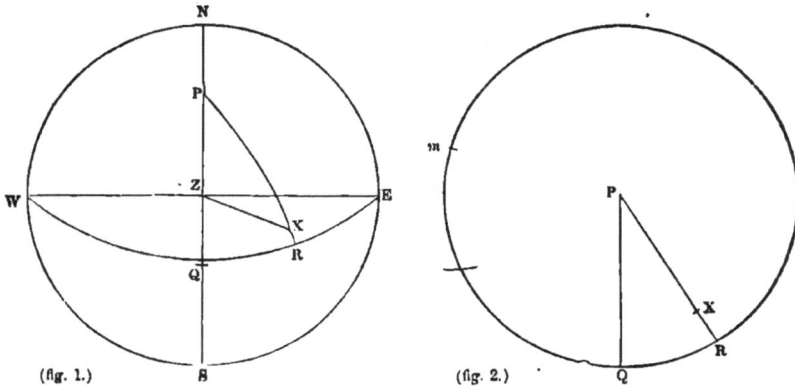

(fig. 1.)   (fig. 2.)

Fig. 2. Take Q n=star's hour-angle, R A=star's right ascension, and A Q m =R A of mean sun. Then Q m=ship mean time to be found, as follows :

Calculation of $24^h - z P x$, *and ship mean time.*

41°	20′	0″ N	....	0·124429
5	35	43 N	....	0·002075
35	44	17		4·823167
47	42	33		4·018181
83	26	50		8·967852
11	58	16		∴ 21^h 38^m $4'=21^h - z P x$, or arc B A Q.

$$7 \quad 41 \quad 40 = A R$$
$$29 \quad 9 \quad 44$$
$$21 \quad 21 \quad 30 = A Q m$$

Ship mean time= 7 48 14 =Q m
Greenwich mean time= 3 31 26
∴ longitude in time= 4 16 48 =64° 12′ 0″ E.

163. In latitude 50° 50′ N., given the altitude of α Cygni=49° 33′ 25 (east of meridian), declination=44° 46′ 6″ N., R A=20^h 36^m 31·3ˢ, R A mean sun=11^h 32^m 46·4ˢ, and Greenwich mean time=7^h 19^m 51·5ˢ. Construct the figures, and find by calculation the longitude.   *Ans.* 35° 22′ 0″ W.

LONGITUDE BY LUNAR OBSERVATIONS.

164. The moon in its motion round the earth is seen continually to change its distance from certain bright stars lying near its path. The angular distances of these heavenly bodies from the moon have been computed and recorded in the *Nautical Almanac* for intervals of every three hours, namely at 3, 6, 9, &c., o'clock at Greenwich. The quantities thus

inserted in the *Nautical Almanac* are the *true* distances; that is, the distances are supposed to be seen from the center of the earth. An observer at any place on the surface of the earth may also measure the angular distance between the moon and one of these stars by means of his sextant; but the distance so observed is the *apparent* distance, and before it can be compared with the true distances in the *Nautical Almanac*, in order to determine the mean time at Greenwich when the observation was taken, it must be corrected for the effects of parallax and refraction (see Chap. V.), so as to obtain the true distance of the two bodies at the moment of observation.

165. The formula about to be investigated will enable us to find the true distance from the apparent or observed distance. This is technically called *clearing the distance*. If the true distance thus computed is exactly the same as the one inserted in the *Nautical Almanac* for some given hour at Greenwich, we have found *Greenwich mean time* when the observation was taken. But this coincidence must rarely happen; the computed true distance will generally come out between two of the recorded distances: to obtain the Greenwich mean time corresponding to that distance a simple proportion only is necessary, as will be seen hereafter.

166. The *mean time at the place* at the instant the observed distance is taken can be readily found in the same manner as in the rule for finding the longitude by chronometer (p. 77), namely by observing the altitude of a heavenly body and computing its hour-angle, and thence ship mean time.

167. The method of finding the longitude by a lunar observation, therefore, divides itself into two distinct parts.

1. To find Greenwich mean time.

2. To find ship mean time.

The former is deduced from the observed distance; the latter, in the usual way, from an observed altitude.

To clear the distance.

PROBLEM XXVI.

Given the apparent distance, and the apparent and true altitudes of the two bodies; to investigate a formula for computing the *true* distance.

Let s′ be the apparent place of the heavenly body, as the sun or star; then, in consequence of refraction and parallax, the former exceeding the

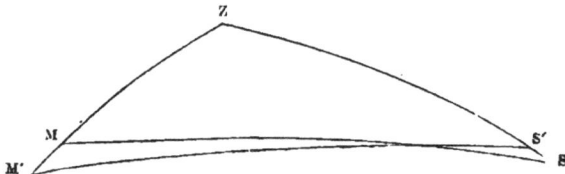

latter, its true place (Chap. V.) will be below s′, as s, in a vertical plane (if we suppose the effect of parallax to take place in a plane passing through the true zenith z, which it does very nearly). Let M′ be the apparent place of the moon; then its true place will be above M′, since the moon is depressed more by parallax than it is raised by refraction: let, therefore, its true place be supposed to be at M, in a plane passing through z. Through M′ and s′ draw the great circle M′s′; this arc will be the *apparent* distance found by observation. Through M and s, the true places of the heavenly bodies, draw the arc Ms. The arc Ms is the *true distance* to be computed.

The true distance Ms may be found by the common rules of Spherical Trigonometry (see seventh method of clearing the distance); but it will be more readily obtained by means of a special formula or rule, which will determine the true distance in *terms of the versines*, since the table of versines is calculated to the nearest second; and this will render it unnecessary to proportion for seconds, which must be done when the distance is expressed in terms of the sines, &c.

Investigation of formula for clearing the distance.

Let z be the zenith of the observer; then, if we suppose the effect of parallax to take place in a vertical circle,* zM′ and zs are circles of altitude.

Let $z = z$M, the *true* zenith distance of the moon,

$z_1 = z$s, the *true* zenith distance of the sun or star,

$a =$ the apparent altitude of moon,

$a_1 =$ the apparent altitude of sun or star.

Then zM′ $= 90° - a$, and zs′ $= 90° - a_1$.

Let $d =$ M′s′, the apparent distance of the centers of the two bodies,

and D $=$ Ms, the true distance of the centers;

it is required to investigate a formula for computing D.

In triangle zMs, cos. $z = \dfrac{\cos. D - \cos. z \, . \, \cos. z_1}{\sin. z \, . \, \sin. z_1}$

,,　　zM′s′, cos. $z = \dfrac{\cos. d - \sin. a \, . \, \sin. a_1}{\cos. a \, . \, \cos. a_1}$

$\therefore \dfrac{\cos. D - \cos. z \, . \, \cos. z_1}{\sin. z \, . \, \sin. z_1} = \dfrac{\cos. d - \sin. a \, . \, \sin. a_1}{\cos. a \, . \, \cos. a_1}$

$\therefore \dfrac{\cos. D - \cos. z \, . \, \cos. z_1}{\sin. z \, . \, \sin. z_1} + 1 = \dfrac{\cos. d - \sin. a \, . \, \sin. a_1}{\cos. a \, . \, \cos. a_1} + 1,$

or $\dfrac{\cos. D - (\cos. z \, . \, \cos. z_1 - \sin. z \, . \, \sin. z_1)}{\sin. z \, . \, \sin. z_1} = \dfrac{\cos. d + \cos. a \, . \, \cos. a_1 - \sin. a \, . \, \sin. a_1}{\cos. a \, . \, \cos. a_1};$

$\therefore \dfrac{\cos. D - \cos. (z + z_1)}{\sin. z \, . \, \sin. z_1} = \dfrac{\cos. d + \cos. (a + a_1)}{\cos. a \, . \, \cos. a_1};$

or cos. D $-$ cos. $(z + z_1) = \{\cos. d + \cos. (a + a_1)\} \cdot \dfrac{\sin. z \, . \, \sin. z_1}{\cos. a \, . \, \cos. a_1}.$

Assume $\dfrac{\sin. z \, . \, \sin. z_1}{\cos. a \, . \, \cos. a_1} = 2$ cos. A (A being an auxiliary angle);

* This is not strictly correct (see Problem XXXVI.).

$$\therefore \cos. \text{D} - \cos. (z+z_1) = 2\cos. \text{A} . \{\cos. d + \cos. (a+a_1)\}$$
$$= 2\cos. \text{A} . \cos. d + 2\cos. \text{A} . \cos. (a+a_1)$$
$$= \cos. (d+\text{A}) + \cos. (d\sim\text{A}) + \cos. (a+a_1+\text{A}) + \cos. (a+a_1\sim\text{A})$$
$$\therefore 1 - \cos. \text{D} = 1 - \cos. (z+z_1) + 1 - \cos. (d+\text{A}) + 1 - \cos. (d\sim\text{A}) + 1$$
$$- \cos. (a+a_1+\text{A}) + 1 - \cos. (a+a_1\sim\text{A}) - 4,$$
$$\therefore \text{vers. D} = \text{vers. } (z+z_1) + \text{vers. } (d+\text{A}) + \text{vers. } (d\sim\text{A}) + \text{vers. } (a+a_1+\text{A})$$
$$+ \text{vers. } (a+a_1\sim\text{A}) - 4.$$

But tabular versine $=1000000 \times$ versine (see *Trigonometry*, Part. I. Art. 32). Reducing the formula to tabular versines and clearing of fractions, we have,

$$\left[\text{since vers. D} = \frac{\text{tab. vers. D}}{1000000}, \text{vers. } (z+z_1) = \frac{\text{tab. vers. } (z+z_1)}{1000000}, \&c.\right]$$

tab. vers. D = tab. vers. $(z+z_1)$ + tab. vers. $(d+\text{A})$ + tab. vers. $(d\sim\text{A})$
$$+ \text{tab. vers. } (a+a_1+\text{A}) + \text{tab. vers. } (a+a_1\sim\text{A}) - 4000000;$$

or, suppressing the word tabular, it being understood that the formula is expressed in terms of tabular versines, we have

vers. D = vers. $(z+z_1)$ + vers. $(d+\text{A})$ + vers. $(d\sim\text{A})$ + vers. $(a+a_1+\text{A})$
$$+ \text{vers. } (a+a_1-\text{A}) - 4000000.$$

From which expression the true distance D may be computed, when the true and apparent altitudes and the apparent distance are known.

Before, however, this formula can be used, the value of the auxiliary angle A must be computed, and formed into a table; this has been done, and may be found in the Nautical Tables of Inman and others.

The auxiliary angle A may be computed as follows.

Construction of table of auxiliary angle A.

Let it be required to compute the value of A, when the zenith distance of the moon $z=32°$ 19′ 50″, the app. alt. of the moon $a=57°$ 11′ 24″, zenith distance of sun $z_1=55°$ 39′ 46″, and app. alt. of sun $a_1=34°$ 21′ 32″.

Since $2\cos. \text{A} = \dfrac{\sin. z . \sin. z_1}{\cos. a . \cos. a_1} = \sin. z . \sin. z_1 . \sec. a . \sec. a_1,$

$\therefore$ in logarithms,

log. cos. A = log. sin. z + log. sec. a + log. sin. z_1 + log. sec. $a_1 - 30·301030.$

Calculation.

log. sin. z............	9·728190
„ sec. a............	10·266120
„ sin. z_1	9·916839
„ sec. a_1	10·083273
	39·994422
	30·301030
„ cos. A............	9·693392 $\therefore$ A=60° 25′ 16″.

And in the same manner may other values of ∆ be computed for any given app. alt. of the moon, the argument of the table.

The true distance D being thus calculated by means of the preceding formula, or otherwise, the mean time at Greenwich corresponding thereto may be found as follows.

PROBLEM XXVII.

Given the true distance of the moon from some other heavenly body, to calculate the corresponding Greenwich mean time.

The Greenwich time corresponding to the true distance may be found by a simple proportion, thus :

As the difference of distances in the *Nautical Almanac* for the three hours between which the computed true distance lies : 3 hours : : difference between the distance in *Nautical Almanac* at the beginning of the 3 hours and the computed distance : t ; t being the hours, minutes, and seconds elapsed from the beginning of the three hours.

The quantity t is, however, usually found by means of proportional logarithms, as follows :

Let d=difference of distances in 3 hours (taken from *Nautical Almanac*),

$\quad d'$=difference between first distance taken out and calculated distance,

$\quad t$=time elapsed from the hour opposite the first distance taken out.

Then $d : d' : : 3$ hours $: t$,   ∴ $td = 3d'$;

∴ log. t + log. d = log. 3^h + log. d', or log. t = log. 3^h − log. d + log. d' ;

∴ log. 3^h − log. t = − (log. 3^h − log. d) + log. 3^h − log. d',

or prop. log. t = prop. log. d' − prop. log. d (Chapter V.).

Then the time t thus found, added to the hour opposite the first distance in the *Nautical Almanac* taken out, will be the Greenwich mean time at the instant when the observation was taken.

168. Given the sun's true distance (calculated from an observed distance) $=73° 51' 58''$, and two distances taken out of the *Nautical Almanac*, namely at $6^h = 73° 23' 44''$, and at $9^h = 74° 54' 7''$; to find Greenwich mean time when the observation was taken.

True distance...	73°	51'	58''
At 6ʰ	73	23	44
	74	54	7
		28	14 = d'
	1	30	23 = d

First method, by proportion.

$1° 30' 23'' : 3^h : : 28' 14'' : t,$

∴ $t = 0^h 56^m 14^s$

Hence Greenwich mean time $= 6^h 56^m 14^s$.

Second method, by proportional logarithms.

prop. log. $t =$ prop. log. d' — prop. log. d.

prop. log. d' .... ·80451

prop. log. d .... ·29918

∴ prop. log. t .... $\overline{50533}$ ∴ $t = 0^h\ 56^m\ 14'$.

Time at first distance in *Nautical Almanac* 6

∴ Greenwich mean time $= \overline{6\ \ 56\ \ 14}$, when the sun's true distance was 73° 51′ 58″.

EXAMPLE FOR PRACTICE.

169. Given the sun's true distance (calculated from an observed distance) $= 110°\ 49'\ 48''$, and two distances taken out of the *Nautical Almanac*, namely at 0^h or Greenwich mean noon $= 110°\ 23'\ 53''$, and at $3^h = 111°\ 47'\ 28''$; to find Greenwich mean time when the observation was taken: first, by proportion; second, by proportional logarithms.

Ans. Greenwich mean time $= 0^h\ 55^m\ 49'$.

(2.) TO FIND SHIP MEAN TIME.

The ship mean time must be calculated for the same instant as the distance was observed which determined the instant of Greenwich mean time. Ship time may therefore be obtained from an altitude of a heavenly body taken at that instant, or at some known interval before or after. It is obtained, in fact, precisely in the same manner as directed in Problem XXIV., for finding the longitude by chronometer.

LASTLY. TO FIND THE LONGITUDE IN TIME.

The difference between Greenwich mean time and ship mean time thus found is the LONGITUDE IN TIME of the ship.

The following sun-lunar, worked out in detail, will indicate the several steps in the rule for finding the longitude by a lunar observation. The method of determining the longitude by a star-lunar differs very little from that by sun-lunar; the construction of the figures, to suit the several cases that may occur, will present no difficulty to the student who is acquainted with the preceding problems.

EXAMPLE OF SUN-LUNAR.   ALTITUDES OBSERVED.

170. Given the necessary observations and quantities taken out of the *Nautical Almanac;* to calculate the several parts of a lunar. These are:

(1.) The true distance between the sun and moon.

(2.) The Greenwich mean time corresponding to that distance.

(3.) The hour-angle of the moon or sun, selecting that body that is farthest from the meridian (Problem XI.).

(4.) The ship mean time deduced from that hour-angle.

(5.) The longitude in time.

(1.) *To calculate the true distance.*

(By formula, p. 86, and auxiliary angle A.)

Quantities required to be given by previous calculation are the following:

Moon's apparent altitude .... $20°$ $57'$ $24'' = a$
Moon's true altitude .... 21 47 26 $= 90° - z$
Sun's apparent altitude .... 13 33 38 $= a_1$
Sun's true altitude .... 13 29 48 $= 90° - z_1$
Apparent distance of sun and moon .... 88 23 12 $= d$
Auxiliary angle A...... $60°$ $10'$ $43''$.

Then, to find the true distance D, we have

vers. $D =$ vers. $(z + z_1) +$ vers. $(d + A) +$ vers. $(d \sim A) +$ vers. $(a + a_1 + A)$
$\qquad\qquad\qquad + $ vers. $(a + a_1 \sim A) - 4000000$.

z	....	$68°$	$12'$	$34''$	
z_1	....	76	30	12	
	$z + z_1 =$	144	42	46	vers.
d	....	88	23	12	
A	....	60	10	43	
	$d + A =$	148	33	55	,,
	$d - A =$	28	12	29	,,
a	....	20	57	24	
a_1	....	13	33	38	
$a + a_1$	....	34	31	2	
A	....	60	10	43	
	$a + a_1 + A =$	94	41	45	,,
	$a + a_1 \sim A =$	25	39	41	,,

Calculation.

vers. $(z + z_1)$........$1816138...130$
,, $(d + A)$........$1853096...140$
,, $(d - A)$........$0118696...\ 66$
,, $(a + a_1 + A)$...$1081649...217$
,, $(a + a_1 \sim A)$...$\underline{0098545...\ 85}$
$\qquad\qquad\qquad 4968124$
$\qquad\qquad\qquad\ \ \underline{638}$
$\qquad\qquad\qquad 4968762$
$\qquad\qquad\qquad \underline{4000000}$
vers. D ....$\cdot0968762$
$\qquad\qquad\qquad 589...173$
$\therefore$ D $= 88°$ $12'$ $36''$.

(2.) *To calculate Greenwich mean time.*

Quantities required to be given are:

True distance at time of observation .... $88°$ $12'$ $36''$
Distance from *Nautical Almanac* at 21^h........... 88 50 57
. ,,　　　　　 ,,　　　 ,,　　24^h........... 87 23 51

Then, to find Greenwich mean time, we have

prop. log. t=prop. log. d'—prop. log. d.

True dist. at obs. ...88°	12′	36″	prop. log. d' ...67151
„ 21ʰ...88	50	57	prop. log. d ...31525
„ 24ʰ...87	23	51	∴ prop. log. t 35626

∴ $d'=$ 38   21   ∴ $t=$ 1ʰ 19ᵐ 15ˢ= { time elapsed
$d=$ 1   27   6   21   { since 21 o'clock.

∴ Greenwich mean time=22   19   15

(3.) *To calculate the moon's hour-angle.* Quantities given are :

    Moon's pol. dist....84° 53′ 19″, found from *Nautical Almanac.*
      „ zen. dist....68 12 34   „   observation.
    Colat. of place......39 22 48 (moon west of meridian).

To find hour-angle z p x.

84°	53′	19″	....	0·001731
39	22	48	....	0·197603
45	30	31	....	4·922892
68	12	34	....	4·294029
113	43	5		9·416255
22	42	3	∴ hour-angle=4ʰ 5ᵐ 40ˢ.

(4.) *To calculate ship mean time.*

Quantities required to be given are :

Moon's hour-angle............ 1ʰ   5ᵐ   40ˢ
   „ right ascension 11   17   1, from *Nautical Almanac.*
R A of mean sun...............17   8   29   „   „

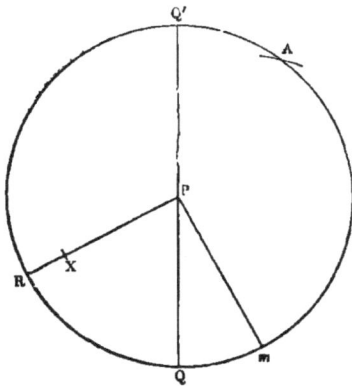

To find ship mean time.

By formula, p. 34, or by fig. (see construction),

Ship mean time=moon's hour-angle +moon's R A—R A mean sun.

Hour-angle	4ʰ	5ᵐ	40ˢ
Moon's R A 11	17	1
	15	22	41
Add 24			
	39	22	41
R A mean sun............17	8	29
∴ ship mean time=22	14	12

(5.) *To calculate the longitude.*

Greenwich mean time 22ʰ 19ᵐ 15ˢ
Ship mean time ....22　14　12
　　　　　　　　　　　　　　　　　───────────
　　　　　∴ long. in time = 　　5　3
　　　　　or in degrees = 1° 15′ 45″ west.

In the above example the ship mean time has been obtained from the *moon's* altitude, and this was done because that heavenly body was the more distant of the two from the meridian (see Problem XI.). If ship mean time is computed from the *sun's* altitude, we must proceed as follows.

(3.) *To calculate the sun's hour-angle.*

Quantities required to be given are :

Sun's pol. dist.........112° 45′ 31″, found from *Nautical Almanac.*
　,,　zen. dist....... 76　30　12　　　,,　　observation.
Colat. of place ...... 39　22　48

The sun is east of meridian (known from observation).

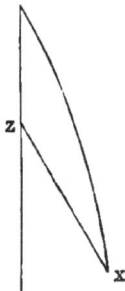

To find hour-angle z ᴘ x,
and thence 24ʰ − z ᴘ x, or app. time.

112° 45′ 31″......................0·035201
　39　22　48 ....0·197603
　────────────　　　　　　　───────
　73　22　43　　　　　　　　　4·984825
　76　30　12　　　　　　　　　3·435643
　────────────　　　　　　　───────
149　52　55　　　　　　　　　8·653272
　3　 7　29
∴ 24ʰ − z ᴘ x = 22ʰ 22ᵐ 1ˢ, or apparent time.

(4.) *To calculate ship mean time.*

Quantities required to be given are :

Ship app. time...22ʰ 22ᵐ 1ˢ.　　Equat. of time...0ʰ 7ᵐ 49·4ˢ sub.

By formula, p. 20 :

Ship mean time = apparent time − equation of time.
Ship apparent time..................22ʰ 22ᵐ 1ˢ
Equation of time.................... 　 4　49·4
　　　　　　　　　　　　　　　　───────────
　　∴ ship mean time = 22　14　11·6

(5.) *To calculate the longitude.*

Greenwich mean time...............22ʰ 19ᵐ 15ˢ
Ship mean time ....22　14　12
　　　　　　　　　　　　　　　　───────────
　　　　　∴ long. in time = 　5　3 W.
　　　　　or long. = 1° 15′ 45″ W.

EXAMPLE OF SUN-LUNAR.   ALTITUDES OBSERVED.

171. Given the following quantities; to construct figures, and compute the longitude.

Moon's app. alt....56° 16′ 40″   True distances, from *Nautical Almanac.*
    „   true alt....56   46   21   At 3ʰ ....77° 13′   6″
Sun's app. alt. ...34   15   42   „ 6ʰ 78   35   22
   „   true alt. ...34   14   24
Apparent dist. ...78   37   16   Sun's pol. dist. 95   58   30
Auxil. angle A ...60   25   12   „   zen. dist. 55   45   36
     Colat......................55   0   0
     Sun west of meridian.

NOTE. Ship time in this ex. is found from sun's altitude.

Equation of time ...11ᵐ 42·9ˢ, additive to app. time.
     Ans. Longitude=36° 40′ W.

EXAMPLE OF STAR-LUNAR.   ALTITUDES OBSERVED.

172. Given the following quantities; to construct figures, and compute the longitude.

Moon's app. alt.......31° 20′   3″   True distances, from *Nautical Almanac.*
   „   true alt. ...35   6   46   At 9ʰ ....96° 32′ 41″
Star's app. alt. 32   57   59   „ 12ʰ 95   7   31
   „   true alt. 32   56   30
Apparent dist. 96   55   26   Star east of meridian.
Auxil. angle A 60   18   2   Star's pol. dist. ...81° 30′ 55″
     „   zen. dist. ...57   3   30
     Colat. 44   39   50

NOTE. Ship time is found from star's altitude.

Star's RA ....19° 43′ 11″
RA mean sun 4   59   47·7
     Ans. Longitude=29° 17′ 15″ E.

LONGITUDE BY LUNAR.   ALTITUDES CALCULATED.

When the altitudes are to be calculated, we must previously know the hour-angles of the heavenly bodies; and these are readily found when the time at the ship is given, which it is supposed to be in this case (note, p. 52). The following examples, and the constructions which accompany them, will sufficiently indicate the several steps to be taken for finding the longitude.

First example. Sun-lunar. Altitudes calculated.

(The distance cleared by means of *auxiliary angle A.*)

173. Dec. 8, 1857, at $10^h 14^m 12^s$ A.M., ship mean time, in lat. 50° 37′ 12″ N., and long. by account 1° 6′ W., the observed distance of the nearest limbs of the sun and moon was 87° 51′ 30″. Construct the figure, and find by calculation the longitude.

Elements from *Nautical Almanac,* computed for the Greenwich date:

(1.) Sun's decl. =22° 45′ 31″ S.; (2.) sun's semi. =16′ 17″; (3.) eq. of time=$7^m 49 \cdot 4^s$ (additive to mean time); (4.) RA mean sun=$17^h 8^m 28 \cdot 8^s$; (5.) moon's RA=$11^h 17^m 1^s$; (6.) moon's decl. =5° 6′ 41″ N.; (7.) moon's semi. (aug.)=15′ 25·5″; (8.) moon's hor. parallax=56′ 15·4″; (9.) distances at 21^h=88° 50′ 57″, at 24^h=87° 23′ 51″.

The several parts of the lunar to be calculated are in order as follows:

(1.) The hour-angles of the sun and moon.

(2.) The true and apparent altitudes of sun and moon.

(3.) The true distance.

(4.) The Greenwich mean time. (5.) The longitude in time.

Construction of figure.

Let $\text{Q} \text{R} \text{Q}'$ represent the celestial equator, $\text{Q} \text{Q}'$ the celestial meridian, P the pole, and z the zenith. Take $\text{Q} \text{Q}'m = 22^h 14^m 12^s$, then m is the place of the mean sun. Let s be the true place of the sun (decl. 22° 45′ 31″ S.), and through s draw zs, a circle of altitude; then since the sun is raised more by refraction than depressed by parallax, its apparent place will be above s, as at s′, in same vertical circle zs (supposing the earth to be a sphere): draw the circle of decl. Ps. Again, take $m\text{Q}\text{A} = 17^h 8^m 28 \cdot 8^s$, the right ascension of mean sun; then A is the place of the first point of Aries. To find the place of the moon, take $\text{A}\text{R}=11^h 17^m 1^s$, the moon's RA, and draw the circle of declination PR; then the moon is in PR. Let M be the true place of the moon (decl. 5° 6′ 41″ N.), and through M draw zM, a circle of altitude; then, since the moon is depressed more by parallax than raised by refraction, its apparent place will be below M, as at M′, in the same vertical circle. Join M′s′ and Ms; then M′s′ is the apparent distance of the two bodies (=87°

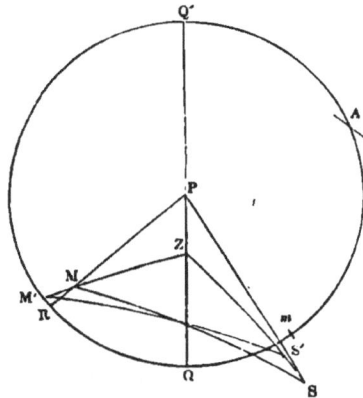

51′ 30″+sum of semidiameters)=88° 23′ 12″, and Ms is the true distance to be computed.

(1.) To find the hour-angles z p s and z p m.

	Sun's hour-angle z p s.			Moon's hour-angle z p m. (See note, p. 52.)		
Ship mean time q q'm ...	22ʰ	14ᵐ	12'	R A mean sun.....17ʰ	8ᵐ	28·8'
Equation of time m p s ...		7	49·4	Ship mean time..22	14	12·0
	22	22	1·4	39	22	40·8
	24			Moon's R A 11	17	1·0
∴ sun's hour-angle z p s...	1	37	58·6	28	5	39·8
				∴ moon's hour-angle z p m = 4	5	39·8

To find the sun's true zenith distance z s.

In the spherical triangle z p s, are given p z the colatitude, p s the polar distance, and z p s the hour-angle; to calculate z s the true zenith distance = 76° 30′ 12″ (*Trig.* Rule IX.).

To find the moon's true zenith distance z m.

In the spherical triangle z p m, are given p z the colatitude, p m the polar distance, and z p m the hour-angle; to calculate z m the true zenith distance = 68° 12′ 34″.

(2.) To find the true and apparent altitudes of the sun.

Sun's true zenith dist. = 76°	30′	12″	= z s	
∴ sun's true alt. = 13	29	48	= 90° − z s	
and by the tables, sun's cor. in alt. =	3	50	= s s′	
∴ sun's apparent alt. = 13	33	38	= 90° − z s′	

To find the true and apparent altitudes of the moon.

Moon's true zen. dist. = 68°	12′	34″	= z m	
∴ moon's true alt. = 21	47	26	= 90° − z m	
and by the tables, moon's cor. in alt. =	50	2	= m m′ (see Chap. V.).	
∴ moon's app. alt. = 20	57	24		
and by the tables, aux. angle A = 60	10	45		

Hence by the formula, p. 86, the true distance m s = 88° 12′ 36″; and the Greenwich mean time corresponding to this distance = 22ʰ 19ᵐ 15' (see p. 88), and the long. in time = 0ʰ 5ᵐ 3' W., or 1° 15′ 45″ W.

174. *Second Example. Sun-lunar. Altitudes calculated.*

(The distance in this example is cleared by the common rules of Spherical Trigonometry; see the Seventh Method.)

Given the following quantities; to construct a figure, and calculate the longitude.

(1.) Lat.=50° 37′ 30″ N., sun's decl.=22° 42′ 28″ N., sun's hour-angle =1ʰ 23ᵐ 24ˢ (west of meridian) ; to calculate sun's true altitude=57° 42′ 16″.

(2.) Lat.=50° 37′ 30″, moon's decl.=11° 43′ 50″ N., moon's hour-angle =3ʰ 6ᵐ 55ˢ (east of meridian); to calculate moon's true altitude=35° 39′ 23″.

(3.) Sun's correction in alt.=32″+, moon's correction in alt.=45′ 8″−; to find sun's app. alt.=57° 42′ 48″, and moon's apparent alt.=34° 54′ 15″.

(4.) Sun's app. zen. dist.=32° 17′ 12″, moon's app. zen. dist.=55° 5′ 45″, and apparent dist. of centers=65° 35′ 17″; to calculate the angle at zenith =99° 15′ 15″.

(5.) Sun's true zen. dist.=32° 17′ 44″, moon's true zen. dist.=54° 20′ 37″, and angle at zenith=99° 15′ 15″; to calculate the true distance=64° 58′ 48″.

(6.) True dist. at observation=64° 58′ 48″, dist. at 0ʰ=64° 15′ 42″, dist. at 3ʰ=65° 45′ 56″; to calculate Greenwich mean time=1ʰ 25ᵐ 58·5ˢ.

(7.) Ship mean time (found as in the rule for longitude by chronometer or otherwise)=1ʰ 21ᵐ 46·5ˢ, and Greenwich mean time=1ʰ 25ᵐ 58·5ˢ, to calculate the LONGITUDE=4ᵐ 12ˢ=1° 3′ W.

Construction of figure from preceding data.

(1.) Let P be the pole, z the zenith, and s the true place of the sun ; then, in the spherical triangle z P s, are given P z=39° 22′ 30″, P s= 67° 17′ 32″, and angle s P z= 1ʰ 23ᵐ 24ˢ ; to calculate the sun's true zenith dist. z s=32° 14′ 44″: hence sun's true altitude s o= 57° 42′ 16″. Let s′ be the apparent place of sun ; then s s′=32″, and ∴ sun's apparent altitude s′o =57° 42′ 48″.

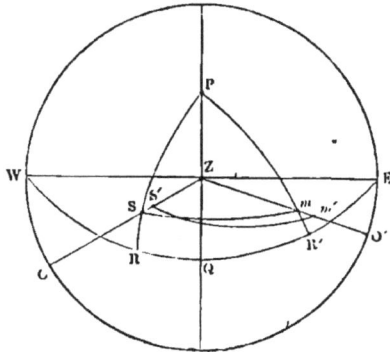

Again, let m be the true place of the moon ; then, in the spherical triangle z P m, are given P z= 39° 22′ 30″, P m=78° 16′ 10″, and angle z P m=3ʰ 6ᵐ 55ˢ; to calculate the moon's true zenith distance z m= 54° 20′ 37″: hence moon's true altitude m o′=35° 39′ 23″. Let m′ be the apparent place of the moon ; then m m′=45′ 8″, and ∴ moon's apparent altitude m′o′=34° 54′ 15″.

(2.) Join s′m′; then, in the spherical triangle z s′m′, are given z s′= 32° 17′ 12″, z m′=55° 5′ 45″, and apparent distance s′m′=65° 35′ 17″; to calculate the angle s′z m′=99° 15′ 15″. Again, in the spherical triangle z s m, are given z s=32° 17′ 44″, z m=54° 20′ 37″, and angle s z m=

99° 15′ 15″; to calculate the true distance $sm = 64^\circ$ 58′ 48″: hence the longitude is found as pointed out above.

PRACTICAL METHODS OF CLEARING THE LUNAR DISTANCE.

It may be necessary sometimes to be acquainted with other methods of clearing the lunar distance besides the one made use of in the preceding examples, since that may not be adapted to the tables in the student's hands. If, for instance, his tables do not contain the auxiliary angle A or a table of natural versines, the rule derived from the formula on p. 86 cannot be used by him. We will therefore now show how the fundamental formulæ on p. 85 may be altered, so as to obtain a rule adapted to other tables; we shall thus form, in fact, several distinct methods of clearing the distance, such as those found in Norie's, Riddle's, and Raper's works on navigation. The investigation of these several methods will not be without its use to the student, even if he should have no occasion to make a practical application of them.

SECOND METHOD OF CLEARING THE DISTANCE.

By means of versines, but not requiring the auxiliary angle A.

Since, as in former method, p. 85,

$$\frac{\cos. \; D - \cos. \; z \; . \; \cos. \; z_1}{\sin. \; z \; . \; \sin. \; z_1} = \frac{\cos. \; d - \sin. \; a \; . \; \sin. \; a_1}{\cos. \; a \; . \; \cos. \; a_1},$$

$$\therefore \; \frac{\cos. \; D - \cos. \; z \; . \; \cos. \; z_1}{\sin. \; z \; . \; \sin. \; z_1} - 1 = \frac{\cos. \; d - \sin. \; a \; . \; \sin. \; a_1}{\cos. \; a \; . \; \cos. \; a_1} - 1;$$

$$\text{or} \; \frac{\cos. \; D - (\cos. \; z \; . \; \cos. \; z_1 + \sin. \; z \; . \; \sin. \; z_1)}{\sin. \; z \; . \; \sin. \; z_1}$$

$$= \frac{\cos. \; d - (\cos. \; a \; . \; \cos. \; a_1 + \sin. \; a \; . \; \sin. \; a_1)}{\cos. \; a \; . \; \cos. \; a_1};$$

$$\therefore \; \frac{\cos. \; D - \cos. \; (z \sim z_1)}{\sin. \; z \; . \; \sin. \; z_1} = \frac{\cos. \; d - \cos. \; (a \sim a_1)}{\cos. \; a \; . \; \cos. \; a_1};$$

$$\therefore \; \cos. \; D = \cos. \; (z \sim z_1) + \frac{\cos. \; d - \cos. \; (a \sim a_1)}{\cos. \; a \; . \; \cos. \; a_1} \; . \; \sin. \; z \; . \; \sin. \; z_1$$

$$= \cos. \; (z \sim z_1) - 2 \sin. \tfrac{1}{2}(d + \overline{a \sim a_1}) . \sin. \tfrac{1}{2}(d - \overline{a \sim a_1}) . \frac{\sin. \; z \; . \; \sin. \; z_1}{\cos. \; a \; . \; \cos. \; a_1};$$

$$\therefore \; \text{vers.} \; D = \text{vers.} (z \sim z_1) + 2 \sin. \tfrac{1}{2}(d + \overline{a \sim a_1}) . \sin. \tfrac{1}{2}(d - \overline{a \sim a_1}) . \frac{\sin. \; z \; . \; \sin. \; z_1}{\cos. \; a \; . \; \cos. \; a_1}$$

$$= \text{vers.} \; (z \sim z_1) + \text{vers.} \; \theta. \quad \text{By assuming}$$

$$\text{vers.} \; \theta = 2 \sin. \tfrac{1}{2}(d + \overline{a \sim a_1}) . \sin. \tfrac{1}{2}(d - \overline{a \sim a_1}) \; . \; \frac{\sin. \; z \; . \; \sin. \; z_1}{\cos. \; a \; . \; \cos. \; a_1}.$$

These formulæ determine the true distance D without using the auxiliary angle A, as in the first method.

EXAMPLE OF CLEARING THE DISTANCE BY SECOND METHOD.

175. Find the true distance D of the sun and moon, having given apparent distance $d = 35° 47' 24''$, moon's app. alt. $a = 57° 11' 25''$, sun's app. alt. $a_1 = 34° 21' 32''$, moon's true zenith distance $z = 32° 19' 50''$, and sun's true zenith distance $z_1 = 55° 39' 46''$.

$$\text{vers. } D = \text{vers.}(z \sim z_1) + \text{vers. } \theta,$$

$$\text{vers. } \theta = 2 \sin. \tfrac{1}{2}(d + \overline{a \sim a_1}) . \sin. \tfrac{1}{2}(d - \overline{a \sim a_1}) . \frac{\sin. z . \sin. z_1}{\cos. a . \cos. a_1}$$

$$\therefore \text{log. vers. } \theta = 6\cdot301030 + \text{log. sin. } \tfrac{1}{2}(d + \overline{a \sim a_1}) +$$

$$\text{log. sin. } \tfrac{1}{2}(d - \overline{a \sim a_1}) + \text{log. sin. } z + \text{log. sin. } z_1 - (\text{log. cos. } a + \text{log. cos. } a_1) - 20.$$

Calculation.

a	57°	11'	25''	6·301030	
a_1	34	21	32	9·689792	
$a \sim a_1$	22	49	53	9·052480	
d	35	47	24	9·728194	9·733883
$d + \overline{a \sim a_1}$	58	37	17	9·916839	9·916727
$d - \overline{a \sim a_1}$	12	57	31	44·688335	19·650610
$\therefore \tfrac{1}{2}(d + \overline{a \sim a_1})$	29	18	38·5	19·650610	
$\tfrac{1}{2}(d - \overline{a \sim a_1})$	6	28	45·5	log. vers. θ... 5·037725	
z	32	19	50	$\therefore$ vers. θ... 109076	
z_1	55	39	46	vers. $\overline{z \sim z_1}$... 81776	
$z \sim z_1$	23	19	56	$\therefore$ vers. D... 190852	

$$\therefore D = 35° 59' 14''.$$

This work may be somewhat shortened by using the table of haversines, as in the following method.

THIRD METHOD OF CLEARING THE DISTANCE.

Using haversines, but not requiring the auxiliary angle A.

By second method,

$$\text{vers. } \theta = 2 \sin. \tfrac{1}{2}(d + \overline{a \sim a_1}) . \sin. \tfrac{1}{2}(d - \overline{a \sim a_1}) . \frac{\sin z . \sin.' z_1}{\cos. a . \cos. a_1},$$

$$\therefore \tfrac{1}{2} \text{vers. } \theta \text{ or hav. } \theta = \sin. \tfrac{1}{2}(d + \overline{a \sim a_1}) . \sin. \tfrac{1}{2}(d - \overline{a \sim a_1}) . \sin. z . \sin. z_1 .$$
$$\text{sec. } a . \text{sec. } a_1$$

$$= \sqrt{\text{hav. } (d + \overline{a \sim a}) . \text{hav. } (d - \overline{a \sim a}) . \sin. z_, . \sec. a . \sin. z_1 . \sec. a_1}$$

From which θ may be found and thence D, as in second method.

EXAMPLE OF CLEARING THE DISTANCE BY THIRD METHOD.

176. Find the true distance D of the sun and moon, having given apparent distance $d = 35° 47' 24''$, moon's app. alt. $a = 57° 11' 25''$, sun's app. alt.

H

$a_1 = 34°\ 21'\ 32''$, moon's true zen. dist. $z = 32°\ 19'\ 50''$, and sun's true zen. dist. $z_1 = 55°\ 39'\ 46''$.

$$\text{hav. } \theta = \sqrt{\text{hav. } (d + a \sim a_1) \cdot \text{hav. } (d - a \sim a_1)} \cdot \sin z \cdot \sec a \cdot \sin z_1 \cdot \sec a_1,$$

and vers. $\text{D} = \text{vers. } \overline{z \sim z_1} + \text{vers. } \theta$.

a	57°	11'	25''	hav. $\overline{(d + a \sim a_1)}$	9·379578
a_1	34	21	32	hav. $\overline{(d - a \sim a_1)}$	8·104941
$a \sim a_1$	22	49	53		2)17·484519
d	35	47	24		8·742259
$d + a \sim a_1$	58	37	17	sin. z	9·916838
$d - a \sim a_1$	12	57	31	sec. a	0·083270
z	32	19	50	sin. z_1	9·728177
z_1	55	39	46	sec. a_1	0·266137
$z \sim z_1$	23	19	56	hav. θ	8·736681

$$\therefore \theta = 27°\ 0'\ 35''$$

vers. θ 109069

vers. $\overline{z \sim z_1}$ 81775

vers. D 190844

$$\therefore \text{ true dist. } \text{D} = 35°\ 59'\ 12''$$

FOURTH METHOD OF CLEARING THE DISTANCE.

The previous method simplified, by tabulating a part of the formula.

In the expression, versine $\theta = 2 \sin \frac{1}{2} (d + a \sim a_1) \cdot \sin \frac{1}{2} (d - a \sim a_1) \cdot \frac{\sin z \cdot \sin z_1}{\cos a \cdot \cos a_1}$, the quantity $\frac{\sin z_1}{\cos a_1}$ changes very slowly for all values of z_1 and a_1; since z_1 and a_1 are the true zen. dist. and app. alt. of the sun or star, and therefore $\frac{\sin z_1}{\cos a_1} = \frac{\cos \text{ true alt.}}{\cos \text{ app. alt.}} = 1$ nearly. We may therefore compute beforehand the values of the expression $2 \cdot \frac{\sin z_1}{\cos a_1}$ for all altitudes of the sun or star, and form a small table of the results. This has accordingly been done (see Table B), and we can take the value of $\frac{2 \sin z_1}{\cos a_1}$ out by inspection (entering the table with the app. alt. of the heavenly body); the labour of calculating $\frac{2 \sin z_1}{\cos a_1}$ is thus avoided. The value of the true distance D may then be expressed as follows:

$$\text{vers. } \text{D} = \text{vers. } \overline{z \sim z_1} + \text{vers. } \theta, \text{ and}$$

$$\text{vers. } \theta = \text{B} \sin z \cdot \sec a \cdot \sqrt{\text{hav. } (d + a - a_1) \cdot \text{hav. } (d - a - a_1)},$$

where log. B = quantity in following table.

Construction of Table B.

Since vers. $\theta = 2 \sin. \frac{1}{2} (d + \overline{a \sim a_1}) . \sin. \frac{1}{2} (d - \overline{a \sim a_1}) \dfrac{\sin. z . \sin. z_1}{\cos. a . \cos. a_1}$

$\therefore$ vers. $\theta = 2 \sqrt{\text{hav. } (d + \overline{a \sim a_1}) . \text{hav. } (d - \overline{a \sim a_1}) . \sin. z . \sec. a . \sin. z_1 . \sec. a_1},$

$\therefore$ log. vers. $\theta - 6 = \cdot 301030 + \log. \sin. z_1 - 10 + \log. \sec. a_1 - 10 + \frac{1}{2} \{ \log. \text{hav.}$
$(d + \overline{a \sim a_1}) + \log. \text{hav. } (d - \overline{a \sim a_1}) \} - 10 + \log. \sin. z - 10 + \log. \sec. a - 10,$

or log. vers. $\theta = 6 \cdot 301030 + \log. \sin. z_1 + \log. \sec. a_1 - 20$

$+ \frac{1}{2} \{ \log. \text{hav. } (d + \overline{a \sim a_1}) + \log. \text{hav. } (d - \overline{a \sim a_1}) \} + \log. \sin. z + \log. \sec. a - 30.$

The quantities $6 \cdot 301030 + \log. \sin. z_1 + \log. \sec. a_1 - 20$, may be computed for all altitudes of the sun or star, and tabulated. Let the sum be denoted by log. B : then log. B $= 6 \cdot 301030 + \log. \sin. z_1 + \log. \sec. a_1 - 20$;
and the formula for finding θ will then become

log. vers. $\theta = \frac{1}{2} \{ \log. \text{hav. } (d + \overline{a \sim a_1}) + \log. \text{hav. } (d - \overline{a \sim a_1}) \} + \log. \sin. z +$
$\log. \sec. a + \log. B - 30.$

177. *Example.* Find the value of log. B, when the sun's apparent altitude $a_1 = 34° 21' 32''$, and sun's true zenith distance $z_1 = 55° 39' 46''$.

log. B $= 6 \cdot 301030 + \log. \sin. z_1 + \log. \sec. a_1 - 20.$

$$6 \cdot 301030$$
log. sin. $z_1 \ldots 9 \cdot 916839$
,, sec. $a_1 \ldots 0 \cdot 083273$
$\therefore$ log. B $\ldots 6 \cdot 301142$ for alt. 34°.

And in the same manner may all the other values be computed.

TABLE B.

FOR THE *SUN*.		FOR A *STAR* OR *PLANET*.	
ARGUMENT.	SUN'S APP. ALT.	ARGUMENT.	SUN'S APP. ALT.
App. alt.	Log. B.	App. alt.	Log. B.
90	6·301134	90	6·301153
75	6·301135	30	6·301152
65	6·301136	20	6·301151
60	6·301137	15	6·301150
55	6·301138	12	6·301149
50	6·301139	10	6·301148
45	6·301140	9	6·301147
40	6·301141	8	6·301146
35	6·301142	7	6·801144
30	6·301143	6	6·301141
25	6·301144	5	6·301137
20	6·301145	4	6·301130
10	6·301144		
8	6·301143		
7	6·301142		
6	6·301139		
4	6·301130		

EXAMPLE OF CLEARING THE DISTANCE BY FOURTH METHOD.

178. Find the true distance D of the sun and moon, having given the apparent distance $d=35°\ 47'\ 24''$, sun's app. alt. $a_1=34°\ 21'\ 32''$, moon's app. alt. $a=57°\ 11'\ 25''$, sun's true zenith distance $z_1=55°\ 39'\ 46''$, and moon's true zenith distance $z=32°\ 19'\ 50''$.

Calculation.

vers. $\theta=D$ sin. z . sec. a . $\sqrt{\text{hav. } (d+a\frown a_1)\ .\ \text{hav. } (d-a\frown a_1)}$

vers. $D=$ vers. $\overline{z\frown z_1}+$ vers. θ.

a_1............34° 21′ 32″		$\frac{1}{2}$ log. hav. $(d+a\overline{-a_1})$...4·689789
a............57 11 25		$\frac{1}{2}$ log. hav. $(d-\overline{a-a_1})$...4·052470
$a-a_1$ 22 49 53		,, sin. z ....9·728177
d 35 47 24		,, sec. a ....0·266137
$d+\overline{a-a_1}$......58 37 17		,, B ....6·301138
$d-\overline{a-a_1}$......12 57 31		,, vers. θ............5·037711
z............32 19 50		∴ vers. $\theta=109072$
z_1............55 39 46		vers. $(z\frown z_1)=\ 81776$
$z\frown z_1$ 23 19 56		∴ vers. $D=190848$
		and ∴ $D=35°\ 59'\ 12''$

FIFTH METHOD OF CLEARING THE DISTANCE.

By rejecting odd seconds.

The log. sines, &c. in most collections of Nautical Tables are given only to the nearest minute, or quarter of a minute ; it will therefore be necessary in all the previous methods (excepting the first, p. 86) to proportion for the odd seconds of the altitudes and distance. Some part of this tedious operation may be avoided by adopting the following arrangement.

Reject the seconds in the apparent altitudes, but add them to the respective true zenith distances ; reject also the seconds in the apparent distance.* Compute the true distance D as before. To the value of D thus found add the seconds rejected from the apparent distance d ; the result will then be the correct true distance required.

EXAMPLE OF CLEARING THE DISTANCE BY FIFTH METHOD.

179. Given the app. dist. $d=35°\ 47'\ 24''$, sun's app. alt. $a_1=34°\ 21'\ 32''$, moon's app. alt. $a=57°\ 11'\ 25''$, sun's true zenith distance $z_1=55°\ 39'\ 46''$, moon's true zenith dist. $z=32°\ 19'\ 50''$; to find the true distance.

* If tables of sines, &c., computed to minutes only are used, this operation must be modified by sometimes *adding* a number of seconds to the apparent distance, so as to make $\frac{1}{2}(d+\overline{a\frown a_1})$ always to consist of degrees and minutes only ; the seconds thus added must be subtracted from the computed distance to get the true distance.

vers. $\theta = $ B . sin. z . sec. a . $\sqrt{\text{hav. } (d + \overline{a \sim a_1}) \text{ . hav. } (d - \overline{a \sim a_1})}$

vers. D $=$ vers. $(z \sim z_1) + $ vers. θ.

a .... 57° 11' (25")	$\therefore z = 32° 19' 50'' + 25'' = 32° 20' 15''$	
a_1 .... 34 21 (32)	$z_1 = 55$ 39 46 $+ 32 = 55$ 40 18	
$a - a_1$ 22 50	$z \sim z_1 = 23$ 20 3	
d 35 47 (24)	$\frac{1}{2}$ log. hav. $(d + \overline{a - a_1})$... 4·689761	
$d + (a - a_1)$... 58 37	$\frac{1}{2}$ log. hav. $(d - \overline{a - a_1})$... 4·052192	
$d - (a - a_1)$... 12 57	„ sin. z .... 9·728277	
	„ sec. a .... 0·266039	
	„ B .... 6·301138	
	„ vers. θ .... 5·037407	
	$\therefore$ vers. $\theta = 108995$	
	vers. $(z \sim z_1) = $ 81790	
	vers. D $= \overline{190785}$	
	or D $= 35° 58' 51'' + 24''$	
	$\therefore$ true distance $= 35$ 59 15	

By this method it appears that we only require to proportion for one quantity, namely sin. z, the moon's true zenith distance.

SIXTH METHOD OF CLEARING THE DISTANCE.

Requiring only the common tables of log. sines, &c.

In triangle zms, p. 84,

$$\cos. z = \frac{\cos. \text{D} - \cos. z \cdot \cos. z_1}{\sin. z \cdot \sin. z_1},$$

$$\therefore \cos. \text{D} = \cos. z \cdot \cos. z_1 + \sin. z \cdot \sin. z_1 \cdot \cos. z$$

$$= \cos. z \cdot \cos. z_1 + \sin. z \cdot \sin. z_1 \cdot \left(2 \cos^2 \cdot \frac{z}{2} - 1\right)$$

$$= \cos. z \cdot \cos. z_1 - \sin. z \cdot \sin. z_1 + 2 \sin. z \cdot \sin. z_1 \cdot \cos^2 \cdot \frac{z}{2}$$

$$= \cos. (z + z_1) + 2 \sin. z \cdot \sin. z_1 \cdot \cos^2 \cdot \frac{z}{2}.$$

But $\cos. \text{D} = 2 \cos^2 \cdot \dfrac{\text{D}}{2} - 1$, and $\cos. (z + z_1) = 2 \cos^2 \cdot \dfrac{z + z_1}{2} - 1$.

Making these substitutions, we have

$$\cos^2 \cdot \frac{\text{D}}{2} = \cos^2 \cdot \frac{z + z_1}{2} + \sin. z \cdot \sin. z_1 \cdot \cos^2 \cdot \frac{z}{2}$$

$$= \cos^2 \cdot \frac{z + z_1}{2} \left\{ 1 + \frac{\sin. z \cdot \sin. z_1 \cdot \cos^2 \cdot \frac{z}{2}}{\cos^2 \cdot \frac{z + z_1}{2}} \right\}$$

$$= \cos^2 \cdot \frac{z + z_1}{2} \{ 1 + \tan^2 \cdot \theta \} = \cos^2 \cdot \frac{z + z_1}{2} \cdot \sec^2 \cdot \theta \quad . . \quad (1)$$

$$\text{by assuming } \tan.^2\theta = \frac{\sin. z \cdot \sin. z_1 \cdot \cos.^2 \frac{z}{2}}{\cos.^2 \frac{z+z_1}{2}}$$

Now to find the value of $\cos.^2 \frac{z}{2}$ in terms of the sides of the triangle $z\,\text{m}'\text{s}'$, p 84, we have (see *Trigonometry*, Part II. p. 58, formula P)

$$\cos.^2 \frac{z}{2} = \sec. u \cdot \sec. a_1 \cdot \cos. \tfrac{1}{2}(a+a_1+d) \cdot \cos. \tfrac{1}{2}(a+a_1-d).$$

Substituting this in the value of $\tan.^2\theta$,

$$\therefore \tan. \theta = \frac{\sqrt{\sin.z.\sec.a.\sin.z_1.\sec.a_1.\cos.\tfrac{1}{2}(a+a_1+d)\cos.\tfrac{1}{2}(a+a_1-d)}}{\cos. \tfrac{1}{2}(z+z_1)} \cdots (2)$$

But $\cos. \tfrac{1}{2}(z+z_1) = \sin. \tfrac{1}{2}(\text{A}+\text{A}_1)$, if A and A_1 represent the true altitudes of the heavenly bodies.

Making these substitutions, we have

$$\tan. \theta = \frac{\sqrt{\cos.\text{A}.\sec.a.\cos.\text{A}_1.\sec.a_1.\cos.\tfrac{1}{2}(a+a_1+d).\cos.\tfrac{1}{2}(a+a_1-d)}}{\sin. \tfrac{1}{2}(\text{A}+\text{A}_1)} \cdots (3)$$

$$= \frac{\text{M}}{\sin. \tfrac{1}{2}(\text{A}+\text{A}_1)}, \text{ if we assume}$$

$$\text{M} = \sqrt{\cos.\text{A} \cdot \sec. a \cdot \cos. \text{A}_1 \cdot \sec. a_1 \cdot \cos. \tfrac{1}{2}(a+a_1+d) \cdot \cos. \tfrac{1}{2}(a+a_1-d)}$$

To simplify these formulæ:

$$\text{From (1), } \cos.\frac{\text{D}}{2} = \frac{\cos. \tfrac{1}{2}(z+z_1)}{\cos. \theta} = \frac{\sin. \tfrac{1}{2}(\text{A}+\text{A}_1)}{\cos. \theta}$$

and since

$$\tan. \theta = \frac{\text{M}}{\sin. \tfrac{1}{2}(\text{A}+\text{A}_1)} = \frac{\sin. \theta}{\cos. \theta} \therefore \frac{\text{M}}{\sin. \theta} = \frac{\sin. \tfrac{1}{2}(\text{A}+\text{A}_1)}{\cos. \theta} \therefore \cos. \frac{\text{D}}{2} = \frac{\text{M}}{\sin. \theta}$$

whence it appears that the true distance D may be found by means of the following formulæ:

$$\text{M} = \sqrt{\cos. \text{A} \cdot \sec. a \cdot \cos. \text{A}_1 \cdot \sec. a_1 \cdot \cos. (a+a_1+d) \cdot \cos. \tfrac{1}{2}(a+a_1-d)}$$

$$\tan. \theta = \frac{\text{M}}{\sin. \tfrac{1}{2}(\text{A}+\text{A}_1)}, \qquad \cos. \frac{\text{D}}{2} = \frac{\text{M}}{\sin. \theta}.$$

EXAMPLE OF CLEARING THE DISTANCE BY SIXTH METHOD.

180. Given the sun's app. alt. $a_1 = 18° 22' 13''$, sun's true alt. $\text{A}_1 = 18° 19' 30''$, moon's app. alt. $a = 13° 21' 21''$, moon's true alt. $\text{A} = 14° 15' 24''$, and apparent distance $d = 61° 19' 49''$; to find the true distance D.

$$\text{M} = \sqrt{\cos. \text{A}_1 \cdot \sec. a_1 \cdot \cos. \text{A} \cdot \sec. a \cdot \cos. \tfrac{1}{2}(a+a_1+d) \cdot \cos. \tfrac{1}{2}(a+a_1-d)}$$

$$\tan. \theta = \frac{\text{M}}{\sin. \tfrac{1}{2}(\text{A}+\text{A}_1)} \qquad \cos. \frac{\text{D}}{2} = \frac{\text{M}}{\sin. \theta}.$$

a..................	13°	21'	21"	$\frac{1}{2}(a+a_1+d)$	46° 31'	41·5"
a_1	18	22	13	$\frac{1}{2}(a+a_1-d)$	14 48	7·5

$a+a_1$............	31	43	34	A	14 15 24
d..................	61	19	44	A$_1$	18 19 30

$a+a_1+d$ 93	3	23	A+A$_1$......................	32 34 54
$a+a_1-d$ 29	36	15	∴ $\frac{1}{2}$(A+A$_1$)......16	17 27

log. cos. A$_1$ .... 9·977398

„ sec. a_1 .... 0·022716

0·000114

„ cos. A .... 9·986414

„ sec. a .... 10·011908

„ cos. $\frac{1}{2}(a+a_1+d)$ ...... 9·837587

„ cos. $\frac{1}{2}(a+a_1-d)$ ...... 9·985343

39·821366

19·910683 .... 19·910683

„ sin. $\frac{1}{2}$(A+A$_1$) .... 9·447953   log. sin. θ...... 9·975639

„ tan. θ .... 10·462730   „ cos. $\frac{D}{2}$...... 9·935044

∴ θ=70° 59' 17"   ∴ $\frac{D}{2}$...... 30° 33' 42"

2

∴ true dist.=61   7 24

The preceding method also involves the necessity of proportioning for seconds; the labour, however, may be diminished by neglecting the seconds in the apparent altitudes and distance, and by applying them to the true altitudes and calculated distance, as pointed out in the fifth method. Moreover, since the apparent altitude of the sun or star is nearly equal to the true altitude, therefore cos. A$_1$. sec. a_1 (two of the factors in the formula for computing M) may be calculated beforehand and formed into a small table : this has accordingly been done (see Table C). Adopting these simplifications, the preceding example may be worked out as follows :

EXAMPLE OF CLEARING THE DISTANCE BY SIXTH METHOD, SUPPRESSING SECONDS AND USING AN AUXILIARY TABLE.

181. Given sun's app. alt. a_1=18° 22' 13", sun's true alt. A$_1$=18° 19' 30", moon's app. alt. a=13° 21' 21", moon's true alt. A=14° 15' 24", and apparent distance d=61° 19' 49"; to find the true distance D.

$$\text{M} = \sqrt{\cos. \text{A} . \sec. a . \cos. \tfrac{1}{2}(a+a_1+d) . \cos. \tfrac{1}{2}(a+a_1-d) . \text{c}}$$

$$\tan. \theta = \frac{\text{M}}{\sin. \tfrac{1}{2}(\text{A}+\text{A}_1)}, \qquad \cos. \frac{\text{D}}{2} = \frac{\text{M}}{\sin. \theta}.$$

a	13°	21′	(21″)		A−21″=14°	15′	3″
a_1...........	18	22	(13)		A₁−13 =18	19	17
	31	43				32	34	20
d	61	19	(49)		½(A+A₁) ...16	17	10
Sum.........	93	2	....	½ sum	46	31	
Diff..........	29	36	....	½ diff.	14	48	

log. cos. A............... 9·986426
 ,,   sec. a...............10·011897
 ,,   cos. ½ sum......... 9·837679
 ,,   cos. ½ diff......... 9·985347   log. M............19·910731
 ,,   c 114   ,, sin. θ ...... 9·975657
                        39·821463   ,, cos. ½ D ... 9·935074
 ,,   M ....19·910731   ∴ ½D=30° 33′ 18″
 ,,   sin. ½(A+A₁)...... 9·447831   ∴ D=61 6 36+49″
 ,,   tan. θ...............10·462900   or true dist.=61 7 25

TABLE C.

FOR THE *SUN.*		FOR A *STAR* OR *PLANET.*	
ARGUMENT. SUN'S APP. ALT.		ARGUMENT. STAR'S APP. ALT.	
App. alt.	Log. C.	App. alt.	Log. C.
90	0·000104	90	0·000123
75	0·000105	30	0·000122
65	0·000106	20	0·000121
60	0·000107	15	0·000120
55	0·000108	12	0·000119
50	0·000109	10	0·000118
45	0·000110	9	0·000117
40	0·000111	8	0·000116
35	0·000112	7	0·000114
30	0·000113	6	0·000111
25	0·000114	5	0·000107
20	0·000115	4	0·000100
10	0·000114	3	0·000090
8	0·000113		
7	0·000112		
6	0·000109		
4	0·000100		

Construction of Table C.

In the expression,

$$\text{M} = \sqrt{\overline{\cos. \text{A}_1 \sec. a_1 . \cos. \text{A} . \sec. a . \cos. \tfrac{1}{2}(a+a_1+d) . \cos. \tfrac{1}{2}(a+a_1-d)}},$$

A_1, the true altitude of the sun or star, is nearly equal to a_1, its apparent alt. ;
the quantity $\cos. \text{A}_1 . \sec. a_1 = \dfrac{\cos. \text{A}_1}{\cos. a_1} = 1$ nearly for all altitudes: it therefore
may be computed and formed into a table.

Let $c = \cos. \text{A}_1 . \sec. a_1$, $\therefore$ log. $c = $ log. cos. $\text{A}_1 + $ log. sec. $a_1 - 20$.

182. Find by calculation the value of log. c, when the sun's true altitude $\text{A}_1 = 18^\circ \, 19' \, 30''$, and app. alt. $a_1 = 18^\circ \, 22' \, 13''$.

$$\begin{aligned}
\text{log. cos. } \text{A}_1 &\ldots \; 9{\cdot}977398 \\
\text{,, \quad sec. } a_1 &\ldots 10{\cdot}022716
\end{aligned}$$

$$\therefore \text{ log. } c = \; 0{\cdot}000114 \text{ for alt. } 18^\circ.$$

It is also evident that Table C may be formed from Table B, by subtracting log. $2 = {\cdot}301030$.

SEVENTH METHOD OF CLEARING THE DISTANCE.

By the Common Rules of Spherical Trigonometry.

The true distance MS may be computed as follows (fig. p. 84) :

1. In triangle M_1ZS_1 the three sides are given, namely, the two apparent distances ZM_1 and ZS_1 and the observed distance M_1S_1 ; to find the angle z.

2. In triangle MZS are given the two sides MZ and SZ, the true zenith distances of the heavenly bodies, and the included angle z just found, to compute the third side MS, *the true distance required.*

The practical inconvenience of this method arises from the necessity of taking out the log. sines, &c., to the nearest second, a work of considerable labour with the common tables of logarithmic sines, &c., which seldom give the arcs nearer than 15''. To obviate this the true distance is now usually found in terms of the versines, the arcs in the table of versines being given to the nearest second.

Other rules have been proposed from time to time for clearing the distance. The methods investigated above have all been derived from the same fundamental formulæ, and from them may be deduced the rules given in most works on Navigation. There are, however, several approximate methods which might be examined with advantage by the student, not so much on account of their practical value, as for the ingenuity they exhibit in the attempt to shorten the labour of calculation ; but these and other matters connected with lunar observations must be left for future discussion in a Third Part. (See Admiral Shadwell's Nautical Works.)

THE VARIATION OF THE COMPASS.

INVESTIGATION OF RULES FOR FINDING THE VARIATION OF THE COMPASS.

The true bearing of a heavenly body can be calculated by either of the three following methods.

First. By computing the true bearing of a heavenly body at its rising or setting, called an *amplitude.*

Second. By observing its altitude, and thence with its declination and the latitude of the observer determining the angle at the zenith or true bearing of the body : this angle, or the corresponding arc of the horizon which it subtends, is called the *azimuth* of the heavenly body.

Third. By noting the time by means of a chronometer, and thence determining the hour-angle (note, p. 52) ; then, knowing the hour-angle and the declination and latitude, the angle at the zenith or azimuth may be computed as before.

If the compass bearing is observed at the time of the observation, the *variation of the compass,* that is, the difference between the true and compass bearing, may be determined.

THE AMPLITUDE.

PROBLEM XXVIII.

Given the latitude of the ship, and the declination of a heavenly body ; to find the amplitude of the heavenly body.

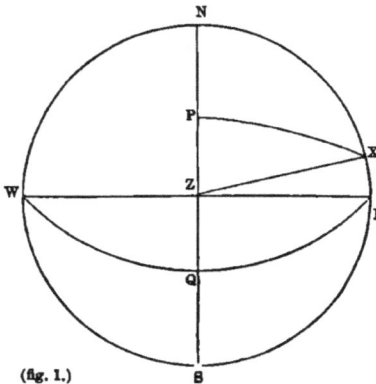
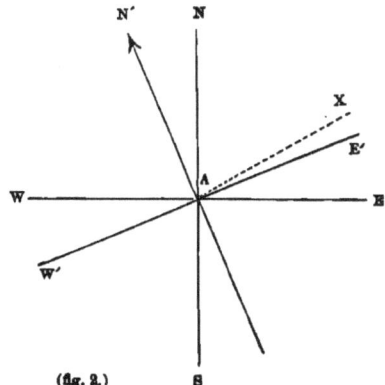

(fig. 1.)   (fig. 2.)

Let N W S E represent the horizon, N Z S the celestial meridian, P the pole,

z the zenith, and x a heavenly body at rising; then the arc ʙ x, or angle
ʙ z x, is its amplitude.

In the quadrantal triangle ᴘzx, are given ᴘz the colatitude, ᴘx the pol.
dist. or co-decl., and zx=90°; to calculate the angle ᴘzx, and thence its
complement ʙzx, or the *amplitude.*

By Rule XIV. *Trig.* Part I., for quadrantal triangles,
   (fig. 1) cos. ᴘx=sin. ᴘz . cos. ᴘzx,
   or sin. decl.=cos. lat. . sin. amplitude;

$$\therefore \text{ sin. ampl.} = \frac{\text{sin. decl.}}{\text{cos. lat.}} = \text{sin. decl. . sec. lat.}$$

EXAMPLE.

183. Given the lat. of ship=47° 50′ N., and the decl. of sun at rising=
17° 54′ 44″ N.; to find the true bearing or amplitude. If the compass
bearing at the same time was E. 7° 10′ N., find also the variation of the
compass.

(1.) To find the true bearing (fig. 1).

Construction. Let x be the place of the sun at rising, and complete the
figure; then, in the quadrantal triangle zᴘx, are given the colat. ᴘz=42° 10′,
the pol. dist. ᴘx=72° 5′ 16″, and the zen. dist. zx=90°; hence the angle
ᴘzx, and therefore its complement ʙzx, may be found by Trigonometry,
Rule XIV., or otherwise by the above formula.

sin. ampl.=sin. decl. . sec. lat.
log. sin. decl......9·487936
  ,, sec. lat. ......0·173090
_____
  .: ,, sin. ampl. ...9·661026
  .: true bearing=E. 27° 16′ 15″ N.=ʙx.

(2.) To find the variation of the compass (fig. 2).

Let ɴAs represent the true meridian; draw wAʙ at right angles to ɴs,
then ʙ and w are the true east and west points. Make the angle ʙAx=
27° 16′ 15″=the true bearing of sun; then x will represent the place of the
sun with respect to the true east point ʙ. To represent the position of the
sun with respect to the magnetic east point, take xAʙ′=7° 10′, and draw
ʙ′Aw′; then ʙ′ and w′ will represent the east and west points of the compass,
and xAʙ′ will=E. 7° 10′ N. by compass. Now draw Aɴ′ at right angles to
ʙ′w′; then the line Aɴ′ will indicate the position of the magnetic meridian
with respect to the true meridian Aɴ, which is in this case to the west
of Aɴ, the true meridian, and therefore the variation of the compass is said
to be westerly: hence

True bearing ......E. 27° 16′ 15″ N.
Compass bearing...E. 7 10 0 N.
_____
  .: variation=20 6 15, which by the figure is evidently westerly;
  .: var. of compass=20 6 15 W.

EXAMPLES FOR PRACTICE.

184. Given the lat.$=56°$ 40′ N., and sun's decl. at setting$=10°$ 58′ 8″ N.; required the true bearing. If the compass bearing at the same time was observed to be W. 5° 50′ S., required also the variation of the compass. Construct figures to show the true bearing and compass bearing.

Ans. Variation$=26°$ 5′ 15″ E.

185. Given the lat.$=49°$ 12′ N., and the sun's decl. at setting$=16°$ 44′ 16″ S.; required the true bearing. If the compass bearing at the same time was observed to be W. 10° 42′ S., required also the variation of the compass. Construct figures to show the true bearing and the compass bearing.   *Ans.* Variation$=15°$ 27′ W.

186. Given the lat.$=50°$ 48′ N., and the sun's decl. at rising$=16°$ 25′ 9″ N.; required the true bearing. If the compass bearing at the same time was observed to be E. 2° 10′ S., required also the variation of the compass. Construct figures to show the true bearing and compass bearing.

Ans. Variation$=28°$ 44′ W.

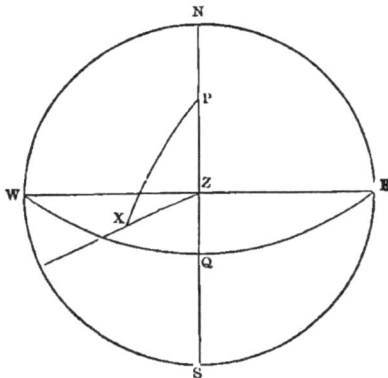

THE ALTITUDE AZIMUTH.

PROBLEM XXIX.

Given the altitude and declination of a heavenly body, and latitude of the place; to find the azimuth.

Let N S W E represent the horizon, N Z S the celestial meridian, P the pole, and Z the zenith. Then, if X be the place of the heavenly body when its altitude was observed, we have given, in the spherical triangle X Z P, the three sides, to find an angle; namely the zenith distance Z X, the polar distance P X, and the colatitude P Z, to find the angle P Z X=the azimuth or true bearing of X.

187. Given the lat.$=51°$ 10′ S., the true alt. of sun$=5°$ 25′ 37″, and its decl.$=16°$ 41′ 42″ S.; required the true bearing, the sun being east of meridian. If the compass bearing was observed at the same time to be N. 87° 50′ E., required also the variation of the compass.

(1.) To find the true bearing (fig. 1, p. 109).

Let N W S E represent the horizon, Z the zenith, and P the south pole; and let X be the place of the sun. Then, in the spherical triangle Z P X, are given Z X the zenith dist.$=84°$ 34′ 33″, P X the pol. dist.$=73°$ 18′ 18″, and P Z the colat.$=38°$ 50′; to find the angle P Z X, the true bearing of the sun.

ZX.........	84°	34'	33"...............	0·001950
PZ.........	38	50	0	0·202693
	45	44	23	4·935422
PX.........	73	18	18	4·377030
	119	2	41	9·517095
	27	33	55	∴ PZX= 69° 59' 30"

or true bearing=S. 69 59 30 E.

(2.) To find the variation of the compass (fig. 2).

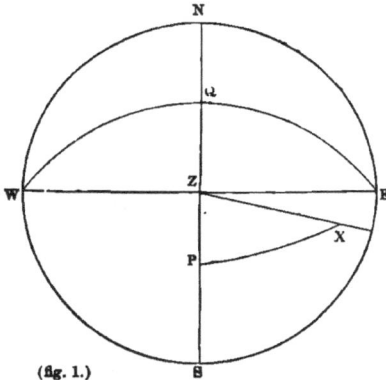
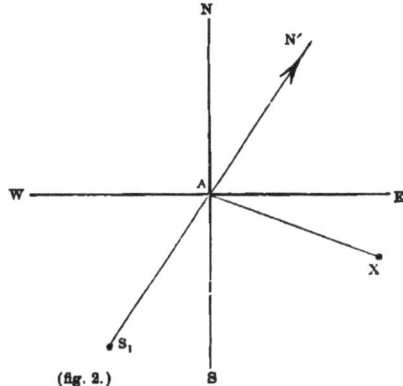

(fig. 1.)   (fig. 2.)

Let NAS represent the true meridian, and make the angle SAX=69° 59' 30", the true bearing; then X will represent the position of the sun with respect to the true meridian. But by the observation with the compass, the compass bearing=N. 87° 50' E., or S. 92° 10' E. Make, therefore, the angle XAS₁=92° 10'; then the angle S₁AX will represent the compass bearing, and the line S₁AN' will be the position of the magnetic meridian; the variation is manifestly easterly: thus

$$\text{True bearing} = \text{S. } 69° \ 59' \ 30''\text{E.}$$
$$\text{Compass bearing} = \text{S. } 92 \ 10 \ \ 0 \ \text{E.}$$
$$\therefore \text{ variation of compass} = \ 22 \ 10 \ 30 \ \text{E.}$$

EXAMPLES FOR PRACTICE.

188. Given the lat. =50° 15' N., the altitude of sun=7° 17' 51" (east of meridian), and its decl. =22° 21' 48" S.; required the true bearing. If the compass bearing was observed at the same time to be S. 61° 15' E., required also the variation of the compass. Construct the figures to show the true bearing and variation.   *Ans.* Variation=20° 11' E.

189. Given the lat. =50° 30' N., the altitude of sun=8° 28' 14" (west of meridian), and its decl. =23° 9' 26" N.; required the true bearing. If the compass bearing was observed at the same time to be N. 89° 40' W.,

required also the variation of the compass. Construct the figures to show the true bearing and variation. *Ans.* Variation=26° 3′ 15″ E.

190. Given the latitude=52° N., the altitude of sun=12° 35′ 38″ (east of meridian), and its decl.=12° 52′ N.; required the true bearing. If the compass bearing was observed at the same time to be E. 36° 30′ S., find also the variation of the compass. Construct figures to show the true bearing and variation. *Ans.* Variation=41° 21′ 30″ W.

THE TIME AZIMUTH.

PROBLEM XXX.

Given the hour-angle and declination of sun, and the latitude; to find the azimuth.

(1.) To find true bearing of sun.

Let N W S E (fig. 1) represent the horizon, N Z S the celestial meridian, P the pole, Z the zenith, and X the place of the sun when the time by chronometer was noted; then, if the error of the chronometer on mean time at the place is known, the hour-angle P may be computed; we have then, in the triangle P Z X, two sides and the included angle given to find an angle, namely, the co-declination P X, the colat. P Z, and the included angle P, to find the angle P Z X, the true bearing or azimuth (*Trig.*, Part I., Rule XI.).

(2.) To find the variation of the compass.

If the compass bearing of the sun was also observed when the time was noted, the variation of the compass may then be determined as before.

191. Given the hour-angle of the sun=1ʰ 12ᵐ 34ˢ (west of meridian), the decl.=13° 44′ 21″ N., and the latitude of the ship=50° 48′ N.; to find the true bearing. If the compass bearing at the time of observation for finding the hour-angle was S. 51° 55′ W., required also the variation of the compass.

(1.) To find true bearing P Z X (fig. 1, p. 111).

Let X be the place of the sun. Then, in the spherical triangle P Z X, are given P X=76° 15′ 39″, P Z=39° 12′, and the angle P=1ʰ 12ᵐ 34ˢ; to calculate the angle P Z X, the true bearing or azimuth.

Calculation (Rule XI. *Trig.* Part I.).

			1.	2.
39°	12′	0″		
76	15	39	10·796826	10·796826
115	27	39	9·502137	9·976882
37	3	39	10·072869	10·272522
57	43	49	10·371832	11·046230
18	31	49	66° 59′ 15″	84° 51′ 45″
P=1ʰ	12ᵐ	34ˢ		66 59 15
½P=0	36	17	∴ true bearing=N.	151. 51 0 W.

(2.) To find the variation of the compass.

Let NAS (fig. 2) represent the true meridian, and make the angle N A X = 151° 51′, the true bearing; then x will represent the place of the sun when

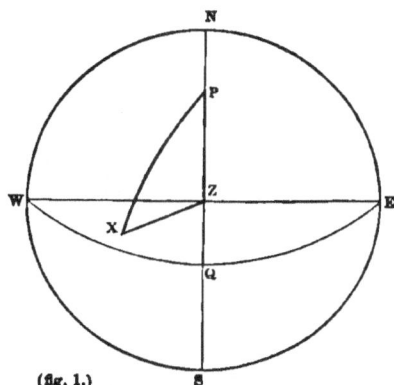

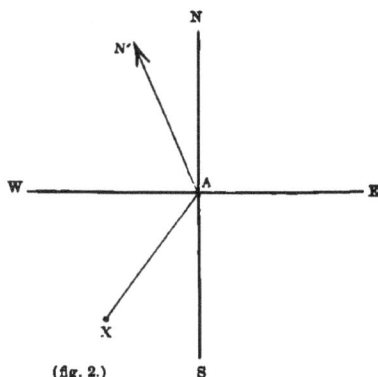

(fig. 1.)   (fig. 2.)

the compass bearing was taken. But the compass bearing was observed to be S. 51° 55′ W., or N. 128° 5′ W., the north point of the needle must therefore be 128° 5′ from x. Make X A N′ = 128° 5′; then N′A will represent the position of the needle, which is evidently to the west of NS, the true meridian; therefore the variation is westerly: thus

$$\text{True bearing} = \text{N. } 151° \ 51′ \text{ W.}$$
$$\text{Compass bearing} = \text{N. } 128 \quad 5 \text{ W.}$$

- .˙. variation of compass = 23   46 W.

EXAMPLES FOR PRACTICE.

192. Given the hour-angle of the sun = 1^h 33^m 33^s (east of meridian), the decl. = 23° 12′ 22″ S., and the latitude of ship = 52° 10′ N.; to find the true bearing. If the compass bearing at the time of observation was N. 170° 20′ E., required also the variation of the compass. Construct figures to show the true bearing and variation of the compass.

Ans. Variation = 12° 13′ W.

193. Given the hour-angle of the sun = 2^h 4^m 56^s (west of meridian), the decl. = 23° 12′ 55″ S., and the latitude of ship = 48° 50′ N.; to find the true bearing. If the compass bearing at the time of observation was S. 51° 40′ W., required also the variation of the compass. Construct figures to show the true bearing and variation of the compass.

Ans. Variation = 22° 25′ 15″ W.

194. Given the hour-angle of the sun = 0^h 36^m 21^s (east of meridian), the decl. = 23° 15′ 18″ S., and the latitude of ship = 39° 40′ N.; to find the

(2.) To find the angle ozy.

In the spherical triangle ozy, are given oz=90°, zy=41° 42' 44", and oy=75° 33' 13"; to calculate the angle ozx=67° 59' 0".

Then ozx−xzs=S. 26° 19' 0" W., the bearing of o.

EXAMPLE (BY TIME AZIMUTH).

196. 20th Oct., at 10^h 42^m 29^s A.M. ship mean time, in lat. 50° 48' N. and long. by account 1° 6' W., observed the angular distance of the sun's nearest limb from an object o to the left of it (see fig. p. 113) to be 78° 31' 45"; the altitude of the object was 4° 28'. Construct a figure, and find by calculation the true bearing of the object.

Elements from *Nautical Almanac* for Greenwich date, Oct. 19, 22^h 47^m.

(1.) Sun's decl. 10° 17' 5" S.; (2.) Sun's semi. 16' 5"; (3.) Equation of time, 15^m 4·5ˢ (additive to mean time); (4.) Correction in altitude, 1' 44".

Construction.

Let N W S E represent the horizon, P the pole, z the zenith, x the true place of the sun, and y its apparent place. Let o be the projected place of object at an altitude=40° 28'. Draw the circles of altitude zo and zx and circle of declination Px, and join oy. Then the true bearing of o is the angle N z o to be computed.

(1.) To find the true zenith distance zx and azimuth Pzx.

In the spherical triangle Pzx, are given Pz=39° 12', Px=100° 17' 5", and the included angle zPx=24^h−(22^h 42^m 29^s+15^m 4·5ˢ)=1^h 2^m 26·5ˢ; to calculate the true zenith distance zx=62° 34' 31", and true bearing of sun, namely Pzx=162° 39' 0".

(2.) To find the angle ozy.

In the spherical triangle ozy, are given oz=85° 32', zy=62° 34' 31" −1' 44"=62° 32' 47", and oy=78° 47' 50"; to compute the angle ozy= 79° 41' 15".

Hence the angle Pzo=Pzx−ozy = N. 82° 57' 45" E., the true bearing of terrestrial object required.

EXAMPLES FOR PRACTICE.

197. August 13, 1858, at 10^h 55^m 19ˢ A.M. ship mean time, in latitude 50° 48' N. and longitude 1° 6' W., observed the angular distance of the sun's farthest limb from an object o in the horizon to the right of sun and west of meridian to be 69° 4' 45". Construct a figure, and find by calculation the true bearing of the object.

Elements from the *Nautical Almanac* for Greenwich date, August 12, at 22ʰ 59ᵐ 43ˢ.

(1.) Sun's decl. 14° 43′ 0″ N.; (2.) Sun's semi. 15′ 50″; (3.) Equation of time, 4ᵐ 39ˢ (subtractive from mean time); (4.) Correction in altitude, 0′ 41″.

Ans. S. 27° 14′ 15″ W.

198. August 13, 1858, at 1ʰ 9ᵐ 21ˢ P.M. ship mean time in lat. 50° 48′ N. and long. 1° 6′ W., observed the angular distance of the sun's nearest limb from an object o in the horizon to the left of sun and east of meridian to be 125° 31′ 32″. Construct a figure, and find by calculation the true bearing of the object.

Elements from *Nautical Almanac* for Greenwich date, August 13, at 1ʰ 13ᵐ 45ˢ.

(1.) Sun's decl. 14° 41′ 18″ N.; (2.) Sun's semi. 15′ 50″; (3.) Equation of time, 4ᵐ 38ˢ (subtractive from mean time); (4.) Correction in altitude, 0′ 41″.

Ans. N. 45° 22′ E.

The true bearing of a terrestrial object (as the spire of a church, seen from some given station, or a ship at anchor) being determined by the above problem, the variation of the compass on board is easily found by simply observing the compass bearing of the object; the difference between the bearings will evidently be the variation of the compass for that position of the ship. Moreover, if it is suspected that the iron on board has some important influence on the compass, this may in some measure be discovered by taking successive observations of the compass bearing of the terrestrial object while the ship is swinging, or whenever she happens to lie with her head in different positions with respect to the magnetic meridian. This method of ascertaining what is called the *deviation of the compass* may often be of use when more accurate methods of forming a table of deviations cannot be adopted.

To obtain the true bearing of the terrestrial object with the greatest accuracy, the heavenly body observed should be at a considerable distance from the meridian; and the angular distance of the terrestrial object from the heavenly body should also be large (not less than 80° or 90°). If these limitations are made, the effects of errors of observation will be considerably diminished; but these and other directions connected with this subject belong rather to marine surveying, where this problem is of great utility.

CHAPTER V.

Magnitude and figure of the earth.

In all the common rules and problems of Nautical Astronomy investigated in the preceding pages, the form of the earth has been considered to be that of a sphere. On this supposition the meridians would be great circles, and the *length* of a degree of latitude in every part of the earth would be equal. But observations and actual measurements of arcs of a meridian, made in different parts of the world, have made it apparent that the lengths of a degree of the meridian are not invariable, but that they *increase from the equator to the poles*, suggesting to us the figure of an oblate spheroid.

The following table contains the results of such measurements·

	Country.	Latitude of middle point of arc measured.			Length of 1° in feet.
1	Sweden	66°	20′	10″	365782
2	Russia	58	17	37	365368
3	England......................	52	35	45	364971
4	France	46	52	2	364872
5	France	44	51	2	364535
6	Rome........................	42	59	0	364262
7	United States	39	12	0	363786
8	India	16	8	22	363044
9	India	12	32	21	363013
10	Peru	1	31	0	362808

The observations recorded in the above table prove that the curvature of the earth must diminish from the equator to the pole: this is sufficient to show that the earth is not a sphere, and that, in fact, it must approach in form to that of an oblate spheroid.

Assuming it to be such a figure, we can compute the lengths of the equatorial and polar diameters from the measured lengths of a degree of the meridian in two places differing considerably in latitude.

As this knowledge of the dimensions of the earth is necessary in several important problems in Nautical Astronomy, we will here investigate an expression for calculating the major and minor diameters qq_1 and pp_1, and also the distance c_A of any point A from the center of the earth.

PROBLEM XXXII.

To calculate the lengths of the equatorial and polar diameters of the earth.

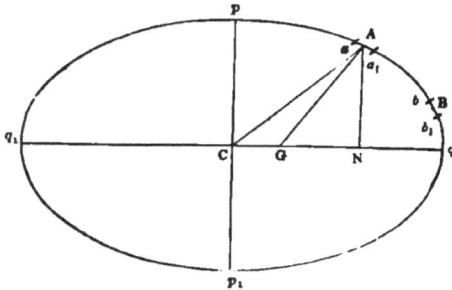

Let pqp_1q_1 represent a section of the earth passing through the poles pp_1; A and B the middle points of two degrees aa_1, bb_1, the lengths of which are supposed to be known from the preceding table.

Draw AG, a normal at A; then the angle AGq is called the *latitude of A* (p. 9).

Let l and l_1 represent the latitudes of A and B;

a and b the semi-major and semi-minor axes;

n the normal AG at A;

D and D_1 the lengths of aa_1 and bb_1 the degrees in latitudes A and B, known from observation.

Then circle of curvature at $A=360 \times D$ nearly$=2\pi r$;

$\therefore$ radius of curvature $r=\dfrac{360 \times D}{2\pi}=57\cdot29577 \times D.$

Similarly, radius of curvature r_1 at $B=57\cdot29577 \times D_1$.

The difference of the equatorial and polar radii is small : call this difference c; then an approximate value of c may be found (by neglecting terms involving $\dfrac{c^2}{a^2}$, &c., which are small), and thence a and b. This may be done by expressing the normal n in terms of a and c and the latitude l; and then, by the differential calculus, n in terms of r. A similar expression being found for r_1, one of the unknown terms a and c may be eliminated between the two expressions, and the other determined as in the following investigation.

Let the coördinates ON and AN of the point A be denoted by x and y; then

(by Conic Sections), $y^2=\dfrac{b^2}{a^2}(a^2-x^2)$, and $GN=\dfrac{b^2x}{a^2}.$

In triangle AGN, $y=n$ sin. l, and $GN=n$ cos. l ;

$$\therefore \frac{b^2}{a^2}(a^2-x^2)=n^2 \text{ sin.}^2 l, \text{ and } \frac{b^2x}{a^2}=n \text{ cos. } l \; ; \; \therefore \; x^2 = \frac{n^2a^4}{b^4}\text{cos.}^2 l.$$

Substituting this value of x^2 in former equation,

$$\frac{b^2}{a^2}\left(a^2-\frac{n^2a^4}{b^4}\text{cos.}^2 l\right)=n^2 \text{ sin.}^2 l, \; \therefore \; n^2=\frac{b^4}{a^2\left(1-\dfrac{a^2-b^2}{a^2}\text{sin.}^2 l\right)}$$

$$\therefore \; n=\frac{b^2}{a}\left(1-\frac{a^2-b^2}{a^2}\text{ sin.}^2 l\right)-\tfrac{1}{2}. \quad \text{But } c=a-b, \therefore \; b=a-c \; ;$$

$$\therefore \; n=\frac{b^2}{a}\left(1-\frac{2ac-c^2}{a^2}\text{ sin.}^2 l\right)-$$

$$=\frac{b^2}{a}\left(1-\frac{2c}{a}\text{sin.}^2 l\right)-\tfrac{1}{2} \text{ nearly, since }\frac{c^2}{a^2}\text{ is small}$$

$$=\frac{b^2}{a}\left(1+\frac{c}{a}\text{sin.}^2 l\right) \text{ neglecting terms involving higher powers of } \frac{c}{a}$$

Again, $r=\dfrac{\left\{1+\left(\dfrac{dy}{dx}\right)^2\right\}^{\frac{3}{2}}}{-\dfrac{d^2y}{dx^2}}=\dfrac{\{b^2x^2+a^2(a^2-x^2)\}^{\frac{3}{2}}}{a^4b}$ (by diff. calc.)

and $n^2=GN^2+AN^2$

$$=\frac{b^4x^2}{a^4}+\frac{b^2}{a^2}(a^2-x^2)=\frac{b^2}{a^4}\{b^2x^2+a^2(a^2-x^2)\}$$

$$\therefore \; n^3=\frac{b^3}{a^6}\{b^2x^2+a^2(a^2-x^2)\}^{\frac{3}{2}} \therefore \; \frac{n^3a^6}{b^3}=\{b^2x^2+a^2(a^2-x^2)\}^{\frac{3}{2}}$$

Substituting this value of the numerator in r, we have

$$r=\frac{n^3a^3}{b^4} \; \therefore \; r=\frac{a^2}{b^4}\cdot\frac{b^6}{a^3}(1+\frac{c}{a}\text{ sin.}^2 l)^3$$

$$=\frac{b^2}{a}(1+\frac{3c}{a}\text{ sin.}^2 l) \text{ nearly} \quad =\frac{(a-c)^2}{a}(1+\frac{3c}{a}\text{sin.}^2 l) \text{ nearly}$$

$$=\left(a-2c+\frac{c^2}{a}\right)\cdot\left(1+\frac{3c}{a}\text{ sin.}^2 l\right) \text{ nearly}$$

$$r=(a-2c)\cdot(1+\frac{3c}{a}\text{ sin. }^2 l) \text{ nearly} \quad =a+3c\text{ sin. }^2 l-2c \text{ nearly.}$$

Similarly $r_1=a+3c$ sin.$^2 l_1-2c$ $\therefore$ $r-r_1=3c$ (sin.$^2 l-$sin.$^2 l_1$)

$$\therefore \; c=\frac{r-r_1}{3\text{ (sin.}^2 l-\text{sin.}^2 l_1)}=\frac{57\cdot29577\text{ (D}-\text{D}_1)}{3\text{ (sin.}^2 l-\text{sin.}^2 l_1)}$$

or c (in miles)$=\dfrac{57\cdot29577\text{ (D}-\text{D}_1)}{3\times1760\times3}$ cosec. $(l+l_1)$ cosec. $(l-l_1)$.

Hence this practical rule to compute c, the difference between the semi-major and semi-minor axes of the earth.

Add together the constant log. $\overline{3}\cdot558367$ (the log. of $\dfrac{57\cdot29577}{3\times1760\times3}$), the log. of the difference of the lengths (in feet) of a degree in the two latitudes, and the log. cosecants of the sum and difference of the two latitudes : the natural number of the sum (rejecting 20 in the index) will be the value of c (the difference between the semi-major and semi-minor axes) in miles.

199. Find the value of c from the measurements (2 and 8) of a degree in Russia and India (see table, p. 116).

$$l = 58^\circ\ 17'\ 37''$$
$$l_1 = 16\ \ \ \ 8\ \ \ 22$$
$$l+l_1 = 74\ \ 25\ \ 59$$
$$l-l_1 = 42\ \ \ \ 9\ \ \ 15$$

$$\text{D} = 365368$$
$$\text{D}_1 = 363044$$
$$\text{D}-\text{D}_1 = \ \ \ 2324$$

const. log............$\overline{3}\cdot558367$
log. $(\text{D}-\text{D}_1)$..........$3\cdot366236$
,, cosec. $(l+l_1)$ ...$0\cdot016230$
,, cosec. $(l-l_1)$ ...$0\cdot173195$
log. $c = 1\cdot114028$
$\therefore c = 13\cdot00$ miles.

With the measurements in England and India (3 and 8), the value of
$$c = 12\cdot58$$
Measurements 1 and 10 $c = 12\cdot83$
,, 3 ,, 9 $c = 12\cdot13$
,, 2 ,, 9 $c = 12\cdot59$
mean value of $c = \overline{12\cdot64}$ nearly.

The value of c being found, we may compute a by the formula
$$r = a - 2c + 3c\ \sin.^2 l,\ \therefore\ a = r + 2c - 3c\ \sin.^2 l.$$
$$= 57\cdot29577\ \text{D} + (2 - 3\sin.^2 l)\cdot\frac{57\cdot29577\ (\text{D}-\text{D}_1)}{3\ (\sin.^2 l - \sin.^2 l_1)}$$
$$= \frac{57\cdot29577}{3}\cdot\frac{2(\text{D}-\text{D}_1) + 3(\text{D}_1\ \sin.^2 l - \text{D}\ \sin.^2 l_1)}{\sin.^2 l - \sin.^2 l_1}$$
$$= \frac{57\cdot29577}{9\times1760}\Big(2\ \overline{\text{D}-\text{D}_1} + 3\ \overline{\text{D}_1\ \sin.^2 l - \text{D}\ \sin.^2 l_1}\ \text{cosec.}\ \overline{l+l_1}\ \text{cosec.}\ \overline{l-l_1}\Big)$$

Hence this practical rule to compute the semi-major axis of the earth.

To the log. of the length of a degree (D_1) in feet, in one latitude, add twice the log. sine of the other latitude. Again, to the log. of the length of a degree (D) in one latitude, add twice the log. sine of the other latitude (rejecting 20 from the index in each case). Take the difference between the natural numbers of the resulting logarithms, and multiply by 3. Add thereto twice the difference between the lengths of a degree in each latitude, and take out the logarithm of the result. To this logarithm add const. log. $\overline{3}\cdot558367$, and the log. cosecants of the sum and difference of the two latitudes. The sum (rejecting the tens in index) is the log. of the semi-major axis of the earth; which find in the tables.

200. Find the value of a (the semi-major diameter of the earth) from the measurements (3 and 9) of a degree in England and India (see table, p. 116).

$$l = 52° \ 35' \ 45''$$
$$l_1 = 12 \quad 32 \quad 21$$
$$l + l_1 = 65 \quad 8 \quad 6$$
$$l - l_1 = 40 \quad 3 \quad 24$$

log. D_1......5·5599222	log. D.........5·5622584
,, sin. l...9·9000231	,, sin. l_1...9·3366737
,, sin. l...9·9000231	,, sin. l_1...9·3366737
5·3599684	4·2356058
229070	17203
17203	
211867	
3	

$$D = 364971$$
$$D_1 = 363013$$
$$\overline{1958}$$
$$2$$
$$2(D - D_1) = \overline{3916}$$

const. log. ....$\bar{3}$·5583670
log. cosec. $(l + l_1)$...0·0422486
,, cosec. $(l - l_1)$...0·1914211
,, 639517.........5·8058521
log. a...3·5978888

635601
$2(D - D_1)$... 3916
639517

∴ semi-major diameter $a = 3961·7$ miles.

EXAMPLES.

201. Find the value of a, the semi-major diameter, from the measurements (2 and 8), p. 116.   *Ans.* $a = 3961·8$ miles.

202. Find the value of a, the semi-major diameter, from the measurements (2 and 9), p. 116.   *Ans.* $a = 3962·6$ miles.

From the last three results we find the mean value of semi-major diameter $a = 3962$ miles, and $c = 12·64$ miles; ∴ b the semi-minor diameter$= 3962 - 12·64 = 3949$ miles nearly, and $\frac{c}{a}$ (called the compression)$= \frac{12·64}{3962} = \frac{1}{313·4}$ nearly.

The distance cA of any point A on the surface of the earth from the center c will be useful hereafter in finding the correction of the moon's equatorial horizontal parallax. It may be computed by the following problem.

PROBLEM XXXIII.

To calculate the distance of a place on the surface of the earth from the center.

Let A (fig. p. 117) be the place, then its distance AC from the center may be investigated as follows :

Since $CA^2 = CN^2 + AN^2$ and subnor. $GN = \dfrac{b^2}{a^2} CN$

$$\therefore \ CN = \frac{a^2}{b^2} \cdot GN = \frac{a^2}{b^2} n \cos. \ l = \frac{a^2}{b^2} \cdot \frac{b^2}{a} \left(1 + \frac{c}{a} \sin.^2 l\right) \cos. \ l \ \ldots \ldots (p. \ 118)$$

$$= a \cos l \left(1 + \frac{c}{a} \sin.^2 l\right) \text{ and } AN = n \sin. \ l = \frac{b^2}{a} \left(1 + \frac{c}{a} \sin.^2 l\right) \sin. \ l$$

$$= \frac{(a-c)^2}{a} \left(1 + \frac{c^2}{a} \sin.^2 l\right) \sin. \ l$$

$$= \left(a - 2c + \frac{c^2}{a}\right) \cdot \left(1 + \frac{c}{a} \sin.^2 l\right) \sin. \ l$$

$$= (a - 2c) \cdot \left(1 + \frac{c}{a} \sin.^2 l\right) \sin. \ l \text{ nearly.}$$

Substituting these values of CN and AN,

$$CA^2 = \left\{a^2 \cos.^2 l + (a - 2c)^2 \sin.^2 l\right\} \cdot \left(1 + \frac{c}{a} \sin.^2 l\right)^2$$

$$= (a^2 \cos.^2 l + a^2 \sin.^2 l - 4 \ ac \sin.^2 l + 4 \ c^2 \sin.^2 l) \cdot \left(1 + \frac{2c}{a} \sin.^2 l\right) \text{ nearly}$$

$$= a^2 + 2ac \sin.^2 l - 4ac \sin.^2 l \text{ nearly (neglecting other terms, which are small)}$$

$$= a^2 + 2ac \sin.^2 l$$

$$\therefore \ CA = a \left(1 - \frac{2c}{a} \sin.^2 l\right)^{\frac{1}{2}} = a \left(1 - \frac{c}{a} \sin.^2 l\right) \text{ nearly}$$

$$= a - c \sin.^2 l = 3962 - 12.64 \sin.^2 l.$$

Hence this practical rule to find the distance of any place from the center of the earth :

To the constant log. 1·101747 (the log. of 12·64) add twice the log. sine of the latitude of the place : the natural number of the result (rejecting the tens in index), subtracted from the semi-major diameter, 3962, is the distance required.

203. Required the distance of a place in latitude 50° 48′ from the center of the earth.

$$\begin{aligned}
&\text{const. log.} \ldots\ldots\ldots\ldots\ldots 1\cdot 101747 \\
&\text{log. sin. lat.} \ldots\ldots\ldots\ldots 9\cdot 889271 \\
&\quad\quad\text{,,} \quad\quad\ldots\ldots\ldots\ldots 9\cdot 889271 \\
&\text{log. cor.}\ldots\ldots\ldots\ldots\ldots 0\cdot 880289 \\
&\therefore \text{ cor.} \ldots\ldots\ldots 7\cdot 6 \text{ miles.}
\end{aligned}$$

Hence distance from center $= 3962 - 7\cdot 6 = 3954\cdot 4$ miles.

204. Required the distance of a place in latitude 30° from the center of the earth. *Ans.* 3958·84 miles.

From these investigations it is manifest that the earth, although not an exact sphere, is very nearly so; accordingly, in the common rules of Navigation, the earth is considered as a sphere, and a meridian a *great* circle (and not an ellipse); and the *latitude* of a place an arc of the meridian intercepted between the place and the equator. The correct definition of the latitude, namely taking into consideration the true figure of the earth, will be considered in the next problem.

PROBLEM XXXIV.

Given the true latitude of a place, to find the reduced or central latitude.

Let true lat. $AGq = l$, and reduced lat. $ACq = l_1$ (fig. p. 117), then $l - l_1$ = the correction required.

$$\tan. l = \frac{AN}{GN} \qquad \tan. l_1 = \frac{AN}{CN}$$

$$\therefore \frac{\tan. l}{\tan. l_1} = \frac{CN}{GN} = \frac{a^2}{b^2} \text{ (since } GN = \frac{b^2}{a^2} CN, \text{ p. 117)}$$

$$\therefore \tan. l_1 = \frac{b^2}{a^2} \tan. l = \frac{(a-c)^2}{a^2} \tan. l = \left(1 - \frac{2c}{a}\right) \tan. l \text{ nearly.}$$

$$\text{Now } \tan. (l - l_1) = \frac{\tan. l - \tan. l_1}{1 + \tan. l \cdot \tan. l_1}$$

$$= \frac{\tan. l \cdot \left(1 - 1 + \frac{2c}{a}\right)}{1 + \tan.^2 l \left(1 - \frac{2c}{a}\right)}$$

$$= \frac{2c}{a} \cdot \frac{\tan. l}{1 + \tan.^2 l} \text{ (since } 1 - \frac{2c}{a} = 1 \text{ nearly)}$$

$$= \frac{c}{a} \sin. 2l = \frac{1}{313 \cdot 4} \sin. 2l$$

Since $l - l_1$ is only a few minutes, the circular measure $\left(\frac{\text{arc } (l - l_1)}{\text{rad.}}\right)$ may be substituted for $\tan. (l - l_1)$;

$$\therefore \text{arc } (l - l_1) \text{ in min.} = \frac{57° \cdot 29577 \times 60}{313 \cdot 4} \cdot \sin. 2l = 11 \sin. 2l \text{ nearly.}$$

Hence this practical rule for finding the correction of the latitude for the spheroidal figure of the earth, and thence the reduced latitude:

To the constant log. 1·041393 (the log. of 11) add log. sine of twice the latitude (rejecting 10 in the index): the natural number of the result will be the correction of latitude, which being subtracted from the given latitude,. the remainder is the reduced or central latitude required.

205. Given the true latitude of a place $A = 50°\ 46'$ N.; to find the reduced latitude.

$$\text{Reduced lat.} = l - \text{cor.} = l - 11 \sin. 2l$$

true lat. $l = 50°\ 46'$	log. 11............1·041393
$\therefore 2l = 101\quad 32$	„ sin. $2l$ ......9·991141
	„ cor. ....$\overline{1·032534}$
$\therefore$ cor. ......$= 10\cdot8' =$	$10'\ 48''$
	true lat. $l = \underline{50\ 46\quad 0}$ N.
	$\therefore$ reduced lat. $= \underline{50\ 35\quad 12}$ N.

206. Given the true latitude $= 62°\ 10'$ N.; calculate the reduction of the latitude for the spheroidal figure of the earth, and thence find the reduced latitude.   *Ans.* Red. lat. $= 62°\ 0'\ 55\cdot2''$ N.

Construction of Table h *in Inman's Nautical Tables.*

By means of the above formula, table h, for finding the reduced latitude, may be computed.

PROBLEM XXXV.

Given the true zenith distance of a heavenly body, and its azimuth or bearing ; to calculate the reduced zenith distance.

Let z be the true zenith, z' the reduced zenith, and x a heavenly body whose azimuth is x z q. Then z x is the true zenith distance, and z'x is the

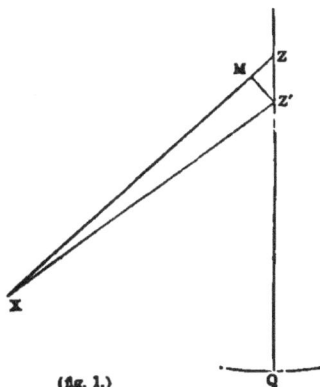

(fig. 1.)

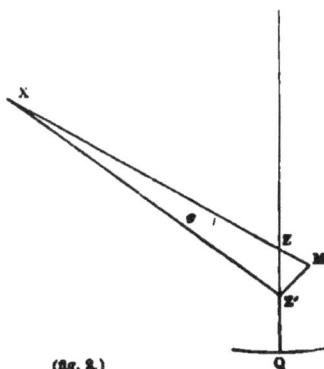

(fig. 2.)

reduced zenith distance. With center x and distance x z' describe the arc z'M : then the triangle z z'M, being very small, may be considered as a right-angled

· plane triangle, and z M is the difference between the true and reduced zenith distances, or z′x=zx±z M; the sign to be used being easily determined by a diagram, remembering that z′ is always nearer the equator than z.

To calculate the correction z M we have, in the right-angled triangle z M z′, z M=z z′ cos. M z z′=11 sin. 2l cos. azimuth; since z z′=reduction of latitude (Prob. XXXIV.), and the angle M z z′ is the azimuth or bearing of the body.

When the bearing is nothing, or the heavenly body is on the meridian, it is evident that z M=z z′ the reduction of the latitude.

207. Calculate the reduced zenith distance of a heavenly body when the true zenith distance (corrected for refraction) is 35° 34′; the latitude of the place being 50° 46′ N., and the azimuth of the body S. 10° W.

$$l= 50°\ 46′ \qquad z M=z z′ \cos. z$$
$$∴ 2l=101\ \ 32 \qquad =11 \sin. 2l \cos. az.$$

Calculation:

$$\log. 11\ ....1·041393$$
$$„\ \sin. 2l.....................9·991141$$
$$„\ \cos. azimuth\ ....9·993351$$
$$„\ z M\ ....1·025885$$
$$∴ z M=10·6′=\qquad 10′\ \ 36″$$
$$and\ zx=35°\ \ 34\quad 0$$
$$∴ reduced\ zenith\ distance=35\quad 23\quad 24$$

208. Calculate the reduced zenith distance of a heavenly body when the true zenith distance (corrected for refraction) is 35° 34′; the latitude of the place being 50° 48′ N., and the bearing or azimuth N. 10° W.

By means of the figure we see that z M, the correction (found as in last example), must be added to the true zenith distance to get the reduced zenith distance.

$$∴ reduced\ zenith\ distance=35°\ 34′\quad 0″+10′\ 36″$$
$$=35\quad 44\quad 36$$

209. Calculate the reduced zenith distance of a heavenly body when its true zenith distance (corrected for refraction) is 22° 30′; the latitude of the place being 40° 48′ S., and the bearing or azimuth S. 50° E. *Ans.* 22′ 37″.

Parallax.

The place of a heavenly body as seen, or supposed to be seen, from the center of the earth, is called its *true* or geocentric place; the place of a heavenly body as seen from any point on the surface is called its apparent place. Thus, let A be any point on the surface of the earth, X a heavenly body, as the moon. Through X draw the straight lines AXm', CXm, from the surface and center to the celestial concave. Then mm' being produced, will pass through the reduced zenith z'. The arc m'm or angle AXC is called the *diurnal parallax*; and if H be the same heavenly body in the horizon of the spectator at A, then the angle AHC is called the *horizontal parallax* of the heavenly body. The circle m'z' coincides very nearly with m'z, a circle through the true zenith z, since the figure of the earth differs very little from that of a sphere; and therefore m'z' is very nearly a vertical circle.

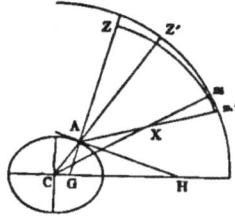

PROBLEM XXXVI.

Given the apparent reduced zenith distance and horizontal parallax; to calculate the diurnal parallax, and thence the true reduced zenith distance.

Let A be the spectator on the surface of the earth, z' his reduced zenith, X a heavenly body. Through X draw AX, CX, and produce the lines to m' and m. Then the arc z'm' is the distance of the body from the reduced zenith z', as seen from the surface of the earth, or it is the apparent reduced zenith distance (corrected for refraction); and z'm is the reduced zenith distance as seen from the center, or it is the true reduced zenith distance; and the arc mm' is the diurnal parallax.

To compute mm'.

Let $AC = r$ the radius of the earth, $CX = CH = d$ the distance of body,

$\quad z'Am' = z'$ the apparent reduced zenith distance,

$\quad AXC = p = mm'$ the diurnal parallax, $AHC = H$ the hor. par.

In triangle AXC we have $\dfrac{r}{d} = \dfrac{\sin. AXC}{\sin. CAX} = \dfrac{\sin. AXC}{\sin. z'Am'}$

$\qquad = \dfrac{\sin. \text{diurnal parallax}}{\sin. \text{app. red. zen. dist.}} = \dfrac{\sin. p}{\sin. z'} \dots\dots\dots\dots (\alpha)$

Again, in AHC, considered as a right-angled triangle,

$\qquad \dfrac{r}{d} = \sin. H = \sin. \text{horizontal parallax} \dots\dots\dots\dots (\beta)$

$\therefore$ equating (α) and (β) $\dfrac{\sin. p}{\sin. z'} = \sin. H$; or $\sin. p = \sin. H \sin. z'$

But since the parallax is small even for the moon, the nearest of the heavenly bodies, we may substitute the circular measure of the angles $\left(\dfrac{\text{arc}}{\text{rad.}}\right)$ for their sines;

$\qquad \therefore \dfrac{p}{\text{rad.}} = \dfrac{H}{\text{rad.}} . \sin. z'$, or $p = H . \sin. z'$;

that is, diurnal par.=hor. par. × sin. apparent reduced zenith distance (corrected for refraction).

210. Given the apparent reduced zenith dist. of the moon=77° 45′ 36″, and the horizontal parallax=59′ 33·2″. Calculate the diurnal parallax, and thence find the true reduced zenith distance.

$$p = \text{H sin. } z'$$

H=59′ 33·2″	log. H............3·553057
60	„ sin. z'......9·990014
3573·2″	„ p............3·543071
	∴ p=3492′= 58′ 12″
	app. red. zen. dist.=77 45 36
	∴ true red. zen. dist.=76 47 24

211. Calculate the diurnal parallax of the moon when the apparent reduced zenith distance is 45° 30′, and the horizontal parallax 55′ 42·5″.

Ans. 39′ 44″.

In the preceding problem the diurnal parallax p is determined when the *apparent* reduced zenith distance is given. If the true reduced zenith distance (as $z'm$ in the figure) is only given, to find the diurnal parallax, the above formula requires to be modified. This may be done as follows.

PROBLEM XXXVII.

Given the *true* reduced zenith distance and horizontal parallax; to calculate the diurnal parallax.

Let the true reduced zen. dist. $z'cm=z$ (see last fig.)

Then $z'=z+p$; and since sin. p=sin. H . sin. z',

∴ sin. p=sin. H . sin. $(z+p)$

    =sin. H {sin. z . cos. p+cos. z . sin. p}

or tan. p=sin. H {sin. z+cos. z . tan. p}

∴ tan. p $(1-$sin. H . cos. $z)$=sin. H . sin. z

∴ tan. p=sin. H . sin. $z \dfrac{1}{1-\text{sin. H . cos. } z}$

    =sin. H. sin. z {1+sin. H . cos. z+ . . . . . }

Substituting $\dfrac{\text{arc}}{\text{rad.}}$ for tan. p, sin. H, &c., we have

$$\frac{p}{\text{rad.}}=\frac{\text{H}}{\text{rad.}} . \text{ sin. } z \left\{1+\frac{\text{H}}{\text{rad.}} . \text{ cos. } z+ . . . .\right\}$$

∴ p=H . sin. $z+\dfrac{\text{H}^2}{\text{rad.}}$. sin. z . cos. z+ . . . .

    =H . sin. $z+\dfrac{\text{H}^2}{57·29577 \times 60 \times 60} . \dfrac{\text{sin. } 2z}{2}+ . . .$

∴ p=H . sin. $z+\dfrac{\text{H}^2 . \text{sin. } 2z}{57·29577 \times 7200}$ nearly (by neglecting terms involving the higher powers of H).

By means of this formula the diurnal parallax may be computed when z, the true reduced zenith distance, and the horizontal parallax are given.

212. Given the true reduced zenith distance of the moon $= 76° 47' 24''$, and the horizontal parallax $= 59' 33\cdot2'' = 3573\cdot2''$; to calculate the diurnal parallax.

$$p = \text{H} \cdot \sin z + \frac{\text{H}^2 \cdot \sin 2z}{57\cdot29577 \times 7200}$$

log. H ....3·553057	........3·553057	3·857333
,, sin. z............9·988353	3·553057	1·758119
8·541410	9·648322	5·615452
	6·754436	
Nat. No. 3478''	5·615452	
14 ....1·138984		

$\therefore p = 3492 = 58' 12''$, as before (see p. 126).

Parallax in altitude.

The reduced zenith z′ is very near the true zenith z, since the figure of the earth is very nearly a sphere (Prob. XXXII.); and the plane passing through z′ and the heavenly body is nearly a vertical plane. In the common problems of Nautical Astronomy, therefore, it is usual to suppose the true zenith z to coincide with the reduced zenith z′, and the diurnal parallax to take place in a circle of altitude. The diurnal parallax then becomes the *parallax in altitude.*

The formula (p. 125), namely,

diurnal parallax = hor. par. × sin. app. red. zen. dist. (corr. for refraction),

may therefore be written thus,

parallax in altitude = hor. par. × sin. zenith distance (corrected for refraction),

or,

parallax in altitude = hor. par. × cos. alt. (corrected for refraction).

A table of the correction in altitude (which combines correction for parallax and refraction) for different altitudes has been computed from the last formula, and is used in navigation as sufficiently accurate for correcting the apparent altitude for the effects of parallax and refraction.

The method of computing this quantity may be seen in the following example :

213. Given the apparent altitude of the moon's center $= 42° 45' 30''$, refraction $= 1' 3''$, and the horizontal parallax $= 55' 45'' = 3345''$; to calculate the true altitude.

Par. in alt.=hor. par. × cos. alt. (corrected for refraction).

obs. alt. ....42° 45′ 30″	log. cos. alt. ......9·865945	
refraction .... 1 3—	„ hor. par......3·524396	
42 44 27	„ par. in alt....3·390341	
40 56+	∴ par. in alt. =2456″, or 40′ 56″	
∴ true alt. 43 25 23		

which result agrees with that found in the table : thus, by Inman's table,

Obs. alt. ....42° 45′ 30″

hor. par. ....55′.........39′ 20″			
45″...... 33			
Cor. for refract. and par. ...39 53 ....39 53			

True alt.=43   25   23

214. Given the apparent altitude of moon's center=72° 42′ 15″, refraction =18″, and horizontal parallax=58′ 49″; required the true altitude (by calculation and by table). *Ans.* By calculation, 72° 59′ 26″. By table, 72° 59′ 26″.

The horizontal parallax$=\dfrac{\text{rad. of earth}}{\text{dist. of body}}$ (p. 125), therefore the horizontal parallax will vary with the radius of the earth ; and as the earth is an oblate spheroid (p. 120), its radius diminishes from the equator to the pole : hence the equatorial horizontal parallax must be the greatest. This may be seen by a figure (fig. 1) ; for if P be the pole of the earth, and EQ the equator, H a heavenly body in the horizon of a spectator at Q, and H′ the same heavenly body in the horizon of P, then it is evident that the angle H is greater than H′, since CQ is greater than CP; that is, the equatorial horizontal parallax H is the greatest, and the horizontal parallax of a heavenly body H′, as seen from the pole, is the least.

The horizontal parallax of the moon, put down in the *Nautical Almanac*, is the equatorial horizontal parallax; to find the horizontal parallax for any other place, a correction must be applied to that taken from the *Nautical Almanac*, and this correction is evidently subtractive.

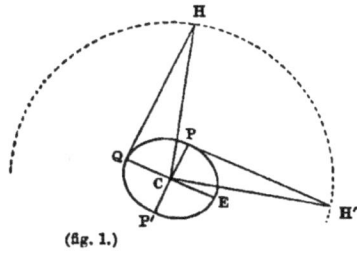

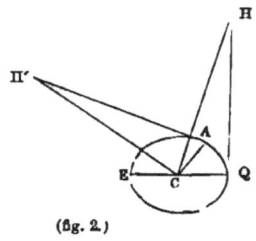

(fig. 1.)   (fig. 2.)

PROBLEM XXXVIII.

Given the horizontal parallax of the moon at the equator; to find the horizontal parallax at any other place on the surface of the earth.

Let A (fig. 2) be the given place, QH, AH′, tangents at Q and A. If equatorial horizontal parallax$=$H, horizontal parallax at A$=$H′, the equatorial semi-diameter CQ$=a$, and CA$=r$, CH$=$CH′$=$R, and lat. of A$=l$.

Then, in triangle CHQ, $\dfrac{a}{R}=$sin. H.

,,   ,,   CH′A, $\dfrac{r}{R}=$sin. H′ nearly, since the earth is nearly a sphere.

$\therefore$ sin. H′ : sin. H :: $\dfrac{r}{R} : \dfrac{a}{R}$ :: $r : a$, $\therefore$ sin. H′$=\dfrac{r}{a}$. sin. H ;

and since the angles H and H′ are small, the arcs may be substituted for the sines, $\therefore$ H′$=\dfrac{r}{a}$. H.

Now the value of OA or r is equal to $a-c$. sin. ^2l (p. 121),

$=a\left(1-\dfrac{c}{a} . \sin.^2 l\right),$ (where $\dfrac{c}{a}=\dfrac{1}{313\cdot4}$, p. 120);

$\therefore$ H′$=\left(1-\dfrac{c}{a}. \sin.^2 l\right).$ H$=$H$-\dfrac{c}{a}$. sin.$^2 l$. H.

The quantity $\dfrac{c}{a}$. sin.$^2 l$. H is the correction to be subtracted from H. This correction has been calculated for different latitudes, and forms part of table h in Inman's *Nautical Tables.*

215. Calculate the correction for the equatorial horizontal parallax of the moon ; the horizontal parallax taken out of the *Nautical Almanac* being 59′ 53·7″, and the latitude of the spectator 50° 48′ N.

$$\text{Cor.}=\frac{c}{a} . \sin.^2 l . \text{H},$$

$\dfrac{c}{a}=\dfrac{1}{313\cdot4}$	$\therefore$ log.$\dfrac{c}{a}=\bar{3}\cdot503900$
59′ 53·7″	log. sin. l...9·889271
60	,, sin. l...9·889271
3593·7″$=$H	,, H.......3·555541
	,, cor.....0·837983
	$\therefore$ cor.$=6\cdot89''$

216. Find the correction for the equatorial parallax of the moon ; the horizontal parallax from the *Nautical Almanac* being 54′ 32·5″, and the latitude of spectator 32° 42′ N.   *Ans.* 3·048″.

PROBLEM XXXIX.

Given the meridian zenith distances of a heavenly body observed at the same instant at two distant places on the same meridian ; to calculate the parallax.

K

Let A and B be the two places on the same meridian, m a heavenly body on the meridian of A and B.

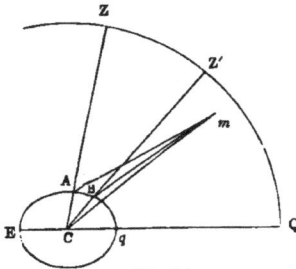

(fig. 1.)

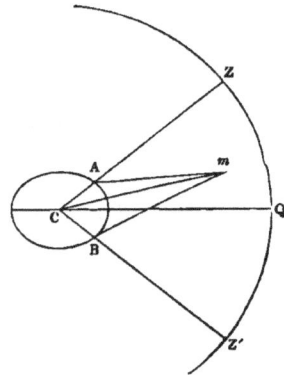

(fig. 2.)

First. Suppose the two places to be on the same side of the equator, as in fig. 1.

Let ZQ the reduced lat. of A $=l$,

       Z'Q   „   „   B $=l'$.

ZAm the reduced zen. dist. of m at A $=z$,

   Z'Bm „   „   B $=z'$.

Horizontal parallax at A $=$ H, and at B $=$ H'.

AC $=r$, BC $=r'$, dist. of body $=d$.

Then diurnal parallax at A $=$ H . sin. $z =$ AmC (p. 125),

     „   „   B $=$ H' . sin. $z' =$ BmC,

and AmC $=$ ZAm $-$ ACm $= z -$ ACm,

BmC $=$ Z'Bm $-$ BCm $= z' -$ BCm ;

$\therefore$ H . sin. z $-$ H' . sin. $z' = z - z' - ($ACm $-$ BC$m)$

$= z - z' -$ ZCZ' $= z - z' - (l - l')$.

But hor. parallax $= \dfrac{\text{rad. of earth}}{\text{dist. of body}}$ (p. 125),

$\therefore$ H $=\dfrac{r}{d}$, and H' $=\dfrac{r'}{d}$ $\therefore$ H' $=$ H . $\dfrac{r'}{d}$

$\therefore$ H . sin. z $-$ H . $\dfrac{r'}{r}$. sin. $z' = z - z' - (l - l')$;

and if we assume $\dfrac{r'}{r} = 1$, which it is very nearly,

then H $= \dfrac{z - z' - (l - l')}{\text{sin. } z - \text{sin. } z'} = \dfrac{z - z' - (l - l')}{2 \cos. \frac{1}{2}(z + z') \sin. \frac{1}{2}(z - z')}$

$= \frac{1}{2}\{z - z' - (l - l')\}$. sec. $\frac{1}{2}(z + z')$ cosec. $\frac{1}{2}(z - z')$.

Next. Suppose the heavenly body to be on different sides of Z and Z' ; and the places of observation on different sides of the equator, as in fig. 2.

Then, as before,

$$\text{H} \cdot \sin. z = \text{A} m \text{C} = z - \text{A} \text{C} m$$
$$\text{H}' \cdot \sin. z' = \text{B} m \text{C} = z' - \text{B} \text{C} m$$

$$\therefore \text{H} \cdot \sin. z + \text{H}' \cdot \sin. z' = z + z' - (\text{A} \text{C} m + \text{B} \text{C} m)$$
$$= z + z' - \text{Z} \text{C} z' = z + z' - (l + l').$$

Let $\text{H} = \text{H}'$, which it is nearly :

$$\therefore \text{H} = \frac{z + z' - (l + l')}{\sin. z + \sin. z'} = \frac{z + z' - (l + l')}{2 \sin. \tfrac{1}{2}(z + z') \cdot \cos. \tfrac{1}{2}(z - z')}$$

$$\therefore \text{H} = \tfrac{1}{2}\{z + z' - (l + l')\} \cdot \text{cosec.} \ \tfrac{1}{2}(z + z') \cdot \sec. \ \tfrac{1}{2}(z - z').$$

By means of this formula, the horizontal parallax of a heavenly body may be calculated from the meridian zenith distances observed at two given places on the same meridian at the same instant.

When the heavenly body is not on the same meridian at the same time, the above formula must be modified to suit the case, or more exact methods used.

217. Calculate the horizontal parallax of the planet Mars, supposing the following observations to be taken at two places on the same meridian at the same instant.

In lat. A. ......59° 20′ 30″ N., the zenith distance was 68° 14′ 6″
  „   B. 33   55   5 S.,   „   „   25   2   0

$$\text{H} = \tfrac{1}{2}(z + z' - \overline{l + l'}) \cdot \text{cosec.} \ \tfrac{1}{2}(z + z') \cdot \sec. \ \tfrac{1}{2}(z - z')$$

z ....68° 14′ 6″			l ....59° 20′ 30″ N.		
z' 25 2 0			l' ... .33 55 5 S.		
$z + z'$ 93 16 6			$l + l'$ 93 15 35		
$z - z'$ 43 12 6			$z + z'$ 93 16 6		
$\tfrac{1}{2}(z + z')$ 46 38 3			$\therefore z + z' - (l + l') =$ 31		
$\tfrac{1}{2}(z - z')$ 21 36 3			$\tfrac{1}{2}(z + z' - \overline{l + l'}) =$ 15·5		

log. cosec. $\tfrac{1}{2}(z + z')$ ....0·138481
  „   sec. $\tfrac{1}{2}(z - z')$.....................0·031621
  „   „   15·5.......................1·190332
  „   „   H...............................1·360434
        $\therefore$ hor. par. $= 22·93''$

DIP.

The altitude of a heavenly body observed from a place above the surface of the earth, as from the deck of a ship, will evidently be greater than the altitude observed from the surface, since the spectator brings the image of the body down to his horizon, which is lower than the horizon seen from the earth's surface beneath him. The difference of the altitudes from this cause is called the dip.

Given the height of the eye above the sea, to calculate the dip.

Let a tangent at B, the point directly beneath the spectator at T, meet the celestial concave at H; from T draw the tangent Th, touching the earth at R. Then, if A be the place of a heavenly body, AH is the altitude observed from B, the surface of the earth, and Ah is the altitude from T. The arc Hh is the dip for the height TB of the spectator above the surface of the earth. Through T draw the diameter TA′, and join RC. Then, the triangles TRC, TC′B, being similar, the angles TCR and TC′B, or HC′h, are equal. Let rad. CR$=r$, and TB, the height of spectator in feet$=a$.

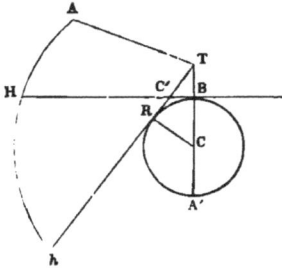

Then tan. TCR$=\dfrac{TR}{RC}=\dfrac{\sqrt{TC^2-RC^2}}{RC}=\sqrt{\dfrac{(a+r)^2-r^2}{r}}=\sqrt{\dfrac{2a}{r}}$ nearly, (since

a^2 is small compared with r), and tan. TCR$=$tan. HC′$h=$HC′h nearly, since the angle is small.

But HC′$h=\dfrac{\text{arc}}{\text{rad.}}=\dfrac{\text{dip}}{57 \cdot 29577° \times 60}$ $\therefore$ $\dfrac{\text{dip}}{57 \cdot 29577° \times 60}=\sqrt{\dfrac{2a}{r}}.$

or dip$=\dfrac{57 \cdot 29577° \times 60 \times \sqrt{2}}{\sqrt{3960 \times 1760 \times 3}}$. $\sqrt{a}=1 \cdot 063 \times \sqrt{a}$

This value of the dip, however, is affected by refraction; the amount of this cannot be very accurately determined on account of the variable nature of refraction at low altitudes. Experiments seem to show that refraction diminishes the amount of dip about $\frac{1}{15}$th of itself; according to Inman its value is about $\frac{1}{13}$th of itself, others make it about $\frac{1}{14}$th. If we take it at $\frac{3}{40}$ths of itself, the results will correspond very nearly with those found in most tables.

Hence dip$=\dfrac{37}{40} \times 1 \cdot 063 \times \sqrt{a}= \cdot 984 \sqrt{a}.$

218. Calculate the dip for the height of the eye above the sea$=20$ feet.

$$\text{dip}= \cdot 984 \sqrt{20}.$$

$$\text{log. } \cdot 984 \ldots \ldots \ldots \ldots \ldots \ldots \ldots \overline{1} \cdot 992995$$
$$\text{,, } \sqrt{20} \ldots \ldots \ldots \ldots \ldots \ldots 0 \cdot 650515$$
$$\text{,, dip} \ldots \ldots \ldots \ldots \ldots \ldots 0 \cdot 643510$$
$$\therefore \text{dip}=4 \cdot 4'=4' \ 24''$$

219. Calculate the dip for the height of the eye above the sea$=110$ feet, and for height$=30$ feet.   *Ans.* 10′ 19·2″, 5′ 23·4″.

AUGMENTATION OF THE MOON'S HORIZONTAL SEMIDIAMETER.

When the moon is above the horizon, as at L' (fig. 1), its distance o L' from a spectator at o is less than its distance o L when in the horizon at L. For the distance c L of the earth's center from the moon is about 60 times the earth's radius, therefore c L $= 60 \times$ c L. But as the horizontal parallax ϖ is small, o L is nearly equal to c L, and therefore L I is less than L O by nearly the earth's radius. Hence if two observers were placed at o and I, one would see the moon when at L in his horizon, and the other in his zenith; but to the spectator at o the moon would be a little more, and to the spectator at I a little less than 60 times its radius, and the diameter would appear to the former about 30″ less than to the latter. It is evident that at any intermediate altitude, as at L', the distance o L' is less than o L, and therefore the moon's diameter at L' would appear to be greater than the true or horizontal diameter at L; that is, the diameter at L' would be *augmented*. To find this correction, or, as it is called, augmentation of the moon's horizontal semidiameter, we must proceed as in the following problem.

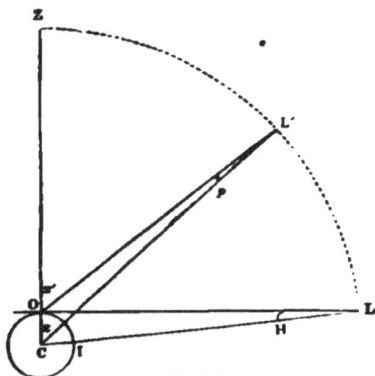

(fig. 1.)

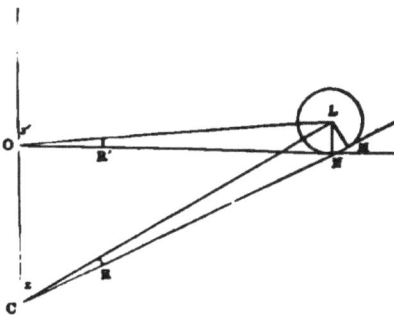

(fig. 2.)

PROBLEM XLI.

Given the horizontal semidiameter of the moon, to calculate its augmentation for a given altitude (fig. 2).

Let o be the place of the spectator, c the earth's center, and L the moon's, o N and c M lines drawn to touch the moon at N and M. Join o L and c L, and draw the perpendiculars L N, L M to the points of contact N and M. The angles L C M, L O N measure respectively the true and apparent semidiameters of the moon, when its zenith distance is z. Let L C M $=$ R, L O N $=$ R'; then R'$-$R is the correction to be calculated.

In the right-angled triangles L O N and L C M we have

$$\sin. \ \mathrm{R}' = \frac{\mathrm{LN}}{\mathrm{LO}}, \ \sin. \ \mathrm{R} = \frac{\mathrm{LM}}{\mathrm{LO}}, \ \therefore \ \frac{\sin. \ \mathrm{R}'}{\sin. \ \mathrm{R}} = \frac{\mathrm{LC}}{\mathrm{LO}}, \ \text{since } \mathrm{LN} = \mathrm{LM}.$$

Let z and z' be the true and apparent zenith distances of the moon. Then in triangle L O C $\dfrac{\sin. \ z'}{\sin. \ z} = \dfrac{\mathrm{LC}}{\mathrm{LO}} = \dfrac{\sin. \ \mathrm{R}'}{\sin. \ \mathrm{R}} = \dfrac{\mathrm{R}'}{\mathrm{R}}$ (since the angles R' and R are small).

$$\therefore \ \frac{\mathrm{R}' - \mathrm{R}}{\mathrm{R}} = \frac{\sin. \ z' - \sin. \ z}{\sin. \ z} = \frac{2 \cos. \ \frac{1}{2} \ (z' + z) \sin. \ \frac{1}{2} \ (z' - z)}{\sin. \ z}$$

Let O L C, the parallax in altitude, $= p$, then $p = z' - z$, and $\therefore \ z = z' - p$, substituting the value of z in the above and reducing, we have

$$\mathrm{R}' - \mathrm{R} = 2 \, \mathrm{R} \ \mathrm{cosec.} \ (z' - p) \cos. \ (z' - \tfrac{1}{2}p) \sin. \ \tfrac{1}{2}p,$$

from which formula $\mathrm{R}' - \mathrm{R}$, the augmentation, may be computed.

220. Calculate the augmentation of the moon's horizontal semidiameter, when the apparent altitude of the center is $60°\ 10'$, the horizontal parallax being $56'\ 1''$, and horizontal semidiameter (as given in the *Nautical Almanac*) $15'\ 16''$.

$$\text{Aug.} = 2 \, \text{n. cosec.} \ (z' - p) \ . \ \sin. \ \tfrac{1}{2}p \ . \ \cos. \ (z' - \tfrac{1}{2}p)$$
$$\text{and } p = \text{hor. par.} \times \cos. \text{ app. alt. (p. 125).}$$

$56'\ 1''$	log. hor. par.........$3\cdot526468$
60	,, cos. alt. ....$9\cdot696774$
$3361'' =$ hor par.	,, p.................$3\cdot223242$
	$\therefore \ p = 1672''$
log. 2 semi. ....$3\cdot262925$	$\therefore \ p = 1672 =$ $27'$ $52''$
,, sin. $\tfrac{1}{2}p$ ....$7\cdot607780$	$z' = 29$ 50 0
,, cos. $(z' - \tfrac{1}{2}p)$ ...$9\cdot939262$	$z' - p = 29$ 22 8
,, cosec. $(z' - p)$ ...$0\cdot309422$	$\tfrac{1}{2}p =$ 13 56
,, aug...............$1\cdot119389$	$z' = 29$ 50 0
$\therefore$ aug. $= 13\cdot16''$	$z' - \tfrac{1}{2}p = 29$ 36 4

221. Calculate the augmentation of the moon's horizontal semidiameter, when the apparent altitude of the center is $32°\ 42'$, the horizontal parallax being $54'\ 42\cdot5''$, and horizontal semi. (in *Nautical Almanac*) $14'\ 56''$:

Ans. $7\cdot86''$.

PROBLEM XLII.

Given the altitude of the moon when near the horizon, and the inclination to the horizon of a line joining the moon's center and that of a distant object; to calculate the contraction of semidiameter on account of refraction.

When the moon is near the horizon its disc assumes an elliptical form, as A BB′, in consequence of the unequal effect of refraction at low altitudes. If, therefore, a contact is made between some distant object in the direction D and the moon's limb at the point P, the semidiameter PC, to be added to the distance to get the distance of the centers, A must be less than CA, the semidiameter as found in the *Nautical Almanac*. If CA=a and CP=r, then $a-r$ is the correction to be computed, or (as it is called) the contraction of the moon's semi. on account of refraction.

To calculate the contraction, let the angle PCA=θ, then by the polar equation to the ellipse we have

$$r=\frac{b}{\sqrt{1-\epsilon^2\cos.^2\theta}}=b\,(1-\epsilon^2\cos.^2\theta)^{-\frac{1}{2}}=b\{1+\tfrac{1}{2}\epsilon^2\cos.^2\theta+\ \ldots\}$$
$$=b\{1+\tfrac{1}{2}\epsilon^2\cos.^2\theta\}$$

omitting all the other terms, since $\epsilon^2\,\epsilon^4$ &c. are small.

$$=b\{1+\tfrac{1}{2}\epsilon^2-\tfrac{1}{2}\epsilon^2\sin.^2\theta\}\ \ldots\ (1)$$

Now, by conic sections, $\dfrac{b^2}{a^2}=1-\epsilon^2,\ \therefore\ a^2=\dfrac{b^2}{1-\epsilon^2}$

$$\therefore a=\frac{b}{\sqrt{1-\epsilon^2}}=b(1-\epsilon^2)^{-\frac{1}{2}}$$
$$=b\{1+\tfrac{1}{2}\epsilon^2+\ \ldots\}=b\{1+\tfrac{1}{2}\epsilon^2\}\ \text{nearly.}$$

Substituting this value of $b(1+\tfrac{1}{2}\epsilon^2)$ in (1) we have

$$r=a-\tfrac{1}{2}b\epsilon^2\sin.^2\theta,\ \therefore\ a-r=\tfrac{1}{2}b\epsilon^2\sin.^2\theta\ \ldots\ (2)$$

We may eliminate $\tfrac{1}{2}b\epsilon^2$ as follows:

Let $\theta=90°$, then $r=$BC and $a-$BC=difference of refractions between the points B and C. This may be found in the Table of Refractions. Call this difference c;

$$\text{then } c=\tfrac{1}{2}b\epsilon^2\sin.^2 90=\tfrac{1}{2}b\epsilon^2.$$

Substituting this value of $\tfrac{1}{2}b\epsilon^2$ in (2) we have

$$a-r \text{ or the correction}=c\sin.^2\theta,$$

where c=difference of refraction for center C and vertex B, and θ=inclination of line joining the centers to the horizon.

222. Calculate the correction or contraction of moon's semidiameter when the altitude is 5° and the inclination of the line joining the centers is 60°, the moon's semidiameter being 16′ 0″.

$$\text{cor.}=c\sin.^2\theta$$

Refraction at 5° 0′...9′ 58″	log. c............1·397940
„ 5 16 ...9 33	„ sin. 60° ...9·937531
$c=$ 25	„ sin. 60° ...9·937531
	„ cor.........1·273002 .·. cor.=18·8″.

223. Calculate the correction or contraction of the moon's semidiameter when the altitude $= 4° 30'$ and the line joining the centers is inclined at an angle of $40°$, the moon's semidiameter being $15' 30''$.   *Ans.* $11·5''$.

Correction of moon's meridian passage.

The time of the transit of a heavenly body can be found by means of Problem VI.; but in the case of the moon, the following approximate method of finding the time of her passage over a given meridian may be sometimes used with advantage.

The mean time of the moon's transit for every day at Greenwich is put down in the *Nautical Almanac.* At any place to the east of Greenwich, the time of the transit, owing to the moon's proper motion to the eastward, must take place sooner (independent of that due to the difference of longitude), and to a place to the westward of Greenwich, later than the time recorded in the *Nautical Almanac.* Thus, if we suppose the moon's daily motion to be 60 minutes : to a place $90°$ to the east of Greenwich the transit will take place 6 hours earlier than that at Greenwich (on account of the difference of longitude) $+ \frac{90}{360}$ of 60^m, or 15 minutes, due to the moon's motion, supposed equable, to the eastward in the 6 hours before she reaches the meridian of Greenwich. To a place west of the first meridian, a retardation will take place for the same reason.

The moon's daily motion in RA varies between 40^m and 60^m, so that it would not be difficult to construct a small table of the correction of the transits given in the *Nautical Almanac* for any given longitude : this has accordingly been done, and may be found in Inman's *Nautical Tables,* p. 5.

The construction of the table may be explained as follows :

Let the moon's proper motion in 24 hours $= d$.

A meridian, as that of Greenwich, describes in that time one complete revolution, or $360°$.

Let the longitude of some given place $= D°$, and the proper motion of the moon for $D° = x$.

Then $360° : D° :: d : x$ (supposing the moon's motion equable),

$$\text{or } x = \frac{D°}{360°} \times d$$

By assuming different values for $D°$ and d, the correction x may be easily calculated, and a table formed of the results.

Example. Given the daily motion of the moon $= 45·7^m$; required the correction of the meridian passage in the *Nautical Almanac* for a place in longitude $50°$ W.

$$D = 50° \qquad\qquad x = \frac{D°}{360°} \times d = \frac{50}{360} \times 45·7 = 6·3^m.$$
$$d = 45·7$$

And in a similar manner may all the other values in the table be calculated.

The moon's daily motion used should be that found by taking the difference between the two transits at Greenwich that happen before and after the one at the place : that is, if the place be in west longitude, the difference should be taken between the transit on the given day and the one following; if in east longitude, that on the given day and the one preceding. By observing this rule, the error arising from the unequal motion of the moon in RA is diminished.

An example or two of finding the time at Greenwich at the transit of the moon over a given meridian will show the use of the table.

224. April 27, required Greenwich mean time nearly at the transit of the moon over the meridian of a place in longitude 50° W.

By *Nautical Almanac,* mer. pass. on 27th...11^h 46·3^m
,,   ,,   on 28th...12   32·0
Moon's motion in 24 hours...   45·7
Correction (from table, or by calculation)...   6·3 +
∴ time of transit at place...11   52·6
long. in time... 3   20·0 W.
∴ Greenwich mean time...15   12·6

225. April 27th, required Greenwich mean time nearly at the transit of the moon over the meridian of a place in longitude 50° E.

By *Nautical Almanac,* mer. pass. on 27th...11^h 46·3^m
,,   ,,   on 26th...11   2·7
43·6

Cor.$=\frac{50}{360} \times 43 \cdot 6$ Correction...   6·0 −
$=6 \cdot 06^m$ Transit at place... 11   40·3
or by table$=6 \cdot 0$ long. in time... 3   20·0 E.
∴ Time at Greenwich... 8   20·3

Required the mean time at the place of the moon's meridian passage on July 19 (astronomical day), in longitude 60° W., and on July 27 (astronomical day), in longitude 175° E., having given the following quantities from the *Nautical Almanac :*

Gr. mer. pass. on July 19......11^h 24·3^m July 27......17^h 30·1^m
,,   ,, 20......12 19·2   ,, 26......16 49·5
Ans. Mer. pass. at place on July 19 at 11^h 33·3^m
,,   ,,   July 27 at 17 11·1$=$July 28 at 5^h 11·1^m A.M.

CONSTRUCTION OF THE TABLES USED FOR CORRECTING THE QUAN-
TITIES TAKEN OUT OF THE NAUTICAL ALMANAC FOR ANY GIVEN
TIME FROM GREENWICH MEAN NOON.

The declination, right ascension, &c., of the sun and moon are inserted in the *Nautical Almanac* for every day at noon, or 0^h 0^m $0'$ Greenwich mean time. To find the same quantities for any other Greenwich time, we may either multiply the hourly differences by the time elapsed since the preceding noon, or use the common rules of proportion, or, what is in some cases the simplest method, find the quantity to be added or subtracted by means of certain tables called Proportional Logarithms, the principal of which are the following :

(1.) The proportional logarithms (properly so called).

(2.) Greenwich date prop. logarithm for the sun.

(3.) Greenwich date prop. log. for the moon.

(4.) The logistic logarithms.

We will give the construction of the above tables, with a few examples to show their application and use.

(1.) *Construction of table of* PROPORTIONAL LOGARITHMS.

Definition. The logarithm of 180^m (or 3 hours), diminished by the logarithm of any other number of hours and minutes (less than 3 hours), is called the proportional logarithm of that number.

Thus prop. log. 2^h $42^m =$ log. $180 -$ log. $162 = \cdot045757$.

(2.) *Construction of table of Greenwich date* PROPORTIONAL LOGARITHMS FOR
THE SUN.

Definition. The logarithm of 1440^m ($= 24$ hours), diminished by the logarithm of any other number of hours and minutes (less than 24 hours), is called the Greenwich date log. for sun for that number.

Thus Greenwich date log. sun for 17^h $42^m =$ log. $1440 -$ log. $1062 = \cdot132238$.

(3.) *Construction of table of Greenwich date* PROPORTIONAL LOGARITHMS FOR
THE MOON.

Definition. The logarithm of 720^m ($= 12$ hours), diminished by the logarithm of any other number of hours and minutes (less than 12 hours), is called the Greenwich date log. for moon for that number.

Thus Greenwich date log. moon for 9^h $48^m =$ log. $720 -$ log. $588 = \cdot087955$.

(4.) *Construction of table of* LOGISTIC LOGARITHMS.

Definition. The logarithm of $3600'$ ($= 1$ hour), diminished by the logarithm of any other number of minutes and seconds (less than 1 hour), is called the logistic logarithm for that number.

Thus logist. log. 42^m $30' =$ log. $3600 -$ log. $2550 = \cdot149762$.

CORRECTION OF THE DECLINATION, RIGHT ASCENSION, &c., OF THE HEAVENLY BODIES FOR A GIVEN TIME FROM GREENWICH NOON.

The following examples will show the use of proportional logarithms for calculating the declination, right ascension, &c., for a given time from noon called " A GREENWICH DATE."

(1.) Given the declination of the sun at two consecutive noons at Greenwich ; to find the declination at some intermediate time.

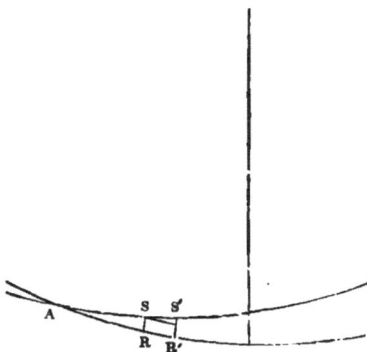

226. Find the sun's declination on April 27, 1846, at $10^h 42^m$ P.M. Greenwich date, $10^h 42^m$.

Sun's decl. by N. A.

$$\text{On 27th} \quad\dots\dots\dots\dots\dots\dots 13^\circ\ 48'\ 3''\ N. = \text{S R} \quad (\text{see fig.})$$
$$\text{„ 28th} \quad\dots\dots\dots\dots\dots\dots 14\quad\ 7\quad 4\ \ N. = \text{s'R'}$$
$$\therefore \text{change in 24 hours}\dots\dots \quad 19\ \ 1\ \ N.$$

Let x=change in $10^h 42^m$,
then (supposing the motion in decl. is equable) we have

$$24^h : 10^h 42^m : : 19'\ 1'' : x,$$
$$\text{or } \log. 24 + \log. x = \log. 10^h 42^m + \log. 19'\ 1'',$$
$$\text{or } \log. x = -\log. 24^h + \log. 10^h 42^m + \log. 19'\ 1''.$$

Adding log. 3^h to both sides, and changing the signs,

$$\log. 3^h - \log. x = \log. 24^h - \log. 10^h 42^m + \log. 3^h - \log. 19'\ 1'',$$
$$\text{or prop. } \log. x = \text{Gr. date log. sun } 10^h 42^m + \text{prop. log. } 19'\ 1''.$$

Greenwich date log. sun for $10^h 42^m$ 35083
Prop. log. for $19'\ 1''$ 97614

Prop. log. x......1·32697

$$\therefore x \text{ or change in } 10^h 42^m = \quad 8'\ 28·6''\ N.$$
$$\text{and declination at noon } \dots\dots\dots 13\quad 48\quad 3\quad N.$$
$$\therefore \text{decl. at Greenwich date} \dots\dots 13\quad 56\quad 31·6\quad N.$$

It is evident the value of x may be found by applying the common rules of proportion or practice, or by multiplying the hourly difference, found in the *Nautical Almanac*, by the time elapsed since noon; and this latter method in some cases will be found the simplest.

(2.) Given the moon's semidiameter at noon and midnight; to find the semidiameter at some intermediate time.

227. Find the moon's semidiameter on April 27, 1846, at $10^h\ 42^m$ P.M. Greenwich date, $10^h\ 42^m$.

Moon's semi.

27th noon......................15′ 27·3″
27th mid. 15   21·8
Change in 12 hours.........   5·5

Let $x=$change in $10^h\ 42^m$, then $12^h : 10^h\ 42^m :: 5·5 : x$,

$\therefore \log. x = -\log. 12^h + \log. 10^h\ 42^m + \log. 5·5''$,

$\therefore \log. 3^h - \log. x = \log. 12^h - \log. 10^h\ 42^m + \log. 3^h - \log. 5·5''$,

or prop. log. $x=$Greenwich date log. moon for $10^h\ 42^m +$ prop. log. $5·5''$.

Greenwich date log. moon for $10^h\ 42^m$ ....   ·04980

prop. log. $5·5''$......................................   3·29306

prop. log. x ....   3·34286

$\therefore x$ or change required$=\ 0'\ \ 4·9'' -$
and semi. at noon$=15\ \ 27·3$
$\therefore$ semi. at Greenwich date$=15\ \ 22·4$

(3.) Given the moon's declination for two consecutive hours; to find the declination for some intermediate time.

228. Find the moon's decl. for April 27, 1846, at $17^h\ 14^m\ 50^s$. Greenwich date, $14^m\ 50^s$.

Moon's decl. on 27th.

At 17^h...........................18° 59′ 30″ N.
„ 18 19   1   58 N.
                                  2   28

Let $x=$change for $14^m\ 50^s$, then $1^h : 14^m\ 50^s :: 2'\ 28'' : x$,

or log. $x+$log. $1^h=$log. $14^m\ 50^s+$log. $2'\ 28''$,

$\therefore$ log. $x = -$log. 1^h+log. $14^m\ 50^s+$log. $2'\ 28''$,

$\therefore$ log. $3°-$log. $x=$log. 1^h-log. $14^m\ 50^s+$log. $3°-$log. $2'\ 28''$,

or prop. log. $x=$logistic log. $14^m\ 50^s+$prop. log. $2'\ 28''$.

logistic log. $14^m\ 50^s$..........................·60691

prop. log. $2'\ 28''$...............................1·86316

prop. log. x.....................2·47007

$\therefore x= 0°\ 0'\ 36·8'' N.$
declination at 17^h ....18 59   30   N.
declination at $17^h\ 14^m\ 50^s$...........19   0   6·8 N.

(4.) The lunar distances of certain stars used in finding the longitude being given in the *Nautical Almanac* for every third hour from Greenwich mean noon, to find the distance at any intermediate time by proportional logarithms.

229. Find the distance of Fomalhaut from the moon on Dec. 1, 1846, at $4^h 30^m$. Greenwich date, $1^h 30^m$.

Star's distance at

$$3^h \ldots\ldots\ldots\ldots\ldots\ldots\ldots\ldots 80° \ 45' \ 54''$$
$$6^h \ldots\ldots\ldots\ldots\ldots\ldots\ldots\ldots 82 \ \ 17 \ \ 51$$
$$\text{change in 3 hours} \ldots\ldots 1 \ \ 31 \ \ 57$$

Let $x=$change in $1^h 30^m$, then $3^h : 1^h 30^m :: 1° 31' 57'' : x$,

$\therefore$ log. $x=$log. 3^h+log. $1^h 30^m+$log. $1° 31' 57''$.

Changing the signs, and adding log. $3°$ to each side, we have

log. $3°-$log. $x=$log. 3^h-log. $1^h 30^m+$log. $3°-$log. $1° 31' 57''$,

prop. log. $x=$prop. log. $1^h 30^m+$prop. log. $1° 31' 57''$.

$$\text{prop. log. } 1^h 30^m \ldots\ldots\ldots\ldots\ldots \cdot 30103$$
$$\text{prop. log. } 1° 31' 57'' \ldots\ldots\ldots\ldots \cdot 29172$$
$$\text{prop. log. } x \ldots\ldots\ldots\ldots\ldots\ldots \cdot 59275$$

$$\therefore x= \ \ 0° \ \ 45' \ \ 58\cdot5''=\text{change in } 1^h 30^m$$
$$\text{distance at } 3^h=80 \ \ \ 45 \ \ \ 54$$
$$\therefore \text{dist. at } 4^h 30^m=81 \ \ \ 31 \ \ \ 52\cdot5$$

The converse of this, namely, to find the time corresponding to some given intermediate distance, may also be found by means of proportional logarithms. This is always required in the rule for finding the longitude by lunar observations (p. 87).

230. Given the lunar distances at 3^h and 6^h to be $80° 45' 54''$ and $82° 17' 51''$ respectively; to find the *time* when the lunar distance will be $81° 31' 52\cdot5''$.

Let $x=$the time elapsed since 3 o'clock.

$$\text{lun. dist. at } x \ldots\ldots\ldots\ldots\ldots 81° \ \ 31' \ \ 52\cdot5''$$
$$\text{,, \quad at } 3^h \ldots\ldots\ldots\ldots 80 \ \ \ 45 \ \ \ 54$$
$$\text{,, \quad at } 6^h \ldots\ldots\ldots\ldots 82 \ \ \ 17 \ \ \ 51$$
$$\text{increase since } 3^h \ldots\ldots\ldots\ldots 0 \ \ \ 45 \ \ \ 58\cdot5$$
$$\text{increase in } 3^h \ldots\ldots\ldots\ldots\ldots 1 \ \ \ 31 \ \ \ 57$$

Then $3^h : x :: 1° 31' 57'' : 45' 58\cdot5''$

$\therefore$ log. $x+$log. $1° 31' 57''=$log. 3^h+log. $45' 58\cdot5''$

log. 3^h-log. $x=$log. $3°-$log. $45' 58\cdot5''-($log. $3°-$log. $1° 31' 57'')$

or prop. log. $x=$prop. log. $45' 58\cdot5''-$prop. log. $1° 31' 57''$.

$$\text{prop. log. } 45' 58\cdot5'' \ldots\ldots\ldots\ldots \cdot 59275$$
$$\text{prop. log. } 1° \ \ 31' 57'' \ldots\ldots\ldots \cdot 29172$$
$$\text{prop. log. } x= \cdot 30103$$

$\therefore x$ or time since 3 o'clock$=1^h 30^m$

$\therefore$ time corresponding to distance $81° 31' 52\cdot5''$ is $3^h+1^h 30^m$, or $4^h 30^m$.

We have supposed in the above examples the motions of the sun and moon in the interval between the given Greenwich times to be *uniform*. This is seldom the case ; and therefore, for very accurate observations, a correction must be used called the equation of second differences, a rule for computing which will be investigated in *Navigation*, Part III.

Other corrections might have been noticed, such as the correction for refraction under different conditions of the air, the correction called the annual variation of the fixed stars, arising from precession, &c. ; but the analytical investigation of these, and others of a similar nature, are not sufficiently elementary for the present volume.

END OF NAUTICAL ASTRONOMY, PART II.

NAVIGATION.

CHAPTER VI.

INVESTIGATION OF RULES IN NAVIGATION OR PLANE SAILING.

THE position of a place on the surface of the earth is found by referring it to two lines drawn on the surface at right angles to one another; these lines are the *terrestrial equator*, and a line continued round the earth passing through its poles and through some well-known place, as *Greenwich*, called the *first meridian*. Thus, if AM represent a portion of the equator, and GA a portion of the first meridian, and P any place on the surface of the earth, and PM an arc of the meridian passing through P; then the position of the point P is said to be determined when the magnitudes of the arcs AM and PM are known. The arc AM is called the *longitude* of P.

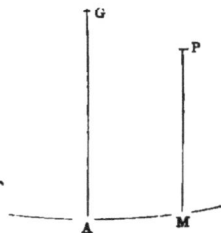

If the earth were an exact sphere, the length of a degree of the meridian in every part of it would be the same; but observations and actual measurements of arcs in different parts of the world have proved to us that the figure of the earth is that of an oblate spheroid, whose equatorial and polar diameters are about 7924 and 7898 miles respectively (see p. 120).

From the above dimensions it is manifest that the earth, although not an exact sphere, is very nearly so; accordingly, in the common rules of Navigation which we are about to investigate, the earth will be considered as a sphere, and on this supposition a meridian is a great circle, and the arc of a great circle PM intercepted between the point P and the equator is the latitude of P.

The following are the principal terms in Navigation: the definitions of these terms, like those in Nautical Astronomy, must be thoroughly understood and committed to memory.

Course.

Distance.

Departure.

True difference of latitude.

Meridional difference of latitude.

Difference of longitude.

Middle latitude.

Definitions of terms in Navigation.

The course is the angle which the ship's track makes with all the meridians between the place left and the place arrived at.

The distance is the spiral line made by the ship's track in describing the course between the place left and the place arrived at.

The departure is the sum of all the arcs of parallels of latitude drawn between the place left and the place arrived at, through points indefinitely near to one another taken on the distance, and intercepted between the meridians passing through those points.

The true difference of latitude is the arc of a meridian intercepted between the parallels of latitude drawn through the place left and the place arrived at.

The meridional difference of latitude is the value in minutes of a great circle of the line on a Mercator's chart, into which the true difference of latitude has been expanded.

The difference of longitude is the arc of the terrestrial equator intercepted between the meridians passing through the place left and the place arrived at.

The middle latitude is the mean of the latitudes (supposed of the same name) of the place left and the place arrived at.

These definitions will be clearly understood by means of the following diagrams.

Let P represent the pole of the earth, T Z an arc of the equator, PT the meridian passing through a known place G, as Greenwich, A and F two other places on the earth (considered as a sphere), PU, PZ, their meridians.

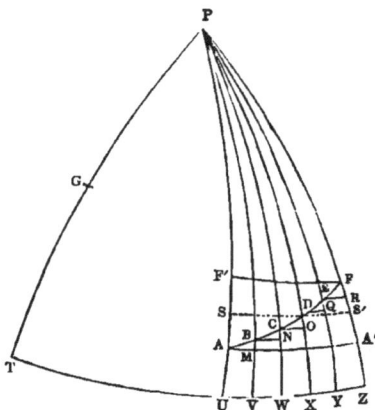

Through the points A and F suppose a curve line AF to be drawn, cutting all the intermediate meridians PV, PW, PX, &c., *at the same angle;* that is, making the angle PAB=PBC= PCD=&c. Then this common angle PAF is called the *course.* The arc AF* is the *distance.* Draw the parallels of latitude AA', FF'; then, since the latitude of A is the arc AU, and the longitude of A the arc TU, and the latitude of F is the arc FZ, and the longitude of F is the arc TZ; therefore the difference, or, as it is usually called, the *true difference of latitude*, between A and F is the arc AF' or A'F, and *the difference of longitude* between A and F is the arc of the equator UZ. Again, suppose the intermediate meridians PV, PW, &c., to be

* AF is sometimes called the *rhumb line*, sometimes the *loxodromic curve*, sometimes the *equiangular spiral*.

drawn through points B, C, D, &c., taken on the line AF indefinitely near to each other; and through the points A, B, C, D, &c., the arcs of parallels of latitude AM, BN, CO, &c. On this supposition (namely that the points A, B, C, &c., are indefinitely near to each other) the elementary triangles ABM, BCN, CDO, &c., may be considered without any error to be right-angled *plane triangles. The departure* between A and F=AM+BN+CO+ ... ER, the points A, B, C, &c., being supposed to be indefinitely near to each other.

If a parallel of latitude ss' be drawn through the middle of AF', then the arc of the meridian su is called the mean or *middle latitude* between A and F. It is manifest that the arc ss' will be nearly equal to AM+BN+ ... DQ+ER, the departure, A and F being supposed to be on the same side of the equator. For short distances ss' is substituted without any practical error for the departure, and one of the principal rules in Navigation deduced from it.

There are two kinds of charts; the Plane chart, and Mercator's chart.

The Plane chart.

The plane chart is a representation of the earth's surface, considering it as a plane. When a small portion of the surface is concerned, this mode of representing it will lead to no practical error; hence coasting charts are usually constructed in this manner, in which the different headlands, light-houses, &c., are laid down according to their bearings.

Mercator's chart.

The chart used at sea for marking down the ship's track, and for other purposes, is called Mercator's chart. It exhibits also the surface of the earth *on a plane;* but the meridians are drawn perpendicular to the equator, and therefore the arcs AM, B'B, &c., of parallels of latitude intercepted between any two meridians are increased to *am, b'b,* &c., and become equal to one another and to line *uv,* and therefore to the intercepted arc UV of the equator. If we wish to make the figures (supposed to be very small) *ambb', b'bcc',* &c., on the chart similar to AMBB', B'BCC', &c., of the globe, it is evident we must increase the sides *bm, bc,* &c., in the same proportion as *am, b'b,* &c. (that represent AM, BB', &c.), have been increased. Let us therefore suppose the straight lines *am, b'b, c'c,* &c., have been drawn at such a distance from each other that the above similarity of figure is preserved (and

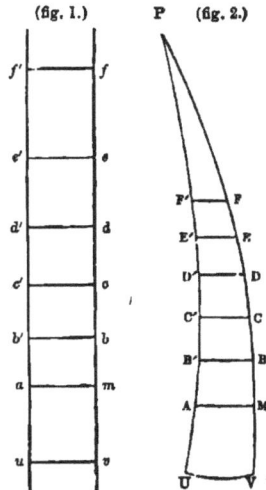

(fig. 1.)   P   (fig. 2.)

L

this can only be done by supposing the surfaces $ambb'$, $b'bcc'$, &c., inde-
finitely small, so that the surfaces AMBB', B'BCC', &c., may be considered as
plane surfaces). Then a representation of the earth's surface, or any part of
it, so constructed, is called a *Mercator's chart.*

The straight line mf, into which MF, the true difference of latitude
between M and F, has been expanded, is called the *meridional difference of
latitude* between M and F, and the values of bv, cv, &c., in minutes, are
called the *meridional parts* of B, C, &c. : hence the *meridional difference of
latitude* between two places is the difference of the meridional parts for the
two places.

*An approximate method of calculating the meridional parts for any
latitude.*

In order to construct a Mercator's chart, we must know the lengths of
the lines corresponding to the latitude of every point on the globe, at small
distances from each other. These lengths, or meridional parts, computed
for every minute of latitude from 0° to 90°, form the table of meridional
parts. It may be obtained with sufficient exactness in the following
manner :

Suppose the meridians PV, PU (p. 145), and parallels AM, BB', &c., are
drawn sufficiently near to each other that the quadrilateral surfaces of the
earth, AMBB', B'BCC', &c., thus formed, may be considered without any
practical error to be *plane surfaces;* this may be done if the arcs MB, BC,
&c., be not taken greater than 1 minute : then these quadrilateral surfaces
being expanded into $ambb'$, $b'bcc'$, similar to them on the chart, we can
prove that

$$bm = \text{BM . sec. lat. M,}$$
$$bc = \text{BC . sec. lat. B,}$$
$$cd = \text{CD . sec. lat. C,}$$
$$de = \text{DE . sec. lat. D,}$$
$$\&c. = \&c.$$

For (by *Trig.* Part II. Art. 69), $\frac{AM}{UV} = \cos. MV = \cos.$ lat. M,

$$\therefore \text{ sec. lat. } M = \frac{UV}{AM} = \frac{uv}{AM} = \frac{am}{AM}.$$

But since the quadrilateral surfaces on the globe and chart are similar,

$$\therefore \frac{bm}{BM} = \frac{am}{AM} = \text{sec. lat. M,}$$
$$\therefore bm = \text{BM . sec. lat. M.}$$
similarly $bc = \text{BC . sec. lat. B,}$
$$cd = \text{CD . sec. lat. C,}$$
$$de = \text{DE . sec. lat. D,}$$
$$\&c. = \&c.$$

Let lat. of M$=l$, and BM$=$BC$=$CD$=$DE$=1$ minute; then lat. B$=l+1'$, lat. C$=l+2'$, lat. D$=l+3'$; $\therefore$ adding, we have

sec. l + sec. $(l+1)$ + sec. $(l+2)$ + sec. $(l+3)=bm+bc+cd+de=me$
$\qquad\qquad$ =meridional difference of lat. between M and E.

If the point M be on the equator, then $l=0$, and the above expression gives the value of the meridional parts for 4 minutes: thus

$$bm+bc+cd+de=\text{sec. }0'+\text{sec. }1'+\text{sec. }2'+\text{sec. }3',$$
or the merid. parts for $4'=$sec. $0°$ + sec. $1'$ + sec. $2'$ + sec. $3'$.

A nearer approximation to the value of the meridional parts for 4 minutes would of course be obtained by taking the parts MB, BC, CD, &c., still smaller, as 1 second; for then the meridional parts for 4 minutes would be found from the expression,

M P for $4'=$sec. $0''$ + sec. $1''$ + sec. $2''$ + sec. $3''$ + . . + sec. $3'\ 59''$.

Hence, generally, we have for any lat. l

Mer. parts for $l°=$sec. $0''$ + sec. $1''$ + sec. $2''$ + . . + sec. $(l°-1'')$.

And from a similar expression to this was the first table of meridional parts computed. (A more correct method is given in p. 171.)

231. *Example.* If we suppose the meridional parts for $70°$ to have been found to be$=5965\cdot92$ (by the above or any other method), let it be required to calculate the meridional parts for $70°\ 10'$.

M P for $70°10'=$sec. $0'$ + sec. $1'$ + sec. $2'$ + sec. $3'$ + . . . + sec. $69°59'(=5965\cdot92)$
$\qquad\qquad\qquad\qquad$ + sec. $70°$ + sec. $70°\ 1'$ + . . . + sec. $70°\ 9'$
$\qquad\quad =5965\cdot92+$sec. $70°$ + sec. $70°\ 1'$ + . . . sec. $70°\ 9'$.

log. sec. $70°$	$0'$	......$0\cdot465948$......$\therefore$	nat. sec.$=$	$2\cdot9238$
"	70	1 ......$0\cdot466296$......	" $=$	$2\cdot9262$
"	70	2 ......$0\cdot466643$	" $=$	$2\cdot9285$
"	70	3 ......$0\cdot466991$	" $=$	$2\cdot9308$
"	70	4 ......$0\cdot467339$	" $=$	$2\cdot9332$
"	70	5 ......$0\cdot467688$	" $=$	$2\cdot9355$
"	70	6 ......$0\cdot468037$	" $=$	$2\cdot9379$
"	70	7 ......$0\cdot468386$	" $=$	$2\cdot9403$
"	70	8 ......$0\cdot468735$	" $=$	$2\cdot9426$
"	70	9 ......$0\cdot469085$	" $=$	$2\cdot9450$

$$\qquad\qquad\qquad\qquad\qquad\qquad\qquad\qquad 29\cdot3438$$

$$\text{mer. parts for } 70°=5965\cdot92$$

$$\therefore \text{ mer. parts for } 70°\ 10'=5995\cdot2638 \cdot$$

232. Given the meridional parts for $70°\ 10'=5995\cdot2638$; calculate the meridional parts for $70°\ 15'$. $\qquad$ *Ans.* Mer. parts$=6010\cdot03$.

PROOF OF THE RULES IN NAVIGATION.

From the definitions and principles in pp. 144, 145, are deduced the following formulæ or equations, and these expressed in words constitute the common rules of Navigation for finding the *place* of a ship, that is, its latitude and longitude.

FUNDAMENTAL FORMULÆ IN NAVIGATION.

Departure=distance × sin. course . . . . . . (1)

True diff. lat.=distance × cos. course . . . . (2)

Diff. long.=meridional diff. lat. × tan. course . (3)

$\left\{\begin{array}{l} \text{In parallel sailing,} \\ \text{Distance=diff. long.} \times \text{cos. lat.} \end{array}\right.$ . . . . . . (4)

$\left\{\begin{array}{l} \text{In middle latitude sailing,} \\ \text{Departure (nearly)=diff. long.} \times \text{cos. mid. lat.} \end{array}\right.$. (5)

PROOF OF THE ABOVE FORMULÆ.

(1.) *Proof that* DEPARTURE=DIST. × SIN. COURSE.

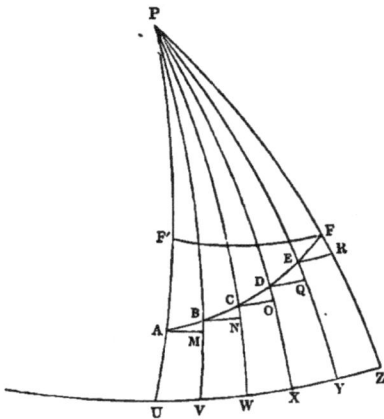

Suppose the ship to sail from A to F, cutting all the intermediate meridians at the same angle PAF, PBF, &c., then this common angle is the course (p. 144), the arc AF is the distance (p. 144); and if in the figure the triangles ABM, BCN, &c., be taken so small that they may be considered as *right-angled plane triangles*, then AM+BN+... +ER is the departure (p. 145): it is also manifest that the angles BAM, CBN, DCO, &c., are equal to one another, each being the complement of the course.

∴ in triangle BAM, AM=AB . cos. BAM=AB . sin. course

,, BCN, BN=BC . cos. CBN=BC . sin. course

,, CDO, CO=CD . cos. DCO=CD . sin. course

&c. = &c.

adding $AM+BN+CO+ .. =(AB+BC+CD+ ..)$ sin. course.

But $AM+BN+CO+ \; ... \; +ER=$departure,

and $AB+BC+CD+ \; ... \; +EF=$distance,

∴ departure=dist. × sin. course.

(2.) *Proof that* TRUE DIFF. LAT.=DIST. × COS. COURSE.

The same construction being made as in the last figure, we have in the elementary triangles ABM, BCN, CDO, &c.

$$BM=AB \; . \; sin. \; BAM=AB \; . \; cos. \; course,$$
$$CN=BC \; . \; sin. \; CBN=BC \; . \; cos. \; course,$$
$$DO=DC \; . \; sin. \; DCO=DC \; . \; cos. \; course,$$
$$\&c.=\&c. \quad ∴ \; adding$$
$$BM+CN+DO+ .. =(AB+BC+DC+ ..) \; cos. \; course.$$

But $BM+CN+DO+ ... +FR=AF'=$true diff. lat.

and $AB+BC+DC+ ... +EF=AF=$distance,

∴ true diff. lat.=dist. × cos. course.

(3.) *Proof that* DIFF. LONG.=MERIDIONAL DIFF. LAT. × TAN. COURSE.

We have already seen that in the construction of a Mercator's chart the meridians are drawn parallel to each other and perpendicular to the equator, and therefore the parts of parallels of latitude AM, BN, CO, &c. (see last figure), are increased, and become equal to the corresponding parts UV, VW, WX, &c., of the equator; and that in order to preserve the similarity of the parts on the chart that correspond respectively to the triangles ABM, BCN, &c., the sides DM, CN, DO, &c., must be increased in the same proportion as AM, BN, CO, &c., have been increased: when, therefore, AM+BN+CO, &c., the departure between A and F, has been increased, and become equal in length to UZ, the difference of longitude, the arcs BM+CN+DO, &c.=AF', the true diff. lat. between A and F, will be expanded, and become equal to a straight line called *the meridional difference of latitude between* A and F. Moreover the lines into which each of the parts AB, BC, CD, &c., is expanded, will be inclined to the parallels at the same angle as the course, since the triangular surfaces on the chart into which the triangles ABM, BCN, &c., are expanded are made similar to ABM, BCN, &c., in every respect; and as AB, BC, &c., cut the meridians at the same angle, lines corresponding to them on the chart must be in one and the same straight line. Hence the meridional difference of latitude, difference of longitude, and the line into which the distance AF is expanded, form the sides of a right-angled plane triangle, as AMN; where if A represents the course, corresponding to the angle PAF on the globe, AM

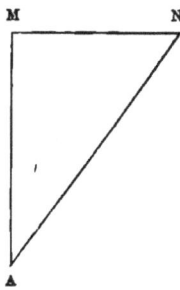

will be the meridional difference of latitude, and MN the difference of longitude between A and F.

Since, in the right-angled triangle AMN, $\dfrac{MN}{AM}$=tan. A,

$$\text{therefore } \frac{\text{diff. long.}}{\text{mer. diff. lat.}}=\text{tan. course,}$$

or diff. long.=mer. diff. lat. × tan. course.

(4.) PARALLEL SAILING.

Proof that DIST.=DIFF. LONG. ×COS. LAT.

When the course is 90°, that is, when the ship is sailing on a parallel of latitude, tan. course=∞, and for this value of the angle formula (3) gives no assistance in finding the difference of longitude: but since the course in

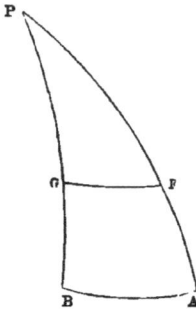

parallel sailing is due east or due west, the distance is in fact the *arc of a parallel of latitude* intercepted between the meridians passing through the two places; we may therefore find the relation between the distance, difference of longitude and latitude, by means of the well-known property that

$\dfrac{GF}{AB}$=cos. FA (*Trig.* Part II. p. 76), in which FG is the distance, AB the diff. long., and AF the latitude of the ship.

Hence we have $\dfrac{\text{dist.}}{\text{diff. long.}}$=cos. lat.,

or dist.=diff. long.×cos. lat.: a formula that will enable us to solve all the common problems in parallel sailing.

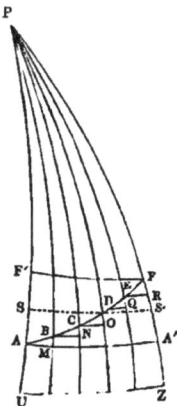

(5.) MIDDLE LATITUDE SAILING.

Proof that DEPARTURE=DIFF. LONG. × COS. MID. LAT.

Another formula, giving results sufficiently correct when the distance run is not great (such as in an ordinary day's sailing), is obtained by considering the parallel ss', drawn through the middle of AF' (the difference of latitude between A and F), as equal to AM+BN+OO+ ..., the departure; for then we have

(*Trig.* Part II. Art. 69) $\dfrac{ss'}{uz}$=cos. su,

in which ss'=departure nearly,

uz=difference of long. between A and F,

and su=the latitude of the middle point between A and F, and is therefore called the *middle latitude.*

Hence $\dfrac{\text{dep.}}{\text{diff. long.}}$=cos. mid. lat.

or departure=diff. long. × cos. mid. lat.

RULES IN NAVIGATION DERIVED FROM PLANE TRIGONOMETRY.

The principal rules in Navigation may be deduced from the formulæ just proved; but as they can be shown to depend also on the solution of a right-angled triangle (the proof of which will be given in the following problem), we will adopt the latter method of developing the rules, as being more practically useful.

PROBLEM XLIII.

The DISTANCE, TRUE DIFFERENCE OF LATITUDE, DEPARTURE, and COURSE, may be correctly represented by the three sides and one of the angles of a *right-angled plane triangle*; also the MERIDIONAL DIFFERENCE OF LATITUDE, and DIFFERENCE OF LONGITUDE, may be represented by two sides of a triangle which is similar to the same right-angled plane triangle.

Suppose the arc AF, the distance between A and F, to be divided into n *equal* parts AB, BC, CD, &c.; then $AF=n$AB, and the true difference of

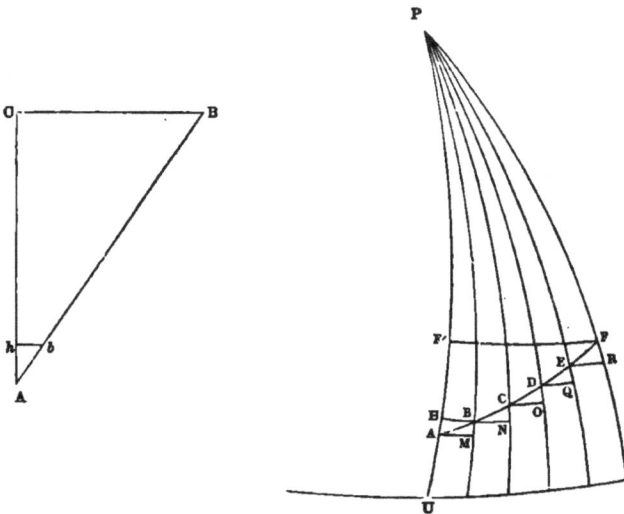

latitude AF' ($=$BM$+$CN$+ \ldots$)$=n$BM, since the elementary triangles ABM, BCN, &c., are on this supposition equal to each other in every respect, and also to the triangle ABH, formed by drawing BH parallel to AM. Let now a small plane triangle Abh be supposed to be constructed equal and similar to the triangle ABH (and this can only be done by supposing ABH to be in its evanescent condition, that is, indefinitely small); produce Ab to B, and Ah to C, so that AB$=$AF, the distance, and AC$=$AF', the true difference of latitude : join BC, then BC shall be at right angles to AC, and equal to AM+BN+CO+ . . . the departure.

For since in triangle ABC, AB=nAb, and AC=nAh,

∴ AB and AC are cut proportionally in b and h,

∴ BC is parallel to bh, and the angle at C is therefore a right angle.

Again, in the same triangle, ABC,

$$BC=AB \cdot \sin A=\text{dist. sin. course.}$$

But by formula (p. 148), departure=dist. sin. course,

$$\therefore BC=\text{departure.}$$

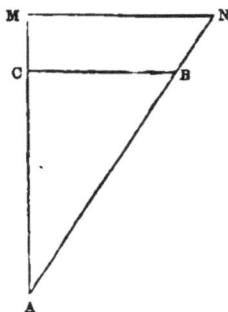

Again, let the side AC be produced to M, so that AM may be equal to the meridional difference of latitude between A and F, and let the right-angled triangle AMN be completed ; then the side MN will equal the difference of longitude UZ (fig. p. 148) between A and F.

For in triangle AMN,

$$MN=AM \tan A=\text{mer. diff. lat.} \times \text{tan. course.}$$

But by formula (p. 149),

$$\text{diff. long.}=\text{mer. diff. lat.} \times \text{tan. course,}$$

$$\therefore MN=\text{difference of long.}$$

We thus see that questions in Navigation or plane sailing may be much simplified by representing the distance, true difference of latitude, departure, meridional difference of latitude, difference of longitude, and course, by the several parts of two similar right-angled plane triangles connected together in the form given in the above figure. In fact, all the principal rules (except in the case when the course is due east or west, and then we must use formula (4), p. 150) can thus be made to depend on the COMMON RULE IN PLANE TRIGONOMETRY FOR SOLVING RIGHT-ANGLED PLANE TRIANGLES.

A few examples will illustrate the method of working questions in Navigation by the common rules of Plane Trigonometry.

To find the course and distance (using meridional parts).

EXAMPLES.

233. Required the course and distance from A to B, having given

Lat. A.........56° 45′ N.   Long. A.........39° 5′ W.

,,   B.........49 10 N.   ,,   B.........29 17 W.

and the merid. diff. lat. (from the table of meridional parts)=758′. The course is evidently towards the south and east, the true diff. lat. is 455', and diff. long. 588'. Construct, therefore, the figure as follows :

Let A represent the place sailed from, and from A draw AC due south=455', the true diff. lat. ; produce AC to M, and make AM=758', the mer. diff. lat. From M draw MN towards the east, at right angles to AM, and make it equal to 588', the diff. long. ; join AN, and through C draw CB parallel to MN; then AB will represent the distance, and the angle A the course required.

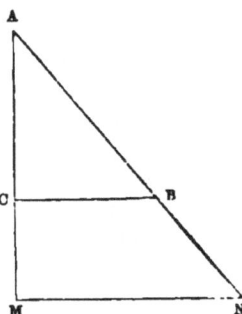

(1.) In right-angled triangle AMN, given AM the mer. diff. lat., and MN the diff. long. ; to find angle A=37° 48': .·. the course=S. 37° 48' E. (see *Trigonometry*, Part I. Rule V.).

(2.) In triangle ABC, given AC the true diff. lat., and angle A ; to find AB=576', the distance (see *Trig.* Rule V. for right-angled triangles).

234. Given lat. A......70° 10' S.   Long. A......54° 40' W.
      „ B......74   40 S.   „ B......57 10 W.

Construct a figure, and find the course and distance from A to B.

Ans. Course=S. 9° 30' W., dist.=274 miles.

To find latitude and longitude in (using meridional parts).

235. Sailed from a place A, in latitude 56° 45' N. and long. 39° 5' W., S. 37° 48' E., 576 miles to another place B ; required the latitude and longitude of B.

Construction.

Let A (last fig.) represent the place sailed from. Draw AM, a part of the meridian, and at A in the straight line AM make the angle A=S. 37° 48' E.; take AB=576', the distance, and through B draw BC at right angles to AM ; then AC will represent the true diff. lat. between A and B. Produce AC to M, and make AM=meridional diff. lat., and complete the triangle AMN ; then MN is the diff. long.

To find true diff. lat. and diff. long. (*Trigonometry*, Rule V.).

(1.) In triangle ABC, are given the angle A the course, and AB the distance ; to calculate AC the true diff. lat.=455'=7° 35' S. Hence the lat. of B=49° 10' N., and the mer. diff. lat. (found from table)=758'=AM.

(2.) In triangle AMN, are given the course A, and AM the mer. diff. lat. ; to find MN the diff. long.=588'=9° 48' E. Hence the long. of B= 29° 17' W.

236. Sailed from A in latitude 20° 30′ N. and longitude 1° 40′ W., N. 45° 33′ 30″ E., 400 miles to another place B. Construct a figure, and find the lat. and long. of B.   *Ans.* Lat. B=25° 10′ N., long. B=3° 30′ E.

To find the course and distance (using middle latitude).

If in the right-angled triangle A B C (p. 153) there is given the true diff. lat. A C, and we also know or can find the *departure* B C, then the other parts of the triangle, namely the course A and distance A B, may also be found directly, and without the aid of the triangle A M N.   Now it has been shown (p. 150) that the arc s s′ of the parallel of latitude, drawn through the middle latitude between two places A and F, may be considered equal to the departure for small distances; and on this assumption we have (formula, p. 148) dep.=diff. long. × cos. mid. lat.*   When, therefore, the latitudes and longitudes of the two places are given, we can compute the departure by the above formula; and then, with the true diff. lat. already known, the course and distance may be found by the common rule in Trigonometry for right-angled plane triangles.

237. Find the course and distance from A to B (using middle latitude), having given

Lat. A	14° 40′ N.	Long. A	56° 40′ E.
„ B	18 20 N.	„ B	60 10 E.

Construction.

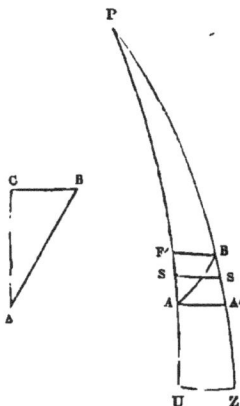

Through A and B, the two places, draw the meridians P U and P Z; draw the parallel B F′, and through s′, the middle of A F′, draw the parallel of latitude s s′; then s s′=departure nearly, and
$\frac{s\,s'}{U\,z}=$cos. s′ U, or $\frac{\text{dep.}}{\text{diff. long.}}=$cos. middle latitude
(p. 150);

∴ dep.=diff. long. cos. mid. lat.

Again, in the right-angled triangle A B C, let A C represent the true diff. lat., and C B the departure (found from the above formula); then A B=distance, and angle A the course required.

* Mid. lat.=½ {lat. from+lat. in} ;
for (fig. p. 150) s U=A U+A s,
and s U=F′ U−F′ s ;
∴ 2 s U=A U+F′ U (since A s=F′ s),
or s U=½ {A U+F′ U} ;
that is, mid. lat.=½ {lat. from+lat. in}.

Calculation.

In triangle CAB, tan. $A = \dfrac{BC}{AC}$,

or tan. course $= \dfrac{\text{dep.}}{\text{true diff. lat.}} = \dfrac{\text{diff. long. cos. mid. lat.}}{\text{true diff. lat.}}$

diff. long. $= 210'$; mid. lat. $= 16° \ 30'$; true diff. lat. $= 220'$

$\therefore$ course $= $ N. $42° \ 28'$ E.

In triangle CAB, $\dfrac{AB}{AC} =$ sec. A; $\therefore$ $AB = AC$. sec. A,

or dist. $=$ true diff. lat. sec. course; $\therefore$ dist. $= 298$ miles.

⁎ In the following examples the student is supposed to know the points of the compass.

CORRECTIONS IN NAVIGATION.

Three corrections are sometimes necessary to be applied to the course steered by compass, to reduce it to the true course; and the converse. These are called :

      (1.) The correction for variation of the compass.

      (2.) The correction for deviation of the compass.

      (3.) The correction for leeway.

(1.) *The Correction for Variation of the Compass.*

The magnetic needle seldom points to the true north. Its deflection to the east or west of the true north is called the *variation of the compass*; it is different in different places, and it is also subject to a slow change in the same place. The variation of the compass is ascertained at sea by observing the magnetic bearing of the sun when in the horizon, or at a given altitude above it. From this observation the true bearing is found by rules given in nautical astronomy. The difference between the true bearing and the observed bearing is the variation of the compass.

The method of correcting the course for variation will be more readily understood by means of a few examples.

Suppose the variation of the compass is found to be two points to the east, that is, the needle is directed two points to the right of the north point of the heavens; then the N.N.W. point of the compass card will evidently point to the true north, and every other point on the card will be shifted round two points. If, therefore, a ship is sailing *by compass* N.N.W., or, as it is expressed, the compass course is N.N.W., her true course will be north; that is, *two points to the right of the compass course*. In a similar manner it may be shown that, when the variation is two points westerly, the true course will be *two points to the left of the compass course*. Hence this rule (the method *by construction* is given at p. 160):

To find the true course, the compass course being given.

    Easterly variation allow to the right.

    Westerly   ,,   ,,   left.

From the preceding considerations it will be easy to deduce the converse rule, namely :

    To find the compass course, the true course being given.

    Easterly variation allow to the left.

    Westerly   ,,   ,,   right.

EXAMPLES.

1. Find the true course, having given the compass course N.W.$\frac{1}{2}$W. and variation $3\frac{1}{4}$ W.

                                  pts. qrs.

    Compass course......4   2 left of N.

    variation...............3   1 left.*

    true course............7   3 left of N.$=$W.$\frac{1}{4}$N.

2. Find the compass course, having given the true course W.$\frac{1}{4}$N. and variation $3\frac{1}{4}$ W.

                                    pts. qrs.

    True course............7   3 left of N.

    variation...............3   1 right.

    compass course........4   2 left of N.$=$N.W.$\frac{1}{2}$W.

Find the true course in each of the following examples :

	Compass course.	Var.	Answers.
3.	N.N.E.	$2\frac{1}{4}$W.	N.$\frac{1}{4}$W.
4.	N.W.	$1\frac{3}{4}$E.	N.N.W.$\frac{1}{4}$W.
5.	S.W.$\frac{3}{4}$W.	$1\frac{1}{2}$E.	W.S.W.$\frac{1}{4}$W.
6.	S.	2W.	S.S.E.
7.	W.	$2\frac{1}{2}$E.	N.W.b.W.$\frac{1}{2}$W.

Find the compass course in each of the following examples :

	True course.	Var.	Answers.
8.	N.N.E.$\frac{1}{2}$E.	$\frac{1}{4}$W.	N.N.E.$\frac{3}{4}$E.
9.	N.	$1\frac{1}{2}$E.	N.b.W.$\frac{1}{2}$W.
10.	S.S.W.	2W.	S.W.
11.	S.W.	0	S.W.
12.	N.b.W.$\frac{1}{4}$W.	$1\frac{1}{4}$W.	N.

    (2.) *The Correction for Deviation of the Compass.*

This correction of the compass arises from the effect of the iron on board ship on the magnetic needle, in deflecting it to the right or left of the magnetic meridian. The increased quantity of iron used in ships has caused this correction to be attended to now more than formerly, as its effects and

---

   * When names are alike (that is, both left or both right), *add :* when unlike, *subtract,* marking remainder with the name of the greater.

magnitude have become more perceptible. The amount of the deviation arising from this local cause varies as the mass of iron changes its position with respect to the compass. When a fore and aft line coincides with the direction of the magnetic meridian, the iron in the ship may be supposed to be nearly equally distributed on both sides of the needle, and its effect in deflecting the needle may be inappreciable. In other positions of the ship with respect to the magnetic meridian, the iron may produce a sensible deflection of the needle; and this deflection or deviation will in general be the greatest when the ship's head points to the east or west.

Various methods are used to determine this correction. The one usually adopted is to place a compass on shore, where it may be beyond the influence of the iron of the ship, or any other local disturbing force, and to take the bearing of the ship's compass, or some object in the same direction therewith; at the same time, the observer on board takes the bearing of the shore compass; then if 180° be added to the bearing at the shore compass, so as to bring it round to the opposite point, the difference between the result and the bearing at ship's compass will be the amount of the deviation of the compass for that position of the ship.

The ship is then swung round one or two points, and a similar observation made; and thus the local deviation found for a second position of the ship. This being repeated for every point or two points of the compass, the deviation is thus known for all positions of the ship. A table, similar to the one below, is then formed, and the courses corrected for this deviation by the following rules; which resemble those already given for correcting for variation.

Deviation of Compass of H.M.S. ——, for given positions of the ship's head.

Direction of ship's head.		Deviation of compass.	Direction of ship's head.		Deviation of compass.
		nearly			nearly
N.	E.	2° 45′ or ¼ pt.	S.	W.	3° 0′ or ¼ pt.
N.b.E.	E.	4 57 or ½ „	S.b.W.	W.	4 20 or ½ „
N.N.E.	E.	7 30 or ¾ „	S.S.W.	W.	5 0 or ½ „
N.E.b.N.	E.	9 0 or ¾ „	S.W.b.S.	W.	6 7 or ½ „
N.E.	E.	10 0 or ¾ „	S.W.	W.	7 0 or ½ „
N.E.b.E.	E.	10 55 or 1 „	S.W.b.W.	W.	7 27 or ¾ „
E.N.E.	E.	10 40 or 1 „	W.S.W.	W.	7 50 or ¾ „
E.b.N.	E.	9 55 or ¾ „	W.b.S.	W.	8 20 or ¾ „
E.	E.	8 50 or ¾ „	W.	W.	8 50 or ¾ „
E.b.S.	E.	7 15 or ¾ „	W.b.N.	W.	8 10 or ¾ „
E.S.E.	E.	5 35 or ½ „	W.N.W.	W.	6 50 or ½ „
S.E.b.E.	E.	3 40 or ¼ „	N.W.b.W.	W.	5 40 or ½ „
S.E.	E.	1 50 or ¼ „	N.W.	W.	4 50 or ½ „
S.E.b.S.	E.	0 20 or 0 „	N.W.b.N.	W.	3 20 or ¼ „
S.S.E.	W.	0 56 or 0 „	N.N.W.	W.	1 40 or 0 „
S.b.E.	W.	2 20 or ¼ „	N.b.W.	E.	1 10 or 0 „

To find the true course, having given the compass course and the deviation.

Easterly deviation allow to the right.

Westerly   „   „   left.

EXAMPLES.

13. Correct the compass course W.b.S. for deviation ¾W. (known from table, p. 157).

```
                        pts. qrs.
Compass course.......7   0 right of S.
deviation...............0   3 left.
                        ─────
true course............6   1 right of S.
        or W.S.W.¼W.
```

14. Correct the compass course N.W.½W. for deviation ½W. (from deviation table, p. 157), and also for variation of compass 3¼W.

```
                            pts. qrs.
Compass course...............4   2 l. N.
deviation..............0   2 l.
variation...............3   1 l.
                    ──   3   3 l.
                        ─────
true course................... 8   1 l. N.
                        16
                        ─────
or true course............... 7   3 r. S.=W.¼S.
```

Find the true course in each of the following examples, by correcting for deviation from table, p. 157, and for variation :

	Compass course.	Var.	Answers.
15.	N.N.E.	2¼W.	N.½E.
16.	N.W.	1¾E.	N.N.W.¾W.
17.	S.W.¾W.	1½E.	S.W.b.W.¾W.
18.	S.	2W.	S.S.E.¼E.
19.	W.	2½E.	W.N.W.¼W.
20.	W.¾N.	1½W.	W.S.W.½W.

To find the compass course, having given the true course and deviation.

Easterly deviation allow to the left.

Westerly   „   „   right.

NOTE.—The true course should first be corrected for variation (if any) by Rule, p. 156 (which is similar to the above), so as to get a compass course nearly, and then this course for deviation, from table, p. 157.

EXAMPLES.

21. Required the compass course, the true course being W.S.W.¼W., variation 0, and deviation ¾W. (see table).

<table>
<tr><td></td><td>pts.</td><td>qrs.</td><td></td></tr>
<tr><td>True course ....6</td><td>1 r.</td><td>S.</td><td></td></tr>
<tr><td>deviation ....0</td><td>3 r.</td><td></td><td></td></tr>
<tr><td>compass course............7</td><td>0 r.</td><td>S., or W.b.S.</td><td></td></tr>
</table>

22. Required the compass course, the true course being S.W., variation of compass 2¼E., and deviation as in table, p. 157.

<table>
<tr><td></td><td>pts.</td><td>qrs.</td></tr>
<tr><td>True course ....4</td><td>0 r.</td><td>S.</td></tr>
<tr><td>variation..................2</td><td>1 l.</td><td></td></tr>
<tr><td>compass course nearly...1</td><td>3 r.</td><td>S., or S.b.W.¾W.</td></tr>
<tr><td>deviation ....0</td><td>2 r.</td><td></td></tr>
<tr><td>compass course............2</td><td>1 r.</td><td>S.=S.S.W.¼W.</td></tr>
</table>

Required the compass course in each of the following examples (for deviation, see table, p. 157):

	True course.	Var.	Answers.
23.	N.½E.	2¼W.	N.N.E.
24.	N.N.W.¾W.	1¾E.	N.W.
25.	S.W.b.W.¾W.	1½E.	S.W.¾W.
26.	S.S.E.¼E.	2W.	S.
27.	W.N.W.¼W.	2½E.	W.
28.	W.S.W.½W.	1½W.	W.¾N.

(3.) *The Correction for Leeway.*

This correction is the angle which the ship's track makes with the direction of a fore and aft line: it arises from the action of the wind on the sails, .&c., not only impelling the ship forwards, but pressing against it sideways, so as to cause the actual course made to be to *leeward* of the apparent course, as shown by the fore and aft line.   The amount of leeway differs in different ships, depending on their construction, on the sails set, the velocity forwards, and other circumstances.   Experience and observation, therefore, usually determine the amount of leeway to be allowed.

The method of correcting for leeway will be best seen by the following example:

Suppose the apparent course is S.S.W.½W., and leeway two points, the wind being S.E., required the correct course.

Draw two lines at right angles to each other towards the cardinal points

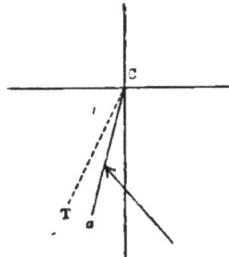

of compass, and a line, as ca, to represent (roughly) the course of the ship, and another to represent the direction of the wind (as the arrow in fig.); then it will be seen that the corrected course, as cT, will be to the *right* of the apparent course; *the observer being always supposed to be at the center* c, *and looking towards the cardinal point from whence the course is measured;* hence

pts. qrs.

Apparent course......2 2 r. S.

leeway.,................2 0 r.

corrected course ......4 2 r. S. $=$ S.W. $\frac{1}{2}$ W.

EXAMPLES.

Correct the following courses for leeway, so as to find the true courses :

	Apparent course.	Wind.	Leeway.	Answers.
29.	N.N.E.	W.N.W.	$1\frac{1}{2}$	N.E. $\frac{1}{2}$ N.
30.	N.W.	N.N.E.	2	W.N.W.
31.	E.S.E.	S.	$2\frac{1}{2}$	E. $\frac{1}{2}$ N.
32.	E.	N.b.E.	$\frac{3}{4}$	E. $\frac{3}{4}$ S.

Correct the following compass courses for deviation, variation, and leeway, so as to find the true courses. The deviation is found in table, p. 157, and the variation of compass is supposed to be in each example $2\frac{1}{2}$ W.

	Course.	Wind.	Leeway.	Answers.
33.	N.W. $\frac{1}{4}$ W.	W.S.W.	$2\frac{1}{2}$	N.W. $\frac{3}{4}$ W.
34.	S.E. $\frac{1}{2}$ E.	E.N.E.	$2\frac{1}{4}$	S.E. $\frac{1}{2}$ E.
35.	W. $\frac{1}{4}$ S.	S.S.W.	2	W.S.W. $\frac{1}{2}$ W.
36.	N. $\frac{3}{4}$ W.	W.b.N.	$1\frac{1}{2}$	N.b.W. $\frac{1}{2}$ W.

These examples may be worked out in the following manner :

                                   pts. qrs.

Ex. 33.   Compass course..........4 1 l. N.

                 deviation ....0 2 l.

                 variation ....2 2 l.

                      —— 3 0 l.

                        7 1 l. N.

                 leeway.....................2 2 r.

                 true course ....4 3 l. N. $=$ N.W. $\frac{3}{4}$ W.

In the preceding examples the courses, both true and compass, are corrected for variation and deviation by a formal rule. The student, however, should also know how to make these corrections by means of a construction, as in the following examples :

238. Given the true course=N. 42° 28' E., and the variation of the compass=1½ points easterly; construct a figure to show the compass course.

Construction.

Let N S represent the true meridian; and since the variation of the com-
pass is 1½ points E., draw N's' 1½ points, or 16° 52', to the east of the true meridian; then N's' will represent the direction of the magnetic meridian, and the angle N O N' the variation of the compass. At the point o, in the straight line N O, make the angle N O F=42° 28'; then N O F will represent the true course, N. 42° 28' E., and N' O F will therefore be the compass course; and it is evident by the figure that

$$N'OF = NOF - NON',$$

or compass course=true course−variation
$$=42° 28' - 16° 52' = 25° 36';$$

and since this angle is to the right of the magnetic meridian,

∴ the compass course=N. 25° 36' E.

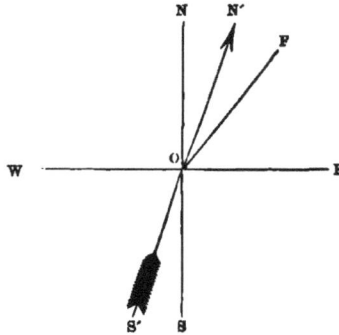

239. Given the true course=N. 25° 36' E., the variation=2 points W., and deviation on account of local attraction=7° 20' E.; to find the corrected compass course (by construction).

Construction.

Let N S represent the true meridian; and since the variation of the compass is 2 points westerly, draw n s 2 points, or 22° 30', to the west of the true meridian; then n s will represent the direction of the magnetic meridian, and the angle N O n the variation of the compass. But the needle is deflected 7° 20' to the east of the magnetic meridian; draw, therefore, N's' 7° 20' to the right of n s; then N'o n=deviation of compass, and N's' will represent the position of the needle. At the point o, in the straight line N O, make the angle N O F=25° 36'; then N O F will represent the true course, N. 25° 36' E., and N' O F the corrected compass course required.

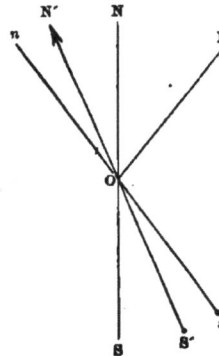

M

By the figure, N′OF = NOF + (NOn − nON′)
$$= 25° \; 36′ + (22° \; 30′ − 7° \; 20′)$$
$$= 25 \quad 36 + 15 \quad 10$$
$$= 40 \quad 46$$
∴ the corrected compass course = N. 40° 46′ E.

By practical rule (p. 158),

True course	.....................25° 36′ r. N.
Variation, 22° 30′ r.	
Deviation, 7 20 l.	......15 10 r.
Compass course	............40 46 r. N. = N. 40° 46′ E.

EXAMPLE FOR PRACTICE.

240. Given lat. A......30° 15′ N.   Long. A......19° 20′ E.
       „   B......26   40 N.   „   B......20 30 E.

variation of compass 2¼ points W.; deviation, 4° 20′ W. Construct figures, and find by calculation the true and compass course, and the distance (by middle lat. method).   *Ans.* True course, S. 15° 58′ E., dist. 223·6′.
                        Compass course, S. 13° 40′ W.

To find latitude and longitude in (middle latitude method).

241. Sailed N. 42° 31′ E., 298·5 miles from a place A, in lat. 14° 40′ N. and long. 56° 40′ E., to another place B; required the latitude and longitude of B.

Construction.

Let AC (see fig. p. 154) represent a part of the meridian, and A the place sailed from. At the point A, in the straight line AC, make the angle CAB = 42° 31′ = the course, and let AB = 298·5 = distance, and from B drop a perpendicular BC on AC: then, in the right-angled triangle CAB, we have given the angle A and side AB; to find AC = true diff. lat., and CB = departure.

Again, let PU, PZ, be the meridians passing through the two places A and B; draw the parallel BF′, and bisect AF′ in S′, and through S′ draw SS′, an arc of a parallel; then SS′ = dep. nearly, and UZ = diff. of long. between A and B.

Calculation.

(1.) In right-angled triangle ABC,

AC = AB cos. A, ∴ true diff. lat. = dist. cos. course . . . (1)

Given dist. = 298·5	true diff. lat. ... 3° 40′ N.
course = 42° 31′	lat. from.........14 40 N.
to find true diff. lat. = 220′ = 3° 40′	∴ lat. of B ....18 20 N.
	∴ mid. lat..........16 30

(2.) Fig. p. 154, $\dfrac{s s'}{u z}=$ cos. $s'u$, or $\dfrac{\text{dep.}}{\text{diff. long.}}=$ cos. mid. lat.

$\therefore$ dep. $=$ diff. long. cos. mid. lat. . . . . (2)

In triangle ABC, $\dfrac{CB}{AB}=$ sin. A, or $\dfrac{\text{dep.}}{\text{dist.}}=$ sin. course,

$\therefore$ dep. $=$ dist. sin. course (3) $\therefore$ equating (2) and (3),

diff. long. cos. mid. lat. $=$ dist. sin. course,

$\therefore$ diff. long. $=$ dist. sin. course sec. mid. lat.

Given, dist. $=298\cdot5$ $\therefore$ diff. long. $=210\ 4=\ \ 3°\ \ 30'$ E.

course $=42°\ \ 31'$ long. from............56   40 E.

mid. lat. $=16\ \ \ 30$ $\therefore$ long. of B............60   10 E.

EXAMPLE FOR PRACTICE.

242. Sailed from a place A, in latitude 42° 15′ N. and longitude 18° 30′ E., N. 46° 23′ E., 195·7 miles. Construct figures, and find by calculation the latitude and longitude in.   *Ans.* Lat. in $=44°$ 30′ N., long. in $=21°$ 45′ E.

PARALLEL SAILING.

In parallel sailing the course is evidently due east or due west, and the distance is the arc of the parallel of latitude intercepted between the two places concerned. Questions in this kind of sailing must therefore be solved by means of the property in Spherical Trigonometry mentioned in p.·150, or by formula (4), p. 148, deduced from it. A few examples will show the use of formula (4).

243. Required the course and distance from a place A, lat. 80° N. and long. 3° 50′ E., to another place B, lat. 80° N. and long. 6° 10′ W. The course is evidently due west.

To find the distance.

Let PZ, PU, represent the meridians of the two given places A and B, UZ the arc of the equator intercepted between them $=$ diff. long. between A and B $=10°$ 0′ $=600'$, and AB the arc of the parallel of latitude through A and B; then AB $=$ distance.

In the fig. $\dfrac{AB}{UZ}=$ cos. AZ, or AB $=$ UZ cos. AZ,

that is, dist. $=$ diff. long. cos. lat. A; $\therefore$ dist. $=104\cdot2$ miles.

EXAMPLE FOR PRACTICE.

244. Required the course and distance from a place B, lat. 41° 30′ N. long. 30° 45′ W., to another place A, lat. 41° 30′ N. and long. 27° 45′ W.

       Ans. Course due east, dist. $=134\cdot8$ miles.

245. Sailed from a place A, due west 492·5 miles, to a place B ; required the latitude and longitude of B.   Lat. A=52° 10′ N., long. A=0° 29′ E.

To find diff. long.

In last fig. $\dfrac{AB}{UZ}$=cos. A U, ∴ . A B=U Z cos. A U,

or dist.=diff. long. cos. lat. A, ∴ diff. long.=dist. sec. lat. A.
Given dist.=492·5 ; lat. A=52° 10′ N.

∴ diff. long.=803′=13° 23′ W.
long. from 0   29 E.
∴ long. B=12   54 W.

and the lat. of B is evidently 52° 10′ N.

EXAMPLES FOR PRACTICE.

246. Sailed from A, due east 226·5 miles, to a place B ; required the latitude and longitude of B.   Lat. A=19° 20′ N., long. A=17° 30′ E.
Ans. Lat. B=19° 20′ N., long. B=21° 30′ E.

247. Two places in the same latitude north, whose difference of longitude is 900 miles, are distant from each other 600 miles ; required the latitude they are in.
In last figure are given AB=600, UZ=900; to find AU the latitude.

cos. A U=$\dfrac{AB}{UZ}$ $\dfrac{600}{900}$=$\dfrac{2}{3}$ ∴ latitude=48° 11′ N.

248. Two places in the same latitude north, whose difference of longi-tude=150 miles, are distant from each other 130·8 ; required the latitude they are in.   *Ans.* Lat.=29° 18′ N.

249. How many miles are there in a parallel of latitude in latitude 80°?
In last fig. are given UZ=1°=60′, and BU=80°; to find AB=length of a degree in lat. 80°.

$\dfrac{AB}{UZ}$=cos. BU, or AB=UZ . cos. BU=60 cos. 80°.∴ AB=10·4 ;

∴ number of miles in parallel=360 × 10·4=3744.

250. How many miles are there in a parallel of latitude in latitude 50° 48′ N.?   *Ans.* 13652 miles.

CONSTRUCTION OF THE TRAVERSE TABLE.

This table contains the true difference of latitude and departure corresponding to every course from 0° to 90°; and for every distance, from 1 nautical mile to about 300.

It is constructed as follows. A course and distance being assumed, the true difference of latitude and departure may be computed for that course and distance: thus, in the triangle ABC, right-angled at C, if the angle CAB represents a given course, and AB a given distance, the side AC will be the true difference of latitude, and CB the departure corresponding to that course and distance.

Given course $=25°$ and distance $=26$ miles; compute corresponding true diff. lat. and departure.

In the triangle CAB, let A $=25°$ and AB $=26$, then true diff. lat. $=$ AC $=$ AB . cos. A $=26$. cos. 25°, $\therefore$ true diff. lat. $=23\cdot56'$, and dep. $=$ BC $=$ AB sin. A $=26$ sin. 25°; $\therefore$ dep. $=10\cdot99$.

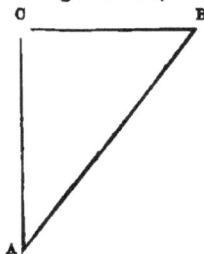

When the difference of latitude and departure are computed in this manner *up* to 45°, the diff. lat. and dep. for courses *above* 45° may be found by interchanging the titles to the columns. Thus, let it be required to find the diff. lat. and dep. for course 65°, and distance 26 miles :

diff. lat. for 65° $=26$ cos. 65° $=26$ sin. 25° $=dep.$ for 25°
dep. for 65° $=26$ sin. 65° $=26$ cos. 25° $=diff. lat.$ for 25°.

Thus it appears that the diff. lat. and dep. for any course are the dep. and diff. lat. respectively for the *complement* of that course; this will easily explain the reason the quantities are tabulated in the following manner.

Form of traverse table.

DISTANCE TWENTY-SIX MILES.			
Course.	Diff. lat.	Dep.	Course.
1°	... ...	... ...	89°
2	... ...	... ...	88
25	23·56	10·99	65
45	... ...	... ...	45
Course.	Dep.	Diff. lat.	Course.

Application of the traverse table.

This table may be applied to a vast variety of problems that depend for their solution on the relation to the several parts of a right-angled triangle. Thus, suppose we have given the middle latitude and departure, and it is required to find the difference of longitude, we may take this quantity out of the traverse table by inspection as follows :

since diff. lat.=dist. cos. course (p. 148),

and dep.=diff. long. cos. mid. lat. (p. 148);

if, therefore, we enter the traverse table with the mid. lat. as a course, and the dep. as a diff. lat., the corresponding distance will be equal to the diff. long. required : thus,

Given the mid. lat.=50° 20' N., and the dep.=14·5' ; find by traverse table the diff. long.   *Ans.* Diff. long.=22·7'.

APPROXIMATE GREAT CIRCLE SAILING.

PROBLEM XLIV.

251. The shortest distance between two places on the surface of the earth is the arc of a great circle passing through them. If the latitudes and longitudes of the two places are known, this arc may be readily calculated by the common rule in Spherical Trigonometry for finding the third side of a spherical triangle, when the other two sides and the included angle are given (see *Trig.* Rule IX.). The practical inconvenience of sailing on a great circle arises from the necessity of continually altering the course. It is for this reason that the rules given in the preceding pages for sailing from one place to another, in which the course is constant, are usually adopted, although the distance run on such a course is not the shortest between the two places.

When the distance between the two places is considerable, as between New York and Liverpool, the following method of approximating to the shortest distance may be adopted with advantage :

(1.) Compute the shortest distance between the two places by *Trig.* Rule IX.

(2). Take two or more convenient points on this arc, and find the latitude and longitude of those points (see following Examples).

(3). Find the course and distance from the place of departure to the nearest point marked on the arc : from thence find the course and distance to the next point, and so on, by the common rule for finding the course and distance from one place to another (p. 152). The sum of the distances described on these several courses will not differ much from the shortest distance.

The points should not be farther apart than 1000 miles. The nearer they are taken to each other, the less will be the difference between the sum of the distances run and the shortest distance.

By proceeding in this manner as far as it is practicable, the advantage of sailing close to the arc of the great circle, and thus of shortening the distance, may be obtained without any difficulty.

The following example, worked out at length, will explain more fully the method of proceeding:

252. Let it be required to find the shortest distance between New York (lat. 40° 42′ N., long. 73° 59′ W.) and Liverpool (lat. 53° 25′ N., long. 2° 59′ W.), also the latitudes and longitudes of certain points taken on the arc of shortest distance, also the course and distance from New York to the nearest point marked on the arc, the course and distance from thence to the second point, &c., and from the last point to Liverpool. Find also the course and distance from New York to Liverpool, in order to determine the distance saved by sailing in this manner near the arc of the great circle.

(1). Let P Y, P L (fig. 1), be the meridians of New York and Liverpool, and L Y the arc of the great circle passing through the two places: then, in the spherical triangle P Y L, are given P Y the colatitude of Y=49° 18′, P L the colatitude of L=36° 35′, and the difference of longitude Y P L=71° 0′; to calculate the arc Y L, the shortest distance between Y and L (=47° 52′, or 2872 minutes, or nautical miles: see calculation, p. 168).

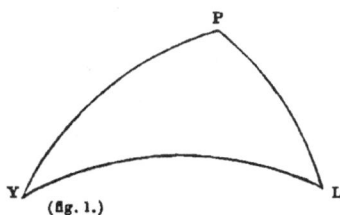

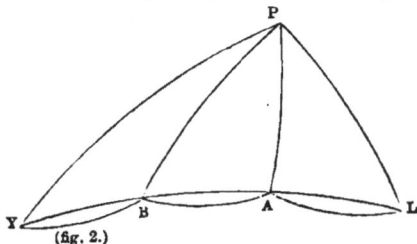

(fig. 1.)   (fig. 2.)

Divide the arc L Y (fig. 2) into three parts at B and A, so that Y B=1000 miles, B A=1000 miles, and therefore the remainder A L=872 miles.

To find the latitudes and longitudes of the points A and B, proceed as follows:

(2.) In the triangle P Y L are given the three sides P L=36° 35′, P Y=49° 18′, and Y L=47° 52′; to compute the angle P Y L (=49° 27′ : see calculation).

(3.) In triangle P Y B are given P Y=49° 18′, Y B=1000′=16° 40′, and the angle P Y B=49° 27′; to compute the arc P B (=40°), the colatitude of B; ∴ the latitude of B=50° N.

(4.) In same triangle P Y B are given the three sides P Y=49° 18′, P B= 40°, and Y B=16° 40′; to compute the angle Y P B (=19° 49′), the difference of longitude between New York and the point B; and thence the longitude of B is 54° 10′ W.

(5.) Again, in the triangle Y P A are given P Y=49° 18′, Y A=2000′=

33° 20′, and the included angle PYL=49° 27′; to compute the arc PA, the colatitude of A (=35° 21′); ∴ the latitude of A = 54° 39′ N.

(6.) In same triangle PYA are given the three sides PY=49° 18′, YA= 33° 20′, and PA=35° 21′; to compute the angle YPA (=46° 12′), the difference of longitude between New York and the point A; and thence the longitude of A is 27° 47′ W.

The latitude and longitude of the points B and A being thus found on the line of shortest distance, the course and distance from New York to B, thence from B to A, and lastly, from A to Liverpool, may be calculated by the common rule for finding the course and distance given in p. 152.

Calculation of the above arcs and angles.

(1.) To find arc YL (*Trigonometry*, Rule IX.).

PY......49° 18′	6·301030		304736
PL......36 35	9·879746		24529
12 43	9·775240		329265
YPL......71 0	9·527908	∴ YL=47° 52′=2872 nautical miles.
	5·483924		

(2.) To find the angle PYL
(*Trig.* Rule VIII.).

PY......49° 18′	0·120254
YL......47 52	0·129782
1 26	4·512826
PL......36 35	4·479940
38 1	9·242802
35 9	∴ PYL=49° 27′

(3.) To find arc PB
(*Trig.* Rule IX.)

PY......49° 18′	6·301030
YB......16 40	9·879746
32 38	9·457584
PYB....49 27	9·242900
	4·881260
	76090
	157861
∴ PB=40° 0′...	233951

and lat. of B=50 0 N.

(4.) To find the angle YPB
(*Trig.* Rule VIII.).

PY......49° 18′	0·120254
PB......40 0	0·191933
9 18	4·351540
YB......16 40	3·807819
25 58	8·471546
15 22	∴ YPB=19° 49′=diff. long.

long. of Y=73 59

∴ long. of B=54 10 W.

(5.) To find arc PA
(*Trig.* Rule IX.).

PY......49° 18′	6·301030
YA......33 20	9·879746
15 48	9·739976
PYL....49 27	9·242900
	5·163652
	145800
	38578
∴ PA=35° 21′...	184378

and lat. of A=54 39 N.

(6.) To find the angle Y P A (*Trigonometry*, Rule VIII.).

P Y	49° 18′	0·120254
P A	35 21	0·237644
	13 57	4·603161
Y A	33 20	4·226203
	47 17	9·187262
	19 23	∴ Y P A = 46° 12′ = diff. long.

long. of Y = 73   59

∴ long. of A = 27   47 W.

By Rule (p. 152), the course and distance from

New York to point B = N. 56° 11′ E., dist. 1002 miles.

From B to A = N. 73 53 E., ,,   1006 ,,

From A to Liverpool = S. 85 10 E., ,,   877 ,,

2885 ,,

From New York to Liverpool = N. 75 10 E., ,, 2982 ,,

∴ distance saved =   97 ,,

We thus see that the distance from New York to Liverpool, by Rule (p. 152), is 2982 miles; but the shortest distance between the two places is 2872 miles, and the sum of the distances described by the ship sailing on the three loops as above is 2885 miles; so that the distance saved by altering the course only twice is about 97 miles.

By taking a greater number of points than two on the arc Y L, so as to bring the ship oftener to the line of shortest distance, the sum of the distances actually sailed will approximate nearer to the shortest distance. It will be seen, however, that even on the assumption of only altering the course once in a thousand miles, the sum of the distances run exceeds by about 12 miles only the shortest distance, and that the absolute saving between the two places, New York and Liverpool, is on this supposition nearly 100 miles.

EXAMPLE FOR PRACTICE.

253. Let it be required to find the shortest distance between Rio de Janeiro, lat. 22° 53′ S. and long. 43° 12′ W., and Java Head, lat. 6° 48′ S., and long. 105° 11′ E.; also the latitudes and longitudes of two points, A and B, taken on the arc of shortest distance so as to divide the arc into three equal parts; also the course and distance from Rio to A, from thence to B, and from B to Java Head. Find also the course and distance from Rio to the Cape, and also from the Cape to Java Head, in order to determine the difference of distances by the two methods.

Ans. The shortest distance between Rio and Java Head = 137° 8′ 15″ or 8228·25 miles.

Latitude and longitude of point A, 44° 6′ S., 6° 32′ E.

,,   ,,   B, 35 40 S., 66 28 E.

The required distances and courses are :

From Rio to A, S. 62° 38′ E.=2772 miles.
From A to B, N. 79   35 E.=2799   „
From B to Java Head, N. 50   56 E.=2751   „

Distance by approximate great circle sailing=8322   „
From Rio to Cape, S. 77   50 E.=3305   „
From Cape to Java Head, N. 70   57 E.=5090   „

Distance by common method=8395   „

The above calculation shows us that the route by the great circle from Rio to Java Head is only about 70 miles shorter than that by the Cape. If we subdivide the arc into more than three parts, we shall be enabled to sail closer to the great circle, and thus approximate still nearer to the shortest distance : the greatest saving will not exceed 160 miles. This route, however, will not take us farther south than lat. 45° 10′,* which is generally sufficiently far to meet the westerly winds, and thereby secure the double advantage of the shortest distance with fair winds.

The above method of calculating the length of the arc of a great circle passing through two places, and of determining a certain number of convenient points thereon to which the ship may be directed so as to keep her as near to the great circle as possible, is direct and general. By dropping a perpendicular from P upon the arc passing through the two places, or the arc produced, or by means of special tables, the labour of the computation may be somewhat diminished, but with the disadvantage of sometimes introducing a distinction of cases, and thereby rendering the problem complicated. As the attempt at great circle sailing can only be of rare occurrence, and as we see that a simple application of the common rules in Spherical Trigonometry is sufficient to obtain all the essential parts of the problem, it does not seem advisable to increase the tabular part of our books of navigation by introducing special rules and tables for the purpose.

* The highest latitude reached by sailing on the arc of a great circle may be found by calculating the value of the perpendicular from the pole upon the arc. See Problems 145 and 146 in the volume of Astronomical Problems by the author.

The value of the meridional parts for any latitude may be correctly found by the following problem :

PROBLEM XLV.

To calculate the meridional parts for any latitude.

Let $l = v\text{D}$, lat. of any point D (fig. 2, p. 145),

M = corresponding angular measure of vd the meridional parts for l,

dl, $d\text{M}$, the contemporary increments of l and M, as the angular measures of DE, de, in fig.

$$\text{Now } \frac{de}{\text{DE}} = \frac{d\,d'}{\text{DD}'} = \frac{\text{U V}}{\text{DD}'} = \text{sec. } l,$$

$$\text{or } \frac{d\text{M}}{dl} = \text{sec. } l ;$$

$$\therefore \text{ M} = \int \text{sec. } l\, dl = \int \frac{\text{cos. } l\, dl}{\text{cos. } {}^2 l} = \int \frac{\text{cos. } l\, dl}{1 - \text{sin. } {}^2 l}$$

$$= \int \frac{\frac{1}{2}\text{cos. } l\, dl}{1 + \text{sin. } l} - \int \frac{-\frac{1}{2}\text{cos. } l\, dl}{1 - \text{sin. } l.}$$

$$= \tfrac{1}{2}\log._\epsilon(1 + \text{sin. } l) - \tfrac{1}{2}\log._\epsilon(1 - \text{sin. } l) + \text{cor. } (= 0)$$

$$= \log._\epsilon \sqrt{\frac{1 + \text{sin. } l}{1 - \text{sin. } l}}$$

Let $l_1 = 90 - l$, the colatitude of l ;

$$\therefore \text{M} = \log._\epsilon \sqrt{\frac{1 + \text{cos. } l_1}{1 - \text{cos. } l_1}} = \log._\epsilon \sqrt{\frac{2 \text{ cos.}^2 \frac{1}{2}l_1}{2 \text{ sin.}^2 \frac{1}{2}l_1}} = \log._\epsilon \text{ cot. } \tfrac{1}{2}l_1$$

Reducing this expression to common logarithms (by *Trig.* Part II. p. 98),

$$\text{M} = 2 \cdot 3025851 \log._{10} \text{ cot. } \tfrac{1}{2}l_1$$

$$\text{Now } \text{M} = \frac{\text{arc}}{\text{rad.}} = \frac{v\,d \text{ (in min.)}}{\text{rad. (in min.)}}$$

$$= \frac{\text{mer. part for lat. } l}{57 \cdot 29577° \times 60}$$

$\therefore$ meridional parts for lat. l

$$= 57 \cdot 29577 \times 60 \times 2 \cdot 3025851 \times \log._{10} \text{ cot. } \tfrac{1}{2}l_1$$

and since log. $57 \cdot 29577 \times 2 \cdot 3025851 = 3 \cdot 8984895$

$\therefore$ log. mer. parts $= 3 \cdot 8984895 + \log \{\log._{10} \text{ cot. } \tfrac{1}{2}\text{colat.} - 10\}.$

EXAMPLE.

Find the meridional parts for lat. 10° and lat. 60°.

For lat 10°.   For lat. 60°.

	For lat 10°	For lat. 60°
log. cot. $\frac{1}{2}$colat. 10	0·076187	0·571948
log. ·076187	$\overline{2}$·8818810 ...... log. ·571948	$\overline{1}$·7573561
const. log.	3·8984895	3·8984895
log. mer. parts	2·7803705 — log. mer. parts	3·6558456
$\therefore$ mer. parts for 10° = 603·07		$\therefore$ mer. parts for 60° = 4527·37

CHAPTER VII.

Transit observations.

THE transit telescope is a meridional instrument for observing, with the assistance of a clock or chronometer, the time when a heavenly body passes the meridian. In the focus of the object-glass is placed a vertical wire, and on each side two or more wires parallel to and at equal distances from it; and the observation consists in noting the second and fractional part of a second when a heavenly body passes these wires: the middle wire is placed in the plane of the meridian. When the *error* of a chronometer or clock is to be determined by a transit of a heavenly body, the instrument should be placed accurately in the plane of the meridian, and to do this some nicety must be observed in making the usual adjustments; but when it is only require1 to obtain a *useful rate* for the chronometer, and the observation is confined to one particular star, any ordinary telescope furnished with a vertical wire may be made use of, and the position of the instrument may be only approximately close to the plane of the meridian.

The principal use to a seaman of transit observations is to enable him to obtain the error and rate of his chronometer; and this he may very often have an opportunity of doing, either by comparing it with the clock in a fixed observatory, or by deducing the error himself from observations made with his own portable transit. The few problems following on transit observations have reference to this latter object.

TO FIND THE ERROR OF A CHRONOMETER ON MEAN TIME AT A GIVEN PLACE BY COMPARING IT WITH A SIDEREAL CLOCK WHOSE ERROR IS KNOWN.

The clock of an observatory used for noting the transits of a heavenly body is generally regulated to show sidereal time, that is, to go 24 hours during one complete revolution of the earth, or of a fixed star about the earth; it is therefore called a sidereal clock (p. 15). This clock enables us to find sidereal time at any instant; and since we can, by Problem V. p. 27, determine the mean time corresponding to any sidereal time, we have thus the means of finding the error of a chronometer, supposed to be regulated to mean time by simply comparing it with the sidereal clock.

The practical rule deduced from Problem V. is as follows:

Compare the chronometer with the sidereal clock at some coincident beat; correct right ascension of mean sun at mean noon at Greenwich (as found in the *Nautical Almanac*) for the difference of longitude (p. 30); subtract right ascension of mean sun so corrected from sidereal time (adding 24 hours if necessary): the result will be mean time nearly at the place. Approximate to correct mean time, as pointed out in Problem V. The difference between mean time thus found and the time shown by the chronometer is the error of the chronometer on mean time at the place.

To find the error of chronometer on mean time at *Greenwich*, proceed as follows:

To the mean time at the place found as above add long. in time if west, and subtract if east; the result will be Greenwich mean time when the comparison was made, the difference between which and the time shown by the chronometer is the error of the chronometer on mean time at *Greenwich*.

254. Oct. 27, 1857, at 11^h 30^m A.M. mean time nearly, in longitude $1°$ $6'$ $3''$ W., a chronometer showed 11^h 32^m 11^s when a sidereal clock showed 13^h 52^m 18^s: the error of the sidereal clock was 0^m 19.28^s slow; required the error of the chronometer on mean time at Greenwich.

Element from *Nautical Almanac*:

Oct. 26, RA mean sun at mean noon ....14^h	19^m	13.49^s	
Correction for long. 4^m 24^s W.*		$.70$	
∴ RA mean sun at mean noon at place = 14	19	14.19	
Sidereal clock showed 13	52	18.00	
Error		19.28	
∴ sidereal time = 13	52	$37.28 + 24^h$	
RA mean sun at noon 14	19	14.19	
Mean time at place nearly 23	33	23.09	
Cor. for 23^h .....3^m 46.70^s			
„ 33^m 5.42			
„ 23^s ...... $.06$	3	52.18	
1st approx. mean time = 23	29	30.91	
Cor. for 23^h ......3^m 46.70^s			
„ 29^m 4.76			
„ $.31^s$ ...... $.08$	3	51.54	
2d approx. mean time = 23	29	31.55	

* This correction may be found thus: 24^h : 4^m 24^s :: 3^m 56^s (change of RA of mean sun in 24 hours): $x = .7^s$; or the correction may be found in the common way by using a table, or by proportional logarithms.

```
Cor. for 23ʰ......3ᵐ 46·70ˢ
  „   29ᵐ......    4·76
  „   31·6ˢ ...    ·09....................    3ᵐ 51·55ˢ
            ∴ Correct mean time at place=23  29  31·54
Longitude in time.............................    4  24·20
Greenwich mean time, Oct. 26...............23  33  55·74
Chronometer showed...........................23  32  11·00
            ∴ error of chronometer=     1  44·74 slow.
```

NOTE. If the comparison is made a little *after* noon, the several approximations above made would be unnecessary, as may be seen in the second and third examples following.

EXAMPLES FOR PRACTICE.

255. Aug. 24, 1858, at 11ʰ 45ᵐ A.M. nearly, in long. 1° 6′ 3″ W., a chronometer showed 11ʰ 41ᵐ 40ˢ when a sidereal clock showed 9ʰ 43ᵐ 4ˢ: the error of the sidereal clock was 7ᵐ 46·92ˢ slow; required the error of the chronometer on Greenwich mean time.

Element from *Nautical Almanac:* RA of mean sun at mean noon at Greenwich, Aug. 23, 10ʰ 5ᵐ 57·07ˢ.

Ans. Error of chronometer, slow 3ᵐ 43·92ˢ.

256. Aug. 24, 1858, at 0ʰ 16ᵐ 19ˢ P.M. mean time nearly, in long. 1° 6′ 3″ W., a chronometer showed 0ʰ 17ᵐ 0ˢ when a sidereal clock showed 10ʰ 18ᵐ 30ˢ: the error of sidereal clock was 7ᵐ 46·92ˢ slow; required the error of the chronometer on Greenwich mean time.

Element from *Nautical Almanac:* RA of mean sun at mean noon at Greenwich, Aug. 24, 10ʰ 9ᵐ 53·62ˢ.

Ans. Error of chronometer, slow 3ᵐ 44·12ˢ.

257. Oct. 27, 1857, at 0ʰ 30ᵐ P.M. mean time nearly, in long. 90° W., a chronometer showed 6ʰ 21ᵐ 42ˢ when a sidereal clock showed 14ʰ 46ᵐ 22ˢ: the error of the sidereal clock was 0ᵐ 21·28ˢ slow; required the error of the chronometer on Greenwich mean time.

Element from *Nautical Almanac:* RA of mean sun at mean noon at Greenwich, Oct. 27, 14ʰ 23ᵐ 10·04ˢ.

Ans. Error of chronometer, slow 0ᵐ 48·4ˢ.

TO FIND ERROR OF SIDEREAL CLOCK BY STAR'S TRANSIT.

PROBLEM XLVI.

Given the time shown by a sidereal clock at the transit of a star; to determine the error of the clock.

Let PQ represent the celestial meridian, AQ the equator, A the first point of Aries, and x a heavenly body on the meridian.   Then AQ represents sidereal time at the transit of the star ; and this is evidently the same as the star's right ascension at that instant, a quantity found in the *Nautical Almanac.* Hence the difference between the star's right ascension and the time shown by the sidereal clock will be the error of the sidereal clock required.

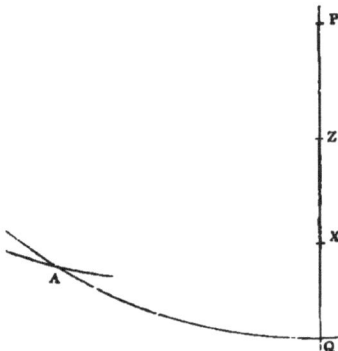

EXAMPLES FOR PRACTICE.

258. Oct. 3, 1846, observed the transit of α Leonis over the three wires of the telescope at the following times, as noted by a sidereal clock ; to determine the error of the sidereal clock, the right ascension of α Leonis by the *Nautical Almanac* being $10^h 0^m 12 \cdot 00^s$.

1st wire	9^h	58^m	$28 \cdot 0^s$
2d, or meridian wire	9	59	13 0
3d	9	59	$58 \cdot 2$
			$39 \cdot 2$
Time by clock	9	59	$13 \cdot 06$
Sidereal time	10	0	$12 \cdot 00$
$\therefore$ Error of sidereal clock $=$		0	$58 \cdot 94$ slow.

259. Aug. 19, 1857, observed the transit of μ Sagittarii over the five wires of a transit telescope at the following times as noted by a sidereal clock ; to determine the error of clock, the right ascension of μ Sagittarii by the *Nautical Almanac* being $18^h 5^m 15 \cdot 8^s$.

1st wire......................	18^h	5^m	12^s
2d ,, ...		5	31
3d ,, ...		5	50
4th ,, ...		6	9
5th ,, ...		6	$28 \cdot 5$

Ans. Error of sidereal clock, fast $34 \cdot 3^s$.

TO FIND ERROR OF A MEAN SOLAR CLOCK OR CHRONOMETER BY STAR'S TRANSIT.

PROBLEM XLVII.

Given the time shown by a mean solar clock, or chronometer, at the transit of a star; to determine the error of the timekeeper.

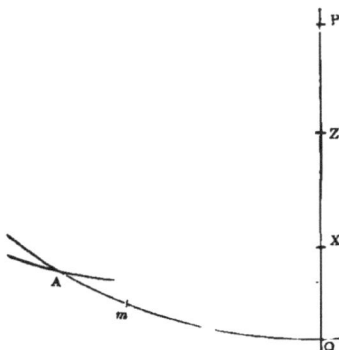

Let P Q represent the celestial meridian, A Q the celestial equator, A the first point of Aries, *m* the mean sun, and X the place of the heavenly body on the meridian. Then the right ascension of the meridian, or sidereal time Q A, is known, since it is the same as the right ascension of the star when on the meridian. The problem reduces itself, therefore, to finding mean solar time, having given sidereal time; and this can be done by means of Problem V. p. 27. The difference between mean time thus found and the time shown by the chronometer is the error of the chronometer on mean time at the place.

EXAMPLES FOR PRACTICE.

260. Oct. 3, 1846, in long. 1° 6' W., observed the transit of Antares over the wires of the telescope at the following times, as noted by a chronometer regulated to mean time; to determine its error on mean time at the place.

Elements from *Nautical Almanac:* RA of mean sun at mean noon at Greenwich, $12^h 47^m 13.80^s$; RA of Antares, $16^h 20^m 1.00^s$.

Times of transit.

1st wire	3^h	25^m	47.5^s
2d ,, ...	3	26	35.8
3d ,, ...	3	27	23.8
	3)	19	47.1
∴ chronometer showed...	3	26	35.7
RA of Antares or sidereal time	16^h	20^m	1.00^s
RA mean sun at mean noon at place...	12	47	14.50^*
Mean time nearly=	3	32	46.50

---

* 0.7ˢ is added to the right ascension of mean sun in *Nautical Almanac,* to adapt it to the meridian of the place; it is, in fact, equivalent to correcting the right ascension for a Greenwich date=long. of place. See Example (254).

Cor. for 3ʰ ....29·57'

   ,,   32ᵐ............ 5·26

   ,,   46·5' .... 0·13............   34·96'

      Mean time, 1st approximation=3   32   11·54

Cor. for 3ʰ ....29·57'

   ,,   32ᵐ............ 5·26

   ,,   11·54'......... ·03............   34·86'

          Correct mean time 3   32   11·64

          Chronometer showed 3   26   35·70

   ∴ error of chronometer (slow)=   5   35·94

261. Aug. 27, 1857, in long. 1° 6' 3" W., observed the transit of μ Sagittarii over the wires at the following times, as noted by a mean solar clock; find the error of clock on *Greenwich* mean time.

Elements from *Nautical Almanac:* R.A of mean sun at mean noon at Greenwich, 10ʰ 22ᵐ 40·24'; R.A of star, 18ʰ 5ᵐ 15·73'.

          1st wire ....7ʰ 42ᵐ 12'

          2d ,, ... 42 32

          3d ,, ... 42 51

          Ans. Error, slow 3ᵐ 11·56' on Greenwich mean time.

TO FIND ERROR OF SIDEREAL CLOCK BY SUN'S TRANSIT.

PROBLEM XLVIII.

Given the time shown by a sidereal clock at the transit of the sun; to determine the error of the clock.

Let PQ (fig. p. 176) represent the celestial meridian, ᴀQ the celestial equator, ᴀ the first point of Aries, and x the place of the sun on the meridian: then ᴀQ represents sidereal time at that instant; and this is known, since it is the same as the right ascension of the true sun x at apparent noon at the place of observation, and this can be deduced from the right ascension of the true sun, as given in the *Nautical Almanac* for apparent noon *at Greenwich.** The difference between the right ascension of true sun at apparent noon at the place thus found and the time shown by the sidereal clock will be the error of sidereal clock required.

262. Oct. 3, 1846, in long. 1° 6' W., observed the transit of the sun's limbs over the wires at the following times, as noted by a sidereal clock; to determine the error of the clock.

---

* The right ascension of the sun for apparent noon at any other place whose longitude is given can be found as pointed out in *Navigation*, Part I. . . .

Times by sidereal clock.

Western limb, 1st wire	....	12ʰ	31ᵐ	8·0ˢ
,,	2d	,,	....	32	2·5
,,	3d	,,	....	32	47·5
Eastern limb, 1st	,,	....	33	28·5
,,	2d	,,	....	34	13·0
,,	3d	,,	....	34	58·0

$$\begin{array}{r} 6)\ \overline{\quad 18\quad 37·5} \end{array}$$

Sidereal time by clock $= 12 \quad 33 \quad 6·25$

RA of sun at app. noon at Greenwich, $\Big\}$ ... $12^{h}\ 36^{m}\ 19\cdot06^{s}$
by *Nautical Almanac*................

Cor. for long. 1° 6′ W...................... $\qquad\qquad\qquad ·70+$

∴ RA of sun at apparent noon at place $= 12 \quad 36 \quad 19·76$

Clock showed................................ $12 \quad 33 \quad 6·25$

∴ error of sidereal clock $= \qquad 3 \quad 13·51$ slow.

263. Aug. 24, 1858, at noon, in long. 1° 6′ 3″ W., observed the transit of the sun's limbs over the wires at the following times, as noted by a sidereal clock ; to determine the error of sidereal clock.

Element from *Nautical Almanac :* RA of sun at app. noon at Greenwich, 10ʰ 12ᵐ 6·75ˢ.

Times of transit by sidereal clock.

Western limb, 1st wire	....	10ʰ	2ᵐ	38·0ˢ
,,	2d	,,	....	2	57·0
,,	3d	,,	....	3	15·0
,,	4th	,,	....	3	33·5
,,	5th	,,	....	3	52·0
Eastern limb, 1st	,,	....	4	49·0
,,	2d	,,	....	5	8·0
,,	3d	,,	....	5	25·5
,,	4th	,,	....	5	44·0
,,	5th	,,	....	6	2·0

Ans. Error on Greenwich mean time, slow 7ᵐ 47·05ˢ

TO FIND ERROR OF MEAN SOLAR CLOCK, OR CHRONOMETER, BY SUN'S TRANSIT.

PROBLEM XLIX.

Given the time shown by a chronometer, or mean solar clock, at the transit of the sun ; to determine the error of the timekeeper.

Let PQ represent the celestial meridian, AQ the celestial equator, x the place of the sun on the meridian, and *m* or *m′* the mean sun. Then apparent time at the instant the sun is on the meridian $= 0^h\ 0^m\ 0^s$, or $24^h\ 0^m\ 0^s$, and mean time $=$ $0^h + Qm$, or $24^h - Qm'$. Now mQ or $m'Q$ represents the equation of time as given in the *Nautical Almanac*, corrected for the place of observation ; by applying which to the apparent time, 0^h or 24^h (adding it to 0^h, or subtracting it from 24^h, according to the sign in the *Nautical Almanac*), it is evident that mean time is obtained for the instant the sun is on the meridian of the observer. The difference between mean time thus found and the time shown by the chronometer, is the error of the chronometer on mean time at the place.

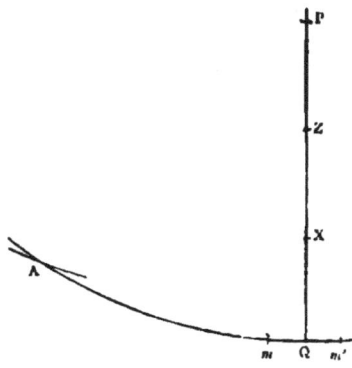

264. May 18, 1846, in longitude 90° W., observed the transit of the sun's limbs over the wires at the following times, as noted by a chronometer regulated to mean time ; to determine the error of the chronometer on mean time at the place.

$$\text{May 18, at} \dots\dots\dots\dots\dots\dots 0^h\ 0^m$$
$$\text{Long. in time} \dots\dots\dots\dots\dots 6\ 0\ W.$$
$$\text{Greenwich, May 18} \dots\dots\dots\dots 6\ 0$$

Equation of time from Nautical Almanac.

$$\left.\begin{array}{l}\text{At 18} \dots\dots\dots\dots 3^m\ 51{\cdot}96^s \\ \text{,, 19} \dots\dots\dots\dots 3\ \ 49{\cdot}73\end{array}\right\}\text{subtractive from apparent time.}$$
$$\overline{2{\cdot}23}$$

Cor. for 6^h .... $\underline{{\cdot}56}$

$3\ \ 51{\cdot}40 =$ equation of time at transit.

Apparent time of transit	24^h	0^m	$0{\cdot}00^s$
Equation of time		3	$51{\cdot}40$
Mean time at place	23	56	$8{\cdot}60$

Times of transit.

Western limb, 1st wire	11^h	54^m	$13{\cdot}2^s$
,, 2d ,,		55	$0{\cdot}0$
,, 3d ,,		55	$46{\cdot}6$
Eastern limb, 1st ,,		56	$27{\cdot}8$
,, 2d ,,		57	$14{\cdot}2$
,, 3d ,,		58	$1{\cdot}0$
		36	$42{\cdot}8$
Mean time by clock	11	56	$7{\cdot}13 + 12^h$
,, at place	23	56	$8{\cdot}60$
$\therefore$ error of chronometer $=$			$1{\cdot}47$ slow.

265. August 24th, 1858, at noon, in long. 1° 6' 3" W., observed the transit of the sun's limbs over the wires at the following times, as noted by a chronometer; to determine the error of chronometer on Greenwich mean time.

Elements from *Nautical Almanac:* equation of time at app. noon on 24th, 2" 12·77'; on 25th, 1" 56·65', additive to apparent time.

Times of transit by chronometer.

Western limb,	1st wire	......	0ʰ	1"	11·0'
,,	2d ,,	......	1	29·7	
,,	3d ,,	......	1	47·0	
,,	4th ,,	......	2	8·0 uncertain, from cloud.	
,,	5th ,,	......	2	24·5	
Eastern limb,	1st ,,	......	3	21·4	
,,	2d ,,	......	3	40·7	
,,	3d ,,	......	3	58·4	
,,	4th ,,	......	4	16·0	
,,	5th ,,	......	4	34·7	

Adding together first and last contact, second and last but one, &c., and taking the mean (rejecting the defective observation) for transit of center, we have

1st and last contact	....	5"	45·7'	
2d and last but one	....	5	45·7	
3d ,, ,, two	....	5	45·4	
4th ,, ,, three	....	5	48·7 lost.	
5th ,, ,, four	....	5	45·9	
		4)	2·7	
		5	45·67	

∴ transit of center by chron. = 2 52·84

				Equation of time.		
Aug. 24	....	0ʰ	0"	Aug. 24	.........2"	12·77'
Long. in time	....	4		,,	1	56·65
Greenwich, Aug. 24	...0	4				16·11
				Cor.		·04
					2	12·73

App. time of transit	....	0ʰ	0"	0'
Equation of time	....		2	12·73 +
Mean time at place	....	0	2	12·73
Longitude in time	....		4	24·20
Mean time at Greenwich	....		6	36·93
Chronometer showed	....		2	52·84
∴ error of chron. on Greenwich mean time =			3	44·09 slow.

Investigation of the correction called the EQUATION OF EQUAL ALTITUDES.

PROBLEM I.

Given the elapsed time, and change of declination in the interval; to investigate an expression for the equation of equal altitudes.

Let P be the pole, z the zenith, and x and Y the places of the sun at the times of the observations; therefore the zenith distances zx and zy are equal.

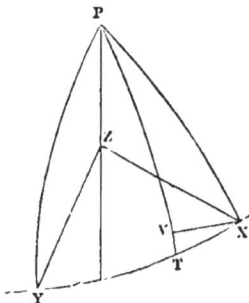

First. Suppose the polar distance to be increasing; then the polar distance PY will be greater than PX, and therefore the hour-angle zPX is greater than the hour-angle zPY, or greater than half the interval xPY. Make zPT=zPY, and call the angle xPT=2x, and the elapsed time, namely the angle xPY=ϵ.

Then ϵ=YPT+xPT=2zPT+2x,

∴ $\frac{1}{2}$ϵ=zPT+x: add x to both sides,

∴ $\frac{1}{2}$ϵ+x=zPT+2x=zPX=hour-angle from noon, or the time that must elapse (by chronometer) before the sun is on the meridian Pz.

Second. Suppose the polar distance to be decreasing, or PY less than PX, then in a similar manner it may be shown that the time to noon or the angle zPX=$\frac{1}{2}$ϵ−x.

The value of x in seconds of time is called the EQUATION OF EQUAL ALTITUDES.

Investigation of the EQUATION OF EQUAL ALTITUDES.

About z as a center, describe the arc xTY, and join TZ. Then PT=PY (for yz=Tz, and zPT was made equal to zPY, and Pz is common to the two triangles). Again, about P as a center describe the arc of a parallel of decl. xV: then TV is the difference between PX and PT or PY; that is, it is the change of declination in the interval of elapsed time ϵ.

Let TV, the change of decl. in elapsed time=d',

PX, the polar dist. at 1st observation=p,

and latitude of place=l.

The triangle xVT, being very small, may be considered as 'a right-angled plane triangle, V being the right angle.

∴ xV=TV . cot. VxT=d' . cot. VxT=d' . cot. PXz,

and {*Trig.* Part II. p. 75} xV=xPV . sin. PX=2x . sin. p,

∴ 2x . sin. p=d' . cot. PXz.....................(1)

To eliminate cot. PXz from this expression. Since

cot. PXz . sin. zPX=cot. Pz . sin. PX−cos. zPX . cos. PX,*

Cot. A . sin. B=cot. a . sin. c−cos. B . cos. c (*Trig.* Part II. p. 58).

or cot. PXZ . sin. ($\frac{1}{2}\epsilon$ nearly)=tan. l . sin. p—cos. ($\frac{1}{2}\epsilon$ nearly) . cos. p

∴ cot. PXZ=tan. l . sin. p . cosec. $\frac{1}{2}\epsilon$—cot. $\frac{1}{2}\epsilon$. cos. p nearly.

Substituting this value of cot. PXZ in (1), we have

$2x$ sin. $p=d'$ tan. l . sin. p . cosec. $\frac{1}{2}\epsilon$—d' cot. $\frac{1}{2}\epsilon$. cos. p,

or $2x=d'$ tan. l . cosec. $\frac{1}{2}\epsilon$—d' . cot. $\frac{1}{2}\epsilon$. cot. p.

From this formula x may be computed, and the practical rule in *Navigation*, Part I., deduced as follows :

Since $2x=d'$. tan. l . cosec. $\frac{1}{2}\epsilon$—d' . cot. $\frac{1}{2}\epsilon$. cot. p.........(in arc)

∴ $=\frac{d'}{15}$tan. l . cosec. $\frac{1}{2}\epsilon$—$\frac{d'}{15}$cot. $\frac{1}{2}\epsilon$. cot. p............(in time)

∴ $x=\frac{1}{30}d'$. tan. l . cosec. $\frac{1}{2}\epsilon$—$\frac{1}{30}d'$. cot. $\frac{1}{2}\epsilon$. cot. p.

Let M$=\frac{1}{30}d'$. tan. l . cosec. $\frac{1}{2}\epsilon$

„ N$=\frac{1}{30}d'$. cot. $\frac{1}{2}\epsilon$. cot. p ;

then the equat. of equal alt. $x=$M$-$N.

To determine the algebraic signs of M and N.

The algebraic signs of M and N will evidently depend on the magnitudes of the angles in the trigonometrical ratios which make up M and N. We will therefore now proceed to determine the algebraic signs of M and N for all conditions of the factors in M and N.

First. Let the declination be *increasing*, and of the *same name* as the latitude.

By making a figure (fig. 1) to suit this case, it will be seen that ZPX is less than half-elapsed time ; ∴ equation of equal alt. x must be subtracted, or

$$ZPX=\tfrac{1}{2}\epsilon-x=\tfrac{1}{2}\epsilon-M+N ;$$

and since the latitude and pol. distance are both less than 90°, the proper signs of M and N are not on this account altered.

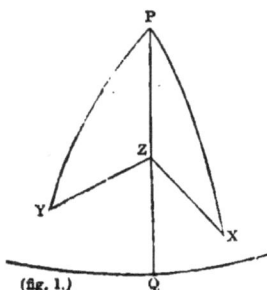
(fig. 1.)

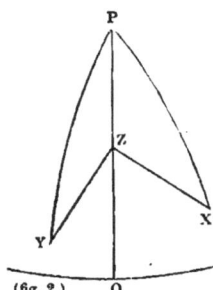
(fig. 2.)

Second. Let the declination be *decreasing*, and of the same name as the latitude.

By making a figure (fig. 2) to suit this case, it will be seen that ZPX is greater than ZPY, and ∴ the equation of equal alt. x must be added, or

$$z\,\mathrm{P\,X} = \tfrac{1}{2}\iota + x = \tfrac{1}{2}\iota + \mathrm{M} - \mathrm{N}\ ;$$

and since, as in the first case, the quantities in M and N are positive, the proper signs of M and N are not altered.

Third. Let the declination be *increasing*, and of a *different* name from the latitude.

By making a figure (fig. 1) to suit this case, it will be seen that z P x is greater than z P Y, and ∴ the equation of equal alt. *x* must be added, or

$$z\,\mathrm{P\,X} = \tfrac{1}{2}\iota + x = \tfrac{1}{2}\iota + \mathrm{M} - \mathrm{N}.$$

But the proper sign of N is negative, since one of its factors, namely cot. *p*,

$$\overset{+}{}\qquad\overset{-}{}$$

is negative, *p* being greater than 90° (for in this case $\mathrm{N} = \tfrac{1}{30}d'$ cot. $\tfrac{1}{2}\iota$. cot. *p*, or $-\mathrm{N} = \tfrac{1}{30}d'$ cot. $\tfrac{1}{2}\iota$. cot. *p*),

$$\therefore\ z\,\mathrm{P\,X} = \tfrac{1}{2}\iota + \mathrm{M} - (-\mathrm{N})$$
$$= \tfrac{1}{2}\iota + \mathrm{M} + \mathrm{N}.$$

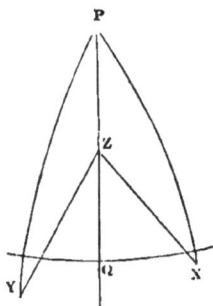

(fig. 1.)

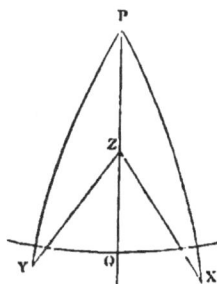

(fig. 2.)

Lastly. Let the declination be *decreasing*, and of a *different* name from the latitude.

By making a figure (fig. 2) to suit this case, it will be seen that z P x is less than z P Y, and ∴ the equation of equal alt. *x* must be subtracted, or

$$z\,\mathrm{P\,X} = \tfrac{1}{2}\iota - x = \tfrac{1}{2}\iota - \mathrm{M} + \mathrm{N}.$$

But, as in the last case, the proper sign of N is negative, since cot. *p* is negative,

$$\therefore\ z\,\mathrm{P\,X} = \tfrac{1}{2}\iota - \mathrm{M} + (-\mathrm{N}) = \tfrac{1}{2}\iota - \mathrm{M} - \mathrm{N}.$$

By inspecting these four results, it appears that M is *positive* when decl. is decreasing and of same sign as latitude, or increasing and of a different sign.

M is *negative* when decl. is decreasing and of a different name from the latitude, or increasing, and of the same name.

N is positive when declination is increasing.

N is negative when declination is decreasing.

In the rule given in *Navigation*, Part I., the values of м and ɴ are expressed in proportional logarithms, and a small table is used for facilitating the computation. We will now show how м and ɴ are adapted to proportional logarithms, and also the construction of the table contained in the following page.

First. To adapt $м = \frac{1}{30} d'$ tan. l . cosec. $\frac{1}{2}ε$ to proportional logarithms:

Let $δ =$ change of declination in 24 hours;

$$\text{then } d' : δ :: ε : 24, \therefore d' = \frac{ε δ}{24}$$

$$\therefore м = \frac{1}{30} . \frac{ε δ}{24} . \text{tan.} \, l . \text{cosec.} \, \tfrac{1}{2} ε$$

$\therefore$ log. м $= -$log. 30 $+$ log. $ε -$log. $24^h +$ log. $δ +$ log. tan. $l +$ log. cosec. $\frac{1}{2}ε - 20,$

subtracting both sides from log. 3^h,

log. $3^h -$ log. м $=$ log. 30 $+$ log. $24^h -$ log. $ε +$ log. $3^h -$ log. $δ -$ log. tan. $l -$

log. cosec. $\frac{1}{2}ε + 20$

$=$ log. 30 $+$ (log. $24^h -$ log. $ε$) $+$ (log. $3^h -$ log. $δ$) $+$ log. cot. $l +$ log. sin. $\frac{1}{2}ε - 20$;

or prop. log. м $=$ log. 30 $+$ Gr. date log. sun for $ε +$ prop. log. $δ +$ log. cot. $l +$

log. sin. $\frac{1}{2}ε - 20$ (p. 138).

Second. To adapt ɴ to proportional logarithms:

$$ɴ = \tfrac{1}{30} d' \, \text{cot.} \, \tfrac{1}{2} ε . \, \text{cot.} \, p.$$

Proceeding in a similar manner to the above, we have

prop. log. ɴ $=$ log. 30 $+$ Gr. date log. sun for $ε +$ prop. log. $δ +$ log. cot. decl. $+$ log. tan. $\frac{1}{2}ε - 20.$

By these two formulæ the values of м and ɴ may be found in proportional logarithms, and thence the equation of equal altitudes x.

The labour of finding м and ɴ is considerably diminished, if we calculate the quantities

log. 30 $+$ Gr. date log. of sun for $ε +$ log. sin. $\frac{1}{2}ε$ in м,

and log. 30 $+$ Gr. date log. of sun for $ε +$ log. tan. $\frac{1}{2}ε$ in ɴ,

for every ten minutes of elapsed time, and form the results into a table.

This Table is constructed as follows:

Calculate the values of A and B when the elapsed time $ε = 5^h 30^m$.

A $=$ log. 30 $+$ Gr. date log. sun for $ε +$ log. sin. $\frac{1}{2}ε$

B $=$ log. 30 $+$ Gr. date log. sun for $ε +$ log. tan. $\frac{1}{2}ε$

A.	B.
log. 30 . . . . 1·47712	log. 30 . . . . 1·47712
Gr. date log. sun for $5^h 30^m$ 0·63985	. . . . 0·63985
log. sin. $2^h 45^m$ . . . . 9·81911	log. tan. $2^h 45^m$ . . . . 9·94299
A $=$ 1·93608	B $=$ 2·05996

And in a similar manner may the values of A and B in the following table be computed for other values of the elapsed time.

VALUES OF A AND B FOR COMPUTING THE EQUATION OF EQUAL ALTITUDES.

Elapsed time.	A	B	Elapsed time.	A	B	Elapsed time.	A	B
1 30	1·97148	1·97991	4 30	1·94886	2·02901	7 30	1·90212	2·15738
1 40	1·97082	1·98123	4 40	1·94692	2·03356	7 40	1·89876	2·16854
1 50	1·97009	1·98272	4 50	1·94490	2·03833	7 50	1·89531	2·18033
2 0	1·96930	1·98435	5 0	1·94281	2·04334	8 0	1·89177	2·19280
2 10	1·96843	1·98614	5 10	1·94064	2·04861	8 10	1·88815	2·20602
2 20	1·96750	1·98808	5 20	1·93840	2·05414	8 20	1·88444	2·22003
2 30	1·96649	1·99017	5 30	1·93608	2·05996	8 30	1·88064	2·23493
2 40	1·96541	1·99243	5 40	1.93368	2·06605	8 40	1·87676	2·25081
2 50	1·96426	1·99484	5 50	1·93122	2·07246	8 50	1·87278	2·26775
3 0	1·96305	1·99743	6 0	1·92866	2·07918	9 0	1·86870	2·28587
3 10	1·96176	2·00019	6 10	1·92604	2·08624	9 10	1·86454	2·30531
3 20	1·96040	2·00312	6 20	1·92333	2·09365	9 20	1·86029	2·32623
3 30	1·95897	2·00623	6 30	1·92054	2·10143	9 30	1·85593	2·34882
3 40	1·95747	2·00954	6 40	1·91767	2·10961	9 40	1·85148	2·37334
3 50	1·95589	2·01303	6 50	1·91473	2·11821	9 50	1·84692	2·40003
4 0	1·95424	2·01671	7 0	1·91170	2·12725	10 0	1·84427	2·42928
4 10	1.95252	2·02060	7 10	1·90859	2·13678	10 10	1·83752	2·46152
4 20	1·95073	2·02470	7 20	1·90539	2·14680	10 20	1·83267	2·49733

The equation of equal altitudes x being thus found, is applied to ½s, and thus the angle z P x is obtained as measured by the chronometer. This being added to the time shown by the chronometer at the first observation, the result will be the time of apparent noon as shown by the chronometer.

The correct mean time of apparent noon is then to be found by applying the equation of time to 0ʰ or 24ʰ (see last problem); the result is the true time of apparent noon.

The difference between this and the time of apparent noon, as shown by the chronometer, will evidently be the error of the chronometer on mean time at the place.

The annexed blank form will facilitate the working out examples for finding the error and rate of a chronometer by equal altitudes.

FORM FOR FINDING ERROR AND RATE OF CHRONOMETER BY EQUAL ALTITUDES.

A.M. time.			(1.)			(2.)	P.M. time.	(3.)		
h	m	s	h	m	s	Equation of time.		A.M. time.	m	s
								h		

App. time at noon (0ʰ or 24ʰ).

Equation of time nearly

Mean time at noon

Longitude in time

Gr. date on ( ) for eq. of time . . . . . On

(4) Diminished by half elapsed time . . . . On

Gr. date on ( ) for decl.

☉'s decl. (5)

On

On

Cor. . . . . . = D

☉'s decl. . . . . .

M

A (from table) . . . . N.

Prop. log. D . . . . . . B (from table)

Log. cot. latitude . . . . . Prop. log. D

Prop. log. N . . . . . . Cot. decl.

„ M* . . . . . . Prop. log. N

„ N . . . . . . „ N

Equation of equal altitudes

Equation of time.

Sub. from P.M. + 12ʰ.

Elapsed time

½ elapsed time

A.M. time + ½ elapsed time

Approx. time of ap. noon

Equation of equal altitude

Time by chronometer at noon.

Ap. time at noon (0ʰ or 24ʰ).

Equation of time.

Mean time at noon

Longitude in time

Greenwich mean time

Time by chronometer

Error on Gr. mean time on ( )

Error on Gr. mean time on ( )

Diff. divided by number of days.

Rate.

* M is *pos.* when decl. is decreasing and of same name as lat., or increasing and of a different name; otherwise *neg.* N is *pos.* or *neg.* according as decl. is increasing or decreasing (*Rule*, p. 182).

LONGITUDE BY OCCULTATION.

The longitude of a place is determined with great accuracy by noting the time of an occultation of a star or planet by the moon; the mean time at the place, at the instant of the disappearance or reappearance of the star, being supposed to be accurately known. The corresponding mean time at Greenwich can be computed by a rule deduced from the following problem.

PROBLEM LI.

Given the mean time at the place at the instant of an occultation of a star by the moon; to determine the longitude of the place of observation.

(fig. 1.) Immersion (east of meridian).   (fig. 2.) Emersion (east of meridian).

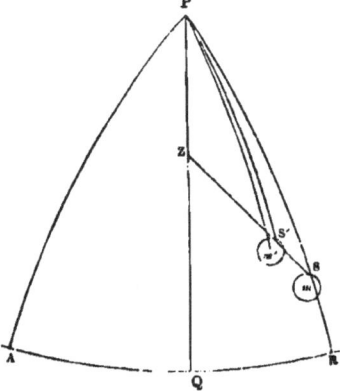

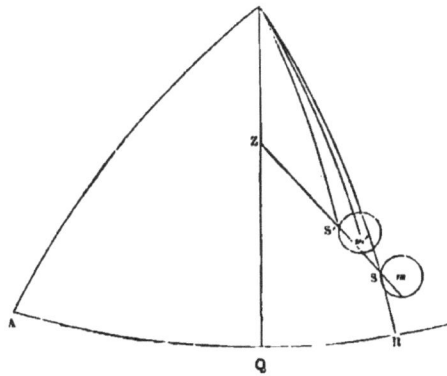

Let p be the pole, z the reduced zenith of the spectator, and s the observed place of the point of contact of the star with the moon.   Then,

(fig. 3.) Immersion (west of meridian).   (fig. 4.) Emersion (west of meridian).

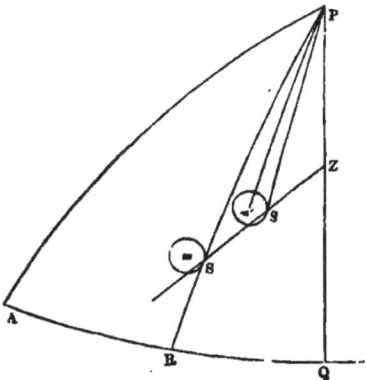

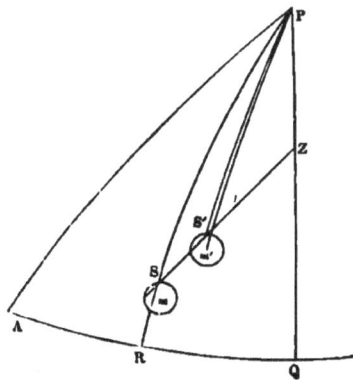

since the moon is depressed by parallax, the true place of the point of contact is above s in a great circle passing through the reduced zenith z.

Suppose s' the true place of the point of contact, m the apparent place of the moon's center, and m' the true place.

Let AQ be the equator, and A the first point of Aries.

At the instant of the occultation, the apparent right ascension AR and decl. SR of the point of the moon's limb in contact with the star are known, since they must be the RA and decl. of the star itself ; and these quantities are readily found in the *Nautical Almanac*. The object of the following investigation is to find the *right ascension of the true place of the moon's center m', namely the angle* ARm' ; for by comparing this right ascension with the right ascensions found in the *Nautical Almanac* for two given times, the *Greenwich mean time* corresponding to the right ascension ARm' can be determined in the same manner as in the lunar.

In order to determine the angle ARm', we must investigate formulæ for computing the following quantities:

(1.) The angle SPS' approximately.

(2.) The parallax in declination, namely PS—PS'.

(3.) The parallax in right ascension, namely the angle SPS', accurately.

(4.) The angle m'PS', the semidiameter in right ascension.

(5.) The Greenwich mean time corresponding to the right ascension ARm'.

The angle APS, or arc AR, is known, since it is the right ascension of the star. It is therefore manifest that, if we can compute the two angles SPS' and s'Pm', and apply them to the angle APS, we obtain the angle ARm', the right ascension of the moon's center at the time of the occultation.

In the figure, let A represent the pole, B the reduced zenith, c and c' the apparent and true places of the point of contact, corresponding to s and s' in the previous figures.

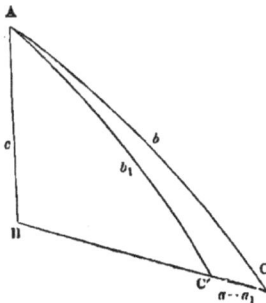

Let BAc the star's hour-angle=A,

BAc' the angular dist. of c' from the meridian=A';

∴ the parallax in right ascension CAc'=A—A'

Let AB the reduced colatitude=c,

AC the polar distance of c, and therefore of the star=b,

AC' the polar distance of c'=b',

∴ the parallax in declination=b—b'

Let BC=a, and BC'=a'. Then, in the two triangles BAC, BAC', are given the colatitude c, the polar distance of star b, and the star's hour-angle A (found from the time at the place, which is supposed to be known accurately) ; to find the parallax in right ascension A—A₁, and parallax in declination b—b₁.

(1.) *To find* $A-A'$, *the* PARALLAX IN RIGHT ASCENSION, *approximately.*

Since c is the apparent place, and c the true place of the point of contact of the moon, the arc cc', or $a-a_1$, is the diurnal parallax.

$$\therefore \sin.(a-a_1)=\sin. H . \sin. a \text{ (p. 125)}.$$

In triangle Acc', $\dfrac{\sin. cAc' \text{ or } \sin.(A-A_1)}{\sin. c}=\dfrac{\sin. cc'}{\sin. b_1}$

$$=\dfrac{\sin.(a-a_1)}{\sin. b_1}=\dfrac{\sin. H . \sin. a}{\sin. b_1}$$

$$\therefore \sin.(A-A_1)=\dfrac{\sin. H . \sin. a . \sin. c}{\sin. b_1} \dots (1)$$

In triangle ABC, $\dfrac{\sin. C}{\sin. A}=\dfrac{\sin. c}{\sin. a}$ $\therefore \sin. c=\dfrac{\sin. c . \sin. A}{\sin. a}$

Substituting this value of sin. c in (1), we have

$$\sin.(A-A_1)=\dfrac{\sin. H . \sin. c . \sin. A}{\sin. b_1} \dots (\alpha)$$

From this expression the value of $A-A'$ cannot be found exactly, since we do not know b_1, the polar distance of the true point of contact; but if we use b, the star's polar distance, for b_1, we shall get an approximate value of $A-A'$, and therefore of A', since A is already known. This value of A' will be required, and will be sufficiently correct to enable us to calculate the parallax in declination $b-b_1$, as follows.

(2.) *To find the* PARALLAX IN DECLINATION $b-b_1$.

In triangle ABC', $\cot. b_1 . \sin. c=\cot. B . \sin. A'+\cos. c . \cos. A'$

,, ABC, $\cot. b . \sin. c=\cot. B . \sin. A+\cos. c . \cos. A$.

Multiply the first equation by sin. A, and the second by sin. A', and subtracting the results, we have

$\cot. b_1 . \sin. c . \sin. A=\cot. B . \sin. A' . \sin. A+\cos. c . \cos. A' . \sin. A$

$\cot. b . \sin. c . \sin. A'=\cot. B . \sin. A . \sin. A'+\cos. c . \cos. A . \sin. A'$

$(\cot. b_1 . \sin. A-\cot. B . \sin. A') . \sin. c=\cos. c . (\cos. A' . \sin. A-\cos. A . \sin. A')$

$$=\cos. c . \sin.(A-A')$$

$$\therefore \cot. b_1 . \sin. A-\cot. b . \sin. A'=\dfrac{\cos. c . \sin.(A-A')}{\sin. c}$$

or $\cot. b_1-\dfrac{\cot. b . \sin. A'}{\sin. A}=\dfrac{\cos. c . \sin.(A-A_1)}{\sin. c . \sin. A}$

By (α) $\sin.(A-A')=\dfrac{\sin. H . \sin. c . \sin. A}{\sin. b_1}$

Substituting this value of sin. $(A-A')$, we have

$$\cot. b_1-\dfrac{\cot. b . \sin. A'}{\sin. A}=\dfrac{\sin. H . \cos. c}{\sin. b_1}$$

Subtracting cot. b from each side,

$$\cot. b_1 - \cot. b = \frac{\sin. \text{II} . \cos. c}{\sin. b_1} - \cot. b + \frac{\cot. b . \sin. \Lambda'}{\sin. \Lambda}$$

$$= \frac{\sin. \text{II} . \cos. c}{\sin. b_1} - \frac{\sin. \Lambda - \sin. \Lambda'}{\sin. \Lambda} . \cot. b.$$

But $\cot. b_1 - \cot. b = \dfrac{\sin. (b - b_1)}{\sin. b . \sin. b_1}$ (*Trig.* Example 180),

and $\sin. \Lambda - \sin. \Lambda' = 2 \cos. \frac{1}{2}(\Lambda + \Lambda') \sin. \frac{1}{2}(\Lambda - \Lambda')$ (*Trig.* p. 31).

Making these substitutions, we have

$$\frac{\sin. (b - b_1)}{\sin. b . \sin. b_1} = \frac{\sin. \text{II} . \cos. c}{\sin. b_1} - \frac{2 \cos. \frac{1}{2}(\Lambda + \Lambda') . \sin. \frac{1}{2}(\Lambda - \Lambda')}{\sin. \Lambda} . \frac{\cos. b}{\sin. b}$$

$$\therefore \sin. (b - b_1) = \sin. \text{II} . \cos. c . \sin. b - \frac{2 \cos. \frac{1}{2}(\Lambda + \Lambda') . \sin. \frac{1}{2}(\Lambda - \Lambda') . \cos. b . \sin. b_1}{\sin. \Lambda}$$

and $2 \sin. \frac{1}{2}(\Lambda - \Lambda') = \Lambda - \Lambda'$ nearly (since $\Lambda - \Lambda'$ is small)

$$= \sin. (\Lambda - \Lambda') = \frac{\sin. \text{II} . \sin. c . \sin. \Lambda}{\sin. b_1} \text{ by } (\alpha)$$

$\therefore \sin. (b - b_1) = \sin. \text{II} . \cos. c . \sin. b - \sin. \text{II} . \cos. \frac{1}{2}(\Lambda + \Lambda') . \cos. b . \sin. c \ldots (\beta)$

☞ Although we do not know Λ' exactly, we have an approximate value of Λ' in (α) (p. 189) sufficiently near to give the value of $b - b_1$ without any sensible error.

(3.) *To find the* PARALLAX IN RIGHT ASCENSION $\Lambda - \Lambda_1$.

As neither of the formulæ (α) and (β) gives a direct value of the quantity sought, we must begin with the one most likely to obtain the nearest approximation to the truth. We shall find it better to calculate a near value of Λ' by (α) *using* b *for* b_1: we shall then be able to find $\frac{1}{2}(\Lambda + \Lambda')$, a quantity that occurs in (β), sufficiently correct to compute $b - b_1$, and thence b_1.

We can then compute $\Lambda - \Lambda'$ by means of formula (α) with the correct value of b_1 thus obtained ; the result will be the parallax in right ascension $\Lambda - \Lambda'$ required.

(4.) *To find the* SEMIDIAMETER IN RIGHT ASCENSION, *namely the angle* m′Ps′.

Let s′ be the true point of contact, m′ the true place of moon's center, and P the pole ; then m′Ps′ is the moon's semidiameter in right ascension,

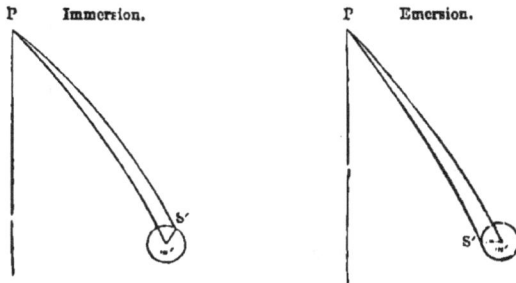

P　　Immersion.　　　　　　　　P　　Emersion.

which may be computed by considering m'PS$'$ as a spherical triangle, whose three sides are given, namely,

PS$'$=polar distance b_1 of true point of contact, found by (β),
m'S$'$=moon's horizontal semi. for the Greenwich date,
Pm'=polar distance of moon, taken from the *Nautical Almanac* for a Green-
     wich date not differing much from Greenwich mean time.

The values of $\text{A}-\text{A}'=$SPS$'$, and m'PS$'$, found as above, being applied with the proper sign to the angle APS, the right ascension of the star, the result will be the angle APm', or the true right ascension of the moon's center at the instant of the observation.

(5.) *To find the instant at Greenwich corresponding to the right ascension of the moon's center* APm'.

Let M be the place of the moon's center at the hour of the Greenwich date, M$'$ the place one hour afterwards, and m' the place of the moon's center when the observation was taken, AR$'$ the celestial equator, and A the first point of Aries. Then RR$'$ is the change of moon's right ascension in 1 hour, or 3600 seconds, and Rr' the difference of right ascension between the calculated right ascension of Ar' of moon's center and AR the right ascension for the hour of Greenwich date, taken from the *Nautical Almanac*.

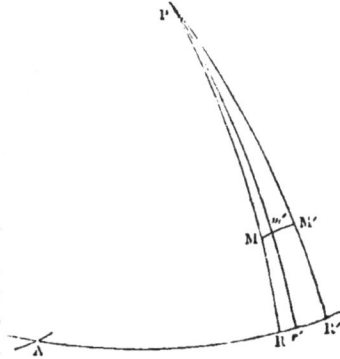

Let $a=$RR$'=$change of RA in 1 hour, or 3600 seconds,
    $b=$Rr',
and $x=$seconds in time, corresponding to b seconds of right ascension.
        Then $a : b : : 3600 : x$.

The value of x thus found, and turned into minutes and seconds, and added to the hour of the Greenwich date, will evidently give the Greenwich mean time when the moon's right ascension was Ar'; that is, the Greenwich time at the instant of the observation. The difference between which and the mean time at the place, found in the usual manner (as in the lunar or chronometer), is the *longitude in time*.

From the formulæ investigated above may be deduced a direct and practical rule for finding mean time at Greenwich at the instant of the observation. To do this, the formulæ must be reduced to logarithms, and the several parts arranged in the manner pointed out in the following investigation.

PRACTICAL RULE FOR FINDING THE LONGITUDE BY AN OCCULTATION OF A FIXED STAR.

Reduction of Formulæ.

Since in $\sin. (\Lambda - \Lambda') = \dfrac{\sin. H \cdot \sin. c \cdot \sin. \Lambda}{\sin. b_1}$ . . . . (a)

and $\sin. (b - b_1) = \sin. H \cdot \cos. c \cdot \sin. b - \sin. H \cdot \cos. \frac{1}{2}(\Lambda + \Lambda') \cdot \cos. b \cdot \sin. c$ (β)
the angles $\Lambda - \Lambda'$, H, and $b - b_1$ are small; the circular measures of these

angles may be substituted for their sines, that is, $\dfrac{\text{arc } (\Lambda - \Lambda')}{\text{rad.}}$ for $\sin. (\Lambda - \Lambda')$,

$\dfrac{\text{arc } H}{\text{rad.}}$ for $\sin. H$, and $\dfrac{\text{arc } (b - b_1)}{\text{rad.}}$ for $\sin. (b - b_1)$.

Making these substitutions, the formulæ become

$$b - b_1 = H \cdot \cos. c \cdot \sin. b - H \cdot \cos. b \cdot \sin. c \cdot \cos. \tfrac{1}{2}(\Lambda + \Lambda'),$$

and $\Lambda - \Lambda' = H \cdot \sin. c \cdot \sin. \Lambda \cdot \text{cosec.} b_1$ . . . . (in arc)

or $\Lambda - \Lambda' = \frac{1}{15} H \cdot \sin. c \cdot \sin. \Lambda \cdot \text{cosec.} b_1$ . . . . (in time)

Before we can compute $b - b_1$, we must know $\frac{1}{2}(\Lambda + \Lambda')$.

Now $\frac{1}{2}(\Lambda + \Lambda') = \Lambda - \frac{1}{2}(\Lambda - \Lambda') = \Lambda - \frac{1}{30} H \cdot \sin. c \cdot \sin. \Lambda \cdot \text{cosec.} b_1$ . . (γ)
This expression would determine $\frac{1}{2}(\Lambda + \Lambda')$, supposing we knew cosec. b_1
(the cosec. of polar distance of the *true* place of point of contact) ; but since
b_1 is not yet found, we can use cosec. b or sec. star's decl. instead (for $b = b_1$
very nearly, and therefore this may be done without any practical error).

Assuming at present that cosec. b_1 = cosec. b = sec. star's decl., we have
$\frac{1}{2}(\Lambda - \Lambda') = \frac{1}{30} H \cdot \sin. c \cdot \sin. \Lambda \cdot \text{sec. star's decl.}$

$\quad = \frac{1}{30} \cdot$ red. hor. par. × cos. red. lat. × sin. star's hour-angle × sec. star's
decl.

To reduce this formula to logarithms, we shall find it convenient to add
together the logarithms that remain unaltered (namely all the quantities
excepting sec. star's decl.), as they will be required hereafter, when $\Lambda - \Lambda'$ is
to be found accurately.

Let ∴ $c = \frac{1}{30}$ H . cos. lat. sin. hour-angle,

∴ $\frac{1}{2}(\Lambda - \Lambda') = c$. sec. star's decl. : reducing to logarithms,

$\quad$ log. c = log. H + log. cos. lat. + log. sin. hour-angle − log. 30 − 20

$\quad\quad$ = log. H + log. cos. lat. + log. sin. hour-angle + ar. co. log. 30 − 30

$\quad\quad$ = log. H + log. cos. lat. + log. sin. hour-angle + 8·522879 − 30.

log. $\frac{1}{2}(\Lambda - \Lambda')$ = log. c + log. sec. star's decl. − 10, ¿
and $\frac{1}{2}(\Lambda + \Lambda') = \Lambda - \frac{1}{2}(\Lambda - \Lambda')$.

Hence this practical rule for finding $\frac{1}{2}(\Lambda + \Lambda')$:
Rule (a), under head (1). (See Example, p. 197.)
Add together log. red. hor. par. (in seconds),

$\quad\quad\quad$ log. cos. reduced lat.,

$\quad\quad\quad$ log. sin. star's hour-angle,

$\quad\quad$ and constant log. 8·522879 ;
the result, rejecting the tens in the index, will be log. c.

To log. σ add log. sec. star's decl.; the sum, rejecting the tens in index, will be log. $\frac{1}{2}(A-A')$ in seconds, the nat. number corresponding to which will be $\frac{1}{2}(A-A')$ in seconds. Turn this into minutes and seconds, and subtract it from the star's hour-angle A; the result will be $\frac{1}{2}(A+A')$ nearly.

This approximate value of $\frac{1}{2}(A+A')$ will be sufficiently correct to enable us to compute $b-b_1$, the parallax in declination.

Reduction of $b-b_1$ (p. 192), the parallax in declination, to logarithms.

$b-b_1 = $ H . cos. c . sin. $b - $ H . cos. b . sin. c . cos. $\frac{1}{2}(A+A')$.

Let M$=$H . cos. c . sin. $b = $H . sin. red. lat. cos. star's decl.

,, N$=$H . cos. b . sin. c . cos. $\frac{1}{2}(A+A')$

$=$H . cos. red. lat. sin. star's decl. cos. $\frac{1}{2}(A+A')$.

In logarithms,

log. M$=$log. red. hor. par. $+$ log. sin. red. lat. $+$ log. cos. star's decl. -20.

,, N$=$log. red. hor. par. $+$ log. cos. red. lat. $+$ log. sin. star's decl.

$\qquad\qquad\qquad\qquad\qquad +$ log. cos. $\frac{1}{2}(A+A') - 30$.

Hence this practical rule for finding the parallax in declination:

Rule (b). Under heads (2) and (3) (see Example, p. 198), put down the following quantities:

Under (2) }
and (3) } ......log. red. horizontal parallax (in seconds).

,, (2).........log. sin. reduced lat.

,, (3).........log. cos. reduced lat.

,, (2).........log. cos. star's decl.

,, (3).........log. sin. star's decl.

,, (3).........log. cos. $\frac{1}{2}(A+A')$.

Add together the logarithms under (2) and (3); the natural numbers corresponding to which, turned into minutes and seconds, will be the values of M and N, the two parts of the parallax in declination.

To determine the algebraic signs of M and N.

The algebraic sign of M is always positive, since the trigonometrical factors which make up M are all positive. In the value of N $\{=$H . cos. b . sin. c . cos. $\frac{1}{2}(A+A')\}$ it may be seen that two of the factors, namely the polar dist. b and $\frac{1}{2}(A+A')$, may be greater or less than 90°, since b is reckoned from the elevated pole. If they are both greater, or both less than 90°, their signs will not affect the negative value of N. (This will be evident by putting the signs $+$ and $-$ over the quantities in N, according to the Rule in *Trig.* Part I. p. 31.) But if one be greater and the other less than 90°, the value of N will be rendered positive.

Hence this practical rule for determining the signs of the two parts of the parallax in declination.

o

Rule (c). If the polar distance and $\frac{1}{2}(\text{A}+\text{A}')$, or rather the hour-angle itself of the star, be one greater and one less than 90°, the second part N must be added to the first part M of the parallax in declination; but if the polar distance and hour-angle be both greater or both less than 90°, the second part N must be subtracted from the first part M. The result will be the parallax in declination $(b-b_1)$. Apply this to the star's declination, so as to diminish its polar distance (b), and we obtain the true declination (namely the complement of b_1) of the point of the moon's limb observed, subject to the small error arising from using in the calculation of $\frac{1}{2}(\text{A}+\text{A}')$, p. 193, the secant of star's declination instead of the secant of true declination of the point of contact. This error, however, will be always very small, and, if necessary, could be entirely removed by recomputing $\frac{1}{2}(\text{A}+\text{A}')$ with the approximate value of b_1 just found.

Reduction of A — A', *the parallax in right ascension to logarithms.*

With the value of b_1, determined as above, we may now find the exact value of A — A', the parallax in right ascension.

For $\frac{1}{2}(\text{A}-\text{A}')=\frac{1}{3}\sigma$. II . sin. c . sin. A . cosec. b_1 (p. 192),

$=c$. cosec. b_1 (see p. 192, where log. c is already found),

and $b_1=$ true pol. dist. of point of moon's limb in contact with the star found above;

$\therefore$ A — A'$=2c$. cosec. b_1;

$\therefore$ log. (A — A')$=$log. $c+$log. sec. decl.$+$·301030$-$10.

Hence this practical rule to determine the parallax in right ascension:

Rule (d). (See Example p. 198.)

Under head (4) put down log. c already found, log. secant of the true decl. of point observed, and constant log. ·301030. The sum (rejecting 10 in index) will be the log. parallax in right ascension $(=$A — A'$)$.

When the star is west of meridian, *add* the parallax in right ascension (A — A') to the star's right ascension.

When east of meridian, *subtract.*

The result will be the true right ascension of the point of the moon's limb in observed contact with the star.

See figures p. 187, where it is evident that to find the angle A P s' we must subtract s P s' $(=$A — A'$)$ from A P s when east of meridian, and add it when west, to obtain the angle A P s'.

To calculate the moon's semidiameter in right ascension m' P s'.

Let moon's hor. semi. for Gr. date$=m's'$ (fig. p. 190),

$$\text{P} s'=90\mp\delta,\ \text{P} m'=90\mp\text{D}$$

($-$ or $+$ according as the decl. and lat. are the same or different names) where $\delta=$decl. of true point of contact,

D$=$decl. of moon's center, taken from *Nautical Almanac* for the Gr. date.

Then, considering $\mathrm{P}m's'$ a spherical triangle, we have (*Trig.* Part II. p. 62),

$$\sin.{}^2\tfrac{1}{2}m'\mathrm{P}\mathrm{S}' = \text{cosec. } \mathrm{P}\mathrm{S}' \text{ . cosec } \mathrm{P}m' \text{ . } \sin.\tfrac{1}{2}(m's' + \overline{\mathrm{P}m'\smallfrown\mathrm{P}\mathrm{S}'})\sin.\tfrac{1}{2}(m\mathrm{S}' - \overline{\mathrm{P}m'\smallfrown\mathrm{P}\mathrm{S}'})$$

But the quantities $m'\mathrm{P}\mathrm{S}'$, $m's' + \overline{\mathrm{P}m'\smallfrown\mathrm{P}\mathrm{S}'}$, and $m's' - \overline{\mathrm{P}m'\smallfrown\mathrm{P}\mathrm{S}'}$ are very small,

and $\therefore$ $\sin. m'\mathrm{P}\mathrm{S}' = \dfrac{\text{arc } m'\mathrm{P}\mathrm{S}'}{\text{rad.}}$, &c.

Making these substitutions, and reducing, we have

$$\frac{(m'\mathrm{P}\mathrm{S}')^2}{4} = \text{cosec. } \mathrm{P}\mathrm{S}' \text{ . cosec. } \mathrm{P}m' \text{ . } \frac{m's' + \overline{\mathrm{P}m'\smallfrown\mathrm{P}\mathrm{S}'}}{2} \text{ . } \frac{m's' - \overline{\mathrm{P}m'\smallfrown\mathrm{P}\mathrm{S}'}}{2}.$$

or $(m'\mathrm{P}\mathrm{S}')^2 = \text{sec. } \delta \text{ . sec. } \mathrm{D} \text{ . } (m's' + \overline{\mathrm{P}m'\smallfrown\mathrm{P}\mathrm{S}'}) \text{ . } (m's' - \overline{\mathrm{P}m'\smallfrown\mathrm{P}\mathrm{S}'})$ (in arc)

and $\therefore$ $(m'\mathrm{P}\mathrm{S}')^2 = \dfrac{1}{15^2} \text{ . sec. } \delta \text{ . sec. } \mathrm{D} \text{ . } (m's' + \overline{\mathrm{P}m'\smallfrown\mathrm{P}\mathrm{S}'}) \text{ . } (m's' - \overline{\mathrm{P}m'\smallfrown\mathrm{P}\mathrm{S}'})$ (in time)

$\therefore$ $2 \log. m'\mathrm{P}\mathrm{S}' = \log. \text{ sec. } \delta + \log. \text{ sec. } \mathrm{D} + \log. (m's' + \overline{\mathrm{P}m'\smallfrown\mathrm{P}\mathrm{S}'})$

$\qquad\qquad + \log. (m's' - \overline{\mathrm{P}m'\smallfrown\mathrm{P}\mathrm{S}'}) + \text{ar. co. log. } 15^2 \ (=7\cdot647818) - 30.$

Hence this practical rule for finding semidiameter in right ascension $m'\mathrm{P}\mathrm{S}'$: Rule (e). (See Example, p. 199.)

Under the declination (δ) of true point of contact put the declination D of moon for Greenwich date; take the difference, which bring into seconds, and under it put moon's horizontal semidiameter in seconds. Take the sum and difference.

To the log. secants of the first two terms in this form add the logs. of the two last terms, and also the constant log. $7\cdot647818$. Half the sum will be the log. of the moon's semidiameter in right ascension $m'\mathrm{P}\mathrm{S}'$ in seconds.

When an immersion is observed, subtract the semi. in right ascension from the right ascension of true point of contact $\mathrm{A}\mathrm{P}\mathrm{S}'$; when an emersion is observed, add it thereto. The result will be the true right ascension of the moon's center $\mathrm{A}\mathrm{P}m'$ at the time of the observation.

See diagrams in p. 187, where it is evident that an emersion takes place to the east of the moon's center, and therefore at a greater distance from the first point of Aries; and the contrary at an emersion.

To calculate the time at Greenwich corresponding to the moon's right ascension $\mathrm{A}\mathrm{P}m'$ just found

If $a =$ change of moon's right ascension in 1^h, or 3600 seconds,

$\qquad b =$ difference of right ascension between the calculated right ascension $\mathrm{A}\mathrm{P}m'$ and the one for hour of Greenwich date, taken out of the *Nautical Almanac*.

and $x =$ seconds of time corresponding to b seconds in RA, it has been shown (p. 191) that $a : b :: 3600 : x$,

$\qquad\qquad \therefore \log. x = \log. b + \log. 3600 - \log. a.$

Hence this practical rule for finding x, the number of seconds that must be added to the hour of the Greenwich date to get the Greenwich mean time at the instant of the observation:

Rule (*f*). From the true right ascension of moon's center, found by Rule (*e*), subtract the right ascension for the hour of the Greenwich date, and thus get *b*.

To the log. difference of right ascension in seconds (*b*) add constant log. 3·556302 (the log. of 3600), and from the sum subtract the log. of the hourly change (in seconds) in right ascension. The result will be the log. of a number of seconds (*x*) in time, which take from the tables and, turned into minutes and seconds, add to the hour of the Greenwich date.

The result will be the GREENWICH MEAN TIME* at the instant of the occultation.

To find MEAN TIME at the place.

This is found in the usual manner from an altitude of a heavenly body, as in the chronometer or lunar observation.

Then the difference between Greenwich mean time and the time at the place will be the *longitude in time*.

EXAMPLE (AN IMMERSION, EAST OF MERIDIAN).

266. Aug. 25, 1839, at $8^h 14^m 34·6'$ mean time at the Royal Naval College, in lat. 50° 48' N. and longitude in time $4^m 24·2'$ W., observed the immersion of φ Aquarii.

Before the rule can be applied, we should take out of the *Nautical Almanac* the following quantities and correct them for a Greenwich date: namely (1.) Moon's semi. ; (2.) Moon's hor. parallax (corrected also for spheroidal fig. of earth, p. 129) ; (3.) Moon's declination ; (4.) Right ascension of mean sun; and (5.) Star's right ascension and declination. The lat. must also be corrected for spheroidal fig. of earth, p. 123; and (6.) The moon's right ascension must be taken out for the hour of Greenwich date, and for the hour following, in order to get the horary motion in right ascension (see working form, p. 204).

$$
\begin{array}{lr}
\text{R. N. College, Aug. 25} \dots\dots\dots & 8^h\ 14^m\ 34·6' \\
\text{Long. in time}\dots\dots\dots\dots\dots\dots & 4\ \ \ 24·2\ \text{W.} \\
\hline
\text{Greenwich, Aug. 25}\dots\dots\dots\dots & 8\ \ 18\ \ 58·8
\end{array}
$$

---

* If the Greenwich time differs much from the Greenwich date, it may be necessary to recompute some part of the work, especially the declination of the moon's center, which is an element required to be found with great accuracy, using the Greenwich time last found as a new Greenwich date.

	(1.) Moon's semi.	(2.) Moon's hor. par.		Latitude.

Noon............16 10·6'' ....59 21·9''   50 48 0'' N.

Midnight ......16 14·2   ....59 35·2 Reduction...   11 0

                3·6   13·3 Red. lat. ...50 37 0 N.

    Cor....   2·5   Cor....   9·2

         16 13·1   59 31·1

           60   Reduction...   7·0—(p. 129)

∴ hor. semi.=973·1   59 24·1

                     60

            ∴ red. hor. par.=3564·1

(3.) Moon's decl. (6.) Moon's R A.   (4.) B A mean sun.

8ʰ ......6° 21' 35·2''S.... 23ʰ 2ᵐ 5·47ˢ Aug. 25.........10ʰ 12ᵐ 15·93ˢ

9ʰ ......   4 44·2 S.... 23 4 13·13 Cor. ....   1 21·97

        16 51·0   2 7·66 R A mean sun...10 13 37·90

Cor. ...   5 19·7   60

     6 16 15·5 S.   127·66   (5.) Star's B A and decl.

                     Star's R A......23ʰ 6ᵐ 2·68ˢ

                     Star's decl. ... 6 54 33·8 S.

To find star's hour-angle A.

Mean time at place......... 8ʰ 14ᵐ 34·60ˢ=Qm

R A mean sun ....10 13 37·90 =Am

R A meridian (+24) ....18 28 12·50 =AmQ

Star's R A ....23 6 2·68 =AQR

                19 22 9·82 =RmQ

           24

∴ star's hour-angle= 4 37 50·18 =RPQ

Rule (a). To compute ½(A+A').

(1.)

Log. red. H P............3·551950

  „ cos. red. lat.......9·802435

  „ sin. hour-angle...9·971469

  „ const. log. ......8·522879

      log. c=1·848733

  „ sec. star's decl....0·003165

  „ ½(A—A') ....1·851898

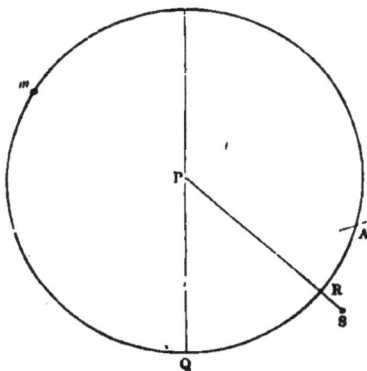

$$\therefore \tfrac{1}{2}(A-A')=71{\cdot}10$$

$$\text{or } \tfrac{1}{2}A-\tfrac{1}{2}A'= \quad 1^{s} \quad 11{\cdot}10'$$

$$\text{and } A=4 \quad 37 \quad 50{\cdot}18$$

$$\therefore \tfrac{1}{2}(A+A')=4 \quad 36 \quad 39{\cdot}08 \text{ nearly.}$$

Rule (*b*). To compute the parallax in declination M − N.

(2.)	(3.)
Log. red. HP ....3·551950	Log. red. HP ....3·551950
„ sin. red. lat. ......9·888133	„ cos. red. lat. ......9·802435
„ cos. star's decl. ...9·996835	„ sin. star's decl. ...9·080198
„ M ....3·436918	„ cos. ½(A+A') ......9·551107
∴ M=2734·7	„ N....................1·985690
	∴ N=96·8

Diagram for this Example.

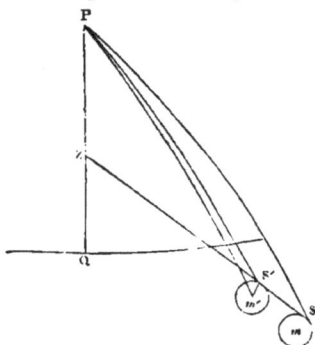

Rule (*c*). To determine the signs of M and N.

$$\overset{+}{M}=\overset{+}{H} . \sin. \text{ red. lat.} . \overset{+}{\sin. \text{ pol. dist.}}$$

$$-N=-\overset{+}{H} . \overset{+}{\cos. \text{ red. lat.}} . \overset{-}{\cos. \text{ pol. dist.}} .$$

$$\overset{+}{\cos. \tfrac{1}{2}(A+A').}$$

Putting the proper signs over the factors (*Trig.* Part I. p. 31), we see that M is +, and − N is + ;

$$\therefore M=2734{\cdot}7 +$$

$$-N= \quad 96{\cdot}8+$$

$$\therefore M-N=2831{\cdot}5$$

$$\text{or par. in decl.}= \quad 47' \quad 11{\cdot}5''=PS-PS'$$

$$\text{star's decl.}=6 \quad 54 \quad 33{\cdot}8 \text{ S.}=PS-90$$

$$\therefore \left.\begin{array}{c}\text{true decl. of point of}\\ \text{moon's limb observed}\end{array}\right\}=6 \quad 7 \quad 22{\cdot}3 \text{ S.}=PS'-90=\delta$$

Rule (*d*). To compute angle SPS'=A − A', the parallax in right ascension, and thence RA of true point of contact S'.

$$\text{(4.)}$$

$$\text{Log. c...........................1·848733}$$

$$\text{„ sec. (PS'−90°) ....0·002484}$$

$$0{\cdot}301030$$

$$\overline{2{\cdot}152247}$$

$$\therefore A-A'= \quad 141{\cdot}98'$$

$$\text{or } A-A'= \quad 2 \quad 21{\cdot}98$$

$$\text{but RA of star S}=23 \quad 6 \quad 2{\cdot}68$$

$$\therefore \text{RA of S'}=23 \quad 3 \quad 40{\cdot}70$$

Rule (e). To calculate the angle m'Ps', the semidiameter in right ascension, and thence RA of moon's center.

$$P s' - 90° \ldots \ldots \ldots 6° \quad 7' \quad 22 \cdot 3'' \text{ S.}$$

$$\text{☾ decl} \ldots \ldots \ldots 6 \quad 16 \quad 15 \cdot 5 \text{ S.}$$

8 53·2	
60	
533·2	sec. P s' — 90°...0·002484
☾ hor. semi. ....973·1	sec. ☾ decl. ......0·002606
1506·3........................3·177911	
439·9........................2·643354	
	7·647818
	2)3·474173
	1·737086

$$\therefore \text{ semi. in RA} = \qquad 54 \cdot 58' = m' \text{P} s'$$

$$\text{RA of } s' = 23^h \quad 3^m \quad 40 \cdot 70$$

$$\left. \begin{array}{l} \therefore \text{ RA of moon's center } m' \text{ at time} \\ \text{of observation} \end{array} \right\} = 23 \quad 2 \quad 46 \cdot 12$$

Rule (f). To calculate the corresponding time at Greenwich, when the moon's RA $= 23^h 2^m 46 \cdot 12'$.

RA at observation..........23^h	2^m	46·12'
RA on Aug. 25, at 8^h ...23	2	5·47
		40·65

Log. 40·65 ....1·609065		
„ const. log. ....3·556302		
5·165367		
„ 127·66 ....2·106055		
„ x........................3·059312		
$\therefore x =$ 1146·4 $= 0$	19^m	6·4'
8		
Greenwich mean time, Aug. 25..............8	19	6·4
R. N. Coll. mean time, Aug. 25..............8	14	34·6
Long. in time $=$ 4	31·8 W.	

EXAMPLE (AN EMERSION, EAST OF MERIDIAN).

267. Aug. 25, 1839, at 9^h 16^m 42·6' mean time at the Royal Naval College, in latitude 50° 48' N. and long. in time 4^m 24·2' W., observed the emersion of φ Aquarii.

Before the rule can be applied, we must take out of the *Nautical*

Almanac the following quantities for a Greenwich date, namely (1.) Moon's semi. ; (2.) Moon's hor. parallax (corrected for spheroidal figure of earth, p. 129) ; (3.) Moon's declination ; (4.) Right ascension of mean sun ; and (5.) Star's right ascension and decl.   The latitude must also be corrected for spheroidal figure of earth (p. 123) ; and (6.) The moon's right ascension must be taken out for the hour of Greenwich date, and for the hour following (see working form, p. 204).

$$
\begin{array}{lr}
\text{R. N. Coll. Aug. 25} & 9^{h}\ \ 16^{m}\ \ 42{\cdot}6^{s} \\
\text{Long. in time} & 4\ \ \ 24{\cdot}2\ \text{W.} \\
\hline
\text{Greenwich, Aug. 25} & 9\ \ \ 21\ \ \ \ 6{\cdot}8 \\
\end{array}
$$

	(1.) ☾ semi.	(2.) ☾ hor. par.		Latitude.
Noon	16′ 10·6″	59′ 21·9″		50° 48′ N.
Midnight	16 14·2	59 35·2	Red	11
	3·6	13·3	Red. lat.	50 37 N.
Cor....	2·8	Cor.... 10·4		
	16 13·4	59 32·3		
	60	Red.... 6·8		
Hor. semi....	973·4	59 25·5		
		60		
	Red. hor. par....	3565·5		

	(3.) ☾'s decl.	(6.) ☾'s R A.		(4.) RA mean sun.
9^h	6° 4′ 44·2″S.	23^h 4^m 13·13^s	25	10^h 12^m 15·93^s
10^h	5 47 50·8 S.	23 6 20·74	Cor.	1 32·17
	16 53·4	2 7·61		10 13 48·10
Cor....	5 56·0	60		
	5 58 48·2 S.	127·61		

(5.) Star's RA and decl.
R A......23^h 6^m 2·68^s
Decl. ....6 54 33·8″S.

To find star's hour-angle (see fig. p. 197).

$$
\begin{array}{lr}
\text{Mean time at place} & 9^{h}\ \ 16^{m}\ \ 42{\cdot}6^{s}\ =\text{Q}m \\
\text{RA mean sun} & 10\ \ \ 13\ \ \ 48{\cdot}1\ =\text{A}m \\
\hline
\text{RA meridian }(+24) & 19\ \ \ 30\ \ \ 30{\cdot}70=\text{A}m\text{Q} \\
\text{Star's RA} & 23\ \ \ \ 6\ \ \ \ 2{\cdot}68=\text{A}\text{Q}\text{R} \\
\hline
& 20\ \ \ 24\ \ \ 28{\cdot}02 \\
& 24 \\
\hline
\therefore\ \text{star's hour-angle}= & 3\ \ \ 35\ \ \ 31{\cdot}98=\text{R}\text{Q} \\
\end{array}
$$

Rule (*a*). To compute $\frac{1}{2}(A+A')$.

(1.)

Log. red. hor. par. ....3·552120

„   cos. red. lat. ....9·802435

„   sin. star's hour-angle ....9·907314

„   const. log. ....8·522879

$\therefore$ log. c$=$1·784748

„   sec. star's decl.................0·003165

„ $\frac{1}{2}(A-A')$ ....1·787913

$\therefore \frac{1}{2}(A-A')=$ 61·37'

or$=$ 1   1·37

A$=$3   35   31·98

$\therefore \frac{1}{2}(A+A')=$3   34   30·61

Rule (*b*). To compute the par. in decl. M$-$N.

(2.)

Log. red. hor. par. ......3·552120

„   sin. red. lat. ......9·888133

„   cos. star's decl. ...9·996835

„   M.....................3·437088

$\therefore$ M$=$2736·1

(3.)

Log. red. hor. par. ......3·552120

„   cos. red. lat. ......9·802435

„   sin. star's decl. ...9·080198

„   cos. $\frac{1}{2}(A+A')$ ......9·773080

„   N.....................2·207833

$\therefore$ N$=$161·4

Diagram for this Example.

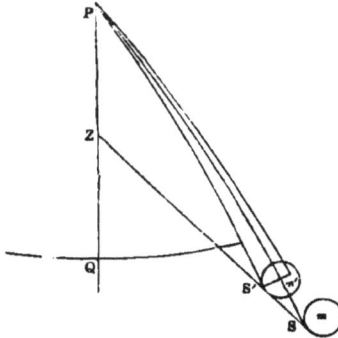

Rule (*c*). To determine the signs of M and N (see *Trig.* Part I. art. 33).

+   +   +

M$-$N$=$hor. par. . sin. lat. . sin. pol. dist.

+   + $-$ +

$-$hor. par. . cos. lat. . cos. pol. dist. . cos. $\frac{1}{2}(A+A')$.

Putting the proper signs over the factors, we see that м is +, and — n
is +.

$$\therefore \text{м} = 2736 \cdot 1 +$$
$$- \text{n} = \underline{161 \cdot 4 +}$$
$$\therefore \text{м} - \text{n} = 2897 \cdot 5$$

or par. in decl. =   48′   17·5″ = ps — ps′ (fig. p. 201)
star's decl. = 6   54   33·8   = ps — 90°

∴ true decl. of point of ⎫ = 6   6   16·3 S. = ps′ — 90°.
moon's limb observed ⎭

Rule (d). To compute s p s′ or a — a′, the parallax in right ascension.
(4.)

Log. c ....1·784748
„ sec. (p s′ — 90°)...0·002469
                       0·301030
log. (a — a′)............2·088247
∴ a — a′ =   122·53′
or =   2ᵐ 2·53 = parallax in r a.
star's r a = 23   6   2·68 = r a of app. point of contact.
r a of s = 23   4   0·15 = r a of true point of contact.

Rule (e). To calculate the angle s′ p m′, the moon's semi. in right ascension.

6° 6′ 16·3″ S.	0·002469
5 58 48·2 S.	0·002366
7 28·1	3·152646
60	2·720407
448·1	7·647818
973·4	2)3·525706
1421·5	1·762853
525·3	

semi. in r a =   57·92′ = m′p s′
r a of point s′ = 23ʰ   4ᵐ   0·15
∴ r a of moon's center m′ at time of observation = 23   4   58·07

Rule (f). To calculate the corresponding time at Greenwich when the
moon's right ascension = 23ʰ 4ᵐ 58·07′.

r a at observation ...23ʰ 4ᵐ 58·07′   Log. 44·94 ....1·652633
r a Aug. 25, at 9ʰ ...23   4   13·13   Const. log. ....3·556302
                      44·94   5·208935
                                          Log. 127·61 ....2·105885
                                          „ x..............................3·103050

$$\therefore x = 1267 \cdot 8', \text{ or} = 0^h \ 21^m \ 7 \cdot 8'$$

		9	
Greenwich, Aug. 25...	9	21	7·8
R. N. Coll. Aug. 25...	9	16	42·6
Long. in time		4	25·2 W.

If the Greenwich time had come out very differently from the assumed Greenwich date, it might have been necessary to recalculate the RA of mean sun, the moon's decl., hor. semi., and hor. par., using the computed date as a new Greenwich date: but this is seldom required; the element that is most affected by an error in the Greenwich date is the moon's declination (see note, p. 196).

The blank form in the next page will considerably reduce the labour of working out an occultation, as the student will see that several of the logarithmic quantities, placed in a horizontal line, can be taken out at the same opening of the tables.

The following is the order in which the form should be filled up:

1. Get a Greenwich date.
2. Find moon's semidiameter in seconds for Greenwich date.
3. „ „ horizontal parallax for Greenwich date (corrected by Inman's tables (h), or by formula, p. 129).
4. Latitude, corrected by table (h), or by formula, p. 122.
5. Right ascension of mean sun for Greenwich date.
6. Moon's horary change in right ascension.
7. Moon's declination for Greenwich date.
8. Take out star's RA and decl. for the day of the month.
9. Find the star's hour-angle, either by a diagram or by the following formula deduced from Prob. IX. p. 34: viz. star's hour-angle= mean time + RA mean sun — star's RA.

Then proceed to compute the several quantities by means of the rules (a), (b), (c), (d), (e), and (f); namely, by rule (a) compute $\frac{1}{2}(A + A')$; by (b) compute parallax in decl. M — N; by (c) determine the algebraic signs of M and N by the rule in the author's *Trigonometry*, Part I.; by (d) compute parallax in RA; by (e) compute semi. in RA; by (f) compute Greenwich mean time at the moment of observation—the difference between which and the mean time at the place at the time of the observation, obtained from an altitude of a heavenly body, as in the rule for chronometer, or otherwise, is the longitude of the observer.

The method of determining the longitude by an occultation will be found, by aid of a form similar to the one in the following page, much easier and more certain than that by a lunar observation.

FORM FOR FINDING LONGITUDE BY OCCULTATION OF STAR.

1.

Greenwich date.

Mean time at place . . . _____

Long. in time . . . _____

Gr. date . . . _____

2.

Moon's semi.

At . . . _____

At . . . _____

In seconds . _____

3.

Hor. par.

At . . . _____

At . . . _____

Red. . . . _____

In seconds . _____

4.

Latitude.

Red. . . . _____

Red. Lat. . _____

5.

RA mean sun.

At . . . _____

Cor. . . _____

6.

Moon's RA.

At ( ) hour . . _____

At ( ) hour . . _____

Hourly change . . _____

In seconds . . _____ $= \delta_1$

7.

Moon's decl.

At ( ) hour . . _____

At ( ) hour . . _____

_____ $= \delta_1$

8.

Star's RA . . . _____

Star's decl. . . _____

9.

To find star's hour-angle .

Mean time at place . _____

$+$

RA mean sun . . . _____

_____

RA meridian . . . _____

$-$

Star's RA . . . _____

_____

∴ star's hour-angle . _____ $= A$

1.

Log. hor. par. (in seconds) . . .
„ cos. red. lat. . . . .
„ sin. ʌ . . . . .
Const. log. . . . . 8·522879
Log. c . . . . .
+
„ sec. star's decl. . .
„ ½(ʌ − ʌ') nearly . .
∴ ½(ʌ − ʌ') . . . .
sub. from ʌ . .
∴ ½(ʌ + ʌ') nearly . .

(a) N has the same sign (N or S) as star's decl., if lat. and star's decl. have the same name; otherwise a different name.
(b) N has a different name from star's decl.; except when star's hour-angle is between 6ʰ and 18ʰ.
(c) Add if like, subtract if unlike, names.
(d) + if west, − if east, of meridian.
(e) + if emersion, − if immersion.

Star's RA . . . .
(d) ± par. in RA . . .
(e) ± semi. in RA . . .
∴ RA ℂ's center . .
−
ℂ's RA at hour of Gr. date .
Diff. . . . . .
In seconds=D' . . .

2.

Log. hor. par. . . .
„ sin. red. lat. . .
„ cos. star's decl. .
„ sum . . . .
Nat. No.=M (a) . .
=N (b) . .
Par. in decl. . . .
Or in min. & sec. (c) ±
Star's decl. . . .
∴ δ . . . .
ℂ's decl. $\bar{\delta}_1$ . .
Diff . . . .
In seconds
ℂ's semi. in seconds
Sum=S . . . .
Diff.=D . . . .
Log. D' . . . .
+Const. log. . . .
Sum . . . . 3·556302
−Log. hourly change .
Log. diff. . . .
Nat. No. . . . .
In min. and sec. . . .
+hour of Gr. date .
∴ Green. mean time .
Ship mean time . .
∴ long. in time . .

3.

Log. hor. par. . . .
„ cos. red. lat. . .
„ sin. star's decl. .
„ cos. ½(ʌ+ʌ'). .
Log. N . . . .
∴ N . . . .
Log. sec. δ . . .
„ sec. δ_1 . .
„ S . . .
„ D . . .
Const. log. . . . 7·647818
2)
Log. . . . .
Nat. No.. . . .
In min. and sec. . .
=semi. in RA . .

4.

Log. c . . . . 0·301030
„ 2 . . . .
„ sec. δ . . .
Sum . . . .
Nat. No. . . . .
In min. and sec. . .
=par. in RA.

EXAMPLES FOR PRACTICE.

268. January 7, 1836, at 10ʰ 45ᵐ 53·2' mean time, in latitude 52° 8' 28″ N. and estimated longitude 1ᵐ W., observed the immersion of *v* Leonis (east of meridian); to determine the longitude.

Elements from *Nautical Almanac*, corrected for a Greenwich date, Jan. 7, at 10ʰ 47ᵐ:

(1.) Moon's hor. semi. 15' 16·1″; (2.) Moon's red. hor. par. 55' 54·9″; (3.) Moon's decl. 15° 49' 40″ N.; (4.) RA mean sun, 19ʰ 6ᵐ 8·5'; (5.) Star's RA, 10ʰ 23ᵐ 26·4', star's decl. 14° 58' 38·8″ N.; (6) Moon's RA at 10ʰ, 10ʰ 18ᵐ 55·5'; at 11ʰ, 10ʰ 20ᵐ 58·5'.   *Ans.* Long. 0ʰ 2ᵐ 7' W.

269. Feb. 12, 1835, at 2ʰ 29ᵐ 40·5' A.M. mean time, in lat. 50° 49' N. and estimated longitude 4ᵐ 17' W., observed the immersion of γ Cancri (west of meridian); to determine the longitude.

Elements from *Nautical Almanac*, corrected for a Greenwich date, Feb. 11, at 14ʰ 34ᵐ:

(1.) Moon's hor. semi. 15' 46·6″; (2.) Moon's red. hor. par. 57' 47·6″; (3.) Moon's decl. 22° 42' 59·1″ N.; (4.) RA mean sun, 21ʰ 25ᵐ 42·61'; (5.) Star's RA, 8ʰ 33ᵐ 44·21', star's decl. 22° 3' 25·8″ N.; (6.) Moon's RA at 14ʰ, 8ʰ 33ᵐ 24·37'; at 15ʰ, 8ʰ 35ᵐ 49·31'·   *Ans.* Long. 0ʰ 3ᵐ 48' W.

270. Feb. 12, 1835, at 2ʰ 10ᵐ 20' A.M. mean time, in lat. 55° 57' 20″ N. and estimated longitude 12ᵐ 43·6' W., observed the immersion of γ Cancri (west of meridian); to determine the longitude.

Elements from *Nautical Almanac*, corrected for a Greenwich date, Feb. 11, at 14ʰ 23ᵐ:

(1.) Moon's hor. semi. 15' 47·6″; (2.) Moon's red. hor. par. 57' 50·6″; (3.) Moon's decl. 22° 44' 12″ N.; (4.) RA mean sun, 21ʰ 25ᵐ 40·8'; (5.) and (6.) see last example.   *Ans.* Long. 0ʰ 12ᵐ 40' W.

By means of the blank form for finding the longitude from an occultation of a star, the labour of working out the observation will be considerably diminished. The rule and form may be used, with some slight alterations, for determining the longitude from the observed occultation of a *planet;* the necessary changes to be observed are the following:

1. Instead of the moon's red. hor. par., use the difference between the moon's red. hor. par. and the planet's horizontal parallax.

2. Instead of the moon's equatorial semi., use the *sum* of the moon's equatorial semi. and the planet's semi. if the interior contact is observed, and their *difference* if the exterior contact is observed. This rule may be neglected if the mean of the two contacts, or the time of contact of the planet's center with the moon's limb, is used, as in the first example following.

EXAMPLES OF AN IMMERSION AND AN EMERSION OF A PLANET.

271. June 20, 1853, observed at Auckland, New Zealand, in lat. 36° 50' 41" S. and estimated longitude 11ʰ 39ᵐ 16ˢ E. (with telescope (G. H. Jones), smallest inverting power, focal length 4 feet, aperture 2·7 inches), the immersion of the eastern and western limbs of Jupiter (west of meridian) at the following times by chronometer:

$$\text{Contact of 1st limb} \ldots \ldots \ldots 11^{h} \; 45^{m} \; 42 \cdot 2^{s} \; \text{P.M.}$$
$$\text{,,} \qquad \text{2d ,,} \quad \ldots \ldots \ldots 11 \quad 47 \quad 40 \cdot 7$$

The error of chronometer on mean time at the place was 23ᵐ 52·37ˢ fast; to determine the longitude.

Elements from the *Nautical Almanac*, corrected for a Greenwich date, June 19, 23ʰ 43ᵐ 33ˢ, corresponding to the mean of the times of contact, namely, June 20, at 11ʰ 22ᵐ 48·88ˢ P.M. at the place:

(1.) Moon's hor. semi. 16' 40·18"; (2.) moon's red. hor. par. – planet's hor. par., 60' 57·29"; (3.) moon's declination, 22° 48' 1·6" S.; (4.) RA mean sun, 5ʰ 54ᵐ 23·62ˢ; (5.) planet's RA, 17ʰ 9ᵐ 32·02ˢ, planet's decl. 22° 26' 9·05" S.; (6) moon's RA at 23ʰ, 17ʰ 6ᵐ 36·54ˢ; at 24ʰ, 17ʰ 9ᵐ 17·83ˢ.

Ans. Long. 11ʰ 39ᵐ 23ˢ E.

272. June 21, 1853, at 0ʰ 25ᵐ 40·83ˢ A.M., observed the emersion of the planet's western limb (west of meridian) at the same place; to determine the longitude.

Elements from *Nautical Almanac*, corrected for a Greenwich date, June 20, 0ʰ 46ᵐ 24·83ˢ.

(1.) Moon's hor. semi. + planet's semi. (16' 41·2" + 21·6") 17' 2·8"; (2.) moon's red. hor. par. – planet's hor. par., 61' 1·1"; (3.) moon's declination, 22° 55' 6·1" S.; (4.) RA mean sun, 5ʰ 54ᵐ 33·86ˢ; (5.) planet's RA, 17ʰ 9ᵐ 30·61ˢ, planet's decl. 22° 26' 7·7"; (6.) moon's RA at 0ʰ, 17ʰ 9ᵐ 17·83ˢ; change to next hour, 161·5ˢ.   *Ans.* Long. = 11ʰ 39ᵐ 25ˢ E.

The above occultation of the planet Jupiter was observed by Captain Byron Drury, while employed in his survey of New Zealand. The longitude of his station determined by other methods was 11ʰ 39ᵐ 16ˢ E.

THE END.

www.ingramcontent.com/pod-product-compliance
Lightning Source LLC
Chambersburg PA
CBHW020902210326
41598CB00018B/1752